MOLECULAR MICROBIAL ECOLOGY MANUAL

MOLECULAR MICROBIAL ECOLOGY MANUAL

Edited by

ANTOON D.L. AKKERMANS
Department of Microbiology, Wageningen Agricultural University,
The Netherlands

JAN DIRK VAN ELSAS
IPO-DLO, Wageningen,
The Netherlands

and

FRANS J. DE BRUIJN
MSU-DOE Plant Research Lab,
NSF Center for Microbial Ecology and
Department of Microbiology,
Michigan State University,
U.S.A.

Springer-Science+Business Media, B.V.

Library of Congress Cataloging-in-Publication Data

Molecular microbial ecology manual / edited by Antoon D.L. Akkermans,
   Jan Dirk Van Elsas, and Frans J. De Bruijn.
       p.   cm.
     ISBN 978-94-010-4156-0   ISBN 978-94-011-0351-0 (eBook)
     DOI 10.1007/978-94-011-0351-0
     1. Molecular microbiology. 2. Microbial ecology.   I. Akkermans,
   A. D. L.   II. Elsas, J. D. van (Jan D.), 1951-   III. De Bruijn,
   F. J. (Frans J.)
   QR74.M64   1995b
   576--dc20   95-31285

ISBN 978-94-010-4156-0

Neither Kluwer Academic Publishers nor any person acting on its behalf is responsible for the use which might be made of the information contained herein.

Printed on acid-free paper

Contents

Preface

For a long time microbial ecology has been developed as a distinct field within Ecology. In spite of the important role of microorganisms in the environment, this group of 'invisible' organisms remained unaccessable to other ecologists. Detection and identification of microorganisms remain largely dependent on isolation techniques and characterisation of pure cultures. We now realise that only a minor fraction of the microbial community can be cultivated. As a result of the introduction of molecular methods, microbes can now be detected and identified at the DNA/RNA level in their natural environment. This has opened a new field in ecology: Molecular Microbial Ecology.

In the present manual we aim to introduce the microbial ecologist to a selected number of current molecular techniques that are relevant in microbial ecology. The first edition of the manual contains 33 chapters and an equal number of additional chapters will be added this year.

Since the field of molecular ecology is in a continuous progress, we aim to update and extend the Manual regularly and will invite anyone to deposit their new protocols in full detail in the next edition of this Manual.

We hope this book finds its place where it was born: at the lab bench!

Antoon D.L. Akkermans, Jan Dirk van Elsas and Frans J. de Bruijn

March 1995

Molecular Microbial Ecology Manual **1.1.2**: 1–11, 1995
© 1995 *Kluwer Academic Publishers.*

Extraction of microbial DNA from aquatic sources: Freshwater

ROGER W. PICKUP[1], GLENN RHODES[1] and
JON R. SAUNDERS[2]

[1] *Institute of Freshwater Ecology, Windermere Laboratory, The Ferry House, Far Sawrey, Ambleside, Cumbria, LA22 0LP, UK;* [2] *Department of Genetics and Microbiology, University of Liverpool, Merseyside, L69 3BX, UK*

Introduction

Although this chapter is principally concerned with DNA extraction other considerations related to sampling and sample preparation are discussed.

Freshwater lake environment

Most deep lakes in temperate regions experience a seasonal cycle of thermal (or density) stratification and destratification and this is a major factor in controlling bacterial activity [8, 9]. Throughout winter and into spring the water column is isothermal and well oxygenated. Increased solar radiation raises the temperature of the surface water and the related changes in density give rise to distinct zones. The surface water (the epilimnion) is warm and well mixed whilst the deep water (the hypolimnion) remains cool. These two layers are separated by a water mass (the metalimnion) which is characterized by steep temperature gradients. The temperature difference, and associated changes in density, ensure that mixing between the epilimnion and hypolimnion is minimal and the latter can be treated essentially as a closed system [8]. The air-water interface has been recognized as an important and active environment which after application of specialized sampling procedures can be treated like other water samples. The surface sediment layers in contact with these zoned water masses exhibit similar temperature cycles and can be broadly defined as shallow (littoral) or deep (profundal) deposits.

A lake can be more or less productive depending on its trophic status (a descriptive term for the nutrient loading of the system with eutrophic and oligotrophic as the extremes). Its status has two components, the cultural (enrichment from the catchment and man's activities) and the morphometric (the volume/area relationship of the lake basin). Generally a large deep lake is less likely to be eutrophic than a shallow lake simply due to the dilution of incoming nutrients and the greater volume of hypolimnetic water which is required to be deoxygenated. The degree to which the hypolimnion becomes deoxygenated will depend on the activity of benthic organisms and the input of suitable electron donors. This in turn affects the redox profiles and subsequent microbial activity in both the water and the sediment [8].

Sampling

DNA extraction is performed for a variety of reasons including studies on biodiversity (reannealing kinetics; [28]), direct detection using specific probes [11] and PCR amplification for both detection [26] and determining diversity [2]. As with all studies in environmental microbiology the sampling procedure chosen is crucial and is often a limiting factor in the detection of specific microorganisms due to their implicit non-representative nature [9, 22, 23]. Therefore, central to the success of any procedure is firstly the strategy used to remove the sample from the environment and secondly, the method employed to extract DNA from the environmental sample.

It should be recognized that any sampling procedure will disturb the natural conditions but it is important to minimize such effects in order to obtain a representative sample [23]. With bulk samples, i.e. those obtained by pumping water into containers where depth is not a consideration or using a sediment grab sampler, the integrity of the sample need not be maintained. However, some sampling strategies require the sample to be maintained as close to *in situ* conditions as possible, or that the sample integrity needs to be maintained, so that precise sub-sampling may be performed in the laboratory. The latter strategy is particularly useful when DNA extraction is directed towards distribution patterns of a particular organism or group of organisms. Maintenance of anaerobiosis is usually a prerequisite in the ·culture of obligate anaerobes and associated activity measurements (e.g. isolation of methanogens and the measurement of methanogenesis), however, its necessity in direct or indirect DNA extraction is of less importance.

The techniques for sampling the water and freshwater sediment have been reviewed elsewhere [see 10]. For normal bottle sampling the containers should be flushed with at least three times their own volume of water prior to stoppering in such a way as to eliminate, or reduce, gaseous exchange or entrapment of air. Gradients within the water column may be steep and conventional sampling devices can straddle different water layers [13]. Under such circumstances the sample should be mixed within the sampler prior to filling the container. If smaller scale resolution is required, pump samplers [5] should be employed with the intake pipes placed horizontally rather than vertically. Moreover, water samples can be obtained at closely defined intervals from the sediment water interface using electronic devices attached to appropriate sampling gear [4]. It is good practice to sample deep water at defined distances from the sediment rather than from the water surface. Unlike sampling the water column, it is difficult to obtain representative samples of sediment because of the existence of steep physical and chemical gradients [3] often associated with specific bacterial activities [14].

The removal of biological particles from water is generally achieved by vacuum filtration using filters of between 0.4 and 0.22 μm pore size often accompanied by prefiltering using a range of pore sizes [9]. However, limitations are imposed on this methods through clogging of the filter and the low volumes that can be processed [1]. These problems can be circumvented using alternative methods that filter whilst maintaining the particles in solution. Cylindrical filter membranes [27], Hollow fibre filtration [25], Tangential flow filtration (TFF, [1]) and vortex flow filtration [12] can be used to this end. The method most frequently employed by our group involves sampling from defined depths, as specified by oxygen and temperature gradients, combined with TFF. With this technique, the final concentrate obtained is termed *retentate* and the filtered water expelled from the TFF unit is called *permeate*. The advantages of this method are that the organisms experience less mechanical stress, that larger volumes can be processed (routinely 100 l by us but over 8000 l with a modification to the pump system [7] and that processing time is short (approximately 40 min to reduce 100 l to 1 l of concentrate). Further processing of the sample can be performed by standard filtration. However, we routinely reduce 1 l samples further to 1–10 ml of concentrate by centrifugation [19].

Sediment sampling can be achieved using a variety of grab and coring devices. In general, corers produce less disturbance than grab samples. The coring devices produce intact cores of defined stratigraphy from a depth of 50 cm (Jenkin corer; [20]) upto 6 m with the Mackereth corer [18]. The processing of sediments and subsequent DNA extraction is detailed in Chapter 1.1.5.

DNA extraction procedures

Methods for DNA extraction from freshwater and freshwater sediments have been reviewed [for example 17, 22] and all share the common feature of cell concentration prior to DNA extraction. The approach taken by most investigators is to concentrate cells on micropore membranes and to lyse the cells on the filter in a small volume of liquid [e.g. 21, 27]. The pore size used ranges from 0.22 to 0.45 μm and often prefiltering is required which permits greater volumes of sample to be processed. Although TFF processing of samples will be described, it is important to note that many workers have used the technique of Somerville *et al.* [27] for cell concentration. This technique requires a Millipore high capacity cylindrical filter membrane through which water is pumped and in which cell lysis is achieved. Lysis is a result of the application of lysozyme, sodium dodecyl sulphate and proteinase K. DNA from the lysate can be purified by precipitation or buoyant-density centrifugation. The technique has been used by Knight *et al.* [16] to detect *Salmonella* spp. in estuarine water.

Statistical handling of data

An impractical degree of replication may be required to work to a given level of statistical significance [9]. Appropriate example calculations relating the mean and variance of a given variable to the appropriate level of significance are provided by Elliott [6]. Such a test is worthwhile performing if only to demonstrate how far the practical limits of replication fall short of what is possible. The degree of variability of microbiological data, on both temporal and spatial scales, is reported by Jones and Simon [15] further emphasizing the caution which must be exercised in interpreting data from a single sample.

Experimental approach

It is apparent from the number of DNA isolation techniques that no single protocol is suitable for the extraction of DNA from all environments. Therefore, we will summarise a technique shown to produce total DNA from freshwater samples of a quality suitable for the restriction endonuclease digestion and DNA amplification by polymerase chain reaction (PCR) and then sequencing.

Procedures

Rapid extraction of bacterial genomic DNA

The following procedure is a modification of the method described by Pitcher *et al.* [24]. Alterations to the published method are outlined in the *Notes* section. This method is a rapid and relatively economical way of extracting total genomic DNA from bacterial cells obtained by TFF from 100 l of lakewater. We have used this method to purify DNA and amplify insertion element sequences directly from lakewater [Rhodes, Saunders and Pickup, unpublished data].

Tangential flow filtration (TFF) and cell concentration

1. Concentrate the particles from 100 l of lakewater to a final volume of 1 l retentate using TFF comprising a Millipore Pellicon cassette holder containing a stack of three 0.22 µm pore size membranes attached to a variable-drive peristaltic pump.
2. Prefilter the retentate through a 3 µm filter (Millipore, type SS) to remove the larger algal cells then through a 0.22 µm filter (Millipore, type GS).
3. Resuspend the cells from the 0.22 µm filter in 50 ml sterile filtered lakewater.
4. Pellet the cells by centrifugation at $20,000 \times g$ for 30 min.

Steps in the extraction procedure
1. Either:
   a) Use the pellet directly and proceed to 2;
   b) Inoculate from a single colony into a 50 ml liquid nutrient medium, and grow cells at the optimum temperature with shaking ($100 \times$ rpm) until approximately 1×10^9 cells ml^{-1} (A_{600} = 1.0–1.5) are present. Harvest cells by centrifugation at 5,000 $\times g$ for 10 min at 10 °C;
   c) It is possible to extract DNA indirectly by filtration. Reconcentrate cellular biomass by filtration on to sterile 0.22 µm pore size micropore membrane(s) and place it on an agar medium to grow into a bacterial lawn or recoverable single colony forming units (CFUs);
2. Resuspend the pellet in 100 µl of fresh lysozyme (50 mg ml^{-1}) in TE buffer (10 mM Tris-HCl; 1 mM EDTA, pH 8.0) and incubate the suspensions at 37 °C for 30 min.
3. Add 0.5 ml of filter sterilized cell lysis solution comprizing 5 M guanidium thiocyanate; 100 mM EDTA and 0.5% sarkosyl, and vortex briefly, then leave on ice for 10 min.
4. To this lysis mixture add 0.25 ml of cold 7.5 M ammonium acetate, mix thoroughly and leave on ice for a further 10 min.
5. Extract DNA by adding 0.5 ml of a chloroform/2-pentanol mixture (24:1) to the lysis mixture, shake well.
6. Transfer samples to 1.5 ml Eppendorf tubes and centrifuge at $25,000 \times g$ for 10 min.

7. Carefully transfer the upper phase of the supernatant fluid into fresh Eppendorf tubes and precipitate the DNA by adding 0.54 volumes of chilled isopropanol (2-propanol). Mix the contents thoroughly and centrifuge at 6,500 × g for 5 min.
8. Remove supernatant and wash the pellet with 1 ml 70% (v/v) ethanol before air drying for approximately 15–30 min.
9. Resuspend the DNA in 100 µl sterile distilled water and store at −20 °C.
10. Examine the DNA by electrophoresis in a 0.7% agarose gel at 100 V for 1–2 h.

Notes
1. If filtration requires several filters then cells should be resuspended from each, harvested, then pooled prior to DNA extraction.
2. Serial dilution of sample material is usually necessary in order to obtain CFUs on the membrane filters.
3. The resuspension of cells in 100 µl of TE buffer is not necessary for lysis of cells. Resuspension directly in 0.5 ml lysis solution is generally sufficient.
4. Use of sterile membrane filtered lakewater prevents premature lysis of cells during harvesting.
5. The quantity and quality of the DNA is assessed visually directly after gel electrophoresis. The quality and quantity depends on several factors including the trophic status of the lake, the season and contaminants such as humic acids.

Solutions

- *Lysis solution*
    - 5 M guanidium thiocyanate
    - 100 M EDTA
    - 0.5% (w/v) sarkosyl pH 8.0
- *Salting solution*
    - 7.5 M ammonium acetate
- *Partitioning solution*
    - chloroform:2-pentanol (24:1)
- *TE buffer*
    - 10 mM Tris-HCl
    - 1 mM EDTA pH 8.0

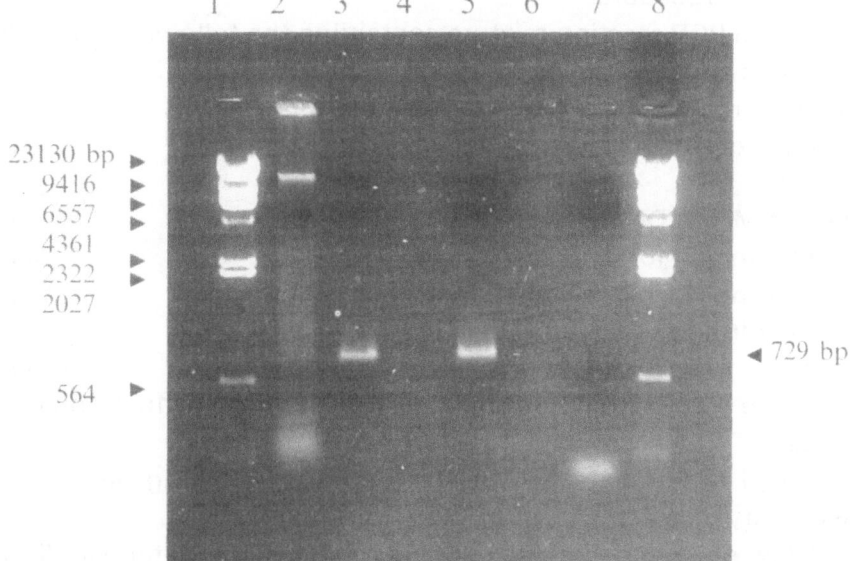

Figure 1. Amplification of insertion element IS*1* from DNA extracted from tangential flow filtration (TFF) processed lakewater. Lanes 1 and 8, *Hin*dIII digested bacteriophage λ DNA; lane 2, TFF/guanidium thiocyanate extracted DNA, undigested (40 μl); lane 3, amplified IS*1* DNA[1]; lane 4, blank; lane 5, positive control, *E.coli* W3110[1]; lane 6, blank; lane 7, negative control[2].

[1]Amplified products purified using QIAEX (Hybaid, Cambridge, UK) according to manufacturer's instructions; [2]PCR negative control as described in text except with no DNA.

Protocol for PCR amplification of genomic DNA obtained from aquatic samples

We have used this procedure to amplify sequences of transposable elements from genomic DNA obtained from bulk extraction bacteria in freshwater obtained by TFF [Rhodes, Saunders and Pickup, unpublished data]. Figure 1 shows the amplification of DNA from a bulk DNA extraction of 100 l of lakewater from Windermere (sample taken 12.12.94) using primers specific for insertion element IS*1* using the procedures described in this chapter.

Steps in the procedure

1. Prepare a PCR reaction mixture containing the following reagents sufficient for 10 samples:

10× PCR reaction buffer	50 µl
Bovine serum albumen (BSA)	10 µl
dNTPs	5 µl
Perfect Match DNA Polymerase Enhancer	1 µl
20–50 pmol 5′ forward reaction primer	10 µl
20–50 pmol 3′ reverse reaction primer	10 µl
Sterile ultra-pure water	384 µl
Total	470 µl

2. From the reaction mixture, aliquot 47 µl into each of 10 0.5 ml PCR tubes.
3. Add 1 µl of template DNA to each tube (at differing dilutions if necessary).
4. Overlay each reaction with 50 µl mineral oil (filter sterilised through 0.22 µm pore size membrane filters).
5. Transfer all tubes to the PCR machine and perform hot start for denaturation at 95 °C for 3 min.
6. During the hot start procedure, dilute the Taq DNA polymerase enzyme 1/5 to obtain a 1 U µl^{-1} stock solution. With 10 tubes, add 4 µl Taq polymerase to 16 µl 1× PCR reaction buffer.
7. Perform the following PCR cycle:

80 °C	Hold, add 2 µl diluted Taq enzyme to each tube	
50 °C	40 s	
72 °C	2 min	
94 °C	1 min	
50 °C	40 s	35 cycles
72 °C	2 min	
94 °C	1 min	
50 °C	40 s	1 cycle
72 °C	10 min	
4 °C	Holding temperature	

8. Freeze samples at −20 °C and remove mineral oil whilst frozen.
9. Thaw the samples and electrophorese 10 µl of reaction product in a 0.7 % agarose gel.

MMEM-1.1.2/8

Notes

1. The addition of both Taq polymerase and mineral oil at different stages in the procedure, i.e. Taq added to the tubes prior to hot start, and mineral oil added after hot start, have resulted in no observable differences in product quality in our experiments. It is therefore a variable part of the procedure which should be carried out within the bounds of the individual's own preferences.
2. More than 1 µl of denatured template DNA may be required for successful amplification after extraction of genomic DNA by the methods described above. If more template is used, then a reduction in the total volume of the reaction mixture must be carried out by excluding the necessary volume of water. This ensures that a constant volume of 50 µl is maintained in all reaction tubes.
3. Perfect match polymerase enhancer is an additive that greatly increases both yield and specificity of primer extension reactions.

Solutions

- *PCR reaction mixture*
  - 10× PCR reaction buffer:
    - 100 mM Tris-HCl
    - 15 mM $MgCl_2$
    - 500 mM KCl pH 8.3
  - BSA (molecular biology grade) 100 µg ml^{-1}
  - 25 mM each dNTP
  - Perfect match DNA polymerase enhancer (Stratagene, Cambridge, UK)
  - 20–50 pmol each of oligonucleotide primers (forward and reverse)
  - sterile ultra-pure water
- Taq DNA polymerase (5 U µl^{-1})

Acknowledgements

The authors wish to thank the Institute of Freshwater Ecology, The University of Liverpool and the Natural Environment Research Council for support. We wish to thank David Chapman for technical assistance.

References

1. Barthel KG, Schneider G, Gradinger R, Lenz J (1989) Concentration of live picoplankton and nanoplankton by means of tangential flow filtration. J Plankton Res 11: 1213–1221.
2. Bej AK, Mahbubani MH, Dicesare JL, Atlas RM (1991) Polymerase chain reaction-gene probe detection of microorganisms by using filter-concentrated samples. Appl Environ Microbiol 57: 3529–3534.
3. Carlton RG, Wetzel RG (1985) A box corner for studying metabolism of epipelic microorganisms in sediment under *in situ* conditions. Limnol Oceanogr 30: 422–426.
4. Cunningham CR, Davison W (1980) An opto-electronic sediment detector and its use in chemical micro-profiling of lakes. Freshwater Biol 10: 413–418.
5. Davison W (1977) Sampling and handling procedure for polarographic measurement of oxygen in hypolimnetic sediments. Freshwater Biol 7: 413–418.
6. Elliot JM (1973) Some Methods in the Statistical Analysis of Samples of Benthic Invertebrates. Freshwater Biological Association Scientific Publication no. 25.
7. Giovannoni SJ, DeLong EF, Schmidt TM, Pace NR (1990) Tangential flow filtration and preliminary phylogenetic analysis of Marine picoplankton. Appl Environ Microbiol 56: 2572–2575.
8. Hall GH (1986) Nitrification in lakes. In: Prosser JI (ed) Nitrification, Special Publication of the Society for General Microbiology, pp. 127–156. Oxford: IRL.
9. Hall GH, Jones JG, Pickup RW, Simon BM (1990) Methods to study the bacterial ecology of freshwater environments. Meth Microbiol 22: 181–210.
10. Herbert RA (1990) Methods for enumerating microorganisms and determining biomass in natural environments. Meth Microbiol 19: 1–40.
11. Holben WE, Tiedje JM (1988) Application of nucleic acid hybridization in microbial ecology. Ecology 69: 561–568.
12. Jiang SC, Thurmond JM, Pichard SL, Paul JH (1992) Concentration of microbial populations from aquatic environments by vortex flow filtration. Marine Ecol Prog Ser 80: 1001–107.
13. Jones JG (1975) Some observations on the occurrence of the iron bacterium *Leptothrix octracae* in freshwater including reference to large experimental enclosures. J Appl Bacteriol 39: 63–72.
14. Jones JG (1979) Microbial activity in lake sediments with particular reference to electrode potential gradients. J Gen Microbiol 115: 19–26.
15. Jones JG, Simon BM (1980) Variability in microbiological data from a stratified eutrophic lake. J Appl Bact 49: 127–135.
16. Knight IT, Scults S, Kaspar CW, Colwell RR (1990) Direct detection of *Salmonella* spp. in estuaries by using a DNA probe. Appl Environ Microbiol 56: 1059–1066.
17. Knight IT, Holben WE, Tiedje JM, Colwell RR (1992) Nucleic acid hybridization techniques for detection, identification and enumeration of microorganisms in the environment. In: Levin MA, Seidler RJ, Rogul M (eds) Microbial Ecology, Principles, Methods and Applications, pp. 65–91. New York: McGraw-Hill Inc.
18. Mackereth FJH (1958) A portable core sampler for lake deposits. Limnol Oceanogr 3: 181–191.
19. Morgan JAW, Pickup RW (1993) Activity of microbial peptidases, oxidases and esterases in lake waters of varying trophic status. Can J Microbiol 39: 795–803.
20. Ohnstad FR, Jones JG (1982) The Jenkin Surface Mud Sampler. User Manual. Occasional Freshwater Biological Association Scientific Publication no. 15.
21. Paul JH, Jeffrey WH, Deflaun MF (1987) Dynamics of extracellular DNA in the marine environment. Appl Environ Microbiol 53: 170–179.
22. Pickup RW (1991) Molecular methods for the detection of specific bacteria in the environment. J Gen Microbiol 137: 1009–1119.
23. Pickup RW (1995) Sampling and detecting bacterial populations in natural environments.

In: Baumberg S, Young P, Wellington E, Saunders J (eds) Population Genetics of Bacteria. Cambridge: Cambridge University Press.

24. Pitcher DG, Saunders NA, Owen RJ (1989) Rapid extraction of genomic DNA with guanidium thiocyanate. Lett Appl Microbiol 8: 151 156.

25. Proctor LM, Fuhrman JA (1990) Viral mortality of marine bacteria and cyanobacteria. Nature 343: 60–62.

26. Smalla K, Van Overbeek LS, Pukall R, Van Elsas JD (1993) Prevalence of *npt*II in kanamycin resistant bacteria from different environments. FEMS Microbiol Ecol 13: 47–58.

27. Somerville CC, Knight IT, Straube WL, Colwell RR (1989) Simple, rapid method for direct isolation of nucleic acids from aquatic environments. Appl Environ Microbiol 55: 548–554.

28. Torsvik VL (1980) Isolation of bacterial DNA from soil. Soil Biol Biochem 12: 15-21.

Molecular Microbial Ecology Manual **1.1.3**: 1–10, 1995
© 1995 *Kluwer Academic Publishers.*

Extraction of microbial DNA from sewage and manure slurries

KORNELIA SMALLA

Federal Institute of Biology for Agriculture and Forestry, Biochemistry and Plant Virology Division, Messeweg 11/12, 38104 Braunschweig, Germany

Introduction

Sewage and manure slurries are environments where rather high numbers of micro-organisms are found. The microbial composition and metabolic activity is strongly dependent on several biotic and abiotic factors (e.g., nutrient availability, duration of storage, pretreatment). Sewage and manure slurries are supposed to play a central role in the natural circulation of pathogens, plasmid-encoded antibiotic or heavy metal resistance genes. Therefore, these habitats are of great interest in view of microbial ecology and hygiene. However, different bacteria of hygienic importance like *Shigella dysenteria* [7], *Salmonella typhimurium* [24], *Salmonella enteritidis* [16], *Campylobacter jejuni* [8,13,14], *Escherichia coli* [2,3], *Vibrio cholerae* [5] or *Vibrio vulnificus* [26] were described to enter the viable but nonculturable state. The potential health hazard presented by pathogens existing in the nonculturable state may be significant because it has been shown that nonculturable bacteria can be resuscitated [11] and remain potentially pathogenic [4]. Furthermore, genetically modified micro-organisms used in industrial settings might be undeliberately released via sewage into the environment. In view of the fact that genetically modified strains like *E. coli* K12 producing biologically active substances might enter the viable but nonculturable state under environmental stress one should not rely on cultivation methods only when tracking the fate of recombinant micro-organisms in the environment. Therefore, microbial ecologists have recently developed methods obviating the need for cultivation by applying molecular techniques to directly extracted nucleic acids from different environmental habitats [6,12,21].

Experimental approach

The methodology of well-established protocols for nucleic acid extraction [18,28] from isolates is applied to the environmental sample directly or after an appropriate concentration step. Directly extracted DNA does not contain only microbial nucleic acids but also DNA of different origin (algae, protozoa). Depending on the environmental matrix, coextraction of substances like humic acids or other organic and inorganic compounds might occur which can disturb subsequent DNA analysis [17,22,23]. Therefore, not only efficient recovery of nucleic acids from an environmental sample but also sufficient purity of the obtained DNA is of importance. PCR-assisted amplification of DNA sequences specific for pathogenic bacteria [1,10], for genetically modified micro-organisms, or antibiotic resistance genes [20] and for genetically modified organisms [25] has considerably improved our ability to detect noncultured organisms (e.g., certain pathogenic bacteria or released genetically modified micro-organisms) in sewage or manure slurries. Furthermore, direct nucleic acid extraction methods are extremely helpful for detection of viruses, e.g., enteroviruses [9] where standard methods are very time-consuming and not applicable for viruses which are not amenable to culture.

For nucleic acid extraction from sewage and manure, different protocols can be adapted. The method of choice will clearly be determined by the sample characteristics, e.g., content of particles and flocks, microorganism density, contaminations with low-molecular substances which have to be removed as well as by the general objective of the investigation. For DNA extraction from manure samples with a very high particle content, protocols originally developed for sediments and soil might be useful. On the other hand, DNA from sewage with low content of particulate matter can be extracted after an appropriate filtration step as applied for nucleic acid extraction from aquatic samples [21]. However, prefiltration steps usually applied to reduce clogging of the filter will diminish the DNA yield when microorganisms are attached to the flocks.

The following protocols were successfully applied to the extraction of genomic DNA from sewage and manure slurries which served as a target for subsequent PCR amplification of antibiotic resistance genes and replication or transfer related sequences of broad host range plasmids. The limit of detection has to be determined separately for each DNA extraction method and the respective target/primer system. In addition, the detection sensitivity of a certain sequence might be influenced by the Taq-polymerase used and by the addition of stabilizing proteins (bovine serum albumin, T4-protein: [15,23]). The following protocols should be regarded just as examples implying that other protocols described in the manual might be useful as well.

Nucleic acid extraction from sewage using the Sterivex protocol [21,27]

The Sterivex protocol was originally developed for nucleic acid extraction from aquatic habitats. The first step of the procedure is filtration of the liquid sample through a bacterial filter (Sterivex, Millipore, 0.22 μm). The volume of sewage which can be pumped through the Sterivex cartridge with a peristaltic pump depends on the content of cellular aggregates. For sewage, volumes of 10 ml (influent of a sewage treatment plant) to 100 ml (sewage treatment plant effluent) should be used. The advantage of the Sterivex procedure is that several steps can be performed in the filter housing thus reducing the probability of contaminations from the laboratory. A further advantage is that the filter can be stored at –20 °C after filtration of sewage.

Procedure

Steps in the procedure

- Sewage with a low content of particulates is pumped through a Sterivex filter (Millipore, 0.22 µm) without prior prefiltration using a peristaltic pump (flow rate will decrease during the filtration due to clogging of the filter).
- Wash the filter with 10 ml sterile SET buffer; close the inlet and the outlet of the filter unit with parafilm and store at –20 °C.
- Thaw filter unit before adding 1.8 ml SET buffer and 67 µl freshly dissolved lysozyme with a syringe; after inverting, incubate on ice for 30 min.
- Add 16 µl SDS solution and incubate the filter on a roller (Spira-mix) to keep the entire filter in contact with the reagents for 1 h (approx. 32 r.p.m.).
- Add 50 µl proteinase K solution and continue incubation on the roller for 3–4 h.
- Remove the lysate from the filter with a syringe; add 1 ml SET buffer to the filter and move on a roller for approx. 5 min.
- Pool the crude lysate with the SET wash buffer in a centrifuge tube.
- Add 0.5 vol 7.5 M ammonium acetate and incubate for 15 min at room temperature after careful mixing to precipitate proteins.
- Centrifuge at 16,000 × g for 15 min at room temperature.
- Place the supernatant in a new centrifuge tube and precipitate DNA by adding 2.5 vol ethanol (at least for 2 h at –20 °C).
- Centrifuge at 26,000 × g for 30 min at 4 °C.
- Resuspend the pellet without drying in 600 µl TE, place the

solution in a microtube and precipitate with 0.6 vol isopropanol for 30 min at room temperature.
- Centrifuge at 26,000 × g for 20 min at 4 °C.
- Wash the pellet with 70% ethanol and air dry it.
- Dissolve the pellet in an appropriate volume of TE.
- In case the DNA extracted is not amplifiable by PCR (testing by addition of a positive control to the target DNA) purification with glassmilk is recommended (Geneclean II, BIO101, Vista, CA).

Notes
Controls. To control for contamination of the resulting DNA during the nucleic extraction procedure, as a negative control one Sterivex filter is processed exactly the same way as described above but with filtration of sterile distilled water instead of the environmental sample; the resulting solution is used as a control for PCR in addition to the water control, which ensures the absence of the respective amplified target sequence in the PCR reagents.

Determination of the detection limit. The detection limit of a naturally occurring sequence can be determined by adding defined cell numbers of the microorganism containing the sequence to be monitored without sewage to the Sterivex procedure. Usually bacterial cells of an overnight culture are harvested by centrifugation. The resulting cell pellet is washed twice with sterile saline before diluting the cells in saline. For an estimate of the cell number a counting chamber is used. The dilutions containing approx. 10^8, 10^5, 10^3, 10^2 and 0 cells are filtered through Sterivex filters (two parallels for each cell density) and are plated in parallel to determine the c.f.u. counts of each dilution. The nucleic acids are recovered as described in the protocol. Usually the limit of detection is determined by Southern blot hybridization of the PCR product with a specific probe. The actual detection limit will depend on the volume used for harvesting bacteria, the copy number of the target sequence per cell and the proportion of recovered DNA used as target in the PCR. We determined the limit of detection for a chromosomally Tn5 labelled *P. syringae* strain applying the Sterivex procedure. When using 1/10 vol of the extracted DNA we obtained strong PCR signals after Southern blot hybridization of amplification products specific for Tn5 for cell numbers corresponding to 4×10^6 and 4×10^3 c.f.u. Faint signals were obtained for the other cell densities used, except for the control without added *P. syringae* (chr::Tn5) (see Fig. 1).
The determination of the detection limit for genetically modified micro-organisms by amplification of a sequence not naturally occurring is done as described above except for resuspending the harvested cell pellet in sewage.

DNA yield. Somerville et al. [21] determined for the protocol described above (with the only modification being 5% sucrose instead of 20%) a yield of about 1 ng high-molecular weight DNA per 10^6 cells. According to Byrd and Colwell [2], free DNA does not contribute to the DNA extracted with the Sterivex protocol.

Figure 1. Determination of the extraction efficiency for chromosomally Tn5 labelled *P. syringae* after PCR amplification and Southern blot hybridization: lanes: 1: Dig-ladder (Boehringer 1218603); 2, 3: 4.2×10^6 c.f.u.; 4, 5: 3.6×10^3 c.f.u.; 6, 7: 3.2×10^1 c.f.u.; 8, 9: 4.5×10^0 c.f.u.; 10, 11: <10 c.f.u.; 12, 13: 0 c.f.u.; 14: water control; 15: Dig-ladder.

Solutions
- SET buffer
  - 5% sucrose.
  - 50 mM EDTA.
  - 50 mM Tris-HCl pH 7.6.
- lysozyme solution
  5 mg lysozyme freshly dissolved in:
  - 10 mM Tris-HCl pH 8.0.
  - 1 mM EDTA.
  - 10 mM NaCl.
- SDS solution
  - 25% in double-distilled water.
- proteinase K solution
  - 20 mg/ml in double-distilled water.

All reagent buffers and water used are to be prepared RNase free as described by Sambrook et al. [18] for the extraction of RNAs.

Genomic DNA extraction from manure slurries or sewage using the Qiagen bacterial DNA isolation kit

The protocol was originally designed for the rapid, easy preparation and purification of genomic DNA from pure cultures using nontoxic reagents. Therefore the following instruction is an adaptation of the

Qiagen protocol to directly extract genomic DNA from manure slurries or sewage. With the protocol described only DNA from microorganisms recovered by the harvesting method is obtained.

At first, the volume of sewage or manure which can be used for genomic DNA extraction without overloading the Qiagen tips has to be determined. According to Qiagen, approximate cell numbers should not exceed 8×10^{10} using the maxi preparation protocol.

The approximate cell number can be determined using a counting chamber or by plating. If the sewage or manure slurry contains flocks, a pretreatment with a blending step to dislodge adsorbed bacteria followed by prefiltration through a Miracloth filter is recommended. Complete resuspension of the bacterial pellet obtained by centrifugation and the absence of particulate matter is important for efficient lysis and to avoid clogging of the Qiagen tips used for purification.

Steps in the procedure
- Centrifuge sewage or manure slurry samples containing a maximum of approx. 8×10^{10} cells. Spin the bacteria down at $5,000 \times g$ for 20 min; higher $\times$ g-forces are recommended to recover also dwarf cells from the environmental sample.
- Resuspend the bacterial pellet thoroughly with 11 ml B1 buffer with dissolved RNase by vortexing at highest speed (no particles should remain).
- Add 300 µl lysozyme solution and 500 µl proteinase K stock solution.
- Incubate at 37 °C for at least 30 min (longer incubation is recommended when particles are still visible).
- Add 4 ml B2 buffer and mix the solutions thoroughly by inverting or vortexing briefly.
- Incubate at 50 °C for at least 15 min; it is necessary that the lysate becomes clear at this stage (if not, extend the incubation time or remove remaining particles by a short centrifugation step).
- Equilibrate a Qiagen tip with 10 ml QBT buffer and drain the tip completely by gravity flow.
- Vortex the lysate for 5–10 sec and load it, if clear and particle-free, onto the equilibrated Qiagen tip (flow can be assisted by a gentle positive pressure with flow rates not exceeding 20–40 drops/min).

- Wash the Qiagen tip 2–3 times with 15 ml QC buffer by gravity flow.
- Elute the genomic DNA with 15 ml QF buffer.
- Add 0.7 volumes of isopropanol equilibrated to room temperature and precipitate by inverting 10–20 times.
- Centrifuge at >5,000 × g at 4 °C for 20 min.
- Resuspend the DNA pellet in TE buffer and transfer to microfuge tube.
- Wash centrifuged DNA in 70% ethanol, air-dry the pellet for 10 min and resuspend in a suitable volume, overnight or for 1–2 h at 55 °C by gentle shaking.

Solutions
- B1 buffer (bacterial lysis buffer) pH 8.0 containing:
    - 50 mM EDTA.
    - 50 mM Tris-HCl.
    - 0.5% Tween-20.
    - 0.5% Triton X-100.
- B2 buffer (bacterial lysis buffer) pH 5.5 containing:
    - 3 M Guanidine Hydrochloride.
    - 20% Tween-20.
- QBT buffer (equilibration buffer) pH 7.0 containing:
    - 750 mM NaCl.
    - 50 mM MOPS (3-(N-Morpholino)propanesulfonic acid).
    - 0.15% Triton X-100.
- QC buffer (wash buffer) pH 7.0
    - 1.0 M NaCl.
    - 50 mM MOPS.
    - 15 % ethanol.
- QF buffer (elution buffer) pH 8.5
    - 1.25 M NaCl.
    - 50 mM Tris-HCl.
    - 15% ethanol.
- RNase A solution
    - 200 µg/ml B1 buffer (stable for 6 months stored at 4 °C).
- lysozyme solution
    - 100 mg/ml distilled water (aliquots should be stored at −20 °C).

- proteinase K solution
  - 20 mg/ml in distilled water (stable for 3–6 months at 4 °C or store aliquots at –20 °C).

Note

Lysozyme and detergents in B1 buffer added to resuspended cells ensure complete lysis of bacterial cells. RNA is degraded by RNase A. Proteinase K and B2 buffer strip all proteins from the DNA. The lysate loaded onto the Qiagen tips should be clear. The salt and pH conditions in the lysate and the selectivity of the Qiagen resin ensure that only DNA is bound while RNA, proteins and other low-molecular compounds are not retained. Wash buffer QC containing 1 M NaCl elutes any residual RNA and proteins bound to nucleic acid. The genomic DNA is efficiently eluted with buffer QF containing 1.25 M NaCl at pH 8.5. The binding, washing and elution conditions for the Qiagen procedure are strongly influenced by pH and salt concentration. Concentrated DNA might diminish the gravity flow rate. Dilution of the DNA with QBT buffer is an alternative. Isopropanol precipitation should be performed at room temperature to prevent salt precipitation. The DNA pellet should be dissolved at slightly alkaline pH (8.0–8.5).

Genomic DNA prepared on Qiagen tips is sufficiently pure to allow restriction enzyme digestion and PCR.

- All buffers should be equilibrated at room temperature before usage.
- Freezing and thawing of enzymes should be avoided.

DNA extraction from manure slurries with a protocol developed for soil DNA extraction

For manure samples with a rather high content of particulate matter the protocols described above might not be suitable due to clogging of the Sterivex filter or the Qiagen tips. The protocol for DNA extraction from soil described in detail in Chapter 1.3.3. is with some modifications applicable to manure slurry. Micro-organisms and particulate matter from 20–100 ml manure slurry are pelleted by centrifugation (16,000 × g, 20 min, 4 °C) . The resulting pellet is washed twice with sterile saline. The pellet is resuspended and processed further as described by Smalla et al. [19]. However, for DNA recovered from manure the purification is much easier than for soil DNA. In most cases, only a one step purification with glass milk (Geneclean II, BIO101, Vista, CA) was needed to obtain PCR amplifiable DNA.

The advantage of this protocol is that the rather harsh cell disruption by applying a bead beating step will result in a more efficient lysis of bacterial cells or spores. Furthermore, a prefiltration step which will definitely reduce the number of cells subjected to DNA extraction is omitted. The resulting DNA is somewhat sheared due to the bead beating treatment which might be an advantage for the amplification of PCR products smaller than 1 kb. However, phenol and phenol/chloroform extraction are applied which is critical with regard to hazardous waste disposal.

Acknowledgement

The work is funded by the EC BIOTECH (grant no. BIO2-CT92–0491).

References

1. Bej AK, Mahbubani MH, Atlas RM (1991) Detection of viable *Legionella pneumophila* in water by polymerase chain reaction. Appl Environ Microbiol 57: 597–600.
2. Byrd JJ, Colwell RR (1990) Maintenance of plasmids pBR322 and pUC8 in nonculturable *Escherichia coli* in the marine environment. Appl Environ Microbiol 56: 2104–2107.
3. Byrd JJ, Leahy JG, Colwell, RR (1992) Determination of plasmid DNA concentration maintained by nonculturable *Escherichia coli* in marine microcosms. Appl Environ Microbiol 58: 2266–2270.
4. Colwell RR (1993) Nonculturable but still viable and potentially pathogenic. Zbl Bakt 279: 154–156.
5. Colwell RR, Brayton PR, Grimes DJ, Roszak DB, Huq SA, Palmer LM (1985) Viable but nonculturable *Vibrio cholerae* and related pathogens in the environment: implications for the release of genetically engineered micro-organisms. BioTechnology 3: 817–820.
6. Holben WE, Jansson JK, Chelm BK, Tiedje JM (1988) DNA probe method for the detection of specific micro-organisms in the soil bacterial community. Appl Environ Microbiol 54: 703–711.
7. Islam MS, Hasan MK, Miah MA, Sur, GC, Felsenstein A, Venkatesan M, Sack RB, Albert MJ (1993) Use of polymerase chain reaction and fluorescent-antibody methods for detecting viable but nonculturable *Shigella dysenteriae* type 1 in laboratory microcosms. Appl Environ Microbiol 59: 536–540.
8. Jones DM, Sutcliffe EM, Curry A (1991) Recovery of viable but nonculturable *Campylobacter jejuni*. J Gen Microbiol 137: 2477–2482.
9. Jothikumar N, Khanna P, Kamatchiammal S, Murugan RP (1992) Rapid detection of waterborne viruses using the polymerase chain reaction and a gene probe. Intervirol 34: 184–191.
10. Koch WH, Payne WL, Wentz BA, Cebolla TA (1993) Rapid polymerase chain reaction for detection of *Vibrio cholerae* in foods. Appl Environ Microbiol 59: 556–560.
11. Nilsson L, Oliver JD, Kjelleberg S (1991) Resuscitation of *Vibrio vulnificus* from viable but nonculturable state. J Bacteriol 173: 5054–5059.
12. Ogram A, Sayler GS, Barkay TJ (1987) DNA extraction and purification from sediments. J Microbiol Meth 7: 57–66.

13. Oyofo BA, Rollins DM (1993) Efficacy of filter types for detecting *Campylobacter jejuni* and *Campylobacter coli* in environmental water samples by polymerase chain reaction. Appl Environ Microbiol 59: 4090–4095.
14. Rollins DM, Colwell RR (1986) Viable but nonculturable stage of *Campylobacter jejuni* and its role in survival in natural environments. Appl Environ Microbiol 52: 531–538.
15. Romanowski G, Lorenz MG, Wackernagel W (1993) Use of polymerase chain reaction and electroporation of *Escherichia coli* to monitor the persistence of extracellular plasmid DNA introduced into natural soils. Appl Environ Microbiol 59: 3438–3446.
16. Roszak DB, Grimes DJ, Colwell RR (1984) Viable but non-recoverable stage of *Salmonella enteritidis* in aquatic system. Can J Microbiol 30: 334–338.
17. Saano A, Kaijalainen S, Lindström K (1993) Inhibition of DNA immobilization to nylon membrane by soil compounds. Microb Releases 2: 153–160.
18. Sambrook J, Fritsch EF, Maniatis T (1989) Molecular cloning. Cold Spring Harbor Laboratory Press, Cold Spring Harbor, New York.
19. Smalla K, Cresswell N, Mendonca-Hagler LC, Wolters A, van Elsas JD (1993a) Rapid DNA extraction protocol from soil for polymerase chain reaction-mediated amplification. J Appl Bacteriol 74: 78–85.
20. Smalla K, van Overbeek LS, Pukall R, van Elsas JD (1993b) Prevalence of *npt*II and Tn*5* in kanamycin-resistant bacteria from different environments. FEMS Microbiol Ecol 13: 47–58.
21. Somerville CC, Knight IT, Straube WL, Colwell RR (1989) Simple, rapid method for direct isolation of nucleic acids from aquatic environments. Appl Environ Microbiol 55: 548–554.
22. Steffan RJ, Atlas RM (1988) DNA amplification to enhance detection of genetically engineered bacteria in environmental samples. Appl Environ Microbiol 54: 2185–2191.
23. Tebbe CC, Vahjen W (1993) Interference of humic acids and DNA extracted directly from soil in detection and transformation of recombinant DNA from bacteria and a yeast. Appl Environ Microbiol 59: 2657–2665.
24. Turpin PE, Maycroft KA, Rowlands CL, Wellington EMH (1993) Viable but non-culturable salmonellas in soil. J Appl Bacteriol 74: 421–427.
25. Van Elsas JD, van Overbeek LS, Fouchier R (1991) A specific marker, pat, for studying the fate of introduced bacteria and their DNA in soil using a combination of detection techniques. Plant Soil 138: 49–60.
26. Weichart D, Oliver JD, Kjelleberg S (1992) Low temperature induced non-culturability and killing of *Vibrio vulnificus*. FEMS Microbiol Lett 100: 205–210.
27. Wendt-Potthoff K, Niepold F, Backhaus H (1992) Fate of plant pathogenic pseudomonads in bean microcosms. In: Stewart-Tull DES, Sussmann M (eds) The release of genetically modified micro-organisms, pp.181–182. Plenum Press, New York.
28. Wilson K (1989) Preparation of genomic DNA from bacteria. In: Ausubel FM, Bent R, Kingston RE, Moore DD, Smith JA, Seidmann JG, Struhl K (eds) Current protocol in molecular biology. Greene and Wiley, New York.

Molecular Microbial Ecology Manual **1.1.4**: 1-7, 1995.
© 1995 *Kluwer Academic Publishers.*

Methods for extracting DNA from microbial mats and cultivated micro-organisms: high molecular weight DNA from French press lysis

MARY M. BATESON and DAVID M. WARD
Department of Microbiology, Montana State University, Bozeman, MT 59717, U.S.A.

Introduction

Only a limited amount of work has been done on extraction of DNA from microbial mat environments. Muyzer et al. [9] have used hot phenolic extraction methods to obtain DNA suitable for PCR amplification. Enzymatic [5] and detergent lysis methods are being used for the same purpose (Paerl, personal communication). As in the case of RNA extraction (see Chapter 1.2.3), little has been published on the comparative analysis of different methods. In the course of our work on the use of small subunit ribosomal RNA (= SSU ~ rRNA) [12, 13] or SSU rRNA genes [6] to characterize inhabitants of hot spring microbial mats, we have tested three methods: French press, mini-bead beater and enzymatic lysis. We have not attempted to optimize these methods but have found that they vary in terms of the quality of the DNA extracted when used as described in the Protocol section. We have also not attempted to compare these methods in terms of DNA recovery. However, if DNA recovery is affected by different lysis methods in the same way as RNA recovery, it may be that French press lysis will give the highest DNA yield (see Section 1.2.4). Little published information is available on French press lysis of natural samples, presumably because it is difficult to process samples that contain mineral material with this technique. The low amount of mineral material in microbial mats has permitted us to evaluate the French press method.

Obtaining high-molecular weight DNA with French press lysis

Even in situations when French press lysis can be used a major concern is the shearing of DNA, since many applications require high-molecular-weight DNA. For instance, in shotgun cloning 10–20 kb sized DNA is advantageous so that partial digestion before cloning leads to a high probability of recovering intact genes [for SSU rRNA gene analysis, see 11]. The same reasoning can be applied in the case of PCR amplification [for SSU rRNA gene analysis, see 1, 2, 4]. On the suggestion of Dr. Norman Pace (Indiana University), we attempted French press lysis under conditions of elevated NaCl and $MgCl_2$ concentrations, reasoning that such conditions would force a "nucleoid" conformation of DNA which might be less subject to shear forces. Under these conditions, a mixture of partially sheared and high-molecular-weight DNA of equivalent size to that obtained by enzymatic lysis was obtained from the Octopus Spring cyanobacterial mat (Fig. 1A). The extent of shearing is apparently related to the pressure used, as the sheared DNA was smaller at 1,405 kg/cm^2 (20,000 lb/in^2) than at 702.5 kg/cm^2 (10,000 lb/in^2). However, at both

A. Octopus Spring 50° C Mat

B. Pure Cultures of Thermophilic Phototrophic Bacteria

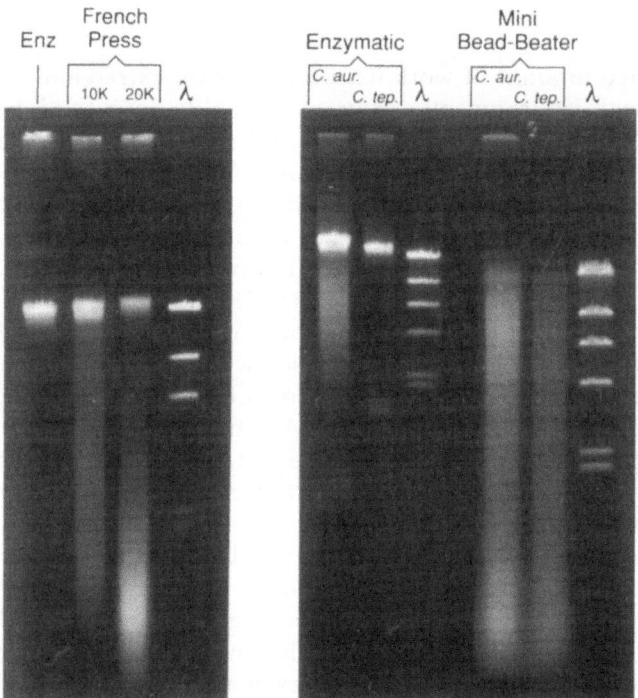

Figure 1. The effect of lysis protocol on the quality of DNA recovered as determined by 0.6% agarose gel electrophoresis.

(A) 50 °C Octopus Spring cyanobacterial mat lysed by passage through French Press at 1,405 kg/cm^2 (20,000 lb/in^2, 20 K), 702.5 kg/cm^2 (10,000 lb/in^2, 10 K), or by the enzymatic protocol (enz).

(B) Pure cultures of *Chloroflexus aurantiacus* (*C.aur.*) Y-400-fl and *Chromatium tepidum* (*C.tep.*), lysed by the enzymatic or mini-bead beater protocol. λ = Hind III digested lambda DNA as size markers (23, 9.4, 6.6, 4.4, 2.3 and 2.0 kilobase pairs, top to bottom).

pressures there was a significant yield of high-molecular-weight DNA. We have applied the same protocol successfully to retrieve high-molecular-weight DNA from the cyanobacterial mat at another Yellowstone hot spring, Clearwater Springs, as well as from every pure culture we have examined to date, including numerous thermophilic *Synechococcus* species, *Bacillus stearothermophilus*, and *Methanobacterium thermoautotrophicum*. In general, DNA from both French press and enzymatic lysis protocols was suitable for PCR amplification of SSU rRNA genes.

Mini-bead beater lysis

We have also investigated another mechanical lysis protocol, mini-bead beating. Under the conditions we employed the DNA was sheared, as illustrated for pure cultures of *Chloroflexus aurantiacus* and *Chromatium tepidum* in Fig. 1B. Performing mini-bead beating under the same conditions of elevated levels of NaCl and $MgCl_2$ as used with French press lysis (above) did not lead to the recovery of high-molecular-weight DNA. this may result from the continuous exposure of DNA to shear forces in the mini-bead beater, as opposed to the limited shearing that occurs as cells pass through the French press orifice. DNA obtained after mini-bead beater lysis was suitable for PCR amplification of SSU rRNA genes.

A warning about formation of chimeras during PCR amplification of SSU rRNA genes

We and others have found evidence of the formation of chimeric artifacts when using PCR to amplify SSU rRNA genes in natural samples [1–3, 6, 7]. Removal of low-molecular weight DNA fragments that may arise from any lysis protocol before PCR amplification may help to reduce chimera formation. The detection of these chimeras is difficult when the parental sequences are unknown [6], as is often the case in analysis of natural samples.

Procedures

French press lysis

Steps in the procedure
1. Transfer frozen sample (≤2 g wet weight) to a sterilized 15 ml Dounce tissue grinder, and homogenize with high salt, $MgCl_2$ lysis buffer.

2. Add the sample to an ethanol-rinsed French press mini-cell (maximum capacity 3.7 ml) and send through the press at a pressure of 702.5 to 1,405 kg/cm^2 (10,000 – 20,000 lb/in^2) following the manufacturer's instructions (SLM Aminco, Urbana, IL, USA).
3. Collect sample and dilute to 0.2 M NaCl and 1 mM MgCl$_2$ in TE (pH 8.0) buffer. Add sodium dodecyl sulfate (SDS) to bring the final concentration to 1% SDS, and add Proteinase K to 60 µg/ml. Incubate for 1 hour at 50 °C.
4. Extract and precipitate nucleic acids and purify DNA as described below.

Solutions
- High salt, MgCl$_2$ lysis Buffer
  - 1 M NaCl.
  - 5 mM MgCl$_2$
  - 10 mM Tris, pH 8.0.
- TE (pH 8.0) buffer
  - 10 mM Tris, pH 8.0.
  - 1 mM EDTA, pH 8.0.

Enzymatic lysis

Modified from the procedure of Marmur [8].

Steps in the Procedure
1. Transfer frozen sample (≤2 g wet weight) to a sterile 15 ml Dounce tissue grinder, and homogenize with TE (pH 8.0) buffer (see above), 200 µg/ml lysozyme and 20 µg/ml RNase A. Incubate at 37 °C for 10 min.
2. Add SDS to 1%, and 200 µg/ml Proteinase K. Incubate at 50 °C for 1 h.
3. Extract and precipitate nucleic acids as described below.

Note
1. RNase A solution should be heated to 80 °C for 10 min to inactivate DNases.

Phenol/chloroform extraction and ethanol precipitation of nucleic acids

Steps in the procedure

1. Add 1 volume TE-buffered phenol to lysed cell suspensions and mix gently for 1 to 3 min in a screw cap polypropylene centrifuge tube. Centrifuge to separate phases (e.g., 2,500 r.p.m. Sorvall RC5B, HB-6 rotor = 1,021 × g) for 5 min.
2. Repeat extraction of aqueous phase as in step 1.
3. Extract aqueous phase as in step 1, but using 0.5 volume TE buffered phenol and 0.5 volume chloroform:isoamyl alcohol (IAA) (24:1).
4. Extract aqueous phase as in step 1, but using 1 volume chloroform:IAA (24:1).
5. Precipitate nucleic acids by adding 0.1 volume 3 M sodium acetate (pH 5.2) and 2.5 volumes absolute ethanol. Place at −20 °C for 2 h or overnight. Centrifuge to pellet nucleic acids (e.g., 10,000 r.p.m., Sorvall RC5B, HB-6 rotor = 16,341 × g) for 15 to 30 min.
6. Wash with 70% ethanol, dry in vacuo, and resuspend in TE buffer (pH 8.0, see above).

Note
1. The reported value of g is the maximum relative centrifugal force for the condition given as an example.

Solution

− TE-buffered phenol–melt distilled phenol at 65 °C, add 8-hydroxyquinoline to 0.1%. Equilibrate phenol with distilled water. Remove aqueous phase and replace several times with TE (pH 8.0) buffer (see above), until the aqueous phase equilibrates to pH 8.0.

Purification of DNA

A standard treatment to remove RNA is often performed following extraction and precipitation. For example, the following method was used for the samples shown in Fig. 1A.

Steps in the procedure
1. Add RNase A to 20 µg/ml and incubate at 37 °C for 10 to 60 min.
2. Phenol/chloroform extract and ethanol precipitate as described above.

Mini-bead beater lysis

This is the procedure described by Pitcher et al., 1989 [10].

Steps in the procedure
1. Add frozen sample (≤0.2 g wet weight) to a 2.0 ml screw cap centrifuge tube with 1.0 ml sterilized zirconium beads (0.1 mm, Biospec Products) and 0.6 ml GES reagent.
2. Insert tube into the holder on the mini-bead beater (Biospec Products) and allow to beat for 5 min.
3. Centrifuge to reduce foam (e.g., 14,000 r.p.m., Eppendorf Microcentrifuge 5415C = 16,000 × g) for 1 min. Add 0.3 ml cold 7.5 M NH$_4$ Ac acetate and cool on ice for 10 min. or longer.
4. Add 0.5 ml of chloroform:2-pentanol (24:1), mix thoroughly, and centrifuge to separate phases (e.g., 14,000 r.p.m., Eppendorf Microcentrifuge 5415C = 16,000 × g) for 10 min.
5. Transfer the aqueous phase to a fresh tube, add 0.54 volumes of cold isopropanol, mix gently for 1 min, then centrifuge to pellet nucleic acids (e.g., 8500 r.p.m., Eppendorf Microcentrifuge 5415C = 6,500 × g) for 20 s.
6. Wash pellet five times with 70% ethanol. Dry in vacuo. Resuspend nucleic acids in water or TE buffer (pH 8.0, see above).

Note
The samples shown in Fig. 1B which were lysed by the mini-bead beater method were not treated with RNase, but this may be performed to purify the DNA further (see above).

Solution
- GES Reagent
  - 5 M guanidium thiocyanate (Sigma).
  - 100 mM EDTA.
  - 0.5% (v/v) sarkosyl.

  Add 60 g guanidium thiocyanate to 20 ml 0.5 M EDTA (pH 8.0) and

20 ml deionized water. Heat at 65 °C with mixing until dissolved. After cooling, add 5 ml of 10% v/v sarkosyl, and bring volume to 100 ml with deionized water. Filter sterilize and store at room temperature.

Acknowledgements

This work was supported by the US National Science Foundation grant number DEB-9209677. We thank the US National Park Service for permission to collect samples in Yellowstone National Park.

References

1. Angert ER, Clements KD, Pace NR (1993) The largest bacterium. Nature 362: 239–241.
2. Barns SM, Fundyga RE, Jeffries MW, Pace NR (1994) Remarkable archaeal diversity detected in a Yellowstone National Park hot spring environment. Proc Nat Acad Sci USA 91: 1609–1613.
3. Choi BK, Paster BJ, Dewhirst FE, Gobel UB (1994) Diversity of cultivable and uncultivable oral spirochetes from a patient with severe destructive periodontitis. Infect Immun 62: 1889–1895.
4. Giovannoni SJ, Britschgi TB, Moyer CL, Field KG (1990) Genetic diversity in Sargasso Sea bacterioplankton. Nature 345: 60–63.
5. Kirshtein JD, Zehr JP, Paerl HW (1993) Determination of N_2 fixation potential in the marine environment: application of the polymerase chain reaction. Mar Ecol Prog Ser 95: 305–309.
6. Kopczynski ED, Bateson MM, Ward DM (1994) Recognition of chimeric small-subunit ribosomal DNAs composed of genes from uncultivated micro-organisms. Appl Environ Microbiol 60: 746–748.
7. Liesack W, Weyland H, Stackebrandt E (1991) Potential risks of gene amplification by PCR as determined by 16S rDNA analysis of a mixed-culture of strict barophilic bacteria. Microb Ecol 21: 191–198.
8. Marmur J (1961) A procedure for isolation of deoxyribonucleic acid from micro-organisms. J Mol Biol 3: 208–218.
9. Muyzer G, de Waal EC, Uitterlinden AG (1993) Profiling of complex microbial populations by denaturing gradient gel electrophoresis analysis of polymerase chain reaction-amplified genes coding for 16S rRNA. Appl Environ Microbiol 59: 695–700.
10. Pitcher DG, Saunders NA, Owen RJ (1989) Rapid extraction of bacterial genomic DNA with guanidium thiocyanate. Lett Appl Microbiol 8: 151–156
11. Schmidt TM, DeLong EF, Pace NR (1991) Analysis of a marine picoplankton community by 16S rRNA gene cloning and sequencing. J Bacteriol 173: 4371–4378.
12. Ward DM, Bateson MM, Weller R, Ruff-Roberts AL (1992) Ribosomal RNA analysis of micro-organisms as they occur in nature. Adv Microbial Ecol 12:219–286.
13. Ward DM, Ferris MJ, Nold SC, Bateson MM, Kopczynski ED, Ruff-Roberts AL (1994) Species diversity in hot spring microbial mats as revealed by both molecular and enrichment culture approaches: relationship between biodiversity and community structure. In: Stal LJ, Coumette P (eds) Microbial mats: structure, development and environmental significance, pp. 33–44. NATO/ASI Series G: Environmental Sciences, vol. 35, Springer Verlag, Heidelberg.

Molecular Microbial Ecology Manual **1.2.3**: 1–14, 1995.
ⓒ 1995 *Kluwer Academic Publishers.*

Methods for extracting RNA or ribosomes from microbial mats and cultivated microorganisms

DAVID M. WARD[1], ALYSON L. RUFF-ROBERTS[1] and
ROLAND WELLER[2]

[1] *Department of Microbiology, Montana State University, Bozeman, MT 59717 USA.;*
[2] *Center for Microbial Ecology, Michigan State University, East Lansing, MI 48824-1325
USA*

Introduction

The division of Section 1 of this manual into many chapters on methods for isolation of DNA or RNA from various natural habitats reflects the opinion that the best method may depend upon which type of nucleic acid is sought from which type of habitat. As the various chapters will undoubtedly exhibit, many methods have been used to extract nucleic acids from different natural habitats. For example, in the specific case of microbial mats, hot phenolic, detergent and enzymatic methods have been used to obtain DNA for PCR amplification [3, 6, and H. Paerl, personal communication]. Further, enzymatic and French press methods have been used to obtain small subunit ribosomal RNA (SSU rRNA) for cDNA synthesis [18, 19] and French press [9] and bead-beater methods (B. Rissati and D. Stahl, personal communication) have been used to obtain SSU rRNA for hybridization probe studies. Very few studies, however, report a comparison of different methods applied to a single system [e.g., 5, 8]. For instance each of the studies reporting RNA extraction methods [e.g., 2, 4, 10, 13] report a single method. While this may result from a tendency to report only the optimized method, we feel that a comparison of different methods is informative, as it is possible that methods will have similar biases when applied to either DNA or RNA recovery from many different habitats.

Here, we report results comparing French press, bead-beater, enzymatic and chemical lysis methods that we used to obtain RNA for different purposes, within the context of our work on hot spring microbial mats and cultivated microorganisms [14, 15, 17 and references therein]. It is important to note, however, that we did not attempt to optimize any of the methods we studied with respect to RNA recovery efficiency. Thus, we are reporting differences in the methods as applied and not ultimate differences that might be obtainable.

Microbial mat samples may present fewer difficulties for nucleic acid recovery than samples from other habitats. For example, their low mineral content means that the removal of microorganisms from such surfaces is unnecessary. For the same reason, it is possible to use one of the most harsh mechanical procedures, French press lysis, which cannot be used directly on samples from many other habitats. As we report below, this method was superior in recovering RNA from microbial mat samples we studied. Further, as reported in Chapter 1.1.4, it is possible to use French press lysis to obtain high-molecular weight DNA from mats and bacteria. This might be worth noting, since French press lysis could in principle be performed on samples from any habitat if cells can be separated from mineral debris.

Comparison of methods for recovery of SSU rRNA from a hot spring microbial mat and cultivated thermophilic bacteria

We examined French press, bead-beater, enzymatic and chemical methods in terms of their efficiency in recovering RNA from 1) a 50 °C region of the Octopus Spring cyanobacterial mat in Yellowstone National Park [16], 2) phototrophic bacteria cultivated from such mats (the cyanobacterium *Synechococcus* sp. Y-7c-S and the green nonsulfur bacterium *Chloroflexus aurantiacus* Y-400-fl), 3) mixtures of the cultivated species and mat samples, and 4) naturally occurring cyanobacterial (A-type) and *Chloroflexus*-like (C-type) community members whose SSU rRNAs we have detected in the mat. The cultivated and uncultivated species serve as added and naturally occurring internal markers, respectively. The lysis methods are detailed in the Procedures section. The total SSU rRNA yield was quantified by the reactivity of a universal hybridization probe. The relative abundance of specific SSU rRNAs of cultivated and uncultivated species was quantified as the relative response of specific and universal hybridization probes [see 9 for probe methods].

The total SSU rRNA yield was significantly ($P<0.05$) higher when samples were lysed by French press as compared with other lysis methods (Fig. 1A), except in the case of *C. aurantiacus* samples where the yield was not significantly different from that obtained with enzymatic lysis. There were no statistically significant differences in recovery of any of the specific SSU rRNAs relative to total SSU rRNA with different lysis methods applied to any of the samples (Fig. 1B). It might be possible to demonstrate that bead beater lysis favors recovery of A-type and C-type SSU rRNA with more replication. With respect to these four SSU rRNA markers the different lysis methods sampled the mat in qualitatively similar ways. We cannot, however, exclude the possibility that some or all of the methods may bias against lysis of species whose SSU rRNAs we did not probe. It is interesting to note that the SSU rRNA of both cultivated species was below detection limits in the mat, but this was obviously not because of their resistance to lysis, since SSU rRNAs of both species could

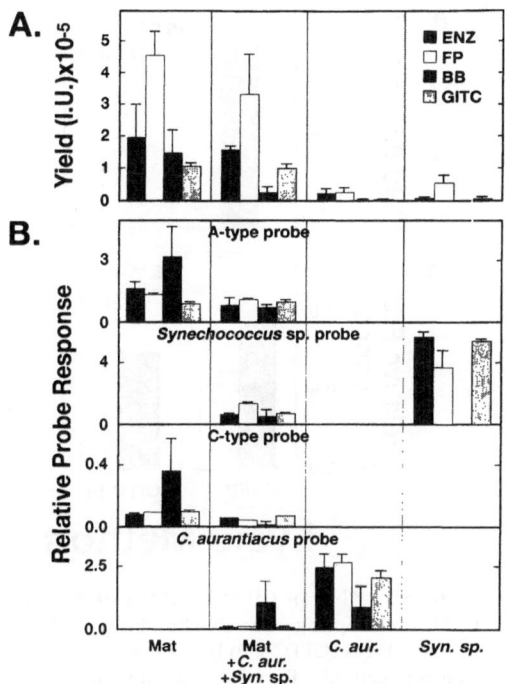

Lysed RNA Sample

Figure 1. Yield of (A) total SSU rRNA or (B) SSU rRNA of A-type and C-type uncultivated members of the Octopus Spring mat, or *Synechococcus* sp. Y-7c-S or *Chloroflexus aurantiacus* Y-400-fl after enzymatic (ENZ), French press (FP), bead-beater (BB) or chemical (GITC) lysis of 50 °C Octopus Spring cyanobacteria mat samples, *Synechococcus* sp., *C. aurantiacus* or mat containing added cells of both species. Total SSU rRNA was quantified using a universal probe which reacts with all bacterial and archaeal SSU rRNAs, and specific SSU rRNAs were quantified as a relative probe response by normalizing the responses of specific probes with the universal probe response (see Ruff-Roberts et al., 1994 [9]). Bars indicate standard error (n=3). Bead-beater lysis of *Synechococcus* sp. failed to recover RNA, so that there was also no *Synechococcus* sp. probe response. IU, integration units.

be detected alone or when added to mat samples before lysis.

The quality of extracted RNA was evaluated by gel electrophoresis. After enzymatic lysis distinct 23S-, 16S- and 5S-sized rRNA bands were visualized [18]. Chemical lysis also resulted in the recovery of RNA with a distinct 16S-sized band. French press lysis resulted in sheared RNA ranging from ca. 120 to 1500 nucleotides in length, which nevertheless sufficed for use as targets for oligonucleotide hybridization probes [9]. Bead-beater lysis also resulted in shearing, with RNA ca. 120 nucleotides in length. We explored combining chemical and French press lysis as a possible means of maximizing both yield and quality of RNA. Though the quality of extracted RNA was improved (i.e., 16S-sized band observed),

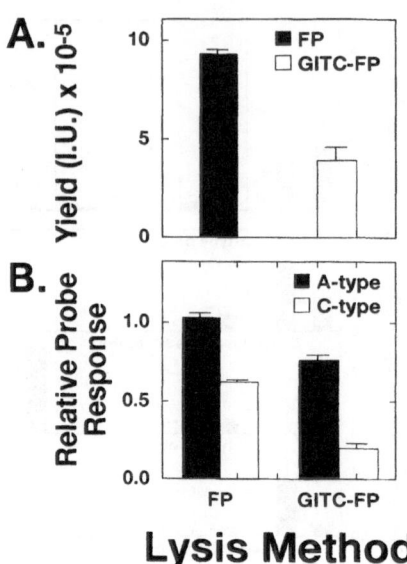

Lysis Method

Figure 2. Yield of (A) total SSU rRNA or (B) SSU rRNA of A-type or C-type uncultivated members of the 50 °C Octopus Spring cyanobacterial mat after French press (FP) or combined chemical and French Press (GITC-FP) lysis of 50 °C Octopus Spring microbial mat samples. See Fig. 1 legend for methods. Bars indicate standard error (n=3). IU, integration units.

the combined method gave lower yield and poorer recovery of both A-type and C-type SSU rRNAs relative to total SSU rRNA, than did French press lysis alone (Fig. 2).

Comparison of methods for recovery of ribosomes from a hot spring microbial mat and cultivated bacteria

In the course of our work on retrieval of SSU rRNA sequences from mats, we developed a method for random priming of cDNA synthesis [19], which depends upon the purification of small ribosomal subunits before RNA extractions, as described in the Procedures section. We have compared the French press lysis method and a Mini-bead-beater lysis protocol in terms of the yield of RNA recovered in this way from some cultivated bacteria. As shown in Table 1, with French press lysis the recovery was higher and less variable than with Mini-bead-beater lysis. This was consistent with gel evidence suggesting that the RNA recovered from ribosomes after Mini-bead-beater lysis was partially degraded. RNA from large or small ribosomal subunits recovered after French press lysis showed distinct 23S- and 16S-sized bands, respectively, and was suitable for cDNA synthesis [19].

Table 1. Recovery of RNA associated with ribosomes obtained after French press or Mini-bead-beater lysis.

Organism	% of total RNA [1] recovered as RNA in ribosomes [2] after lysis by	
	French press	Bead-beater
Bacillus subtilis	64%	58–60%
Corynebacterium flaccumfaciens	64%	32–36%
Pseudomonas sp. DB01	87%	14–22%
Micrococcus flavus	73%	40%

[1] estimated from absorbance after lysis and extraction, assuming that 80% of total RNA is rRNA [7]
[2] estimated assuming that 58% of the weight of ribosomes [11], as determined by absorbance (1 A_{260} unit = 69 ug ribosomes/ml), is rRNA

Protocols

Homogenization of mat samples for French press, Bead-beater, and enzymatic lysis

Steps in the Procedure

1. Place frozen sample (up to ca. 2 g wet weight) in a 15 ml Dounce tissue grinder (Corning) on ice. Add 3.25 ml lysis buffer and thaw quickly at 50 °C. Homogenize until a smooth slurry is obtained.
2. Transfer the homogenate to the lysis chamber of a mechanical lysis device or into a Corex tube for enzymatic lysis.

Solution

- lysis buffer
  - 80 mM NaCl
  - 8 mM Tris-HCl, pH 7.6
  - 0.8 mM EDTA
  - 15 µg/ml lysozyme

French press lysis

Steps in the procedure

1. Lyse sample by three passes through the French press Mini-Cell (SLM Aminco) at 1,405 k/cm^2 (20,000 lb/in^2), collecting each time in a chilled 30 ml Corex tube.
2. Add 0.5 ml STE to the tissue grinder to remove all residual sample. French Press at 1,405 kg/cm^2 (20,000 lb/in^2). Collect in the same Corex tube.
3. Add 1 ml proteinase digest buffer and 2 mg proteinase K and digest at 50 °C for 1 hour.
4. Phenol/chloroform extract nucleic acids, and purify RNA as described below.

Notes

1a. If there is mineral debris in the sample it should be removed before French Press lysis in order to avoid damage to the chamber. For this study, mineral debris was removed after homogenization by low-speed centrifugation (e.g., 3000 rpm, Sorvall RC5B, HB-6 rotor = 1,470 $\times$ g) and the supernatant slurry was removed to a chilled 30 ml Corex tube. The reported value of g is the maximum relative centrifugal force for the example conditions given.

1b. Eliminate contaminating RNase from French press Mini-Cell by soaking 15 minutes in 10% H_2O_2.

1c. Chill Mini-Cell at —20 °C in an airtight container.

2. Change nylon ball in French press valve after lysing each sample.

Solutions

– STE
  – 100 mM NaCl
  – 10 mM Tris-HCl, pH 7.6
  – 1 mM EDTA
– proteinase digest buffer
  – 2.5 M NaCl
  – 5% sodium dodecyl sulfate

Bead-beater lysis

Steps in the procedure

1. Add 2.75 ml lysis buffer (see above) to the homogenate.
2. Place 15 ml sterile 0.1 mm glass beads in the 30 ml chamber of a Bead-beater (Biospec Products). Add 6 ml STE-saturated phenol to chamber. Add sample and close the chamber. To keep the sample from overheating, place an icewater-filled outer chamber over the sample chamber.
3. Homogenize mixture for 2 minutes with the Bead-beater.
4. Transfer mixture to a chilled 30 ml Corex tube, phenol/chloroform extract nucleic acids and purify RNA as described below.

Notes
2a. Heat sterilize glass beads in a 177 °C oven for 12 hours.
2b. Remove RNase from the Bead-beater chamber by soaking 15 minutes in 10% H_2O_2.

Solution

– STE-saturated phenol – Add 8-hydroxyquinoline to 0.1% final volume. Equilibrate phenol with distilled H_2O overnight. Equilibrate phenol with several volumes of STE, pH 7.6 (see above), until pH of aqueous layer is >7.6.

Enzymatic lysis

Steps in the procedure

1. Freeze homogenate quickly by swirling tube vigorously in a dry ice-ethanol (95%) slurry.
2. Thaw the frozen sample quickly in a 50 °C water bath.
3. Repeat the freeze-thaw cycle twice.
4. Add 0.5 ml proteinase digest buffer (see above) and 1 mg proteinase K to the thawed sample and digest at 50 °C for 1 hour.
5. Phenol/chloroform extract nucleic acids and purify RNA as described below.

Chemical lysis

This is essentially the guanidium isothiocyanate (GITC)-phenol-chloroform extraction first described by Chomczynski and Sacchi [1].

Steps in the procedure

1. Place frozen sample (up to 2 g wet weight) in a 7 ml Dounce tissue grinder. Add 3 ml working solution D and thaw at 50 °C.
2. Add 100 µl 2M sodium acetate, 1 ml H_2O-saturated phenol, and 200 µl chloroform:isoamyl alcohol (24:1). Homogenize until a smooth slurry is obtained.
3. Transfer homogenate to a chilled 30 ml Corex tube and centrifuge to separate phases (e.g. 10,000 rpm, Sorvall RC5B, HB-6 rotor = 16,341 g) for 20 minutes at 4 °C. Remove the aqueous fraction to a chilled Corex tube.
4. Precipitate nucleic acids with 3 ml isopropanol for 2 hours at —20 °C.
5. Centrifuge to pellet nucleic acids (e.g., 10,000 rpm, Sorvall RC5B, HB-6 rotor = 16,341 g) for 10 minutes at 4 °C.
6. Discard supernatant and air dry pellet.
7. Resuspend pellet in 500 µl working solution D by heating 5 minutes at 65 °C.
8. Ethanol precipitate nucleic acids and purify RNA as described below.

Solutions

− Working solution D
  − Add 360 µl 2-mercaptoethanol to 5 ml stock solution D
− Stock solution D
  − 4 M guanidium thiocyanate
  − 25 mM sodium citrate, pH 7.0
  − 0.5% sarcosyl

Combined chemical and French press lysis

Steps in the procedure

1. Place frozen sample (up to 2 g wet weight) in a 7 ml Dounce tissue grinder. Add 1 ml working solution D (see above) and thaw at 50 °C. Homogenize until a smooth slurry is obtained.
2. Add 100 µl 2M sodium acetate, 1 ml H_2O-saturated phenol (see above), and 200 µl chloroform: isoamyl alcohol (24:1).
3. Transfer homogenized sample into the chamber of a French press Mini-Cell.
4. Lyse sample by three passes through the French press at 1,405 kg/cm^2 (20,000 lb/in^2), collecting each time in a chilled 30 ml Corex tube.
5. Centrifuge to separate phases (as above). Remove the aqueous fraction to a chilled Corex tube.
6. Precipitate nucleic acids with 3 ml isopropanol for 2 hours at — 20 °C.
7. Centrifuge to pellet nucleic acids (as above).
8. Discard supernatant and air dry pellet.
9. Resuspend pellet in 500 µl working solution D by heating 5 minutes at 65 °C.
10. Ethanol precipitate nucleic acids and purify RNA as described below.

Mini-Bead-beater lysis (to obtain ribosomes)

Steps in the procedure

1. Combine 0.7 ml zirconium beads (100-150 µm, Biospec Products) and 1 ml of a cell suspension in Buffer I in a 2 ml screw-cap centrifuge tube.
2. Insert the tube into the holder of the Mini-Bead-beater (Biospec Products) and beat in 30 second cycles for a total of 6 minutes, chilling on ice for 1.5 minutes between cycles.
3. Purify ribosomes as described below.

Notes

1a. Acid-wash and bake beads before use.

Solutions

- buffer I
    - 20 mM Tris, pH 7.4
    - 150 mM KCl
    - 15 mM $MgCl_2$
    - 2 mM Dithiothreitol (add to buffer after preparation of the first ingredients in diethyl pyrocarbonate (DEPC)-treated water).
- DEPC-treated distilled H_2O – Treat distilled H_2O with 0.1% DEPC with constant stirring for 12 hours, then autoclave.

Purification of ribosomes

The purification of ribosomes is a prerequisite to the isolation of pure SSU rRNA, which can be used for a random priming approach to obtain cDNA from SSU rRNA [19]. The following protocol is a modification of those used for the isolation of ribosomes in research addressing the structure and function of the ribosome [12].

Steps in the procedure

1. Adjust the volume of the cell lysate (after French press or Mini-bead-beater lysis, as above) with buffer I (see above) to less than 18 ml to allow processing in one ultracentrifuge tube and centrifuge to sediment unlysed cells (e.g., 5,000 rpm, Sorvall ultracentrifuge, T865 rotor = 2,541 g).
2. Collect supernatant and centrifuge to pellet cellular debris (e.g., 17,000 rpm, Sorvall ultracentrifuge, T865 rotor = 29,876 × g) for one hour.
3. Collect ribosomes by centrifugation over a 40% sucrose cushion (sucrose in buffer I) for 2.5 hours (e.g., 65,000 rpm, Sorvall Ultracentrifuge, T865 rotor = 429,458 × g).
4. Discard the supernatant above the sucrose cushion, dilute the sucrose to 20% or less with buffer I and add a new 40% sucrose cushion (about 7 ml) under the 20% solution.

5. Pellet the ribosomes using centrifugation (e.g., 65,000 rpm, Sorvall ultracentrifuge, T865 rotor = 429,458 × g) for 17 hours at 4 °C.
6. Remove all liquid and resuspend ribosomes in 1 ml of buffer I.
7. Phenol extract rRNA from a small aliquot (as described below) and check the integrity of the ribosomes by gel electrophoresis [see 19].
8. If small or large ribosomal subunits are desired, resuspend pellet from step 6 in buffer II to dissociate the subunits. Separate on a 5 to 30% sucrose gradient (wt/wt in buffer II) by centrifugation for 2.5 hours (e.g., 49,600 rpm, Sorvall ultracentrifuge, T865 rotor = 250,000 × g). Collect fractions and use absorbance at 260 nm to detect peaks corresponding to SSU or LSU rRNA. Extract RNA from corresponding fractions as described below.

Notes

1. During all steps all solutions and tubes should be kept as cold as possible to avoid RNase activity.
1a. Beads from Mini-bead-beater lysis are removed in the centrifugation step.
8. Gradient makers are available from many manufacturers. Follow manufacturer's instructions for gradient preparation.

Solution

- buffer II
    - 20 mM Tris, pH 7.4
    - 100 mM KCl
    - 1.5 mM MgCl$_2$

Phenol/chloroform extraction of nucleic acids

Steps in the procedure

1. Add one volume STE-saturated phenol to lysed cell suspensions or ribosome preparations (see above), mix using a vortex mixer, and centrifuge to separate phases (e.g., 10,000 rpm, Sorvall RC5B, HB-6 rotor = 16,341 × g) for 10 minutes at 4 °C. Remove the top fraction to a chilled 30 ml Corex tube.

2. Add phenol:chloroform:isoamyl alcohol (25:24:1), mix using a vortex mixer, and centrifuge to separate phases (as above). Remove the aqueous fraction to a chilled 30 ml Corex tube.
3. Add one volume chloroform:isoamyl alcohol (24:1), mix using a vortex mixer, and centrifuge to separate phases (as above). Remove the aqueous fraction to a chilled 30 ml Corex tube.
4. Ethanol precipitate nucleic acids and purify RNA as described below.

Notes
1. Dot not add phenol to a Bead-beater lysed sample before centrifugation, as phenol was added during the lysis protocol. The Beads and phenol will be centrifuged beneath the aqueous layer.

Solution
– Phenol:chloroform:isoamyl alcohol – Mix 25 parts of STE-saturated phenol (see above) with 24 parts chloroform and 1 part isoamyl alcohol.

Ethanol precipitation of nucleic acids

Steps in the procedure

1. Add 0.1 volume 3M sodium acetate and two volumes absolute ethanol to aqueous extracts containing nucleic acids (see above), invert to mix, and precipitate overnight at —20 °C.
2. Pellet the precipitated nucleic acids by centrifugation (as above). Remove supernatant.
3. Wash pellet with 70% ethanol. Centrifuge (as above) to pellet nucleic acids. Remove supernatant. Dry pellet at room temperature under vacuum.
4. Purify RNA as described below.

Purification of RNA

Steps in the procedure

1. Resuspend pellet described in step 4 of the ethanol purification procedure in 400 µl DNase buffer and transfer to a 2 ml Eppendorf tube.

2. Add 100 units RQ1 DNase (Promega) and digest 30 minutes at 37 °C.
3. Phenol/chloroform extract and ethanol precipitate as previously described.
4. Resuspend pellet in DEPC-treated distilled H_2O (see above).

Solution

- DNase buffer
  - 40 mM Tris-HCl, pH 7.9
  - 10 mM NaCl
  - 6 mM $MgCl_2$
  - 10 mM $CaCl_2$

Acknowledgements

This work was supported by the US National Science Foundation (DEB-9209677). We thank the US National Park Service for permission to collect samples in Yellowstone National Park.

References

1. Chomczynski P, Sacchi N (1987) Single-step method of RNA isolation by acid guanidium thiocyanate-phenol-chloroform extraction. Anal Biochem 162: 156–159.
2. Hahn D, Kester R, Starrenburg MJC, Akkermans ADL (1990) Extraction of ribosomal RNA from soil for detection of *Frankia* with oligonucleotide probes. Arch Microbiol 154: 329–335.
3. Kirshtein JD, Zehr JP, Paerl HW (1993) Determination of N_2 fixation potential in the marine environment: application of the polymerase chain reaction. Mar Ecol Prog Ser 95: 305–309.
4. Moran MA, Torsvik VL, Torsvik T, Hodson RE (1993) Direct extraction and purification of rRNA for ecological studies. Appl Environ Microbiol 59: 915–918.
5. More MI, Herrick JB, Silva MC, Ghiorse WC, Madsen EL (1994). Quantitative cell lysis of indigenous microorganisms and rapid extraction of microbial DNA from sediment. Appl Environ Microbiol 60: 1572–1580.
6. Muyzer G, De Waal EC, Uitterlinden AG (1993) Profiling of complex microbial populations by denaturing gradient gel electrophoresis analysis of polymerase chain reaction-amplified genes coding for 16S rRNA. Appl Environ Microbiol 59: 695–700.
7. Neidhardt FC (1987) Chemical composition of *Escherichia coli*, p. 3–6 in *Escherichia coli* and *Salmonella typhimurium* (F.C. Neidhardt, ed.), Cellular and molecular biology, Am Soc Microbiol, Washington, DC.
8. Picard C, Ponsonnet C, Paget E, Nesme X, Simonet P (1992) Detection and enumeration of bacteria in soil by direct DNA extraction and polymerase chain reaction. Appl Environ Microbiol 58: 2717–2722.

9. Ruff-Roberts AL, Kuenen JG, Ward DM (1994) Distribution of cultivated and uncultivated cyanobacteria and *Chloroflexus*-like bacteria in hot spring microbial mats. Appl Environ Microbiol 60: 697–704.

10. Somerville CC, Knight IT, Straube WL, Colwell RR (1989) Simple, rapid method for direct isolation of nucleic acids from aquatic environments. Appl Environ Microbiol 55: 548–554.

11. Spirin AS, Gavrilova LP (1969) The Ribosome. Springer-Verlag, NY.

12. Tapprich WE, Hill WE (1986) Involvement of bases 787–795 of *Escherichia coli* 16S ribosomal RNA in ribosomal subunit association. Proc Natl Acad Sci USA 83: 556–560.

13. Tsai Y, Park MJ, Olson BH (1991) Rapid method for direct extraction of mRNA from seeded soils. Appl Environ Microbiol 57: 765–768.

14. Ward DM, Bateson MM, Weller R, Ruff-Roberts AL (1992) Ribosomal RNA analysis of microorganisms as they occur in nature. Adv Microbial Ecology 12: 219–286.

15. Ward DM, Ferris MJ, Nold SC, Bateson MM, Kopczynski ED, Ruff-Roberts AL (1994) Species diversity in hot spring microbial mats as revealed by both molecular and enrichment culture approaches – relationship between biodiversity and community structure, pp. 33–44. In Microbial mats: structure, development and environmental significance (Stal LJ, Coumette P, eds.), NATO/ASI Series G: Ecological Sciences, Vol. 35, Springer-Verlag, Heidelberg.

16. Ward DM, Weller R, Shiea J, Castenholz RW, Cohen Y (1989) Hot spring microbial mats: anoxygenic and oxygenic mats of possible evolutionary significance. p. 3–15 in Microbial Mats: Physiological ecology of benthic microbial communities (Cohen Y, Rosenburg E, eds.) Am Soc Microbiol, Wash DC.

17. Ward DM, Weller R, Bateson MM (1990) 16S rRNA sequences reveal numerous uncultured microorganisms in a natural community. Nature 345: 63–65.

18. Weller R, Ward DM (1989) Selective recovery of 16S ribosomal RNA sequences from natural microbial communities in the form of cDNA. Appl Environ Microbiol 55: 1818–1822.

19. Weller R, Weller JW, Ward DM (1991) 16S rRNA sequences retrieved as randomly primed cDNA from a hot spring cyanobacterial mat community. Appl Environ Microbiol 57: 1146–1151.

Molecular Microbial Ecology Manual **1.3.1**: 1–15, 1995
© 1995 *Kluwer Academic Publishers.*

Cell extraction method

VIGDIS TORSVIK

Department of Microbiology, University of Bergen, Jahnebakken 5, N-5020 Bergen, Norway

Abbreviations:

CPB	–	cetylpyridinium bromide
DAPI	–	4,6-diamidino-2-phenylindole
EDTA	–	sodium ethylene diamine tetra-acetate
NaP	–	sodium phosphate buffers
NaPPi	–	sodium pyrophosphate
PBS	–	phosphate buffered saline
PVPP	–	polyvinylpolypyrrolidone
SDS	–	sodium dodecyl sulphate
SSC	–	standard saline citrate

Introduction

The major part of soil material consists of organic and inorganic particles, colloids and amorphous organic matter. The fact that bacterial cells make up only a small fraction of the soil and are often strongly attached to surfaces make studies of indigenous soil bacteria a difficult task. Direct observations by epifluorescence and immunofluorescence techniques are impeded by the shadowing effect and nonspecific adsorption of fluorochromes and antibodies to soil particles. Measurements of metabolic activities of soil bacteria are impeded by interference, quenching and inhibition by humic material. Therefore, analysis of the indigenous soil bacteria requires efficient methods for separation and purification of the cells from the soil as a first step. Also when the purpose of the investigation is to study the activity of prokaryotic and eukaryotic organisms separately, mechanical cell separation may be a better alternative than the use of specific inhibitors which may not inhibit the entire target group, or may affect the nontarget microflora in some unpredictable way.

The main fraction of bacteria in soil (≥99%) cannot be isolated on agar plates, and therefore no information about these bacteria can be obtained by classical microbiological techniques. On the other hand all the information about these bacteria is stored in their DNA. Recently, molecular methods have become valuable tools for the study of soil microbial populations and communities. An alternative to phenotypic characterization of the bacterial community would be to isolate DNA from all the soil bacteria and characterize the community genotypically. Specific bacteria or taxonomic groups may be identified by the use of nucleic acid probing or fingerprinting techniques without having to isolate and culture them [7].

When nucleic acids are extracted directly from soil, it consists of a mixture of prokaryotic and eukaryotic nucleic acid. In some investigations, for instance when the purpose is to assess the diversity of the bacterial community by reassociation kinetics [17], it is necessary to ascertain that the isolated DNA only arises from bacteria. In such cases and whenever highly purified bacterial DNA is needed, the bacteria have to be extracted from the soil as the first step, after which the nucleic acids are isolated from them [16].

Extraction of bacteria

The forces binding bacteria to the soil surfaces include bonding between extracellular bacterial polymers and humic colloids, electrostatic forces, hydrogen bonding, and other surface interactions. They are different for clay and humus, and vary for different bacteria. The bacteria may also be physically entrapped in soil aggregates. The fractionation efficiency is dependent on how efficient bacteria can be released from soil aggregates and soil surfaces. Optimal procedures for separating bacteria from soil particles and humic colloidal material should be rapid and simple. They must provide purified samples with reasonably high yields without being selective. Fægri et al. [4] described a differential centrifugation technique for the separation of bacteria from fungi and particles in organic soil. It has later been used by several investigators with some modifications. In the presence of clay the bacteria are not readily released by mechanical treatment [1], and the fractionation efficiency has been improved by using chemical desorbing agents.

In order to break the electrostatic forces between bacteria and colloidal material, chelating cation exchanger has been used. Polymer bridges can be broken by using detergents. Macdonald [10] found that the most efficient extraction method was to disperse the soil by shaking with the sodium salt of an iminodiacetate based resin (Dowex A1, Sigma) combined with treatment with 0.1% (w/v) sodium deoxycholate. Hopkins et al. [8] have recommended a combination of different dispersion methods; shaking the soil with anionic detergent and chelating cationic exchange resin (Chelex 100, BioRad) followed by ultrasonication (<1kJ g^{-1} soil).

Addition of polyvinylpolypyrrolidone (PVPP) before homogenization gives a purer bacterial fraction as it binds to humic material [7]. Steffan et al. [15] investigated the extraction efficiency with various extraction solutions and found that adding 0.1% sodium dodecyl sulphate and lowering the pH of the extraction solution (0.1 M phosphate buffer) apparently resulted in an improved recovery. Addition of PVPP to the first extraction slightly reduced the recovery. In our laboratory we have tried different solutions for homogenization and extraction of bacteria from soil: Winogradsky's salt solution (1:20), 0.1 M sodium phosphate buffer pH 7.5, and distilled water. Our experience with organic and sandy soil (not published) is that distilled water without the addition of chemical dispersal agents and PVPP gives the highest bacterial recovery.

For further purification of the bacterial cells, they can be subjected to isopycnic density gradient centrifugation in colloidal silica (Percoll, Pharmacia LKB Biotechnology, Uppsala, Sweden) with densities 1–1.15 g/ml [1, 11]. This purification method works best with soil low in organic matter content. Humic material covers the range of buoyant densities of bacterial cells and will not be separated from the bacteria in the gradient. The recoveries of bacteria by rate-zonal density gradient centrifugation were much higher for organic soil (82%) than for clay loam and sandy loam (24%) where cells attached to the particles are removed by sedimentation [8].

Estimation of the fractionation yield

The estimation of fractionation yield is based on microscopic count of bacteria in the pooled supernatants after the low speed centrifugation taken as a percentage of the total count of bacteria in the soil. For microscopic counting of bacteria, diluted soil samples are filtered through a Nucleopore filter prestained with Irgalan Black [6]. The bacteria on the filter are stained with DAPI according to Porter and Feig [13]. When counting bacteria stained with a fluorescent dye, care must be taken in order to reduce shadowing by soil debris. With epifluorescence microscopy, both excitation light and the light emitted from the fluorochrome complex can be reduced by shadowing and that will result in a corresponding decrease in fluorescence intensity. Small soil organic matter particles may be distinguished from bacteria by not having a distinct boundary. When counting bacteria in soil samples, a twofold dilution series should be made, and counting carried out in the range where there is inverse proportionality between counts and dilution. This range may be rather narrow, and even within this range counting of bacteria directly in the soil often gives lower values than those from the pooled supernatants with extracted bacteria plus the residue after the last low-speed centrifugation. We consider it correct to calculate the yield from the latter data, and have observed recoveries between 50 and 80% in organic soils with very low clay content [4]. Holben et al. [7] found a recovery of about 35% when counting in this manner and Steffan et al. [15] reported 27.8% recovery.

Isolation of DNA from the extracted bacteria

DNA isolated from the bacterial fraction of soil is free from eukaryotic DNA. Practically all eukaryotic organisms are removed by the fractionated centrifugation, but some free DNA [12] and viral DNA may be left in the bacterial fraction. Free DNA is removed from the bacterial fraction by washing with pyrophosphate or hexametaphosphate prior to lysis of the bacteria. Most of the soil humic matter is also removed from the extracted bacteria prior to DNA extraction. It is therefore easier to obtain ultrapure DNA with relatively high molecular weight from a bacterial fraction than from the total soil [9].

The most critical steps when isolating bacterial DNA from soil are the cell lysis and the separation of DNA from humic substances. A protocol for lysing the bacteria should ensure maximal disruption of the various bacterial types present in the soil without destroying their DNA. Mechanical disruption of cells by a sonicator is not recommended, since it may result in severe shearing of DNA. Bead beating with glass beads (0.2 to 0.5 mm in diameter) is a more gentle disruption method than sonication, and is widely used in direct DNA extraction procedures. Treatment with enzymes and detergents combined with high temperature is preferred when DNA with high molecular weight is needed. Purification of DNA from humic substances is problematic because they are macromolecular and acidic in nature, and thus follow each other in several fractionation steps. Highly purified DNA can be obtained by precipitating most of the humic matter in the presence of high concentrations of KCl or NH_4 acetate, followed by hydroxyapatite chromatography to remove other cell constituents and residual humic acids.

The yield of DNA isolated from extracted bacteria is lower than that obtained by direct isolation from soil. Nevertheless it is conceivable that DNA of the extracted bacteria is representative for the total soil bacterial DNA. DNA of extracted bacteria has a very high heterogeneity, indicating that DNA from most of the different bacterial types in the soil is included. We have also observed that the restriction fingerprints of PCR-amplified rRNA are identical in DNA extracted from the bacterial fraction and directly from the soil (unpublished results).

The purity and quantity of the DNA can be determined by spectral analysis. An absorbance of 1.0 at 260 nm wavelength and 10 mm light path, is equal to 50 μg double stranded DNA per ml. This method is only reliable with quite pure DNA, i.e., only during the last purification steps when the DNA sample is colorless. Pure DNA has a A_{260}/A_{280} ratio higher than 1.5, and a A_{260}/A_{230} ratio higher than 1.9. The increase in A_{260} of a pure DNA sample after heating to complete melting (the hyperchromicity) in 0.1 × SSC is normally 30–40%.

Procedures

Protocol 1

Extraction of bacterial cells from soil

    The protocol consists of careful homogenization of the soil (10 to 30 g portions in 100 ml distilled water in a Waring blender), dilution of the homogenate to 350 ml, centrifugation at low speed (1,000 × g) in a refrigerated centrifuge, collecting the supernatant and homogenizing the precipitate anew, repeating this three times, and finally centrifuging the combined supernatants at high speed (10,000 × g). The pellet from this centrifugation will contain bacteria together with some humic material and other particles of the same density as bacteria. After extracting the bacteria from the soil, extracellular DNA and other impurities are removed with hexamethaphosphate. As there are no indications that the procedure is selective, probably all bacterial types present in the soil are represented in the bacterial fraction. This is corroborated by the finding that plate counts and microscopic counts gave roughly the same yield [4, 7]. Balkwill et al. [2] found little difference when studying the size distribution of soil bacteria using electron microscopy before and after separation by a centrifugation procedure similar to the one described here. The bacterial fractions may be stored for a short time as pellets in the refrigerator after the last centrifugation. For longer periods the pellets may be stored frozen at —20 °C. When the pellets have high contents of humic material, and are going to be used for DNA extraction, storing in the refrigerator with isopropanol [14] seems to give higher yields. The procedure presented has worked well with organic and sandy soil. Time required: 4–5 h.

Steps in the procedure
1. Weigh out 4–6 portions of 30 g soil, previously sieved through a soil sieve, mesh size 2 mm. Homogenize each portion with 100 ml ice-cold sterile distilled water in a Waring blender at low speed three times for 1 min, with 5 min cooling on ice (or 1 min in a freezer) between each run.
2. Transfer the soil suspensions to 500 ml centrifuge flasks with 300 ml sterile water, and centrifuge for 15 min at 5 °C and 1,000 × g.

3. Pool the supernatants and store cold (4 °C) until further processing. Transfer the soil pellets back into the blender with 100 ml sterile water and homogenize anew for 1 min. Then, transfer the homogenate to the centrifuge flasks with 300 ml sterile water and centrifuge as before. The supernatants are combined with the earlier pooled supernatants. The procedure is repeated once more.

4. Centrifuge the combined supernatants from the three low-speed centrifugations batchwise for 30 min at high speed ($10,000 \times g$) and 5 °C (e.g., 9,000 r.p.m. in a Sorvall RC-5 centrifuge with GSA rotor). Discard the supernatants from these runs, and leave the pellets in the flasks until the total volume has been centrifuged.

5. Combine the bacterial pellets and resuspend them in 200 ml cold 2% (w/v) sodium hexametaphosphate, adjusted to pH 8.5 with 0.2% (w/v) Na_2CO_3. Homogenize in the Waring blender for 1 min at low speed and centrifuge at $10,000 \times g$ and 5 °C for 30 min.

6. Resuspend the pellet in 200 ml Crombach buffer (0.033 M Tris HCl, 0.001 M EDTA, pH 8.0 [3]) and centrifuge as before. Transfer the pellet to a 30 ml centrifuge tube with 20 ml Crombach buffer and centrifuge again (SS-34 rotor) at $10,000 \times g$ and 5 °C for 30 min.

7. Add an excess of isopropanol to the pellet, and homogenize with a tissue homogenizer (Ystral, Ballrechten-Dottingen, Germany, speed setting 2) to obtain a homogenous suspension. The bacterial fraction can be stored on isopropanol in a refrigerator.

Notes
- Ad 1. The homogenization is most efficient when the soil is diluted 1/10 in the homogenization solution. Too large samples reduce the cell recovery and the purity of the bacterial fraction.
- Ad 1. Water may be replaced by Winogradsky's salt solution, diluted 1:20 or 0.1 M sodium phosphate buffer for extraction of bacteria. The use of distilled water simplifies the procedure and gives purer bacterial fractions.
- Ad 1. If the soil humic content is high, 10 g of acid-washed polyvinylpolypyrrolidone (PVPP) (Sigma) can be added to the soil suspension before the homogenization. PVPP will remove humic material from the bacterial fraction, but the cell recovery will be slightly reduced.
- Ad 3. The number of homogenizations applied is a compromise between the time used and the loss of bacteria which still adhere to soil particles. This loss will depend on soil type, but with most soils three homogenizations seem to be satisfactory. Increasing the number of homogenizations does not improve the bacterial yield significantly.

- Ad 4. Alternatively, the bacteria can be concentrated by tangential flow filtration (Millipore, Pellicon), using a filter cassette with cut-off at 0.22 µm.
- Ad 5. Sodium hexametaphosphate can be replaced by 0.2% sodium pyrophosphate, but removes humic material more efficiently than pyrophosphate. The pH should be as high as possible without damaging the bacterial cells, in order to extract the humic material as efficiently as possible.
- Ad 6. Crombach buffer is recommended if the bacterial fractions are used for DNA extraction as this buffer promotes bacterial lysis. If the bacterial fractions are used for physiological studies, PBS, Winogradsky's salt solution or soil extract may be used.
- Ad 7. For long-term storage the bacterial fraction may be frozen. When the bacterial fraction is to be used for DNA extraction, it should be stored in isopropanol or frozen. When the content of humic material is high, storing on isopropanol seems to give higher DNA yields than freezing. Bacterial fractions from organic soil which are frozen form aggregates which are difficult to disperse after thawing, thereby restraining the cell lysis and reducing the DNA yield.

Equipment
- Waring blender with 2–4 buckets.
- High capacity refrigerated centrifuge with rotor for 4–6 bottles of 500–1000 ml.
- High-speed refrigerated centrifuge with rotor for 250–500 ml bottles. As an alternative to the high-speed centrifuge, a tangential flow filtration apparatus and filter cassette (cut-off 0.22 µm) can be used.
- Tissue homogenizer with rotor-stator generator.

Chemicals, reagents, solutions
- Crombach buffer [3] (0.33 M Tris hydrochloride, 0.001 M sodium ethylene diamine tetra-acetate (EDTA), pH 8.0).
- Acid washed polyvinylpolypyrrolidone (PVPP).
- 2% (wt/vol) sodium hexametaphosphate adjusted to pH 8.5 with 0.2% (wt/vol) sodium carbonate (Na_2CO_3).
- Phosphate buffered saline (PBS): 0.8 g sodium chloride, 0.2 g potassium chloride, 1.44 g disodium hydrogen phosphate, 0.24 g potassium dihydrogen phosphate, 1000 ml sterile distilled water, pH 7.4.

Protocol 2

Determination of bacterial recovery

For estimating the recovery of bacteria from a soil sample, it is necessary to have satisfactory methods for counting the bacteria in the soil and the bacterial fractions.

For microscopic counting of bacteria, we recommend filtration of a soil homogenate in distilled water through a Nucleopore filter, pore size 0.2 µm, prestained with 0.2% Irgalan Black in 2% acetic acid [6], and staining the bacteria with DAPI (4'6-diamidino-2-phenylindole) according to Porter and Feig [13]. It is recommended to prepare a twofold serial dilution of the soil, and use count values where there is inverse proportionality between count values and degree of dilution.

Steps in the procedure

1. From the soil suspension (1/10 dilution of soil) after the first 3×1 min homogenization take a 1 ml sample and dilute further (1/100 and 1/1000) with filtered (0.2 µm pore size) PBS+2% formaldehyde (filtered using a 10 ml syringe with sterile Acrodisc 13, with 0.2 µm pore size). Do the same with the pellet after the last low-speed centrifugation. From the pooled supernatants after each of the low-speed centrifugations take 1 ml samples.
2. Use a small (<20 mm) Millipore filtration funnel. Place a support membrane (Metricel), then an Irgalan Black stained Nucleopore filter (0.2 µm pore size) on top of that.
3. Add 1 ml diluted sample to 10 ml prefiltered PBS/formaldehyde in the funnel and apply low vacuum to suck the sample slowly onto the filter. Wash once with 10 ml PBS/formaldehyde.
4. Disconnect vacuum and cover the filter with sterile filtered (0.2 µm) DAPI solution (10 µg/ml). Stain for approximately 10 min, and suck the stain through the filter.
5. Wash 3 times with 5 ml filtered PBS/formaldehyde.
6. Suck the filter dry. Mount on an objective glass with a small drop of fluorescence-free immersion oil or liquid paraffin under and over the filter. Cover with a coverslip.
7. Count in an epifluorescence microscope using UV-light at 365 nm for excitation (the emission light is at and above 390 nm). Count with a 100× fluorescence immersion oil objective.

Equipment
- Epifluorescence microscope.
- Membrane filter apparatus.
- Vacuum pump or water suction.
- 10 ml syringe with sterile Acrodisc 13 (Amicon), with 0.2 µm pore size.
- Metricel Membrane Filter GA-6, 25 mm, 0.45 µm Gelman Sciences Inc.
- Polycarbonate filters (Nucleopore or equivalent) stained with Irgalan Black (Ciba Geigy). Let the Nucleopore filters stay in the Irgalan Black solution overnight. Rinse twice with filtered sterile water. Store in distilled water or PBS buffer with 2% formaldehyde at 4 °C.

Solutions
- Phosphate buffered saline (PBS)+2% formaldehyde: 0.8 g sodium chloride, 0.2 g potassium chloride, 1.44 g disodium hydrogen phosphate, 0.24 g potassium dihydrogen phosphate, 54 ml 37% formaldehyde, 946 ml sterile distilled water, pH 7.4.
- Irgalan Black solution; 0.2 g Irgalan Black in 100 ml of 2% acetic acid made up in sterile distilled water. The solution can be stored for several months in the refrigerator, but an old solution has to be filtered before use.

Protocol 3

Extraction of DNA from the bacterial fraction
This protocol for isolation of soil bacterial DNA is a large scale procedure, whereby about 1 mg ultrapure DNA can be prepared from 120–180 g soil containing approximately 1×10^{10} bacteria per g soil dry weight [16,17]. This is at least 100 times more than what is needed for DNA hybridization and PCR techniques, but for the reassociation technique several hundred micrograms of DNA are needed.

For removal of humic substances and extracellular DNA, the extracted bacteria are washed with sodium pyrophosphate or hexametaphosphate prior to lysis. The protocol includes treatment with lysozyme, protease and a detergent (sodium dodecyl sulphate (SDS) or sarkosyl). The lysis efficiency should be controlled by direct microscopy, and is normally about 95% [7,15]. After lysis, the suspension is centrifuged to remove cellular debris. A substantial amount of humic matter can be removed by precipitation in 1 M solutions of KCl or NH_4acetate. Time required: 5 h.

Steps in the procedure
1. If the bacterial fraction has been stored in isopropanol, centrifuge at 10,000 × g for 5 min in order to remove the isopropanol.
2. Resuspend the precipitate in 30 ml Crombach buffer in a 50 ml centrifuge tube and homogenize with a tissue homogenizer to completely disrupt lumps and aggregates.
3. Adjust the volume to 40 ml with Crombach buffer. Add 5 mg/ml lysozyme and incubate the suspension at 37 °C for 1 h.
4. Add 0.2 mg/ml proteinase K and incubate further at 37 °C for 30 min.
5. Bring the suspension to 65 °C, and add SDS to a final concentration of 1% (1 ml of a 25% solution). Incubate at 65 °C for 10 min or until increased viscosity is observed.
6. Cool the suspension on ice. Add NH_4acetate to a final concentration of 1 M, and mix gently. Keep the mixture on ice for 2 h, or in the refrigerator at 4 °C overnight.
7. Centrifuge for 20 min at 12,000 × g and 5 °C. Collect the supernatant containing the DNA.
8. Resuspend the pellet once in 10 ml 1 M NH_4 acetate, and centrifuge. Pool the supernatants, which are now ready for purification on hydroxyapatite.

Note
– Ad 3. If the bacterial fraction is very mucoid, the mucus may be dissolved by adding bromhexine (Sigma) prior to, or at the same time as the lysozyme treatment. Use 100 µg/ml bromhexine and incubate at 37 °C.

Equipment
– High-speed refrigerated centrifuge with rotor for 50 ml tubes.
– Tissue homogenizer (Ystral, Ballrechten-Dottingen, Germany, or equivalent type).
– Thermocontrolled water bath.

Chemicals and solutions
- Crombach buffer [3]: 0.33 M Tris hydrochloride, 0.001 M sodium ethylene diamine tetra-acetate (EDTA), pH 8.0.
- Lysozyme.
- Proteinase K (Sigma).
- 25% (w/v) sodium dodecyl sulphate.
- 0.1 M sodium ethylene diamine tetra-acetate (EDTA), pH 8.0.
- 5 M ammonium acetate, pH 5.4.

Protocol 4

DNA purification by hydroxyapatite chromatography
The DNA solution is loaded on a hydroxyapatite column in the presence of 8 M urea [16,17]. The high urea concentration will disrupt the hydrogen bonds between DNA and humic acids. Humic acids, proteins, polysaccharides and single stranded nucleic acids are washed from the column with low concentration phosphate buffer with 8 M urea. Double stranded DNA is eluted with a buffer with higher phosphate concentration.

The eluted DNA is relatively diluted, and therefore has to be concentrated. This can be done by precipitation or ultrafiltration. In order to precipitate DNA in dilute solutions in the presence of high phosphate concentrations without co-precipitation of phosphate, cetylpyridinium bromide (CPB) is used [5].

The DNA should be stored as concentrated as possible, preferably at 4 °C with a drop of chloroform, or at −80 °C. The amount of DNA obtained can be calculated from the absorbance at 260 nm. The molecular weight of the isolated DNA is high enough for most studies. A molecular size of 10–25 kb pairs is normally obtained. Time required: 2 h; overnight; 3 h the next day.

Steps in the procedure
1. Suspend 40 ml hydroxyapatite (Bio-Gel HT, BioRad) in 120 ml 0.1 M NaP buffer, pH 6.8, and heat to 80 °C for 20 min. Decant to remove the finest particles and repeat the procedure once.
2. Resuspend in 120 ml 0.24 M NaP, 8 M urea (start buffer). Shake and let the suspension settle for 5 min. The supernatant is discarded.

3. Mix the DNA extract and the hydroxyapatite by gentle intermittent swirling for about 30 min at room temperature.
4. Swirl well and pour the suspension into a column (Pharmacia, 25 mm inner diameter). Allow the hydroxyapatite to settle with the column outlet closed. Then open the outlet and let the liquid pass through the column.
5. Wash the column with the start buffer overnight or until the eluate no longer absorbs light at 260 nm. This may be controlled with an LKB UVICORD connected to a fraction collector. The buffer flow rate should be approximately 25 ml/h (LKB Varioperpex II pump, setting 4). During this washing step, proteins, RNA, single stranded DNA, humic material and other impurities are removed.
6. Wash the column with approximately 10 column volumes of 0.014 M NaP (pH 6.8) to remove the urea.
7. Elute double stranded DNA with 0.4 M NaP (pH 6.8), using a flow rate of approximately 25 ml/h. Collect 3 ml fractions.
8. Measure the DNA in the fractions with a spectrophotometer. Pool the fractions which do not contain humic impurities and have DNA concentrations higher than 5 µg/ml (A_{260} > 0.1).

Notes

- Ad 3. The concentration of chelating agents (EDTA, citrate etc) in the DNA extract should be low as they can interfere with the binding of DNA to hydroxyapatite.
- Ad 4. The column should be relatively wide and short to ensure a good flow rate. A 1–2 cm layer of siliconized glass beads (0.2–0.5 mm) in the bottom of the column can also improve the flow rate. The column should never run dry as that can create channels through it and result in poor washing and elution of DNA.

Equipment
- Chromatography column (25 mm internal diameter). It is convenient to use columns with outlet that can be closed (i.e., LKB Pharmacia).
- Peristaltic pump (Varioperpex II, LKB Pharmacia).
- UV detector (UVCORD, LKB Pharmacia).
- Fraction collector (LKB, Pharmacia).
- Thermocontrolled water bath.

Chemicals and solutions
- Hydroxyapatite (Bio-Gel HT, BioRad).

- Sodium phosphate buffers (NaP), pH 6.8; Equimolar concentrations of sodium dihydrogen phosphate and disodium hydrogen phosphate. To make working solutions; mix equal volumes of 1 M stock solutions and dilute with sterile distilled water to the desired concentrations.
- Start buffer, 8 M urea, 0.24 M NaP; Dissolve the urea in half final volume of sterile distilled water. Add NaP to give a final concentration of 0.24 M NaP and adjust the volume to 1000 ml. Do not heat or autoclave. Store at room temperature.

Protocol 5

Concentration by precipitation with cetylpyridinium bromide (CPB)
The precipitation is carried out in Tris buffer with EDTA at pH 6.0. The mixture is frozen at —20 °C for 24 h, thawed slowly and centrifuged. The pellet is washed in water, ethanol-acetate buffer pH 4.5, and finally in analytical grade acetone. After drying in a vacuum desiccator, the DNA can be dissolved in the desired buffer or distilled water. Time required: 5 min; overnight; 2–3 h.

As an alternative to precipitation, DNA can be concentrated by filtration in spin concentrators (Centricon or Centriprep, Amicon, W.R. Grace Co.). The time required for concentration and buffer exchange is 1–2 h.

Purified DNA should be stored as concentrated as possible. Add a drop of chloroform, and store at 4 °C in the refrigerator. DNA can be stored like this for several months.

Steps in the procedure
1. Mix 10 volumes of DNA solution carefully with 1 volume 0.5 M EDTA pH 6.0 and 1.2 volumes of 1% (w/v) CPB in distilled water.
2. Freeze the mixture at —20 °C for 24 h, thaw slowly (room temperature) and centrifuge for 10 min at 6,000 × g in a swing-out rotor. The precipitate is quite voluminous. The supernatant must be sucked off carefully.
3. Wash the pellet once with 5 ml H_2O, twice with 5 ml etanol/sodium-acetate buffer pH 4.5, once with 70% ethanol and twice with 5 ml acetone.
4. Dry the pellet in a vacuum desiccator, and dissolve in the desired buffer (0.1 × SSC or Tris:EDTA (10:0.5) mM), or distilled water.

Equipment

- Centrifuge with swing-out rotor for 30–50 ml tubes.
- Freezer (—20 °C).
- Vacuum desiccator.

Solutions

- 1% (w/v) cetylpyridinium bromide (CPB) in distilled water.
- Ethanol/sodium-acetate buffer: Make 20 ml 0.5 M sodium-acetate; dissolve the sodium-acetate in a small volume of distilled water, adjust pH to 4.5 with glacial acetic acid (in a hood) and sterile filter the solution. Add 80 ml 100% ethanol.
- 0.5 M sodium ethylene diamine tetra-acetate (EDTA) solution; Dissolve the sodium-EDTA in distilled water (heath the solution and stir with a magnetic stirrer to completely dissolve the EDTA). Adjust pH to 6.0 with 0.5 M Tris base. Sterile filter and store at room temperature.
- Standard Saline Citrate (SSC); $1 \times$ SSC is 0.15 M sodium chloride, 0.15 M trisodium citrate in distilled water, pH adjusted to 7.0. Prepare as a $10 \times$ SSC stock solution and dilute as needed.
- 70% ethanol.
- acetone (analytical grade).

References

1. Bakken, LR (1985) Separation and purification of bacteria from soil. Appl Environ Microbiol 49: 1482–1487.
2. Balkwill DL, Labeda DP, Casida LEJ (1975) Simplified procedures for releasing and concentrating micro-organisms from soil for transmission electron microscopy viewing as thin-sectioned and frozen-etched preparations. Can J Microbiol 21: 252–262.
3. Crombach WHJ (1972) DNA base composition of soil arthrobacters and other coryneforms from cheese and sea fish. Antonie van Leeuwenhoek 38: 105–120.
4. Fægri A, Torsvik VL, Goksøyr J (1977) Bacterial and fungal activities in soil: Separation of bacteria and fungi by a rapid fractionated centrifugation technique. Soil Biol Biochem 9: 105–112.
5. Geck P, Nasz I (1983) Concentrated, digestible DNA after hydroxyapatite chromatography with cetylpyridinium bromide precipitation. Anal Biochem 135: 1426–1429.
6. Hobbie JE, Daley RJ, Jasper S (1977) Use of Nucleopore filters for counting bacteria by fluorescence microscopy. Appl Environ Microbiol 33: 1225–1228.
7. Holben WE, Jansson JK, Chelm BK, Tiedje JM (1988) DNA probe method for the detection of specific micro-organisms in the soil bacterial community. Appl Environ Microb 54: 703–711.
8. Hopkins DW, Macnaughton SJ, O Donnell AG (1991) A dispersion and differential centrifugation technique for representative sampling of micro-organisms from soil. Soil Biol Biochem 23: 217–225.

9. Jacobsen CS, Rasmussen OF (1992) Development and application of a new method to extract bacterial DNA from soil based on separation of bacteria from soil with cation-exchange resin. Appl Environ Microbiol 58: 2458–2462.

10. Macdonald RM (1986a) Sampling soil microfloras: Dispersion of soil by ion exhange and extraction of specific micro-organisms from suspension by elutriation. Soil Biol Biochem 18: 399–406.

11. Macdonald RM (1986b) Sampling soil microfloras: Optimization of density gradient centrifugation in Percoll to separate micro-organisms from soil suspensions. Soil Biol Biochem 18: 407–410.

12. Ogram A, Sayler GS, Barkay T (1988) DNA extraction and purification from sediments. J Microb Meth 7: 57–66.

13. Porter KG, Feig YS (1980) The use of DAPI for identifying and counting aquatic microflora. Limnol Oceanogr 25: 943–948.

14. Rake A (1972) Isopropanol preservation of biological samples for subsequent DNA extraction and reassociation studies. Anal Biochem 48: 365–368.

15. Steffan RJ, Goksøyr J, Bej AK, Atlas RM (1988) Recovery of DNA from soils and sediments. Appl Environ Microbiol 54: 2908–2915.

16. Torsvik VL (1980) Isolation of bacterial DNA from soil. Soil Biol Biochem 12: 15–21.

17. Torsvik V, Goksøyr J, Daae FL (1990) High diversity in DNA of soil bacteria. Appl Environ Microbiol 56: 782–787.

Molecular Microbial Ecology Manual **1.3.3**: 1–11, 1995
© 1995 *Kluwer Academic Publishers.*

Extraction of microbial community DNA from soils

JAN DIRK VAN ELSAS[1] AND KORNELIA SMALLA[2]

[1] *IPO-DLO, P.O. Box 9060, 6700 GW Wageningen, the Netherlands;* [2] *Federal Institute of Biology, Biochemistry Division, Messeweg 11/12, Braunschweig, Germany*

Introduction

Soil represents a highly heterogeneous environment consisting of solid, liquid and gaseous phases. The dominating soil solid phase is composed of inorganic (sand, silt and clay) and organic (humic matter) materials, which are to varying degrees complexed with one another. The soil biota, including soil microorganisms such as bacteria, fungi and protozoans, are known to inhabit different sites in the soil pore matrix. Organisms associate in particular with soil solids, e.g. clay/organic matter complexes, in soil pores conducive to their survival. Although a wealth of information is available about many microorganisms isolated from soil by culturing techniques, the nature of a substantial part of the soil microbiota is essentially unknown due to their unculturability.

The extraction and analysis of total microbial community DNA from soil is useful for several purposes [15]. First, it provides insight in the prevalence of specific genes in microbial communities in the soil ecosystem, possibly resulting in a better understanding of natural selection of specific microbial groups under the influence of soil conditions. Secondly, by using 16S/18S or 23S/25S ribosomal DNA sequences as "signature molecules" (biomarkers), overall community DNA analysis may help to describe microbial communities in terms of their population structure. This can be achieved by applying temperature or denaturing gradient gel electrophoresis (TGGE or DGGE) to PCR products generated with sets of conserved primers, resulting in a type of community structure data which were not obtainable previously. Finally, the application of microbial community DNA extraction methodologies allows investigations on the nature of non-culturable cells, which are known to abound in soil. For instance, a new deep-branching group of proteobacteria (Planktomycetales) has recently been described based on the analysis of DNA sequences obtained from total soil communities [2].

After the pioneering work of Torsvik [14], extraction of microbial DNA from soil has been primarily carried out using two different approaches: 1) separation of microbial cells from soil particles, followed by subsequent cell lysis and extraction [1], and 2) direct cell lysis and DNA extraction from a soil slurry in the presence of soil particles [4]. While the former method provides DNA which is considered quite specific for the bacterial fraction of the microbial community in soil, the latter, although less specific due to the presence of eukaryotic (fungal) and extracellular DNA, has been shown to provide much higher yields [12]. The extent to which either of the two methods extracts extracellular DNA from soil is a still unresolved question, which should be addressed in further work. In spite of potential problems with the interpretation of results, i.e. whether positive detection indicates the presence of microbial cells carrying target DNA or merely extracellular target DNA, the direct lysis and extraction method has gained preference over the indirect method in many laboratories working with soil DNA. Further theoretical considerations on these approaches can be found in Chapter 1.5.2.

Direct community DNA extraction from soil, as developed by Ogram *et al.* [4], has been shown to provide a substantial amount of total soil bacterial DNA, however has been found to be prone to unavoidable yield losses. Recently, Moré *et al.* [3] showed that it is likely that many direct protocols extract DNA mainly from easy-to-lyse cells, whereas in particular the minute microbial forms (dwarfs) that are abundant in soil were not included. Since the original protocol encompassed tedious CsCl gradient and/or hydroxyapatite chromatography purification steps, many more recent protocols have attempted to simplify the original Ogram *et al.* protocol [4]. Examples of these simplified protocols can be found in the recent literature [5, 6, 8, 9, 16, 17].

In this chapter, we describe a rapid and simple direct DNA extraction protocol developed in our laboratories, which has been successfully applied to numerous soils of different textural classes (sand through clay). It is shown that restrictable and PCR amplifiable DNA of relatively high molecular weight can be obtained with this protocol, which consists of a few simple successive lysis, extraction and purification steps. The protocol uses some principles which are described in Molecular Biology manuals, e.g. Sambrook *et al.* [7] and the reader is referred to these manuals for background information.

Experimental approach

The approach described here is based on an efficient lysis of bacterial cells in the soil matrix in a slurry, followed by quick removal of soil particles and humic compounds, using different rapid purification steps. Removal of humic material, proteins, RNA and polysaccharides, as well as other soil compounds (minerals), is required to obtain DNA of sufficient purity for hybridization, restriction or PCR amplification analysis, as well as for cloning purposes [10, 12, 13]. The protocol is based on the direct extraction protocol of Ogram *et al.* for the extraction of DNA from sediments [4], as adapted by Smalla *et al.* [10], with omission of the laborious and often inefficient purification via hydroxyapatite chromatography. We adopted bead beating of soil slurries in a Braun's cell homogenizer (Braun, Melsungen) as the method of choice for cell lysis, since it was shown to yield higher quantities of DNA than freeze/thaw lysis [10]. However, in laboratories which do not possess a bead beater, freeze/thaw-assisted cell lysis may be used as an alternative. Careful control of the bead beating time and conditions was found to be essential to obtain DNA of large fragment size. This is important, since severely sheared DNA is unsuitable for PCR based detection of specific genes or analysis of community structure, e.g. using bacterial 16S ribosomal sequences as targets. Following cell lysis, extraction with cold phenol in the presence of soil particles was found to optimally separate DNA from contaminating compounds in the soil matter, while offering protection of the DNA from soil nucleases. Subsequent precipitation steps with CsCl and KAc were included to further remove impurities (proteins, RNA, humic material) from the DNA in the aqueous phase. For some soils, e.g. several silt loam soils, DNA preparations thus obtained are of sufficient purity to serve as targets for restriction, amplification or hybridization analysis. In the case of other soils, e.g. a high organic matter loamy sand, a final clean-up step was found to be required. This was originally performed using spermine-HCl [10], however this method was superseeded by using adsorption/elution over commercially available glassmilk (Geneclean II kit, Bio 101, La Jolla, CA, USA) or resin spin columns (Wizard DNA Clean-Up System, Promega, Madison, WI, USA). In our experience, this flexible protocol allows for the extraction and purification of high quality DNA from virtually any soil type.

Procedures

This procedure is based on the extraction of 10 g of soil. It can be scaled down easily to accommodate 1–5 g soil samples. In our experience, the full procedure is needed to obtain restrictable and amplifiable DNA from a loamy sand soil with high organic matter content, whereas purification up to and including purification step I was often sufficient for DNA from a silt loam soil with low organic matter content.

Lysozyme treatment

1. Resuspend 10 g of soil in 15 ml 120 mM sodium phosphate buffer pH 8.0 [7] in a 50 ml polypropylene tube.
2. Add 75 mg lysozyme to the soil suspension, homogenize and incubate during 15 min at 37 °C.
3. Chill on ice.

Bead beating lysis

4a. Transfer the soil suspension obtained under 3 into a bead beating vial containing 15 g beads (0.09–0.13 mm dia). Homogenize 3 times for 90 s in the bead beater (MSK cell homogenizer, B. Braun Diessel Biotech, Melsungen, Germany) at 4,000 U/min with intervals of 15–30 s. Transfer the lysate to a 45 ml centrifuge or 50 ml polypropylene tube. Add 900 µl of 20% sodium dodecyl sulphate (SDS) and mix well. Leave either on ice for 1 h to enhance lysis, or at room temperature for 15 min.

Freeze/thaw lysis (instead of bead beating)

4b. Add 900 µl of 20% SDS to the soil slurry obtained under 3 and mix. Freeze at —20 °C (or —80 °C) for 1 h, then keep at 37 °C for 30–45 min. Repeat freeze/thaw cycle twice.

Extraction and precipitation

5. Add an equal volume of Tris-buffered phenol pH 8.0 [7] to the lysed cell slurry obtained under 4a or 4b.
6. Mix well (manually) and centrifuge during 5 min at 10,000 xg (or 15 min at 3,000 xg for polypropylene tubes) at room temperature.
7. Recover the aqueous (upper) phase in a new centrifuge (or polypropylene) tube.
8. Back-extract the phenol/soil mixture with 5 ml 120 mM sodium phosphate buffer (pH 8.0).
10. Pool the aqueous phases.
11. Extract the pooled aqueous phases with an equal volume of chloroform/iso-amylalcohol (24:1).

12. Recover the upper aqueous phase. If a heavy interphase is present, extract the aqueous phase again with an equal volume of chloroform/iso-amylalcohol.
13. Add 0.1 volume of 5 M NaCl and 2 volumes of icecold 96% ethanol. Keep at —80 °C for 20 min or at —20 °C for at least 1 h (often overnight).
14. Centrifuge during 5–10 min at 10,000 xg (or 15–20 min at 3,000 xg for polypropylene tubes).
15. Discard the supernatant and wash the pellet with icecold 70% ethanol. Air-dry pellet.
16. Resuspend pellet in 1–1.5 ml (sterile) TE buffer pH 8.0 (10 mM Tris-HCl pH 8.0, 1 mM EDTA pH 8.0; [7]). Due to the volume of the pellet, the final volume may be larger. The solution at this stage is called the crude extract.

Purification step I – CsCl and KAc precipitations

Perform all steps at room temperature, unless stated otherwise.

1. Add 0.5 g CsCl to 500 µl crude extract.
2. Incubate for 1–3 h.
3. Centrifuge during 20 min at maximum speed in an Eppendorf centrifuge.
4. Recover the supernatant ($\approx$ 500 µl) in a 10 ml tube.
5. Add 2 ml deionized water and 1.5 ml iso-propanol. Mix and incubate for minimally 5 min at room temperature.
6. Centrifuge during 15 min at 10,000 xg. When polypropylene tubes are employed, use maximally 3,000 xg for 20 min. Check degree of pelleting.
7. Discard supernatant and resuspend pellet in 500 µl TE buffer (pH 8.0) and transfer the suspension to a new Eppendorf tube.
8. Add 100 µl 8 M potassium acetate (= KAc), mix and incubate for 15 min at room temperature.
9. Centrifuge during 15 min at maximum speed in an Eppendorf centrifuge.
10. Recover supernatant and add 0.6 volume of iso-propanol, mix and incubate for 5 min at room temperature.
11. Centrifuge during 15 min at full speed in an Eppendorf centrifuge.

12. Wash pellet with icecold 70% ethanol, dry and resuspend it in 500 µl TE buffer (pH 8.0).

Purification step IIa – glassmilk (Geneclean II kit, Bio 101, La Jolla, CA, USA) clean-up

The system is based on DNA adsorption to glassmilk in high molar (4M) NaI, followed by washing in the presence of ethanol and elution with TE buffer or deionized water.

Notes

* Before use, vortex the glassmilk stock until all contents are in suspension.
* Add 5 µl glassmilk to solutions containing 5 µg or less of DNA. Add an additional 1 µl for each 0.5 µg above 5 µg.

1. Add 3 volumes of 6 M NaI stock solution (750 µl) to 250 µl partially purified soil DNA extract (after purification step I).
2. Add 5 µl glassmilk, mix and place at room temperature for 5 min to allow binding of the DNA to the silica matrix, mixing every 1–2 min to ensure that the glassmilk stays suspended.
3. Pellet the glassmilk by centrifugation of the suspension for 5 s at maximum speed in an Eppendorf centrifuge. Remove the NaI supernatant.
4. Wash pellet by adding 200 µl of icecold NEW WASH (commercial name used in kit), then pipet the suspension back and forth while digging into the pellet with the pipet tip (resuspension of pellet), spin for 5 s and discard the supernatant. Repeat twice.
5. Resuspend pellet with 125 µl TE buffer (pH 8.0) or deionized water. Incubate the tube at 45–55 °C for 2–3 min. Centrifuge (30 s) to make a solid pellet. Carefully remove the supernatant containing the eluted DNA and place in a new tube. Elute DNA for a second time with 125 µl TE-buffer (pH 8.0) or deionized water and combine the eluates. The purified DNA (250 µl) may be stored at 4 or —20 °C.

If needed (as judged by colour and/or suitability for restriction digestion or amplification): repeat steps 1–5.

Purification step IIb – Wizard DNA clean-up system (Promega, Madison, WI, USA), formerly Magic DNA clean-up system (instead of purification step IIa).

The system is based on DNA adsorption to clean-up resin in 6 M guanidine thiocyanate, washing with 80% iso-propanol and elution with TE buffer or deionized water. The reagents can be kept at room temperature protected from exposure to direct sunlight. See also Chapter 1.3.4 (Saano and Lindström) for a full account of this procedure.

1. Add 1 ml of DNA clean-up resin to 250 μl partially purified DNA extract (after purification step I) in an Eppendorf tube and mix by gently inverting several times.
2. Attach the syringe barrel of a 2.5 ml disposable luer-lock syringe to the extension of the Wizard Minicolumn.
3. Pipet the suspension from the Eppendorf tube into the syringe barrel. Using the plunger, push the slurry slowly into the Minicolumn.
4. To wash the column, pipet 2 ml of 80% iso-propanol into the syringe barrel. Gently push the solution through the Minicolumn.
5. Remove syringe and place the Minicolumn containing the loaded resin on top of an Eppendorf tube. Centrifuge for 20 s at full speed in an Eppendorf centrifuge to dry the resin. Leave the Minicolumn at room temperature for 5–15 min to evaporate traces of iso-propanol still present.
6. Transfer the Minicolumn to a new Eppendorf tube. Apply 125 μl of prewarmed (60–70 °C) TE buffer (pH 8.0) or deionized water and leave for 5–10 min. The DNA will remain intact on the Minicolumn for up to 30 min. Centrifuge for 20 s at full speed in an Eppendorf centrifuge to elute the bound DNA.
7. Repeat step 6 using the same Eppendorf tube. Total volume of eluate will be about 250 μl.
8. Discard Minicolumn. The purified DNA may be stored at 4 or —20 °C.

If needed (as judged by colour and/or suitability for restriction digestion or amplification): repeat steps 1–8.

The soil DNA extraction method has been used in our laboratories for a wide variety of purposes. Routinely, 15 to 20 µg of DNA of 10–40 kb in size has been obtained from the sandy and clayey soils we have studied (Fig. 1). As shown in Fig. 2 (lane 2), the DNA obtained from a loamy sand soil was digestable with *Eco*RI and *Hind*III, resulting in an expected smear of fragments of 10–20 kb down to 1 kb. As commonly found for complex mixtures of DNA, individual fragments were not distinguishable in the digests of total soil microbial community DNA, as opposed to the *Eco*RI digest of genomic DNA obtained from a pure culture of *Pseudomonas fluorescens*, in which discrete bands were visible (Fig. 2, lane 7). The *Hind*III digest was inefficient, as also evidenced in digests of the *Pseudomonas fluorescens* genomic DNA (Fig. 2, lanes 4, 5, 9 and 10). With the protocol, we have successfully detected genetically modified *Pseudomonas fluorescens* as well as *Bacillus amyloliquefaciens* introduced into soil, using PCR amplification systems based on introduced unique marker sequences [10; Smalla *et al.*, in preparation], with a limit of detection of 10^3 added cells per g soil. It

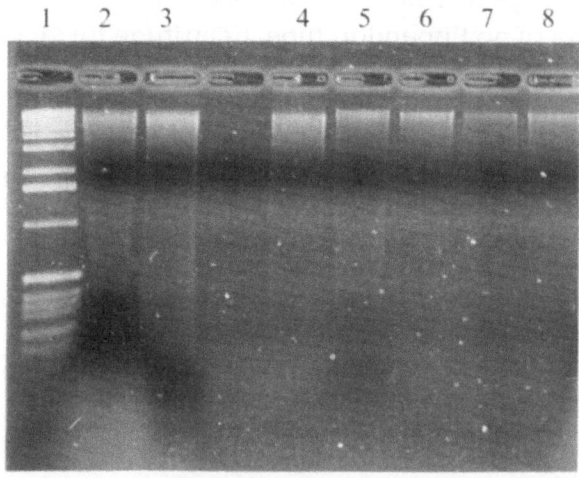

Figure 1. Soil DNA obtained from Ede loamy sand and Flevo silt loam soils using the direct extraction protocol. Lanes: 1, 1-kb ladder (Gibco-BRL, from bottom to top respectively 1, 2, 3, 4, 5, and 6–23 kb in size); 2 to 4, DNA from Ede loamy sand: respectively crude extract, extract after purification step 1 and extract after Wizard clean-up; 5 to 8, DNA from Flevo silt loam: respectively crude extract, extract after purification step 1, extract after glassmilk clean-up and extract after Wizard clean-up.

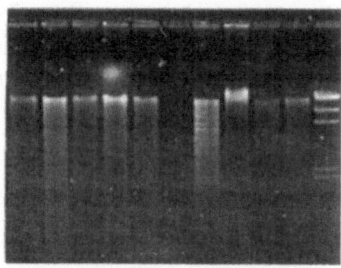

1 2 3 4 5 6 7 8 9 10 11

Figure 2. Soil DNA obtained from Ede loamy sand using the direct extraction protocol digested or not with *Eco*RI or *Hind*III. Lanes: 1, purified DNA from Ede loamy, undigested; 2, idem, digested with *Eco*RI; 3, idem, undigested; 4 and 5, idem, digested with *Hind*III; 6, void; 7, *Pseudomonas fluorescens* R2f genomic DNA digested with *Eco*RI; 8, idem, undigested; 9 and 10, idem, digested with *Hind*III, 11, molecular size marker (λ DNA digested with *Hind*III, bands from top to bottom respectively 23, 9.6, 6.7, 2.2 and 2.0 kb in size).

has also been applied to detect *Mycobacterium chlorophenolicum* with a 16S ribosomal RNA based PCR amplification system, at a detection limit in the order of 10^2 cells per g soil [Briglia *et al.,* submitted]. Further, PCR analysis of total microbial community DNA extracts also enabled us to detect naturally occurring antibiotic resistance (*npt*II) as well as transposon Tn5 sequences in soil communities, without the need to culture organisms [11]. We are further applying it to detect plasmids of different *Inc* groups in soil bacterial communities in a project aimed at unraveling the gene mobilizing capacity of soil microbial communities.

Concluding remarks

The protocol described here is very suitable for the detection of genetically engineered microorganisms and their genes following introduction into soil [10]. It can also aid in detecting naturally occurring DNA sequences in soil microbes, however there may be some doubt as to the localization of such DNA sequences inside cells or extracellularly. This ambiguity makes the method *a priori* less suitable for detection and quantification of microbial cells in soil, unless it is carefully controlled for the co-extraction of extracellular DNA (e.g. by running controls not subjected to lysozyme/bead beating lysis). Furthermore, even though the bead beater is known to efficiently lyse a major part of bacterial species including many Gram-positives, there is no absolute certainty that cell lysis is representative for the soil bacterial community. The findings of Moré *et al.* [3] suggest lysis may well be confined to the larger cell-size fraction of the microbial community. These considerations have to be taken into account when the protocol is to be used for microbial population structure studies in soil. In addition, the protocol not only yields bacterial, but also eukaryotic DNA, which may originate from a variety of source organisms present in the system (e.g. fungi, protozoans, plants). For quantification purposes, a eubacterial probe is necessary to determine the proportion of bacterial DNA in the total extracts obtained.

Acknowledgements

This work was supported by grants from the EC-BIOTECH (grant no. BIO2-CT92–491) and EC-Environment (grant no. EVSV-CT93-0250) Programmes. We thank A.C. Wolters and E. Smit for carefully reading the manuscript.

References

1. Holben WE, Jansson JK, Chelm BK, Tiedje JM (1988) DNA probe method for the detection of specific microorganisms in the soil bacterial community. Appl Environ Microbiol 54: 703–711.
2. Liesack W, Stackebrandt E (1992) Occurrence of novel groups of the domain bacteria as revealed by analysis of genetic material isolated from an Australian terrestrial environment. J Bacteriol 174: 5072–5078.
3. Moré MI, Herrick JB, Silva MC, Ghiorse WC, Madsen EL (1994) Quantitative cell lysis of indigenous microorganisms and rapid extraction of microbial DNA from sediment. Appl Environ Microbiol 60: 1572–1580.
4. Ogram A, Sayler GS, Barkay TJ (1987) DNA extraction and purification from sediments. J Microbiol Meths 7: 57–66.
5. Pillai SD, Josephson KL, Bailey RL, Gerba CP, Pepper IL (1991) Rapid method for processing soil samples for polymerase chain reaction amplification of specific gene sequences. Appl Environ Microbiol 57: 2283–2286.
6. Porteous LA, Armstrong JL (1991) Recovery of bulk DNA from soil by a rapid, small-scale extraction method. Curr Microbiol 22: 345–348.
7. Sambrook J, Fritsch EF, Maniatis T (1989) Molecular Cloning. A Laboratory Manual. Second Edition. New York: Cold Spring Harbor Press.
8. Selenska S, Klingmüller W (1991) Direct detection of nif-gene sequences of Enterobacter agglomerans in soil. FEMS Microbiol Lett 80: 243–246.
9. Selenska S, Klingmüller W (1991) DNA recovery and direct detection of Tn5 sequences from soil. Lett Appl Microbiol 13: 21–24.
10. Smalla K, Cresswell N, Mendonca-Hagler LC, Wolters AC, Van Elsas JD (1993a) Rapid DNA extraction protocol from soil for polymerase chain reaction-mediated amplification. J Appl Bacteriol 74: 78–85.
11. Smalla K, Van Overbeek LS, Pukall R, Van Elsas JD (1993) Prevalence of *npt*II and Tn5 in kanamycin-resistant bacteria from different environments. FEMS Microbiol Ecol 13: 47–58.
12. Steffan RJ, Goksoyr J, Bej AK, Atlas RM (1988) Recovery of DNA from soils and sediments. Appl Environ Microbiol 54: 2908–2915.
13. Tebbe CC, Vahjen W (1993) Interference of humic acids and DNA extracted directly from soil in detection and transformation of recombinant DNA from bacteria and a yeast. Appl Environ Microbiol 59: 2657–2665.

14. Torsvik VL (1980) Isolation of bacterial DNA from soil. Soil Biol Biochem 12: 18 21.

15. Trevors JT, Van Elsas JD (1989) A review of selected methods in environmental microbial genetics. Can J Microbiol 35: 895–902.

16. Tsai Y-L, Olson BH (1992) Detection of low numbers of bacterial cells in soils and sediments by polymerase chain reaction. Appl Environ Microbiol 58: 754–757.

17. Van Elsas JD, Van Overbeek LS, Fouchier R (1991) A specific marker, pat, for studying the fate of introduced bacteria and their DNA in soil using a combination of detection techniques. Plant Soil 138: 49–60.

Molecular Microbial Ecology Manual **1.3.4**: 1-6, 1995.

Small scale extraction of DNA from soil with spun column cleanup

AIMO SAANO and KRISTINA LINDSTRÖM

Department of Applied Chemistry and Microbiology, Division of Microbiology, PO Box 27, FIN-00014 University of Helsinki, Finland

Introduction

A general outline of procedures for isolation of DNA from soil by direct lysis is: 1) lysis of the cells; 2) separation of DNA from other cell components such as polysaccharides and proteins; 3) release of DNA from soil particles and purification of the DNA extract from soil compounds; 4) precipitation of DNA; and 5) further purification of DNA from soil compounds. Steps 2–4 do not proceed in succession in all of the published protocols, and 2 and 3 may have to be repeated after 4. However, in the protocol described below, the procedure follows the outline given above.

The following DNA-extraction procedure was developed by combining and modifying protocols published by Selenska and Klingmüller [7] for soil DNA and Ausubel *et al.* [1] for bacteria. The aim was to combine high yield of DNA pure enough for enzyme manipulations with fast and easy performance. This was reached by exclusion of freezing-thawing [10], sonication [3] or other multi-step treatments of the sample and minimizing the use of harmful reagents. The method easily yields over 10 µg of DNA/g of soil depending on soil type. It is advisable to pass the soil through a 2 mm sieve before measuring the samples.

All published methods on extraction of DNA from soil samples aim at getting a high yield of DNA that is pure enough for molecular analysis by DNA-DNA-hybridization, restriction analysis, or amplification by PCR. However, the chemical properties of the humic impurities from soil render them resistant to separation from DNA. This introduces extensive purification steps into the DNA isolation protocols. In many cases, soil humic compounds present in soil DNA extracts create problems for PCR amplification, because they strongly inhibit polymerase activity [11, 12]. For the solution of this problem, many different attempts to purify microbial DNA extracted from soil have been described [2, 3, 8, 9, 11, 12]. Soil compounds also decrease the sensitivity of DNA-DNA-hybridization [5], because they strongly inhibit the immobilization of DNA onto positively charged nylon membrane [4].

Different methods can be used to purify the crude DNA extracted from soil, e.g. CsCl-EtBr equilibrium density gradient ultracentrifugation, hydroxyapatite column chromatography, phenol-chloroform extraction, agarose gel electrophoresis, and passage through elutips, glass milk or various resins from different manufacturers. However, CsCl-EtBr equilibrium density gradient ultracentrifugation, hydroxyapatite purification, phenol-chloroform extraction and agarose gel electrophoresis are all laborious procedures, which require special skills, care and expensive apparatus and reagents. The spun column clean-up protocol, suggested below, is based on use of DNA-binding resin. It is inexpensive and very simple and fast to perform.

Procedures

Extraction of DNA from soil

Steps in the procedure

1. Add to one gram soil sample in a 15 ml polypropylene (PP) tube 2.5 ml of the following buffer: 120 mM Na_2HPO_4 (pH 8), 1% SDS, 100 µg/ml proteinase K. Mix well and incubate one hour at 37 °C with occasional shaking. This step is for lysing the cells.
2. Add 450 µl of 5 M sodium chloride. Vortex carefully (5 sec). The NaCl concentration is raised to prevent nucleic acids from dissolving in CTAB (cetyl-trimethylammonium bromide), which will be added next.
3. Add 375 µl of 10% CTAB in 0.7 M NaCl. Vortex carefully (10 sec). Incubate at 65 °C for 20 min. CTAB denatures and binds proteins and polysaccharides.
4. Add chloroform in a volume equal to that of the whole mixture. Vortex carefully (5 sec). Chloroform extraction removes CTAB, its complexes, as well as other compounds except nucleic acids from water.
5. Transfer the mixture into a 15 ml PP tube autoclaved with about 1 ml of silicone grease (Wacker-Chemie GmbH, Germany). Centrifuge for 15 min at 9,000 × g at 4 °C. The silicone grease forms a tight interphase between the upper water phase and the lower solvent phase during the centrifugation. It facilitates collection of the water phase remarkably.
6. Collect the water phase into a fresh 15 ml PP tube. Add an equal

volume of isopropanol. Mix and incubate for one hour at —20 °C. Isopropanol precipitates the nucleic acids.
7. Centrifuge for 15 min at 10,000 × g at 4 °C. Pour off the iso- propanol, dry the pellets.
8. Dissolve the DNA pellets in 200 µl of sterile distilled water. This is the crude DNA sample.
    The procedure takes about 3 hours to perform for a small number of samples. The samples can be stored at room temperature for several months without noticable degradation of DNA.

Yield estimation

There are three things with crude DNA samples that should be taken into account before trying to estimate the yield of the extracted DNA. First, crude DNA samples are too contaminated with soil compounds to be evaluated by UV-spectrophotometry (A_{260}). The estimation of DNA quantity and quality can instead be done by running aliquots of samples with quantitative DNA standards in agarose gel electro- phoresis [6]. Second, soil compounds prevent ethidium bromide from binding to DNA during electrophoresis. This is due to the dif- ference in the electrophoretic properties of DNA and ethidium bromide. DNA moves towards the positive electrode, whereas ethidium bromide moves towards the negative electrode. Soil compounds moving faster than DNA block ethidium bromide from binding to DNA in the electrophoresis. This can lead to severe underestimation of the DNA yield. Therefore, it is advisable to stain the gel after electrophoresis. Third, crude DNA as such can easily be loaded into the gel well without any addition of loading buffer. The brown soil compounds also serve as an electrophoresis marker, running approximately as fast as bromophenol blue.

Steps in the procedure

1. Run 10 µl of sample next to DNA standards in 0.8-1.2% agarose gel electrophoresis without ethidium bromide at 70-90 V for about 90 min.
2. Stain the gel with 10 µg/ml ethidium bromide electrophoresis buffer for about 20 min. Destain in distilled water for another 20 min. Compare sample with standards under UV (302 nm).

Spun column purification

For DNA hybridizations and enzymatic manipulations, aliquots of crude DNA should be further purified. Spun columns manufactured by Promega, USA (formerly Magic DNA Clean-up columns, now Wizard DNA Clean-up columns) can be used for rapid purification of small samples, because they fit into tubes used in microcentrifuges. One round of purification takes about 25 min + 5 min for every sample exceeding two. Two rounds of purification are generally needed for efficient removal of humic compounds. The losses of DNA are low (<10%).

Steps in the procedure

1. Pipet 50 µl of crude DNA extract into a sterile microcentrifuge tube.
2. Add 1 ml of well-mixed DNA purification resin, mix by inverting the tube several times.
3. Collect the mixture into a sterile 2 ml syringe using a sterile 0.9-1.2 mm diameter needle. If the mixture has stood for a while before this step, it may appear impossible to get it into the needle. Shake or vortex the tube strongly and try again. Discard the needle.
4. Plug the syringe into a column, push the DNA solution gently through it (approximately one drop/second).
5. Wash the DNA in the column with 2 ml of 80% sterile isopropanol using the same syringe.
6. Plug the column into a microcentrifuge tube, spin at maximum speed in a microcentrifuge for 20 seconds. Discard the tube with isopropanol. Transfer the column to a fresh sterile microcentrifuge tube, let stand >5 min at room temperature to allow the rest of the isopropanol to evaporate.
7. Add 50 µl of sterile water at 65 °C into the column. Let stand for about 1 min.
8. Spin at maximum speed in a microcentrifuge for 20 sec. Discard the used column. The colourless/yellowish/brownish liquid in the microcentrifuge tube is the DNA sample, which probably will work fine in DNA hybridizations. For enzyme manipulations, like restriction or PCR, a second round of spun column purification may be necessary.

Summary

Crude DNA extract can be analyzed for the presence of specific target DNA by dot-blot hybridization or by Southern blotting and hybridization. However, this is possible only when the target DNA is in excess in the sample. For better sensitivity, spun column clean-up of crude DNA should be used. Apparently, spun column purification shears all the DNA molecules larger than 20 kb into shorter fragments (Fig. 1), but this has not affected the efficiency of DNA hybridizations [4] or PCR analyses [10].

The described protocol has worked well with different soils, but especially with 'difficult' soils with high organic matter content, such as peat [10].

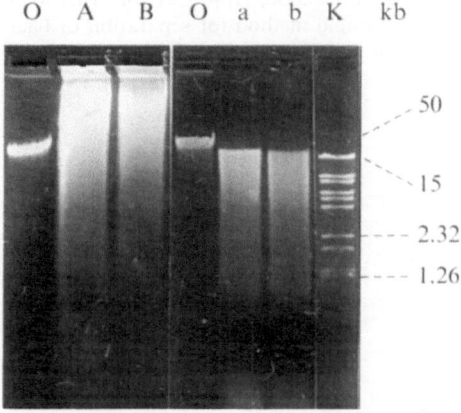

Figure 1. Gel electrophoresis of the total DNA extracted from 1 g of agricultural soil with high organic matter content (Agricultural Research Center, Research Station for Ecological Farming, Juva, Finland). Lane 0, bacteriophage lambda DNA undigested, 200 ng; A and B, crude DNA samples, 1/10 of the total volume; a and b, DNA samples A and B after spun column purification, volume corresponding to 1/10 of the total volume of the crude DNA samples; K, bacteriophage lambda DNA digested with *Bst*EII, 100 ng.

References

1. Ausubel FM, Brent R, Kingston RE, Moore DD, Seidman JG, Smith JA, Struhl K (eds) (1989) Current protocols in Molecular Biology. Wiley, NY.
2. Hilger AB, Myrold DD (1991) Method for extraction of *Frankia* DNA from soil. Agriculture, Ecosystems and Environment 34: 107–113.
3. Picard C, Ponsonnet C, Paget E, Nesme X, Simonet P (1992) Detection and enumeration of bacteria in soil by direct DNA extraction and polymerase chain reaction. Appl Environ Microbiol 58: 2717–2722.

4. Saano A, Kaijalainen S, Lindström K (1993) Inhibition of DNA immobilization to nylon membrane by soil compounds. Microb Releases 2: 153–160.
5. Saano A, Lindström K (1992) Detection of *Rhizobium galegae* from one-gram soil samples by non-radioactive DNA-DNA-hybridization. Soil Biol Biochem 24: 969–977.
6. Sambrook J, Fritsch EF, Maniatis T (1989) Molecular cloning: a laboratory manual. 2nd ed. Cold Spring Harbor Laboratory, Cold Spring Harbor, NY.
7. Selenska S, Klingmüller W (1991) DNA recovery and direct detection of *Tn5* sequences from soil. Lett Appl Microbiol 13: 21–24.
8. Selenska S, Klingmüller W (1992) Direct recovery and molecular analysis of DNA and RNA from soil. Microb Releases 1: 41–46.
9. Smalla K, Cresswell N, Mendonca-Hagler LC, Wolters A, Van Elsas JD (1993) Rapid DNA extraction protocol from soil for polymerase chain reaction-mediated amplification. J Appl Bacteriol 74: 78–85.
10. Tas, É, Saano A, Leinonen P, Lindström K (1995) Identification of *Rhizobium* in peat-based inoculants by DNA hybridization and PCR-application in inoculant quality control. Appl Environ Microbiol (in press).
11. Tsai Y-L, Olson BH (1991) Rapid method for direct extraction of DNA from soil and sediments. Appl Environ Microbiol 57: 1070–1074.
12. Tsai Y-L, Olson BH (1992) Detection of low numbers of bacterial cells in soils and sediments by polymerase chain reaction. Appl Environ Microbiol 58: 754–757.
13. Tsai Y-L, Olson BH (1992) Rapid method for separation of bacterial DNA from humic substances in sediments for polymerase chain reaction. Appl Environ Microbiol 58: 2292–2295.

Molecular Microbial Ecology Manual **1.3.5**: 1-9, 1995.

Gel purification of soil DNA extracts

DAVID D. MYROLD, KENDALL J. MARTIN and
NANCY J. RITCHIE
Department of Crop and Soil Science, Oregon State University, ALS 3017, Corvallis,
OR 97331-7306, USA

Introduction

Many methods have been developed for extracting nucleic acids from soils or other environmental samples. They can be grouped into two general categories: (i) cell separation followed by lysis [3, 11, 12], and (ii) direct lysis of cells *in situ* [1, 2, 5–7, 9–11, 13, 15]. Releasing DNA from cells is only the first step, additional purification must be performed in order to produce DNA of sufficient quality for subsequent enzymatic reactions. Several approaches have been tried, including the inclusion of PVPP [2, 3, 5], selective precipitation of DNA [3, 5], hydroxyapatite columns [12], gel filtration [14], and gel electrophoresis [2, 7, 15].

Gel purification is based on the different electrophoretic mobilities of DNA and humic materials. Humic materials move much faster than high molecular weight DNA when both are run under standard conditions for agarose electrophoresis. Agarose gel purification had been used by Hilger and Myrold [2], Rochelle and Olson [7], and Van Elsas *et al.* [15]. Rochelle and Olson [7] used 1-g soil samples and in-gel lysis to extract DNA, however, the DNA extracted was not of sufficient quality for use with restriction enzymes. Hilger and Myrold [2] and Van Elsas *et al.* [15] used preparative agarose gel electrophoresis and produced DNA that could be cut by restriction enzymes or amplified by the polymerase chain reaction. An alternative electrophoresis approach, which incorporates polyvinylpyrrolidone into the agarose gel, and thereby results in the retarded migration of humic materials, has also been reported [16].

The following procedures are a modification of the direct lysis method reported by Hilger and Myrold [2]. Its main features are: (i) enzymatic cell lysis, (ii) gentle extraction methods to isolate relatively intact DNA of high molecular weight, and (iii) agarose gel electrophoresis for purification of DNA from contaminating soil humic materials. The method has been used successfully to isolate DNA from a wide range of soils. Total DNA extracted has ranged from 5 to 27 µg DNA g^{-1} soil. Extraction efficiency has been estimated to range from nearly 100% in low organic matter, sandy soils to about 10% in soils with significant amounts of clay and organic matter. The isolated DNA is suitable for PCR (polymerase chain reaction) amplification [4].

In principle, the following procedure should work in connection with most direct DNA extraction protocols, however, this should be confirmed on a case by case basis, because soils vary in their amounts and types of humic materials and extraction methods produce DNA of differing sizes and degree of purity.

Procedures

The following procedures have been developed for purifying crude soil DNA extracts of about 1 ml volume generated from 5 g of soil. It is typically most convenient to process six soil samples at a time, because this is the capacity of a standard 20 × 25 cm gel mold. Table 1 lists the materials and Table 2 describes the reagents used in this procedure. Two modifications of the gel purification method have been developed: the gel well method for soils low in organic matter, and the gel block method for high organic matter soils. Figures 1 and 2 show features of the gel purification methods.

Table 1. Materials

Autoclaved or new sterile	per sample
Stirring rods (bamboo cooking skewers work well)	2
Cotton-tipped swabs	1
15-mL screw-cap centrifuge tubes (polypropylene)	2
1.5-mL microcentrifuge tubes (polypropylene)	4
Plastic transfer pipettes	3
Siliconized well-formers (gel well)	1
5-mL plastic beakers (gel block)	1
Gel molds (gel block)	1
250-µL pipette tips	2
1-mL wide-bore pipette tips	6
5-mL pipette tips	2
Dialysis membrane (Spectrapor 132678, 2.5 cm wide, 15 cm long, MW cut-off 14,000-16,000; prepared according to [8])	1
Clips for dialysis tubing	2
Centrifuge and rotors for 15-mL screw-cap tubes	
Microcentrifuge	
70 °C water bath	
Electrophoresis apparatus (20 × 25 cm gel tray) and power supply (250 V, 7.5 amp)	

Table 2. Reagents

Table 2. Reagents

Ethanol (100% and 70%) at −20 °C
TE (10:1) – (**Tris-E**DTA; 1× is **10** m*M* Tris, **1** m*M* EDTA, pH 8.0) [8]
TE (10:0.1) – (**Tris-E**DTA; 1× is **10** m*M* Tris, **0.1** m*M* EDTA, pH 8.0)
Phenol: chloroform: isoamyl alcohol (25:24:1) [8]
2-butanol (100%, also known as *sec*-butanol)
TAE (**Tris-A**cetate-**E**DTA; 50× stock is 2 *M* Tris-acetate, 0.05 *M* EDTA, pH 8.0) [8]
Ammonium acetate (10 *M*) [8]
Glycerol (100%, autoclaved)
Loading buffer (10× Type II; 0.42% bromophenol blue, 0.42% xyelene cyanol FF, 25% Ficoll Type 400) [8]

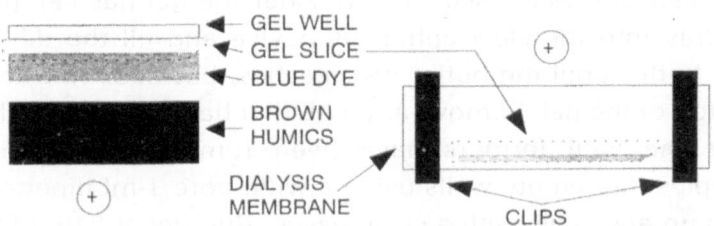

Figure 1. Schematic representation of the separation of humic materials from DNA during agarose gel electrophoresis and subsequent step of electroelution of DNA from an agarose gel slice. The brown humic materials migrate more quickly towards the anode than the blue dye (xylene cyanol) and DNA. The DNA typically migrates only a short distance out of the well (or gel block) with smaller DNA extending just into the light blue dye band. RNA co-migrates with the humic materials.

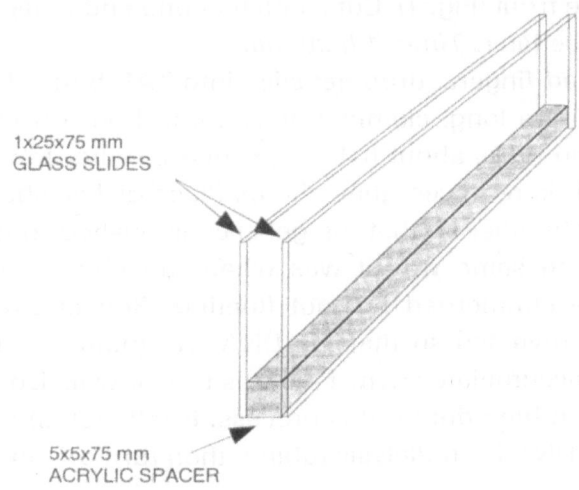

Figure 2. Diagram of a gel block mold, which consists of two 1 × 25 × 75 mm glass microscope slides glued to a 5 × 5 × 75 mm acrylic spacer.

MMEM-1.3.5/3

Gel Well Purification

1. Precipitate the crude soil DNA and dissolve the pellet in 800 µl 1×
   TE (10:1). Set tube in 70 °C water bath until the pellet is
   completely dissolved (15–30 min). Transfer DNA solution to a 1.5-
   ml microcentrifuge tube using wide-bore pipette tip. Add 50 µl
   glycerol (100%) and 20 µl 10× Loading Buffer, and mix gently by
   inversion. Spin down drops and store on ice until loaded on gel.
   Time: 45 min.
2. Prepare an agarose gel (1% agarose with 1× TAE buffer) with six,
   1-ml capacity wells (see Note 1). After the gel has set, place the
   gel tray into the electrophoresis device and fill the device with
   TAE buffer until the buffer just reaches, but does not flow over,
   the top of the gel. Remove any TAE that has moved into the wells
   after the well formers have been removed. Carefully, load
   samples into empty wells using a wide-bore 1-ml pipette tip and
   mop up any spills with a clean tissue. Run gel at 240 V for 5 min
   then at 120 V until brown humic material is separated from the
   blue dye nearest the well and the dye is about 1 cm from the well
   (about 30 min). Turn off power and collect buffer from the well
   with a transfer pipette and place in a labelled 15-ml polypropylene
   tube (store at 4 °C). Cut out gel slice (about 0.8–1.0 cm width)
   including well edge as one side and extending to the blue, xylene
   cyanol dye front (Fig. 1). Cut a notch at one end to aid in orienting
   the gel slice later. *Time: 1 h 30 min.*
3. With gloved fingers, drop gel slice into TAE buffer-filled dialysis
   tubing (15 cm long, clamped at bottom). Pour off excess buffer
   until tubing looks about half full, squeeze buffer towards the top
   until all air is removed, then clamp. Place dialysis bag in electrophoresis chamber so that the gel slice is pushed to one side and
   oriented the same way it was when run (Fig. 1). Dialysis bag
   should be submerged but not floating. Run at 240 V for 1 h,
   terminals reversed so that the DNA will migrate out of the gel
   slice and accumulate against dialysis tubing wall. Do not touch or
   rearrange tubing during this process. Briefly (30 s) reverse leads
   to detach DNA from dialysis tubing, then turn off current. *Time: 1
   h 30 min.*
4. Massage the dialysis tubing wall where the DNA has stuck. Using

a transfer pipette, draw up the buffer and gently rinse the inside of the dialysis membrane. Repeat three times. Collect the buffer from the dialysis tube and pool with the buffer from the gel well. Rinse the gel slice and inside of the dialysis tube with 1 ml of sterile water and pool wash with buffer in the tube. Total volume should be 5 to 8 ml. *Time: 30 min.*

5. Extract with 2-butanol to reduce volume of the DNA solution (see Note 4). Add one volume of 2-butanol, gently invert until emulsion forms (about 20 times), centrifuge (2,000 × *g*, 1 min), carefully pipette off the butanol (upper) phase, and repeat these butanol extractions until the volume of DNA solution is 1.5 ml. Depending upon the sample, three to six extractions may be needed to reduce a 5 to 8 ml sample to the desired 1.5 ml. *Time: 1 h.*

6. Extract with one volume of phenol/chloroform/isoamyl alcohol mixture (25:24:1). Mix gently by inversion until emulsified, and centrifuge (2,000 × *g*, 5 min). Carefully transfer upper phase to a new 15-ml tube using a wide-bore pipette tip, being careful to avoid the white material at the interface. Add 1 ml 1× TE (10:1) to the lower layer, mix gently by inversion, and centrifuge (2,000 × *g*, 5 min) again. Pool the upper phase with the first and extract with one volume of phenol/chloroform/isoamyl alcohol mixture (25:24:1). Mix gently by inversion until emulsified, and centrifuge (2,000 × *g*, 5 min). Transfer upper phase to a new 15-ml tube and re-extract lower phase as described above. Pool upper phases. The volume is now 3 to 4 ml. Reduce volume to about 400 μl with 2-butanol as described in step 5 and transfer to a 1.5-ml microcentrifuge tube. Add 0.25 volume (about 100 μl) 10 *M* ammonium acetate and two volumes (about 1.0 ml) ice-cold 100% ethanol, mix gently by inversion and store at −20 °C for 2 h (or at −70 °C for 20 min). Centrifuge (14,000 × *g*, 30 min, 4 °C). Pour off ethanol and rinse pellet with 70% ethanol. Dry pellet under vacuum at room temperature. Dissolve pellet in 100 to 250 μl of 1 × TE (10:0.1) or sterile water (see Note 5). Store at −20 °C. *Time: 2 to 3 h.*

Gel Block Purification

1. Precipitate the crude soil DNA and dissolve pellet in 800 µl 1× TE (10:1). Set tube in 70 °C water bath until completely dissolved (15–30 min). Transfer DNA solution to 5-ml plastic beaker using a wide-bore pipette tip. Add 60 µl ethidium bromide (200 µg ml^{-1}) and 20 µl Loading Buffer, and dilute to 1.5 ml with 1× TE (10:1). Gently swirl and return mixture to 70 °C water bath. *Time: 1 h.*

2. Coat the bottom of the gel molds (see Note 3) with 3% agarose (low melting point). When firm, transfer 1.5 ml of the molten (70 °C) 3% agarose solution to each of the samples in the 70 °C water bath. (Do one sample at a time.) Mix the solution with a stirring rod and quickly transfer to the gel mold. Let the slice set in the freezer (–20 °C) for 10 min. Gently push the slice out of the mold using four microscope slides glued together. When all slices are in position in the 20 × 25 cm gel mold, gently cover with 1% agarose (about 500 ml) containing 0.5 µg ml^{-1} ethidium bromide which has been cooled to 50 °C. It is important that the agarose solution is cooled to 50 °C, otherwise humic materials from the gel blocks will diffuse into the gel. Run gel at 240 V for 5 min and then at 120 V until the brown humic material is separated from the light blue dye closest to the original gel slice. Different samples will 'clean up' at different rates. View gel under UV illumination and cut out the gel slice containing the fluorescing DNA. In most cases, this slice is the original gel slice that was transferred from the mold. *Time: 2 h.*

3. With gloved fingers, drop gel slice into TAE buffer-filled dialysis tubing (15 cm long, clamped at bottom). Pour off excess buffer until tubing looks about half full, squeeze buffer towards the top until all air is removed, then clamp. Place dialysis tubing in the electrophoresis chamber, orienting it so that the 3% bottom coat is pushed to one side of the membrane and facing the (–) electrode and there is room for elution. Dialysis bag should be submerged but not floating. Run at 240 V for 1 h, terminals reversed. DNA will migrate out of the gel slice and accumulate against dialysis tubing wall. Do not touch or rearrange tubing during this process. Briefly (30 s) reverse leads to detach DNA from dialysis tubing, then turn off current. *Time: 1 h 30 min.*

4. Massage the dialysis tubing wall where the DNA has stuck. Using a transfer pipette, draw up the buffer and gently rinse the inside of the dialysis membrane. Repeat three times. Collect the buffer from the dialysis tube. Rinse the gel slice and inside of the dialysis tube with 1 ml of sterile water and pool wash with buffer in the tube. Total volume should be 5 to 8 ml. *Time: 30 min.*
5. Extract with 2-butanol to reduce volume of the DNA solution. Add one volume of 2-butanol, gently invert until emulsion forms (about 20 times), centrifuge (2,000 × *g*, 1 min), carefully pipette off the butanol (upper) phase, and repeat these butanol extractions until the volume of DNA solution is 1.5 ml. Depending upon the sample, three to six extractions may be needed to reduce a 5 to 8 ml sample to the desired 1.5 ml. *Time: 1 h.*
6. Extract with one volume of phenol/chloroform/isoamyl alcohol mixture (25:24:1). Mix gently by inversion until emulsified, and centrifuge (2,000 × *g*, 5 min). Carefully transfer upper phase to a new 15-ml tube using a wide-bore pipette tip, being careful to avoid the white material at the interface. Add 1 ml 1× TE (10:1) to the lower layer, mix gently by inversion, and centrifuge (2,000 × *g*, 5 min) again. Pool the upper phase with the first and extract with one volume of phenol/chloroform/isoamyl alcohol mixture (25:24:1). Mix gently by inversion until emulsified, and centrifuge (2,000 × *g*, 5 min). Transfer upper phase to a new 15-ml tube and re-extract lower phase as described above. Pool upper phases. The volume is now 3 to 4 ml. Reduce volume to about 400 µl with 2-butanol as described in step 5 and transfer to a 1.5-ml microcentrifuge tube. Add 0.25 volume (about 100 µl) 10 *M* ammonium acetate and two volumes (about 1.0 ml) ice-cold 100% ethanol, mix gently by inversion and store at −20 °C for 2 h (or at −70 °C for 20 min). Centrifuge (14,000 × *g*, 30 min, 4 °C). Pour off ethanol and rinse pellet with 70% ethanol. Dry pellet under vacuum at room temperature. Dissolve pellet in 100 to 250 µl of 1× TE (10:0.1) or sterile water (see Note 5). Store at −20 °C. *Time: 2 to 3 h.*

Conclusions

Gel purification is a gentle procedure that keeps genomic DNA intact and does not noticeably shear the DNA. Thus, DNA of high molecular weight can be obtained. Generally the yield of DNA is quite good, with recoveries approaching 100% in a low organic matter sandy soil [4]. Somewhat lower yields would be expected for samples high in organic matter. DNA purified by gel electrophoresis has been used successfully for hybridization and PCR amplification and is likely to be suitable for other molecular methods, such as restriction enzyme digests and cloning.

Notes

1. All containers, solutions, and supplies should be sterile and DNase free.
2. For the gel well procedure, 1-ml well-formers can be made from three $1 \times 25 \times 75$ mm microscope slides stacked and glued together with epoxy. Siliconize and boil before using, handle with gloves, tape to support to hold in position during use. A 20×25 cm gel can be used to process six samples, with two wells per row and three rows 8 cm apart.
3. For the gel block procedure, gel block molds can be formed from two $1 \times 25 \times 75$ mm microscope slides glued to a 5-mm spacer.
4. Butanol extraction removes water only, salts become concentrated. If EDTA (from TAE buffer) has a final concentration greater than 10 mM, it will co-precipitate with DNA during the subsequent step. The removal of water is not linear; the entire DNA sample can be lost from over-extraction with 2-butanol. This extraction also removes ethidium bromide.
5. DNA should be colorless and free from humic contaminants, however, a slight yellow color usually does not inhibit the PCR reaction following dilution of the DNA sample. If further clean-up is required, precipitate with PEG (polyethylene glycol) as follows: Add an equal volume of PEG/NaCl (20% PEG 8000, 2.5 M NaCl) incubate at 37 °C for 15 min, centrifuge at $10,000 \times g$ for 5 min, pipette off supernatant, wash pellet with 200 µl cold 80% ethanol, vacuum dry and dissolve the pellet in 10:0.1 TE.

References

1. Cresswell N, Saunders VA, Wellington EMH (1991) Detection and quantification of *Streptomyces violaceolatus* plasmid DNA in soil. Lett Appl Microbiol 13: 193–197.
2. Hilger AB, Myrold DD (1991) Method for extraction of *Frankia* DNA from soil. Agric Ecosys Environ 34: 107–113.
3. Holben WE, Jansson JK, Chelm BK, Tiedje JM (1988) DNA probe method for the detection of specific microorganisms in the soil bacterial community. Appl Environ Microbiol 54: 703–711.
4. Myrold DD, Huss-Danell K (1994) Population dynamics of *Alnus*-infective *Frankia* in a forest soil with and without host trees. Soil Biol Biochem 26: 533–540.
5. Ogram A, Sayler GS, Barkay T (1987) The extraction and purification of microbial DNA from sediments. J Microbiol Meth 7: 56–66.
6. Porteous LA, Armstrong JL (1991) Recovery of bulk DNA from soil by a rapid, small-scale extraction method. Current Microbiol 22: 345–348.

7. Rochelle PA, Olson BH (1991) A simple technique for electroelution of DNA from environmental samples. BioTechniques 11: 724–728.
8. Sambrook J, Fritsch EF, Maniatis T (1989) Molecular Cloning: A Laboratory Manual, 2nd Edition. New York: Cold Spring Harbor Laboratory Press.
9. Selenska S, Klingmüller W (1991) DNA recovery and direct detection of *Tn5* sequences from soil. Lett Appl Microbiol 13: 21–24.
10. Smalla K, Cresswell N, Mendonça-Hagler LC, Wolters A, Van Elsas JD (1993) Rapid DNA extraction protocol from soil for polymerase chain reaction-mediated amplification. J Appl Bacteriol 74: 78–85.
11. Steffan RJ, Goksøyr J, Bej AK, Atlas RM (1988) Recovery of DNA from soils and sediments. Appl Environ Microbiol 54: 2908–2915.
12. Torsvik V (1980) Isolation of bacterial DNA from soil. Soil Biol Biochem 12: 15–21.
13. Tsai Y-L, Olson BH (1991) Rapid method for direct extraction of DNA from soil and sediments. Appl Environ Microbiol 57: 1070-1074.
14. Tsai Y-L, Olson BH (1992) Rapid method for separation of bacterial DNA from humic substances in sediments for polymerase chain reaction. Appl Environ Microbiol 58: 2292–2295.
15. Van Elsas JD, Van Overbeek LS, Fouchier R (1991) A specific marker, *pat*, for studying the fate of introduced bacteria and their DNA in soil using a combination of detection techniques. Plant Soil 138: 49–60.
16. Young CC, Burghoff RL, Keim LG, Minak-Bernero V, Lute JR, Hinton SM (1993) Polyvinylpyrrolidone-agarose gel electrophoresis purification of polymerase chain reaction-amplifiable DNA from soils. Appl Environ Microbiol 59: 1972-1974.

8. Kornberg A, Baker TH (1991) DNA Replication, 2nd edition. WH Freeman and Company, New York.

9. Sherratt DJ, Blakeley G, Burke K et al. (1994) Site-specific recombination and circular chromosome segregation. Philos Trans R Soc Lond B 347:37-42.

10. Staudt LM, Lenardo MJ (1991) Immunoglobulin gene transcription. Annu Rev Immunol 9:373-398.

11. Stark GR, Debatisse M, Giulotto E, Wahl GM (1989) Recent progress in understanding mechanisms of mammalian DNA amplification. Cell 57:901-908.

12. Stark GR, Wahl GM (1984) Gene amplification. Annu Rev Biochem 53:447-491.

13. Van Dyke MW, Hertzberg RP, Dervan PB (1982) Map of distamycin, netropsin, and actinomycin binding sites on heterogeneous DNA. Proc Natl Acad Sci USA 79:5470-5474.

14. Wells RD (1988) Unusual DNA structures. J Biol Chem 263:1095-1098.

Molecular Microbial Ecology Manual **1.4.2**: 1–20, 1995
ⓒ 1995 *Kluwer Academic Publishers.*

Extraction and PCR amplification of DNA from the rhizoplane

PENNY A. BRAMWELL[1], RITA V. BARALLON[2], HILARY J. ROGERS[2] and MARK J. BAILEY[1,*].

[1] *Molecular Microbial Ecology, Institute of Virology and Environmental Microbiology, Natural Environment Research Council, Mansfield Road, Oxford, OX1 3SR, UK,* [2] *Analytical Molecular Biology Group. Laboratory of the Government Chemist, Queens Road, Teddington, Middlesex, TW11 0LY, UK*

Introduction

The rhizoplane environment

While the term rhizoplane has been adopted for the plant root surface [8], the root environment represents a plant-soil-microbial continuum [40]. That is, microbial communities are found to transect the root, from the soil under the influence of the root (rhizosphere) through to the cortical tissue. The root surface is a rapidly changing substrate, particularly during the first 10 days following emergence of the radicle from the seed [13]. Root habitats are characterized by environmental gradients, including reduced oxygen tensions, elevated CO_2 levels as a result of plant and microbial respiration, pH differences and negative water potentials caused by evaporation at the leaf surface and subsequent transpiration.

Mucilage, the main plant root excretory product, binds clay particles, minerals, organic matter and bacteria into a layer around the root epidermis. This mucigel layer may be discrete and bound by a distinct pellicle, membrane or cuticle; alternatively it may extend into the adjacent soil. The proliferation of microbial life at the root surface has been attributed to the release of organic and inorganic nutrients. It has been estimated that 25% of the organic matter from maize roots is lost through rhizodeposition [16]. The chemistry of root exudates is complex, but essentially comprises a mixture of sugars, amino acids, glycosides, organic acids, vitamins, enzymes and complex carbohydrates. Root cap and tip cells have been identified to be sites of active exudation. In studies directed at the rhizosphere and rhizoplane, the influence of the whole plant on the root system cannot be excluded. For example, there is a direct relationship between the photosynthetic capacity of the plant and the carbohydrate content in the roots.

---

* Corresponding author.

The rhizoplane microbial community

The root supports microbial communities involved in a number of key processes such as nitrogen fixation, ammonification, nitrification, denitrification, mineralisation and the transformation of nutrient elements. Certain root-associated microbial communities have been identified to be Plant Growth Promoting Rhizobacteria (PGPR) [24,25] and play key roles in the suppression of phytopathogens [52] or enhancement of plant growth. These activities may be attributed to the production of antibiotics [14], aggressive pre-emptive colonisation of available niches and the consequent exclusion of deleterious microorganisms [52]. Root associated bacteria also produce plant growth hormones such as auxins and gibberellins [4,5] as well as siderophores [25].

Bacterial coverage of the root surface has been estimated to be in the range of 5 to 10% [43]. Macro-ecology pattern analysis techniques have been applied to analyze the distribution of rhizoplane bacteria at the root surface. These studies indicate that they are not randomly distributed, but are present in aggregates [38]. The non-random distribution is apparent both at a microscopic and on a larger scale and may in part reflect preferential microsites for colonization. Further studies suggest that the distribution of bacterial cells appears to be related to plant microbe contact rather than bacterial motility [19,1,36]. Bacterial cells have rarely been found associated with the rapidly growing apical tip with the first area to be colonized identified to be just beyond the zone of elongation.

Experimental approach

Sampling, extraction, enumeration and characterization of rhizoplane bacteria

Few studies on the microbial ecology of plant roots have employed molecular techniques, rather the field has been developed using microscopic and plate cultivation procedures. The reluctance to apply molecular techniques to plant microbial ecology is surprising since much work has been done to recover good quality nucleic acids from plants. However, there are still difficulties to be resolved in terms of how to apply the sensitive molecular approaches quickly and accurately in order to process a number of samples that is sufficiently representative of the heterogeneity of the natural environment. These considerations suggest that a combination of molecular and traditional approaches will be fundamental to developing our understanding of this area. In our comparisons, molecular detection techniques appear as sensitive as classical plating methods, although it is likely that the full potential of molecular approaches has not yet been realized especially in the monitoring of nonculturable organisms. Selective plating frequently forms the basis for analyses of individual members of a community, which may be characterised by molecular and phenotypic methods. Microorganisms can be characterised using molecular fingerprinting methods such as DNA restriction endonuclease pattern analysis [47], Restriction Fragment Length Polymorphism (RFLP) [3]; Randomly Amplified Polymorphic DNA (RAPDs) [32,34,35], Polymerase Chain Reaction (PCR) amplification of repetitive DNA sequences [10] and restriction site analysis of PCR amplified 16S ribosomal RNA/DNA [46]. The application of genetic fingerprinting techniques to populations recovered on selective media provides a powerful means to describe the genetic basis of microbial community structure. Selective plating in combination with molecular techniques may also be used for tracking genetically modified microorganisms (GMMs) in the natural environment. Approaches to the monitoring of GMMs have been described elsewhere [2,20].

Sampling

Rhizosphere microbiologists have used a variety of approaches for sampling the culturable component of microbial communities at the root/soil interface. The lack of a universally adopted standard is a reflection of the complexity of the root system, and the limitations imposed by each sampling regime. These include the time available, plant and soil type, the communities under study and the environmental questions being posed.

Initial problems in sampling the rhizoplane or rhizosphere arise from removing the plant root from the soil [6]. This may be easier for pot-grown plants, than for those grown in the field. The amount of root and soil extracted will be dependent on the plant type, method of extraction (usually with a fork or spade), root morphology and soil structure. Separating finer roots from soil structures can be extremely difficult, but has been achieved using artificial growth chambers. Kuchenbuch and Jungk [26] describe a method in which the roots are separated from the soil by a screen of nylon cloth. Root hairs but not roots, penetrate through the screen into the soil. The method, initially developed for the determination of concentration gradients in the vicinity of plant roots, has been applied to investigate the distribution of bacteria at the root surface [11].

In our studies on the microbial ecology of sugar beet, the plants are loosened in the field with a fork and then pulled out by hand. The loosely attached soil is then shaken free. Soil that is not released is then considered to be rhizosphere soil. This can be removed by scraping the roots and washing in deionised water to expose the root surface (rhizoplane).

Bacterial populations at the plant-soil interface have been found to be log–normally distributed [29]. That is, the population densities, expressed in colony forming units per gram (c.f.u. g^{-1}) dry weight tissue are not normally distributed, whereas $\log_{10}$ transformed values are. These findings highlight the problems of taking representative samples. This variation may be compensated for by taking pooled samples from a number of different plants. Bulked samples do have a tendency to overestimate the mean population size. The degree of overestimation is dependent on the variance of the population, which fluctuates with environmental conditions [17,18]. An algorithm can be applied to compensate for overestimated mean values by calculating the variance from individual plants on each sample occasion [18].

Extraction

Several methods have been compared for extracting a biological control strain of *Pseudomonas fluorescens* introduced as an inoculum to cotton seed [23]. Assessments of efficiency of recovery were determined by comparing average c.f.u. g^{-1} root on *Pseudomonas* selective agar, for total roots per plant and 2.0 cm segments of the tap root sample and three root regions. These were a) the top 5 cm of crown roots with attached laterals, trimmed to 1 cm lengths, b) The tip 5 cm of the crown roots with attached laterals trimmed and c) all lateral roots. The five methods that were compared were: (i) agitation with a wrist action shaker at 150 r.p.m. for 30 min in phosphate buffer (PB), (ii) agitation with a wrist action shaker at 150 r.p.m. for 5 min in PB containing 0.5 g ml^{-1} glass beads, (iii) sonication for 1 min in PB using a Microson™ ultrasonic cell disrupter, (iv) maceration for 1 min in a stomacher lab blender and (v) maceration using a pestle and mortar. The sonication conditions applied did not affect viability. The count obtained varied with the method of extraction. Populations recovered using the pestle and mortar, agitation with beads and stomacher blending were higher than those obtained through sonication or agitation. The most consistent results were obtained with the automated paddle-action of the stomacher.

In our laboratory, rhizoplane samples of sugar beet are routinely prepared by homogenizing coarsely chopped, rinsed root peelings (<1 mm) in physiological saline at full speed for 1 min in a Waring blender [28]. We find that homogenization compares favourably in extraction efficiency with either stomacher blending or bead agitation. The extraction method used is dictated by the numbers and size of samples. For example, a minimum of 5 g of root material is required for homogenization, whereas small quantities are suitable for agitation or stomacher blending.

Extraction of bacteria from rhizoplane samples contaminated with soil

The extraction of bacterial DNA from the rhizoplane requires efficient methods for the removal of soil colloids adhering to the root surface. At the simplest level small aggregates of soil may be removed from the root using forceps. More frequently soil is removed by washing roots in a suitable diluent or isotonic solution either with or without glass beads or gravel. A method for disrupting soil aggregates to release soil micro-organisms, has been described using an ion exchange resin in conjunction with a non-biocidal detergent [31]. The resin (the sodium ion of Dowex A1) acts by exchanging cations with the soil and cell, and helps in the disruption of soil particles intimately associated with the rhizoplane [6]. Once removed the microbial component of the rhizosphere soil fraction may be analyzed [6] and bacteria extracted from the soil-free root or rhizoplane material by abrasion or maceration.

Extraction of microbial DNA from the rhizoplane

DNA may be extracted from rhizoplane samples either indirectly, by recovery of the microbial fraction prior to DNA extraction, or directly, by extracting both root and microbial DNA. Cells are most frequently removed from rhizoplane material by agitating soil-free roots in water, with or without glass beads for long periods [15,30,44]. Alternatively, root surface peel or whole fibrous root tissue samples can be homogenised and the bacterial fraction (which contains both surface associated and endophytic microorganisms) recovered by centrifugation. Other techniques have been devised to remove the mucigel layer, present on certain plant roots, such as shaking in 1 M NH_4Cl solution [22]. We routinely remove rhizoplane bacteria from the fibrous roots of sugar beet by simple abrasion in a 10% suspension of fine sand.

Procedures

Indirect extraction of DNA from rhizoplane microbes

Equipment required
- Orbital or wrist shaker.
- Waring blender.

Collection of plants and removal of rhizosphere soil

1. Remove a plant from the field, plant pot or microcosm.
2. Shake away all soil loosely adhering to the root.

3. Remove tightly adhering soil (rhizosphere soil) by washing in sterile water or saline.
4. Collect rhizoplane samples either by abrasive washing or homogenization of whole roots, fibrous roots and/or a fine surface peel.

Abrasive root washing

a. Prepare the plant by washing away all soil and place approx. 10 g of root material or a fine peel of rhizoplane material in a 250 ml conical flask with 100 ml 1/4 strength Ringer's solution and 10 g of autoclaved sand.
b. Agitate on an orbital shaker at 200 r.p.m. for approx. 1 h.
c. Allow sand and root material to settle on the bench for 10–15 min and recover cell suspension.
d. Place 1.5 ml aliquots into Eppendorf tubes and spin at 6,500 r.p.m. for 10–15 min in a microcentrifuge to pellet cells.
e. Enumerate bacteria on selective agar or extract genomic DNA.

Homogenisation of rhizoplane material

a. Prepare the plant root by washing away all soil and place rhizoplane material in a suitable diluent (1/4 strength Ringer's solution or water), 1 part plant material to 10 volumes of diluent and homogenize for 1 min in a sterile Waring blender.
b. Alternatively, mix the root material in a 1:10 ratio of diluent with sufficient beads to fill a screw capped tube when vortexed and agitate vigorously for 10 min on a Griffin wrist-action shaker (Fisons, UK).
c. Enumerate bacteria on selective agar or extract genomic DNA.

Notes
– The choice of extraction method will be dictated by the morphology of the plant root system and size of the sample to be analyzed.
– Rhizoplane material may be sub-divided to assess microbial distribution, these may include: sections of the tap root, fibrous roots, root hairs or root storage organs.
– Alternative approaches for studying plant root microbiology include growing plants in soil or other materials between glass plates. Separation of the plates, allows specific regions of the root at particular depths or age etc. to be sampled by micro-manipulation.

MMEM-1.4.2/6

- Fine rhizoplane peelings can be collected from established roots or from root storage structures by shaving with a safety razor or peeling with a vegetable knife.
- Disruption via homogenization or abrasive bead beating of whole roots or root sections will extract endophytes as well as microbes associated with the root surface.

Isolation of bacteria by plating methods

The rhizoplane cell suspension is diluted in sterile distilled water (or physiological saline) and spread onto agar containing the antifungal agents, cycloheximide at 50–100 µg ml^{-1} and nystatin at 10–25 µg ml^{-1}. The choice of media will depend on the target bacterial population. For example, tryptic soy broth agar (TSBA, Difco) or plate count agar (PCA, Difco) are used to obtain total culturable counts. Similarly, selective agars (Oxoid) have been used to isolate specific components of the bacterial community from the phyllosphere [48,49]. In order to enumerate a community using selective plating, the bacterial species of interest should be present in at least 1×10^2 c.f.u. g^{-1} weight of environmental substrate [50].

Isolation of bacteria from the root surface using agar overlay

The spatial distribution of bacteria on the root surface may be studied by overlaying roots with a selective agar containing a metabolic dye. Reduction of TTC (2,3,5-triphenyltetrazolium chloride) to produce a strong red colour allows metabolically active colonies to be identified. These may then be isolated using a sterile needle or seeker and characterized genotypically and/ or phenotypically.

Equipment required
- Petri dishes or Nunc bioassay dishes (depending on root type and stage of development).
- Binocular microscope with photographic set up (optional).
1. Wash away all soil attached to the root with sterile water or physiological saline.
2. Place roots in the bottom of a Petri or Nunc bioassay dish and overlay with molten agar cooled to approx. 45 °C. (For location of total bacteria, use a general purpose bacterial medium such as tryptic soy broth agar (Oxoid) containing 1% TTC.

3. Incubate plates at 28 °C for 24–48 h.
4. Inspect colonies under a binocular microscope and photograph if required.
5. Isolate individual colonies from the area of interest using a sterile seeker.
6. Streak isolates to single colonies onto fresh plates and analyze as required.

Notes
– Bacterial colonies stain red with TTC. Selective agars or agars containing appropriate additives may be used to target specific populations. We use this approach to localize a fluorescent pseudomonad modified to express *lacZ* (β-galactosidase) by incorporating X-gal (25 µg/ml into *Pseudomonas* selective agar). GMM colonies stain blue.

Extraction of genomic DNA from recovered bacterial cells

This method uses the mild detergent CTAB (hexadecyltrimethyl ammonium bromide, Sigma, Co.) to extract total DNA [1] from colonies recovered from isolation plates, the overlay method, overnight cell cultures and bacterial cell suspensions concentrated via centrifugation from root washings or homogenates.
1. Spin the cells to produce a pellet in 1.5 ml Eppendorf tubes in a microcentrifuge (Heraeus Sepatech, Biofuge A or similar model) for 5 min at 13,000 r.p.m.
2. Wash the pellet, by centrifugation (5 min at 13,000 r.p.m. in a microcentrifuge) with ice-cold 1 M NaCl.
3. Resuspend the cells in 500 µl lysis buffer (LB: 0.5% Sodium Dodecyl Sulphate; 20 µg ml^{-1} Proteinase K) and incubate at 55 °C for 30 min.
4. Add 100 µl of prewarmed (65 °C) 1 M NaCl, mix thoroughly and precipitate the denatured proteins and polysaccharides with 80 µl prewarmed (65 °C) CTAB solution (10% w/v hexadecyltrimethyl ammonium bromide in 0.7 M NaCl) by incubation for 10 min at 65 °C.
5. Add 680 µl isoamyl alcohol:chloroform (1:24), mix thoroughly and extract the top, aqueous phase after centrifuging for 5 min at 13,000 r.p.m. in a microcentrifuge.
6. Precipitate the DNA with 0.6 volumes of isopropanol at room

temperature for 10 min. Recover the DNA by centrifugation (10 min at 13,000 r.p.m.). Wash the pellet with ice-cold 70% ethanol.

7. Remove residual ethanol (Dry the pellet either by placing briefly at 65 °C with the Eppendorf tube caps open, by drying in a desiccation or via air drying on the bench). Do not overdry as problems may be encountered when resuspending DNA contaminated with impurities.

8. Resuspend the pellet in 100 µl TE buffer (10 mM Tris-HCl pH 8.0, 1 mM EDTA) at 4 °C overnight. DNA extracted using this method can be used directly for dot hybridization, restriction endonuclease digestion, Southern hybridization etc., and in PCR without further cleaning steps.

Notes
- Particularly viscous DNA extracts, may require a phenol/chloroform extraction prior to the isoamyl alcohol:chloroform step. To extract DNA from Gram positive bacteria a harsher lysis procedure is necessary such as snap freezing the washed cell pellet in liquid nitrogen and boiling for 2 min or bead beating (agitate for 5 min in a Braun homogenizer with 0.1 mm glass beads [9]).
- Bacterial genomic DNA prepared by this method can be stored at 4 °C.

Direct extraction

The direct extraction of rhizoplane and root samples offers the potential to recover both surface associated micro-organisms and those more intimately associated with the root tissue. The rhizoplane may be sampled by peeling plants with bulbous roots or underground organs, however, plants with fine seminal roots may have to be homogenized or macerated as they may not be amenable to peeling or scraping.

Rapid DNA extraction from root and rhizoplane samples

Root tissue, like phylloplane material is high in phenolic compounds, complex carbohydrates such as lignin and other plant secondary metabolites. In addition, root tissue is often contaminated by soil which contains humic and fulvic acids in addition to clay particles, all of which may strongly bind DNA and prevent its analysis by PCR and/or hybridization. For PCR analysis some additional purification of extracted DNA may be necessary.

A number of protocols for rapidly extracting DNA from plant tissue have been described [7,12,39,51]. These recover DNA of sufficient purity for PCR amplification from small amounts of starting material. An overview of other approaches to the isolation of microbial DNA from plant systems is included in the chapter "Extraction of microbial DNA from the phylloplane"(Chapter 1.4.3).

We have made a comparison of these rapid DNA extraction methods [42] on washed scrapings of the sugar beet storage organ (tap root) and on unwashed oil seed rape roots and conclude that certain methods appear to be advantageous in terms of speed, while others may be more applicable to specific plant species and tissues. All of the methods analyzed required a further purification step and/or 1/100 dilution of the root tissues for PCR analysis of DNA extracted by the direct approach. In surface washings, DNA yield varies with the numbers of microbes present, but surface washings are less contaminated by plant DNA than samples prepared following extraction of homogenates. The method identified to be both rapid [51] and most reliable was adapted as follows:

Equipment required
– Eppendorf grinders.
– Eppendorf tubes.
– microcentrifuge.
– Gilson pipettes.
– thermal cycler.
– gel electrophoresis apparatus.
1. Wash away all soil attached to the root with sterile water or physiological saline.
2. Homogenize 5–10 mg of root tissue or cell pellet from rhizoplane sample with an Eppendorf grinder for approx. 2–3 min in 10 µl of 0.5 M NaOH. The root material may absorb the NaOH, if so add a further 10 to 20 µl of 0.1 M Tris-HCl pH 8.0.
3. Microfuge the homogenate for 1–2 min at full speed on a micro-centrifuge (13,000 r.p.m.), and remove 5 µl of the supernatant, dilute to a final volume of 0.5 ml with 0.1 M Tris-HCl pH 8.0. (If it is not possible to recover 5 µl, then add 10–20 µl of Tris to the homogenate and remove 5µl).

4. Purify DNA solution using a Sepharose CL-6B spin column.
5. Use 1 µl of this solution in a 25 µl PCR assay.

Note
This procedure will take approximately 5–10 min. Extracts can be stored for several months at –20 °C.

Alternative methodology

The following method [39], has been adapted and successful applied to extracted rhizoplane material.

Additional equipment needed
- boiling water bath in a fume hood.
- refrigerated microfuge or cold room.
- – 70 °C freezer (or – 20 °C).
- 60 °C oven or vacuum desiccator (optional).
1. Grind approx. 5–10 mg of root tissue using an Eppendorf grinder in an Eppendorf tube.
2. Add 600 µl of extraction buffer (EB; 100 mM Tris-HCl pH 8.0, 50 mM EDTA, 500 mM NaCl, 10 mM mercaptoethanol), mix and boil the homogenate for 7 min (pierce a hole in the cap of the tube using a syringe needle).
3. Place the tube on ice and then spin at full speed in a microcentrifuge for 15 min at 4 °C.
4. Transfer 450 µl of the supernatant to a fresh tube and precipitate the DNA with 45 µl of 10 M ammonium acetate (pH 4.2) and 1 ml of ethanol at –70 °C for 30 min (or –20 °C for 1 h).
5. Recover the DNA by centrifugation in a microfuge at 13,000 r.p.m. for 15 min at 4 °C.
6. Remove and discard the supernatant, and air dry the pellet (which will usually appear visible and coloured) at 60 °C for 5–10 min (or vacuum dry for 5–10 min, or air dry at room temperature until dry).
7. Resuspend the pellet in 50 µl of TE buffer.
8. Purify DNA suspension using a Sepharose CL-6B column.
9. Use 1 µl in a 25 µl PCR reaction.

– The whole procedure will take approx. 90 min. Extracts can be stored for several months at –20 °C.
– All rapid methods are prone to occasional failure and the inclusion of appropriate controls is advised. If the DNA is not sufficiently pure for PCR, restriction enzyme analysis etc., further purification by CL-6B Sepharose spin columns is recommended.

Removal of inhibitors with sepharose CL-6B

1. Equilibrate 100 g of the Sepharose CL-6B (Pharmacia) with TE pH 8.0 at 4 °C.
2. Construct a spin column. Using a hypodermic needle, pierce a small hole at the bottom of a 0.7 ml Eppendorf tube. Place the pierced tube in a 2 ml Eppendorf tube.
3. Add a drop of 0.4 mm glass beads (autoclaved and stored in TE pH 8.0 buffer at room temperature) to the 0.7 ml Eppendorf tube.
4. Add 500 µl of equilibrated Sepharose CL-6B.
5. Spin, with the caps removed, in a bench top microfuge at 2,000 r.p.m. for 2 min.
6. Replace the 2 ml tube with a fresh, sterile 1.5 ml Eppendorf tube (cap removed). Gently pipette the boiled leaf wash onto the top of the column (maximum volume/column is 50 µl).
7. Spin at 2,000 r.p.m. for 2 min.
8. Store the eluate at 4 ° C or –20 °C until required.

Detection of bacteria or target genes using PCR

The monitoring of genetically modified organisms in the environment necessitates simple and rapid methods for collecting and processing material from environmental samples. PCR offers an appropriate molecular tool for the detection of low levels of bacteria released into the environment either deliberately or accidentally [37], since it is both highly sensitive and amenable to microtitre technology.

Preparation of sample and DNA purification/dilution for successful PCR

Take an aliquot of DNA purified as described above, an aliquot taken directly from a root wash suspension or an isolated bacterial colony. Boil samples for at least 5 min to lyse cells and release DNA. Remove debris by centrifugation in a microfuge, dilute sample as necessary and perform the PCR. Sepharose CL-6B purification can be used for the removal of impurities from root extracts which are inhibitory to PCR amplification. A dilution of 1/100 prior to PCR has also worked in some cases, however, this was not as reliable and should only be used if appropriate controls demonstrate that PCR inhibition is effectively eliminated and sufficient target DNA is present. For PCR detection of bacteria isolated from the sugar beet rhizoplane simple CTAB extraction of homogenates or surface washings is effective in eliminating inhibition by plant and soil material (Fig. 1).

Conditions used for PCR

The reaction conditions for PCR will be dependent on the size and sequence of primers, the Polymerase enzyme and the type of thermal cycler being used. For more specific details about PCR refer to the appropriate texts [21,33] or to chapter 2.7.2.

The conditions described below are those used to demonstrate the presence of eubacterial DNA prepared from a rhizoplane homogenate by targeting highly conserved regions of the 16S ribosomal RNA gene. The following 16S rRNA eubacterial primers are routinely used: 5' CAGCa/cGCCGCGGTAATa/tC 3' and 5' CCGTCAATTCa/cTTTa/gAGTTT 3' [27]. Specific primers have also been applied for the detection of well-characterised target bacteria, e.g., a GMM (Fig. 1).

PCR conditions

1. Prepare approx. 10–100 ng template DNA, bacterial colonies from culture or rhizoplane samples.
2. Add 2.5 µl of each primer (20 µM).
3. Add water to make the volume to 25 µl.

4. Add 50 µl of mineral oil (Sigma) (this step is not required for thermal cyclers with a heated lid).
5. Heat the solution in a thermal cycler at 95 °C for 7 min followed by 80 °C for 30 min (i.e., a 'hot start').
6. While the solution is at 80 °C, add 25 µl volume of the following cocktail to the reaction mixture:
   8 µl dNTP (200 µM final concentration).
   5 µl 10 × PCR buffer.
   0.25 µl *Taq* polymerase (Perkin-Elmer).
   11.75 µl sterile distilled water.
   Final volume 50 µl.
7. Begin thermal cycling.
   The PCR cycle used for this set of primers was:

   1 Cycle
   95 °C for 4 min.
   50 °C for 10 s
   72 °C for 1 min.

   28 Cycles
   95 °C for 1 min.
   50 °C for 10 s
   72 °C for 1 min.

   1 Cycle
   95 °C for 1 min.
   50 °C for 10 s
   72 °C for 8 min.
8. Analyze samples on a 1% agarose gel in 1 × TBE buffer containing 10 µg ml^{-1} ethidium bromide. To 15 µl PCR product add 3 µl Bromophenol blue solution [45] and load onto the agarose gel. If mineral oil was used in the PCR, it is not necessary to remove the oil prior to loading the samples (pipette through the oil meniscus).

Notes
− Controls that should be included: DNA extracted from pure bacterial cultures (a positive control), while sterile water should be included as a negative control to confirm specificity.
− When using Universal primers, a faint product is frequently observed in the negative control. This phenomenon has been reported [41] and is due to the

presence of DNA in commercial preparations of *Taq* polymerase. This problem may be overcome through the use of low DNA *Taq*(*Taq* LD ™ available from Perkin-Elmer).

- A serial dilution of known DNA concentration or bacterial cells numbers can be included for quantitative estimates.
- Following the production of a PCR product of predictable size, confirm specificity of the product by either dot blot or southern blot hybridization [45] or restriction endonuclease analysis [46].

Use of CTAB extraction to remove inhibitors of PCR present in sugar beet root homogenate

Direct PCR amplification (hot start) was performed on pure bacterial cultures diluted in water (Fig. 1a), on rhizoplane homogenate (10% w/v in water) (Fig. 1b) and on DNA extracted by the CTAB method from bacteria diluted in rhizoplane homogenate (Fig. 1c).

A rhizoplane homogenate was prepared by finely peeling fully differentiated sugar beet and homogenizing 30 g of tissue in 270 ml of sterile water at full speed for 1 min in a Waring blender. The homogenate was allowed to stand at room temperature for 30 min, before addition of bacteria. Rhizoplane homogenate, 900 µl, was dispensed into Eppendorf tubes and mixed with a 100 µl aliquot of bacterial cell suspension. The cells and homogenate were allowed to stand at room temperature for 1 h after thorough mixing. DNA was extracted from the diluted cell suspension spiked into the rhizoplane homogenate using the CTAB method, precipitated DNA was resuspended in 10 mM Tris-HCl pH 8.0 to the original volume.

PCR amplification targeted the kanamycin resistance gene (*kan*^{r}, from pUC4K, Pharmacia), introduced as a single copy into the chromosome of a fluorescent *Pseudomonas* strain SBW25 (Bailey unpublished). A 1.1 kbp product was generated with the following forward and reverse primers: 5' TGACTCATACCAGGCCTGAA 3' and 5' TACAAGGGGTGTTATGAGCC 3'. The PCR conditions employed are given in the text. Controls included 10 ng of CTAB–Sepharose CL-6B purified genomic DNA recovered from the recombinant microorganism (positive control), sterile distilled water and rhizosphere homogenate. A 15 µl aliquot of each reaction was loaded on a 1% agarose gel and stained with ethidium bromide after electrophoresis.

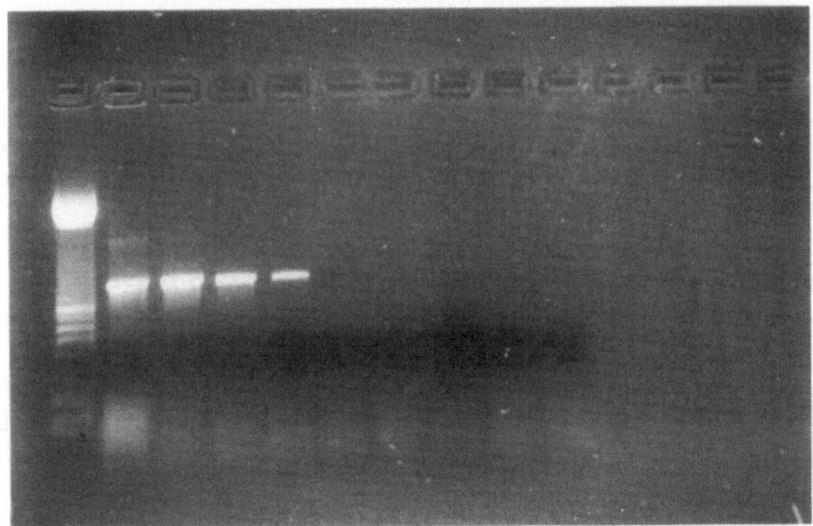

Figure 1a. Direct PCR amplification of pure GM-pseudomonad cells, hot start PCR: Lane 1, 100 bp molecular weight marker ladder; Lanes 2–10, 1.1 kbp kanamycin resistance gene fragment amplified from dilute cell suspension: 2, 10^6; 3, 10^5; 4, 10^4; 5, 10^3; 6, 10^2; 7, 10^1; 8, 1; 9, 10^{-1}; 10, 10^{-2} (c.f.u./reaction).

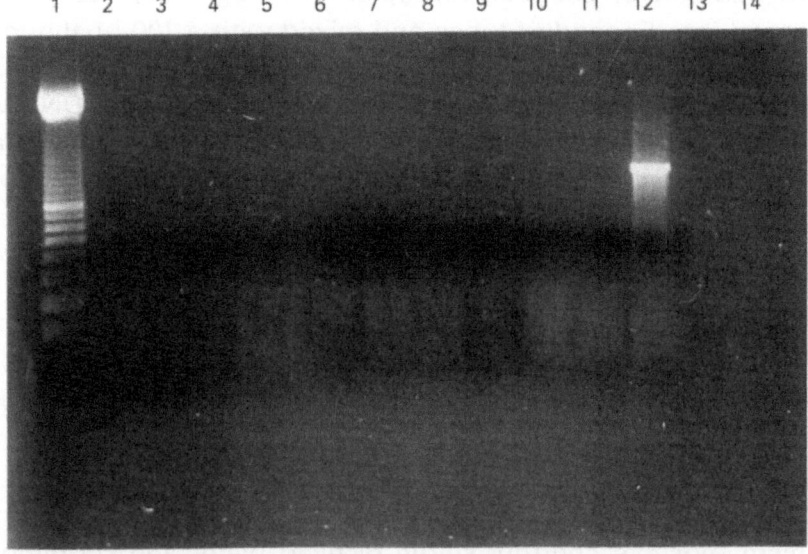

Figure 1b. Direct PCR of a serial dilution of GM-pseudomonad in rhizoplane homogenate, hot start PCR. Lane 1 100 bp molecular weight marker ladder; lanes 2–9 contain products from PCR reaction: 2, 10^5; 3, 10^4; 4, 10^3; 5, 10^2; 6, 10^1; 7, 1; 8, 10^{-1}; 9, 10^{-2} (c.f.u./reaction); 11, rhizoplane homogenate amended with sterile water; 12, 10 ng genomic DNA from the target microorganism.

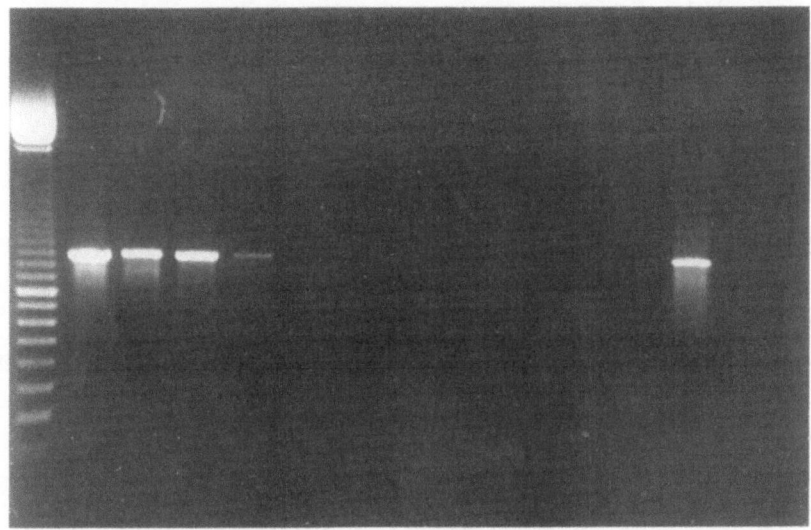

Figure 1c. Removal of rhizoplane inhibitors of PCR, Fig. 1b, by CTAB extraction. Lane 1, 100 bp molecular weight marker ladder; lanes 2–9 products from PCR reaction: 2, 10^5; 3, 10^4; 4, 10^3; 5, 10^2; 6, 10^1; 7, 1; 8, 10^{-1}; 9, 10^{-2}; 10, blank; 11, rhizoplane homogenate and sterile distilled water; 12, blank; 13, 10 ng genomic DNA from the target micro-organism. Note increased sensitivity of detection following (c.f.u./reaction) CTAB preparation even in the presence of rhizoplane homogenate.

Buffers and reagents

TE buffer
– 10 mM Tris-HCl pH 8.0.
– 1 mM Na_2EDTA pH 8.0.

CTAB solution
– 10% w/v hexadecyltrimethyl ammonium bromide in 0.7 M NaCl.

Lysis buffer (LB)
– 0.5% (w/v) Sodium Dodecyl Sulphate; 20 µg ml^{-1} Proteinase K.

Extraction buffer: (EB)
– 100 mM Tris-HCl (pH 8.0), 50 mM EDTA, 500 mM NaCl, 10 mM mercaptoethanol (stock solution is 14 M).

10 × TBE
– 108 g Tris base.

- 55 g Boric acid.
- 9.3 g Na$_2$EDTA.
- Water to 1 l.

10 × PCR buffer
- 500 mM KCl.
- 100 mM Tris-HCl pH 8.3.
- 15 mM MgCl.
- 0.01% w/v gelatin.

Make up to 100 ml with sterile distilled water
Aliquot 1 ml portions into screw cap Eppendorf tubes and sterilise

Bromophenol blue buffer (10×)
- 0.25% bromophenol blue.
- 30% glycerol in TE pH 8.0.
- 0.25% xylene cyanol(ff); optional.

References

1. Ausubel FM, Brent R, Kingston RE, Moore DD, Seidman J D, Smith JA, Struhl K (1989) Current Protocols on Molecular Biology. John Wiley and Sons, New York.
2. Bailey MJ (1994) Extraction of DNA from the Phyllosphere. In: Trevors JT, Van Elsas JD (eds) Nucleic Acids in the Environment: Methods and Applications, pp. 89–110. Springer-Verlag, Berlin.
3. Bloom RA, Mullin BC, Tate RL (1989) DNA restriction patterns and DNA-DNA solution hybridisation studies of *Frankia* isolates from *Myrica pensylvanica* (bayberry). Appl Environ Microbiol 55: 2155–2160.
4. Brown ME (1974) Seed and root bacterization. Annu Rev Phytopath 12: 181–195.
5. Brown ME (1972) Plant growth substances produced by micro-organisms of soil and rhizosphere. J App Bact 35: 443–451.
6. Campbell R, Greaves MP (1990) Methods for studying the microbial ecology of the rhizosphere. Meth Microbiol 22: 447–477.
7. Cheung WY, Hubert N, Landry BS (1993) A simple and rapid DNA microextraction method for plant animal and insect suitable for RAPD and other PCR analyses. PCR Meth Applic 3: 69–70.
8. Clarke FE (1949) Soil micro-orgainsims and plant roots. Adv Agron 1: 241–288.
9. Cresswell N, Saunders VA, Wellington EMH (1991) Detection and quantification of *Streptomyces violaceolatus* plasmid DNA in soil. Lett Appl Microbiol 13: 193–197.
10. De Bruijn FJ (1992) Use of repetitive (Repetitive extragenic palindromic and entero-bacterial repetitive intergenic consensus) sequences and the polymerase chain reaction to fingerprint the genomes of *Rhizobium meliloti* isolates and other soil bacteria. Appl Environ Microbiol 58: 2180–2187.
11. Dijkstra AF, Govaert JM, Scholten GHN, Van Elsas JD (1987) A soil chamber for study-ing the bacterial distribution in the vicinity of roots. Soil Biol Biochem 19: 351–352.
12. Edwards K, Johnstone C, Thompson C (1991) A simple and rapid method for the preparation of plant genomic DNA for PCR analysis. Nucl Acid Res 19: 1349.
13. Fokkema NJ, Schippers B (1986) Phyllosphere versus rhizosphere as environments for

saprophytic colonisation. In: Fokkema NJ, Van den Heuvel, J (eds) Microbiology of the Phyllosphere, pp. 137–159. Cambridge University Press.

14. Fravel DR (1988) Role of antibiosis in the biocontrol of plant diseases. Ann Rev Phytopath 26: 75–91.
15. Greaves MP, Webley DM (1965) J Appl Bacteriol 28: 454–465.
16. Haller T, Stolp M (1985) Quantitative estimation of root exudation from maize plants. Plant Soil 86: 207–216.
17. Hirano SS, Upper CD (1986) Temporal, spatial and genetic variabilty of leaf-associated bacterial populations. In: Fokkema NJ, Van den Heuvel, J (eds) Microbiology of the Phyllosphere. pp. 235–251. Cambridge University Press.
18. Hirano SS, Nordheim EV, Arny DC, Upper CD (1982) Lognormal distribution of epiphytic bacterial populations on leaf surfaces. Appl Environ Microbiol 44: 695–700.
19. Howie W, Cook RJ (1985) The effect of motility on wheat root colonization by fluorescent pseudomonads antagonistic to take-all of wheat. Phytopath 75: 1344.
20. Hwang I, Farrand SK (1994) A novel gene tag for identifying micro-organisms released into the environment. Appl Environ Microbiol 60: 913–920.
21. Innis MA, Gelfand DH, Sninsky JJ, White TJ (1990) In: PCR Protocols; A Guide to Methods and Applications. Academic Press.
22. Jenny H, Grossenbacher K (1963) Root-soil boundary zones as seen in the electron microscope. Proc Soil Sci Soc Am 27: 273–277.
23. Kloepper JW, Mahaffee W, McInroy JA, Backman PA (1990) Comparative analysis of isolation methods for recovering root-colonizing bacteria from roots. In: Keel C, Kroller B, Defago G (eds) Plant Growth-Promoting Rhizobacteria-Progress and Prospects. The second International workshop on Plant Growth-Promoting Rhizobacteria. Interlaken, Switzerland.
24. Kloepper JW, Lifshitz R, Zablotowicz RM (1989) Free-living bacterial inocula for enhancing crop biotechnology. Trends Biotechnol 7: 39–44.
25. Kloepper JW, Leong J, Teintze M, Schroth MN (1980) Enhanced plant growth by siderophores produced by plant growth-promoting rhizobacteria. Nature 286: 885–886.
26. Kuchenbuch R, Jungk A (1982) A method for determining concentration gradients at the soil-root interface by thin slicing rhizopheric soil. Plant Soil 68: 391–394.
27. Lane DJ, Pace B, Olsen GJ, Stahl DA, Sagin ML, Pace NR (1985) Rapid determination of 16S ribosomal RNA sequences for phylogenetic analysis. Proc Natl Acad Sci USA 82: 6955–6959
28. Lilley AK, Fry JC, Day MJ Bailey MJ (1994) In situ transfer of an exogenously isolated plasmid between *Pseudomonas spp.* in sugar beet rhizosphere. Microbiol 140: 27–33.
29. Loper JE, Suslow TV, Schroth MN (1984) Lognormal distribution of bacterial populations in the rhizosphere. Phytopath 74: 1454–1460.
30. Louw HA, Webley DM (1959) The bacteriology of the root region of the oat palnt grown under controlled pot culture conditions. J Appl Bacteriol 22: 216–226.
31. MacDonald RM (1986) Sampling soil microfloras: dispersion of soil by ion exchange and extraction of specific micro-organisms from suspension by elutriation. Soil Biol Biochem 18: 399–406.
32. McMillin DE, Muldrow LL (1992) Typing of toxic strains of *Clostridium-difficile* using DNA fingerprints generated with arbitary polymerase chain reaction primers. FEMS Microbiol Lett 92: 5–10.
33. McPherson MJ, Quirke P, Taylor GR (1991) PCR: A Practical Approach. IRL Press.
34. Mazurier SI, Wernars K (1992) Typing of *Listeria* strains by random amplification of polymorphic DNA. Res Microbiol 143: 499–505.
35. Menard C, Brousseau R, Mouton C (1992) Application of polymerase chain reaction with arbitary primer (AP-PCR) to strain identification of *Porphyromonas- (Bacteroides) gingivalis*. FEMS Microbiol Lett 95: 163–168.
36. Misaghi IJ, Olsen MW, Billotte JM, Sonoda RM (1992) The importance of rhizobacterial motility in biocontrol of bacterial wilt of tomato. Soil Biol Biochem 24: 287–297.

37. Mullis K, Faloona F (1987) In: Wu R (ed) Methods in Enzymology 155: p. 335. Academic Press, New York and London.
38. Newman EI, Bowen HJ (1974) Patterns of distribution of bacteria on root surfaces. Soil Biol Biochem 6: 205–209.
39. Oard JH, Dronovalli S (1992) Rapid isolation of rice and maize DNA for analysis by random primer PCR. Plant Mol Biol Reporter 10: 236–241.
40. Old KM, Nicolson TH (1978) The root cortex as part of a microbial continuum. In: Loutit MW, Miles JAR (eds) Microbial Ecology pp. 291–294. Springer Verlag, Berlin.
41. Rand KH, Houck H (1990) Taq Polymerase Contains Bacterial DNA of Unknown Origin. Mol Cell Probes 4: 445–450.
42. Rogers HJ, Parkes HC (1994) (unpublished observations).
43. Rovira AD (1979) Biology of the soil-root interface. In: Hartley JL, Russell RS (eds) The soil root interface, pp.145–160. Academic Press, New York and London.
44. Rovira AD, Newman EI, Bowen HJ, Campbell R (1974) Quantitative assessment of the rhizoplane microflora by direct microscopy. Soil Biol Biochem 6: 211–216.
45. Sambrook J, Fritsch EF, Maniatis T (1989) Molecular Cloning: A Laboratory Manual, 2nd ed. Cold Spring Harbor Laboratory Publications, New York.
46. Schneider B, Ahrens U, Kirkpatrick BC, Seemuller E (1993) Classification of plant-pathogenic mycoplasma-like organisms using restriction-site analysis of PCR-amplified 16S rDNA. J Gen Microbiol 139: 519–527.
47. Sorensen B, Falk ES, Wisloff-Nilsen E, Bjorvatu B, Kristiansen BE (1985) Multivariate analysis of Neisseria DNA restriction endonuclease patterns. J Gen Microbiol 131: 3099–3104.
48. Thompson IP, Bailey MJ, Ellis RJ, Lilley AK, McCormack PJ, Purdy KJ, Rainey PR (1995) Short term community dynamics in the phyllosphere microbiology of field grown sugar beet. FEMS Microbiol Ecol 16: 205–212.
49. Thompson IP, Bailey MJ, Fenlon JS, Fermor TR, Lilley AK, Lynch JM, McCormack PJ, McQuilken M, Purdy KJ, Rainey PB, Whipps JM (1993) Quantitative and qualitative seasonal changes in the microbial community from the phllosphere of sugar beet (*Beta vulgaris*). Plant Soil 150: 177–191.
50. Trevors JT, Van Elsas JD (1989) A review of selected methods in environmental microbial genetics. Can J Microbiol 35: 895–902.
51. Wang H, Qi M, Cutler J (1993) A simple method of preparing plant samples for PCR. Nucl Acids Res 21: 4153–4154.
52. Weller DM (1988) Biological control of soil borne plant pathogens in the rhizosphere with bacteria. Ann Rev Phytopath 26: 379–407.

Molecular Microbial Ecology Manual **1.4.3**: 1–21, 1995
© 1995 Kluwer Academic Publishers.

Extraction of microbial DNA from the phylloplane

PENNY A. BRAMWELL[1], RITA V. BARALLON[2],
HILARY J. ROGERS[2] and MARK J. BAILEY[1,*]

[1] Molecular Microbial Ecology, Institute of Virology and Environmental Microbiology,
Natural Environment Research Council, Mansfield Road, Oxford, OX1 3SR, UK, [2] Analytical
Molecular Biology Group, Laboratory of the Government Chemist, Queens Road, Teddington,
Middlesex, TW11 OLY, UK

Introduction

The phylloplane environment

The phylloplane represents the largest terrestrial habitat. The leaf offers an exposed, desiccating, radiation-prone, vibrating and potentially crowded environment for microorganisms. It is a habitat characterized by physical and chemical barriers to microbial colonisation. The leaf surface which is protected by non-uniform layers of wax, cutin and sometimes by polyisoprene may also develop trichomes (epidermal appendages). These structures offer the plant protection from infection, excessive water loss and play a key role in the production, metabolism and secretion of a number of natural compounds such as terpenoids, phenolics, mucoproteins and resins [54]. The microbial community within the leaf tissue may be exposed to fungitoxic and bacteriotoxic substances and phytoalexins. Since leaf surfaces are porous these compounds will have a direct effect on microbial communities both at the leaf surface and within the tissue [2]. Leaf leachates also represent a rich source of nutrients which include sugars, amino acids, salts and phenolics. The physiological status of the leaf surface and internal tissue change constantly as the leaf and host plant grow, mature and senesce.

The phylloplane microbial community

Bacterial epiphytes are characteristically diverse and highly adapted to the harsh leaf environment [4,40]. They are adept at scavenging nutrients and attaching to leaf surfaces [37], often having been deposited via rain [16] and aerosols [51]. Phylloplane bacteria are predominantly Gram negative, chromogenic rods many of which may be identified as *Pseudomonas* and *Xanthomonas* [4]. Populations on the leaf surface appear to be clonal with

* Corresponding author.

few bacterial species recovered at any one time. Fluctuations in community composition occur with environmental shifts and the physiological status of the leaf [23]. Microbial communities have been estimated to be in the order of 10^4–10^6 cells per cm^2, or 10^5–10^7 colony forming units (c.f.u.) g^{-1} dry weight tissue [40].

Experimental approach

Sampling, extraction, enumeration and characterization of microbes from leaves

The majority of investigations in phyllosphere microbial ecology have employed methods to physically extract microbes from target tissue followed by their cultivation on selective media. Few reports describe the application of molecular approaches to the microbial ecology of the phylloplane, although plant molecular biologists have addressed problems associated with the recovery of DNA from plant material. In our comparisons, molecular detection techniques appear as sensitive as classical plating methods, although it is likely that the full potential of molecular approaches has not yet been realized. Selective plating may also form the basis to analyses of individual members of a community, through the characterization of isolates using molecular and phenotypic methods. Microorganisms may be analyzed using molecular fingerprinting methods such as DNA restriction pattern analysis [58], Restriction Fragment Length Polymorphism (RFLP) [10], Randomly Amplified Polymorphic DNA (RAPDs) [42,43,47], Polymerase Chain Reaction (PCR) amplification of repetitive DNA sequences [11] and restriction site analysis of PCR amplified 16S ribosomal RNA/DNA [57]. The application of genetic fingerprinting techniques to isolates recovered on media selective for a defined component of the community offers a powerful means to describe the genetic basis of microbial community structure. Selective plating in combination with molecular techniques also provides a powerful approach for tracking genetically modified micro-organisms (GMMs), which may harbour genes that facilitate their selective isolation and also provide unique sequences for the development of probes and primers. Approaches to the monitoring of GMMs have been described elsewhere [6,31].

Sampling

Phyllosphere microbial communities display a log–normal pattern of distribution [28,29]. It has been observed that population densities, expressed in colony forming units per gram (c.f.u. g^{-1}) dry weight tissue, are not normally distributed whereas $\log_{10}$ transformed values are [25]. Population sizes can vary by over 1,000-fold between individual leaves of the same size, collected from a similar position in the canopy, on the same sampling occasion [29]. Consequently, bulked samples tend to overestimate the mean population size. The degree of overestimation is dependent on the variance of the population, which fluctuates with environmental conditions [30]. An algorithm has been successfully applied to compensate for the overestimated mean values by calculating the variance from individual leaves on each sample occasion [30]. Diurnal fluctuations have also been described in some phyllosphere communities. Some populations exhibited similar numbers at the same time on two consecutive days, but varied dramatically throughout the day and night. These observations highlight the dynamic nature of the leaf environment and the importance of adopting a specific daily sample time [5,60].

Studies in our laboratories have focused on the microbial community of the sugar beet phyllosphere. The microbial ecology of three stages of leaf development, designated immature, mature and senescent has been investigated [61]. Consistent samples are obtained by collecting a similar weight or area of each leaf and pooling leaves from several plants. Since mature sugar beet leaves are large, the tips of the leaves are taken. A 5 or 10 g sub-sample of the pooled sample is used for the extraction of micro-organisms.

Extraction

The extraction of bacteria from bulk samples of varying leaf numbers has been adopted widely [20,29,38,39]. Bacteria may be extracted from leaves by washing, stomaching, homogenizing, sonicating, and then recovered by selective plating procedures. As the plant associated microbial community represents a broad spectrum of micro-organisms that lyse differentially, the suitability of each extraction method should be assessed with the particular population under study. Furthermore, plant metabolites may affect the cultivation of specific microbial communities when low dilutions of homogenates are spread onto selective agar.

Studies can focus on the microbial communities loosely associated with the leaf surface and/or those within the internal tissue. Leaf washing removes loosely attached leaf surface microorganisms, while stomaching removes those attached more firmly. Conversely, homogenization or sonication results in the recovery of internal and surface associated microorganisms. Donegan et al. [20] evaluated a number of procedures for recovering microorganisms from the phylloplane. Their findings indicated that stomaching was more reproducible and recovered higher viable counts than blending, sonication or washing. The prompt processing of samples was also recommended, since storage at room temperature, 4 °C and freezing affected counts.

There are difficulties to be resolved in terms of how to apply the sensitive molecular approaches quickly and accurately in order to process sufficient samples that are representative of the heterogeneity of the natural environment. Extraction of DNA from bacteria found in the phyllosphere can be achieved by: indirect extraction, through the separation of intact bacterial cells from plant material by washing the leaves, followed by extraction of the DNA from the recovered cells. Alternatively, direct extraction of microbial DNA from non-sterile plants permits the isolation of all plant associated microbial DNA. A method for DNA amplification direct from leaf tissue without a DNA extraction step, has been reported [8]. However, we have found this approach to be unreliable.

A number of methods have been described for the extraction of DNA from plant material. Some of these have been evaluated for their potential to extract plant and microbe DNA with a view to detecting the target microbe. These approaches can be divided into rapid, small scale methods (requiring 5–100 mg of leaf tissue, equivalent to 0.5 cm^2) and more time consuming methods which require larger quantities of starting material. The comparisons outlined below have focused on the rapid, small scale extractions, since these approaches are most appropriate to environmental sampling. There are cases, however, where the DNA may not be sufficiently pure for PCR analysis as a result of contaminating plant secondary compounds such as polyphenols. These may inhibit PCR and hybridizations leading to false negative results, unless analyzed in the presence of appropriate negative controls. A review of methodologies for extracting DNA from plants and approaches to overcoming contamination problems has been included at the end of the chapter.

Procedures

Indirect extraction of DNA from phyllosphere microbes

Leaf washing

Equipment required
- Eppendorf tubes.
- Microcentrifuge.
- Gilson pipettes.
- Orbital shaker.
1. Remove leaf discs (collect by closing the lid of an Eppendorf tube directly through the leaf).

2. Place the leaf discs singly into sterile Eppendorf tubes containing 50 µl of sterile distilled water.
3. Agitate for 5 min on a bench-top orbital shaker (Fisons, 200 r.p.m.), spin for 1 min at 13,000 r.p.m. in a microcentrifuge.
4. Collect and transfer the wash to sterile Eppendorf tubes.

Enumerate bacteria in the leaf wash by selective plate count methods. Alternatively, bacteria may be detected either by direct PCR amplification of the DNA in the leaf wash or PCR amplification of extracted and purified DNA.

Isolation of bacteria by plating methods

Dilute the bacterial wash in sterile distilled water (or physiological saline) and spread onto agar containing antifungal agents (cycloheximide at 50–100 µg ml^{-1} and nystatin at 10–25 µg ml^{-1}). The choice of media will depend on the target bacterial population. For example, Tryptic soy broth agar (TSBA, Difco) or plate count agar (PCA, Difco) are used to obtain total culturable counts. Similarly, selective agars have been used to isolate specific components of the bacterial community from the phyllosphere [60,61]. In order to enumerate a community using selective plating, the bacterial species of interest should be present at levels of 1×10^2 c.f.u. g^{-1} of environmental substrate [62].

Extraction of genomic DNA from recovered bacterial cells

This method uses the mild detergent CTAB (hexadecyltrimethyl ammonium bromide) to extract total DNA [3] either from colonies picked with a toothpick from agar plates, from an overnight culture, or from cells concentrated via centrifugation from leaf washings.
1. Spin the cells to produce a pellet in 1.5 ml Eppendorf tubes in a microcentrifuge for 5 min at 13,000 r.p.m.
2. Wash the pellet, by centrifugation in an Eppendorf centrifuge (5 min at 13,000 r.p.m.) with ice-cold 1 M NaCl.
3. Resuspend the cells in 500 µl lysis buffer (0.5% sodium dodecyl sulphate; 20 µg ml^{-1} Proteinase K) and incubate at 55 °C for 30 min.
4. Add 100 µl of prewarmed (65 °C) 1M NaCl, mix thoroughly,

precipitate the denatured proteins and polysaccharides after mixing with 80 µl prewarmed (65 °C) CTAB solution (10% w/v hexadecyltrimethyl ammonium bromide in 0.7 M NaCl) and incubate for 10 min at 65 °C.

5. Add 680 µl isoamyl alcohol:chloroform (1:24), mix thoroughly using a whirlimixer and collect the upper aqueous phase after centrifugation, 5 min at 13,000 r.p.m.

6. Precipitate DNA with 0.6 volumes of isopropanol at room temperature for 10 min. Recover the DNA by centrifugation (10 min at 13,000 r.p.m.). Wash the pellet with ice-cold 100% ethanol.

7. Remove all residual ethanol, resuspend the pellet in 100 µl TE buffer and store at 4 °C. DNA extracted using this method can be used directly in PCR or for restriction enzyme digestion and hybridization (RFLP etc.) analysis without further cleaning steps.

Notes
- Particularly viscous DNA extracts may require phenol/chloroform extraction prior to the isoamyl alcohol:chloroform step.
- To extract DNA from Gram positive bacteria a harsher lysis procedure is necessary: i.e., snap freezing the washed cell pellet in liquid nitrogen and following thawing, boiling for 2 min. Alternatively bead beating for 5 min using a Braun homogeniser and 0.1 mm glass beads may be employed [17].
- Yield from extracted leaf washing sufficient to detect ,10^4 target cells by hybridization. DNA yield from 5 mm colony µg.

Direct extraction

A number of protocols for rapidly extracting DNA from plant tissue have been described [12,22,49,64]. These recover DNA of sufficient purity for PCR amplification from small amounts of starting material.

We have made a comparison of these rapid DNA extraction methods [55] and conclude that certain methods appear to be advantageous in terms of speed, while others are more applicable to specific plant species. The method identified to be both rapid [64] and successful with most of the plants tested has been adapted for the detection of plant associated microbial DNA as follows:

Rapid DNA extraction from leaf discs for PCR

Equipment required
- Eppendorf grinders.
- Eppendorf tubes.
- Microcentrifuge.
- Gilson pipettes.
- Thermal cycler.
- Gel electrophoresis apparatus.

1. Homogenise 10–50 mg of leaf disc tissue (collected by closing the lid of an Eppendorf tube directly through the leaf) by using an Eppendorf grinder for approx. 2 min in 10 µl of 0.5 M NaOH.
2. Microfuge the homogenate for 1–2 min at full speed (13,000 r.p.m.), and remove 5 µl of the supernatant, dilute (1/100) to a final volume of 0.5 ml with 0.1 M Tris-HCl pH 8.0. (If it is not possible to recover 5 µl add 10–20 µl of Tris-HCl to the homogenate and then remove 5 µl).
3. Use 1 µl of this solution in a 25 µl PCR assay.

Notes
— This procedure takes approximately 5–10 min. Extracts can be stored for several months at –20 °C.
— PCR was successfully carried out when this method was applied to sea beet, oil seed rape, sugar beet and tobacco leaf material, with little to no inhibition of the PCR. Less reproducible results were obtained using this method with maize and wild brassica. In these cases, more consistent results were obtained using the following boiling method, adapted from Oard and Dronovalli [49].

Alternative methodologies

Where the rapid direct approach is insufficient to extract the target DNA or where extracted DNA is of poor quality more rigorous disruption of the plant tissue may be necessary. Below two methods are described the first of which is more suitable to leafy species.

Extraction of total DNA by boiling

Equipment required
- Boiling water bath in a fume hood.
- Refrigerated microfuge or cold room.
- −70 °C freezer (or −20 °C).
- 60 °C oven or vacuum desiccator (optional).

1. Grind approx. 10–50 mg of leaf tissue (a disc obtained by closing the cap of an Eppendorf tube on a leaf) using an Eppendorf grinder in an Eppendorf tube.
2. Add 600 µl of Extraction buffer I (100 mM Tris-HCl pH 8.0, 50 mM EDTA, 500 mM NaCl, 10 mM mercaptoethanol), mix and boil the homogenate for 7 min (n.b. pierce a hole in the cap of the tube, prior to boiling).
3. Place the tube on ice and then spin at full speed in a microcentrifuge for 15 min at 4 °C.
4. Transfer 450 µl of the supernatant to a fresh tube and precipitate the DNA with 45 µl of 10 M ammonium acetate (pH 4.8) and 1 ml of ethanol at −70 °C for 30 min (or −20 °C for 1 h).
5. Pellet the DNA by centrifugation in a microfuge at 13,000 r.p.m. for 15 min at 4 °C.
6. Remove and discard the supernatant, and air dry the pellet at 60 °C for 5–10 min (or vacuum dry for 5–10 min, or air dry at room temperature until dry).
7. Resuspend the pellet in 50 µl of TE buffer, use 1 µl in a 25 µl PCR reaction.

Notes
- The whole procedure will take approx. 90 min. Extracts can be stored at −20 °C for several months.
- All rapid methods are prone to occasional failure and the inclusion of appropriate controls is advised. If the DNA is not sufficiently pure for PCR, restriction enzyme analysis etc., further purification by CL6B Sepharose spin columns is recommended.

Physical disruption and phenol extraction of total DNA

Equipment required
- Mortar and pestle.
- Eppendorf tubes.

MMEM-1.4.3/8

- Microcentrifuge.
- Gilson pipettes.
1. Collect leaf material and grind under liquid nitrogen (observe appropriate safety precautions, wear gloves and eye protection).
2. Resuspend 1 g tissue in 5 ml Extraction buffer II (100 mM Tris-HCl pH 8.0, 100 mM NaCl, 50 mM EDTA, 2% SDS, 0.1 mg/ml proteinase K, 10 mM 2-mercaptoethanol). Incubate (37 °C, 1.5 h).
3. Extract in an equal volume phenol saturated with Tris-HCl pH 8.0, centrifuge 10 min.
4. Extract aqueous phase twice with phenol:chloroform (24:1).
5. Extract aqueous phase once with chloroform.
6. Precipitate DNA with 0.1 volume 3 M NaAc pH 5.2 and 2.5 volumes ethanol at –20 °C (1 h).
7. Collect DNA by centrifugation, wash in 70% ethanol.
8. Resuspend DNA pellet in TE at 4 °C overnight.
9. Analyze DNA by hybridization or PCR.

Notes
- Although the approach produces DNA of reasonable quality the whole procedure can be time consuming and only permits a limited number of samples to be analyzed.
- Isolated DNA can be analyzed in the normal way, bacterial content approx. 1%.

Review of general approaches to DNA extraction from plants, and strategies for purifying DNA to allow PCR, hybridization and restriction analysis

Total (plant and microbe) DNA, extracted from certain woody and monocotyledonous plant species using the rapid methods described above, may require further purification before it can be analyzed; alternatively modified protocols may need to be devised. Below we have outlined a few simple guidelines which may improve DNA yield or purity.

Methods for DNA extraction generally include a step to physically or chemically disrupt cells. Plant material is often ground in liquid nitrogen, followed by incubation in a buffer containing EDTA, a denaturing agent, and often, a proteinase. The lysed preparation is then extracted with phenol/chloroform and the DNA recovered by ethanol precipitation. These methods generally yield DNA of good quality, suitable for PCR, hybridization or restriction enzyme digestion. Bacterial DNA can represent approximately 1% of total DNA extracted from homogenized leaf tissue. The removal of microbes from leaves by washing will reduce contamination by plant DNA and improve the relative content of bacterial DNA in the sample. However washing steps may not be reliable and reproducible and will not allow the detection of endophytes. Alternative approaches to the preparation of total DNA (plant and microbe) are available at each given step described below. The choice of method is dependent on the plant material, microbial community under study, accessibility of equipment, time available, etc.

Physical methods for cell disruption

Grinding in liquid nitrogen (see above). Many large scale methods advise cell breakage using liquid nitrogen either in a pestle and mortar, or in a coffee grinder (grinders are fast and effective, but not very durable). Liquid nitrogen probably produces the best yields, however, it can be inconvenient, and is probably not necessary when preparing DNA for PCR.

Grinding with sand is used in some methods [14]. Sand may assist the grinding of harder tissues at room temperature using either a pestle and mortar, or an Eppendorf grinder for small scale preparations.

Freeze drying plant tissue [59] can be effective both as a means of breaking down tissue and for long-term storage of plant material.

Chopping plant tissue, followed by incubation with benzyl chloride, has been used frequently [65] where tissue disruption is enhanced by the benzyl chloride step.

Eppendorf grinding, as described above, is favoured as it is rapid, reproducible, easy to apply and suitable for most small scale preparations.

Denaturing agents

Phenol/chloroform extraction forms the basis of many DNA preparation methods both for plant and bacterial DNA. Generally, one phenol, one phenol/chloroform and finally one chloroform extraction are performed, followed by ethanol precipitation. These organic extractions are effective at removing proteins, lipids and some carbohydrate material Both phenol and chloroform are toxic and should be used in a fume hood. These extractions can be time consuming and preferably avoided. Methods using guanidium thiocyanate are used routinely for extraction of DNA from meats and fish [13] and result in high quality DNA, however our attempts to apply this method to plant tissue have not been successful. Benzyl chloride has also been used [65] in conjunction with sodium acetate precipitation. Furthermore, it eliminates the need to grind samples to achieve cell disruption. This method has not yet been evaluated, but may be worthy of investigation.

Digestion of contaminants

Proteinase K is used widely in plant DNA extraction protocols for digesting proteins. Glycoside hydrolase has been used for eliminating polysaccharides from DNA preparations from plant materials [53] and may be particularly suitable for dealing with starchy tissues, e.g., potato tubers.

Buffer systems

One of the key problems associated with the extraction of plant DNA is the release of polyphenols found in vacuoles into the cytoplasm and their oxidation to quinones by polyphenol oxidase. This is followed by covalent coupling, or by oxidation of the proteins by the quinones [41]. The addition of reducing agents such as mercaptoethanol can alleviate this problem. Certain plant tissues such as lotus require high levels (up to 5%) of mercaptoethanol [33], while 10 mM is adequate for most plant tissues. An alternative to mercaptoethanol is DIECA (Diethyl Dithiocarbamic Acid), used at 0.1% final concentration, which also inactivates phenol oxidases [41,50]. Ascorbic acid may also be included as an antioxidant [50].

One reliable and widely used larger scale method uses an extraction buffer containing SDS as the denaturing agent, proteinase K and mercaptoethanol [36], followed by phenol/chloroform extractions and ethanol precipitation. Another popular method uses hexadecyltrimethyl ammonium bromide (CTAB) as the denaturing agent [21] followed by chloroform extraction and isopropanol precipitation. The use of Sarcosyl alone, or in conjunction with Triton, to replace SDS in the extraction buffer [15] has also proved successful, where SDS-extract complexes are inhibitory to the PCR. Small amounts of non-ionic detergents such as these can in certain cases, improve PCR results [32]. Chelex 100 (Biorad) chelates metal ions, and has been found to increase PCR yields with DNA extracted from various sources [1,9] including plants. An acidic buffer system containing sodium acetate, PVP (Polyvinyl Pyrrolidine) and SDS with a final pH of 5.5 and the addition of cysteine (10 mM) avoids ionisation and subsequent oxidation of phenolic compounds and has proved successful in the extraction of high purity DNA from mature tree leaves of various species [26].

Precipitation of contaminants

Removal of contaminants by precipitation with sodium or potassium acetate in conjunction with an SDS based buffer was found to be effective at removing proteins and polysaccharides from the extract as a complex with the insoluble potassium or SDS precipitate. Spermine/spermidine precipitation [18] produces good yields of high quality DNA, but in our experience, inhibits PCR.

Precipitation of DNA

DNA may be precipitated using ethanol or isopropanol with either sodium chloride, sodium acetate, magnesium chloride, or ammonium acetate. Precipitation can either be carried out at –20 °C or –70 °C or at room temperature. Generally precipitation using relatively small volumes of isopropanol (around 0.6 volumes) at room temperature is considered better for DNA purity, leaving the majority of contaminating polysaccharides in solution [18]. However, it is not advisable when DNA yields are likely to be very low, or if the DNA is likely to be highly degraded. Precipitation with magnesium chloride is generally advised when dealing with very low amounts of low molecular weight DNA. Ethanol precipitation of DNA using high sodium chloride concentrations has also been reported as being useful for the removal of polysaccharide [24]. Precipitation with PEG (polyethylene glycol) has also been found to remove substances inhibitory to enzymes from plant DNA preparations [27]. However, removal of the PEG is important as this can in itself be inhibitory to enzymes.

DNA purification

After extraction further steps to improve DNA purity include: use of CF11 cellulose [35] or Elutips (supplied by Schleicher and Schuell) [19], for removal of polysaccharides; use of PVP-40 for removal of phenolics [27,50]. Although CsCl gradients are used less frequently, this time consuming method may be necessary for very "difficult" tissues such as coniferous plants [7].

Chromatography using ion exchange columns (e.g., DEAE cellulose [27]) and size fractionation (e.g., Sepharose CL6B or Sephadex [63] have also been reported. We have found CL6B to be quick, easy and produce DNA of suitable quality for PCR from plants derived from a wide variety of sources (including cotton and ramie leaf; rice and maize seed and several other species and tissue types) and extracted by various methods. RPC-5 columns, originally applied to tRNA isolation, have also been found to improve PCR results on plant derived DNA [26]. In our experience NACS columns (Pharmacia) produce very clean DNA, however, this is accompanied by a considerable loss during purification. In addition, the technique is time consuming. Qiagen columns also require relatively large quantities of DNA, moreover, current Qiagen methods are not tailored to DNA extraction from plant material. The use of Gene-clean/glassmilk has been reported to improve PCR results when used on DNA derived from preserved material and its use with plant material should be investigated. Some companies offer kits specifically designed for the purification of DNA from plants, e.g., Clonetech "GREEN-GENE" plant DNA isolation kit. The kit utilizes a sarcosyl containing buffer, and claims to produce purified DNA in 4 h.

Detection of bacterial DNA extracted from the phyllosphere

The monitoring of genetically modified or indigenous target organisms in the environment necessitates simple and rapid methods for collecting and processing material. Although isolated DNA can be analyzed by hybridization (dot blotting or by Southern analysis) PCR offers an appropriate molecular tool for the detection of low levels of bacteria released into the environment either deliberately or accidentally [44], since it is both highly sensitive and amenable to microtitre technology.

Preparation of sample for PCR analysis

Take an aliquot of the leaf wash suspension or a bacterial colony from an isolation plate and boil for 5 min in 100 µl water to lyse the bacterial cells and release DNA. Remove debris by centrifugation for 5 min at 13,000 r.p.m in a microcentrifuge.

Alternatively take aliquots of DNA purified from bacteria or extracted from leaf samples. As the presence of phenolics from leaves may inhibit PCR, potential inhibitors may be removed by passing the boiled suspension through chromatography columns. A wide range of columns and column material are available. One simple and rapid method [45] is described below.

Removal of PCR inhibitors with Sepharose CL-6B

1. Equilibrate 100 g of the Sepharose CL-6B (Pharmacia) with TE pH 8.0 at 4 °C.
2. Construct a spin column. Pierce a small hole at the bottom of a 0.7 ml Eppendorf tube. Place the pierced tube in a 2 ml Eppendorf tube.

3. Add 50 µl of a suspension of 0.4 mm glass beads (50% w/v in TE pH 8.0) to the 0.7 ml Eppendorf tube.
4. Add 500 µl of equilibrated Sepharose CL-6B.
5. Spin, with the caps removed, in a bench top microfuge at 2,000 r.p.m. for 2 min.
6. Transfer column to a fresh, sterile 1.5 ml Eppendorf tube (cap removed). Gently pipette the boiled leaf wash onto the top of the column (maximum volume/column is 50 µl).
7. Spin at 2,000 r.p.m. for 2 min.
8. Store the eluate at 4 °C or –20 °C until required.

Conditions used for PCR

The reaction conditions for PCR will be dependent on the size and sequence of primers, the Taq polymerase enzyme and the type of thermal cycler being used. For more details about PCR refer to the appropriate texts [32,48].

The conditions described below are those used to demonstrate the presence of eubacterial DNA prepared from a leaf wash. The sequences for the primers used are 5' CAGCa/cGCCGCGGTAATa/tC 3' and 5' CCGTCAATTCa/cTTTa/gAGTTT 3' [34, see also ch. 1.4.2 of this manual].

1. Prepare template DNA as described above.
2. Add 2.5 µl of each primer (primer stock solution 20 µM).
3. Add water to a final volume of 25 µl.
4. Add 50 µl of mineral oil (Sigma) (n.b. this step is not required for thermal cyclers with a heated lid).
5. Heat the solution in a thermal cycler at 95 °C for 7 min followed by 80 °C for 30 min.
6. Hold the solution at 80 °C and add 25 µl volume of the following to the reaction mixture:
    8 µl dNTP (200 µM final concentration).
    5 µl 10 × PCR buffer.
    0.25 U *Taq* polymerase (Perkin-Elmer).
    11.75 µl sterile distilled water.

7. Begin cycle.

    Typical PCR cycling scheme used for this set of primers:

    1 Cycle
    95 °C for 4 min.
    50 °C for 10 s
    72 °C for 1 min.

    28 Cycles
    95 °C for 1 min.
    50 °C for 10 s
    72 °C for 1 min.

    1 Cycle
    95 °C for 1 min.
    50 °C for 10 s
    72 °C for 8 min.

8. Analyze samples on a 1% agarose gel in 1 × TBE buffer containing 10 µg ml^{-1} ethidium bromide. To 5–10 µl PCR product add 3 µl Bromophenol blue solution and load into the agarose gel. If mineral oil was used in the PCR, it is not necessary to remove the oil prior to loading the samples (pipette through the oil meniscus).

Notes

– Heterogeneous samples treated with specific oligonucleotides should be serially diluted, 10-fold, and analyzed to determine target content. 100 ng represents the amount of DNA extracted from approx. 10^7 bacterial cells (<100 mg leaf).

– Include a positive control (DNA extracted from a pure culture) and sterile water as a negative control in any PCR assay to confirm specificity.

– When using Universal primers, a faint product is frequently obtained in the negative control. This phenomenon has been reported by [52] and subsequently confirmed by workers at the Laboratory of the Government Chemist, UK. The problem is due to the presence of DNA in commercial preparations of *Taq*. This problem may be overcome through the use of low DNA *Taq* (*Taq* LD ™ available from Perkin-Elmer) or by treating the enzyme with DNase.

– Following the production of a PCR product of predictable size, it is worthwhile to confirm the specificity of the product by dot or Southern blot hybridisation with a specific probe [56] or by restriction endonuclease analysis [57].

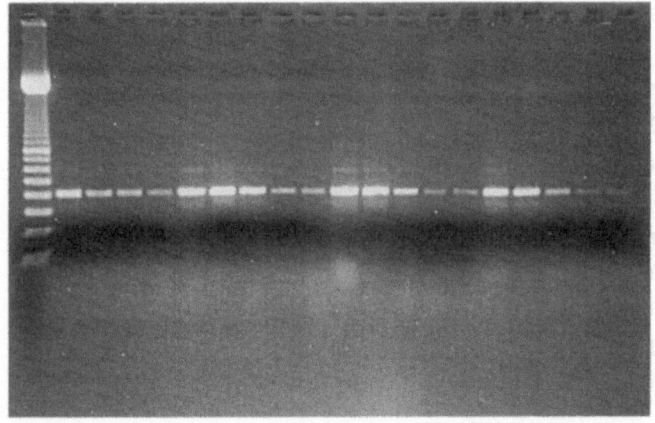

Figure 1. PCR detection of eubacterial 16S sequences in leaf washes. Bacteria were removed from leaf discs by washing in 50 μl of water. 1 μl samples were boiled and PCR amplified using 16S primers [34] as described in the text. Amplified DNA (>400 bp) was stained with ethidium bromide, visualised in 1% agarose under UV illumination and photographed. Lane 1 100 bp ladder (Pharmacy), lane 2 control DNA. Samples arranged in sets of three, one set per individual sugar beet plant, 6 plants compared. Each set of three represented by extract from senescing leaf, mature leaf and immature leaf [61]. Bacterial DNA detected in every sample.

Other approaches to the analysis of environmental microbial community DNA

Recently, a technique described as denaturing gradient gel electrophoresis (DGGE) for analysis of PCR-amplified genes coding for 16S rRNA has been used to estimate the genetic diversity of microbial communities [46]. This procedure separates PCR-amplified 16S rDNA fragments by electrophoresis in polyacrylamide gels containing a linearly increasing gradient of denaturants. Essentially, DNA fragments of the same length with different base-pair sequences migrate differentially. In this study [46] 10 different rDNA fragments (each of which was thought to correspond to a different species) were generated from the genomic DNA fraction recovered from an aerobic bacterial biofilm. Other methods based on DNA analysis such as hybridization or DNA fingerprinting are described in detail in the appropriate laboratory manuals [3, 56].

Buffers and reagents

TE buffer
- 10 mM Tris-HCl pH 8.0.
- 1 mM Na$_2$EDTA pH 8.0.

Extraction buffer I
- 100 mM Tris-HCl pH 8.0.
- 50 mM EDTA.
- 500 mM NaCl.
- 10 mM 2-mercaptoethanol).

Extraction buffer II
- 100 mM Tris-HCl pH 8.0.
- 100 mM NaCl.
- 50 mM EDTA.
- 2% SDS.
- 0.1 mg/ml proteinase K.
- 10 mM 2-mercaptoethanol).

10 × TBE
- 108 g Tris base.
- 55 g Boric acid.
- 9.3 g Na$_2$EDTA.
- Water to 1 l.

10 × PCR buffer
- 500 mM KCl.
- 100 mM Tris-HCl pH 8.3.
- 15 mM MgCl.
- 0.01% w/v gelatin.

Make up to 100 ml with sterile distilled water.
Aliquot 1 ml portions in screw cap Eppendorfs and sterilise.

Bromophenol blue buffer (10 ×)
- 0.25% bromophenol blue.
- 30% glycerol in TE pH 8.0.
- 0.25% xylene cyanol(ff); optional.

References

1. Aldrich J, Cullis CA (1993) RAPD analysis in flax: optimization of yield and reproducibility using Klen*Taq* 1 DNA polymerase, Chelex 100, and gel purification of genomic DNA. Plant Mol Biol Reporter 11: 128–141.
2. Andrews JH, Hirano SS (1991) Microbial Ecology of leaves. Springer Verlag, New York.
3. Ausubel FM, Brent R, Kingston RE, Moore DD, Seidman JD, Smith JA, Struhl K, eds. (1989) Current Protocols on Molecular Biology. John Wiley and Sons, New York.
4. Austin B, Goodfellow M, Dickinson CH (1978) Numerical taxonomy of phylloplane bacteria isolated from *Lolium perenne*. J Gen Microbiol 104: 139–55.
5. Austin B, Goodfellow M (1979) *Pseudomonas mesophilica*, a new species of pink bacteria isolated from leaf surfaces. Int J Syst Bacteriol 29: 373–378.
6. Bailey MJ (1995) Extraction of DNA from the Phyllosphere. In: Trevors JT, Van Elsas JD (eds) Nucleic Acids in the Environment: Methods and Applications, pp. 89–110. Springer Verlag, Berlin.
7. Baker SS, Rugh CL, Kamalay JC (1990) RNA and DNA isolation from recalcitrant plant tissues. BioTechniques 9: 268–272.
8. Berthomieu P, Meyer C (1991) Direct amplification of plant genomic DNA from leaf and root pieces using PCR. Plant Mol Biol 17: 555–557.
9. Berthold DA, Best BA, Malkin R (1993) A rapid DNA preparation for PCR from *Chlamydomonas reinhardtii* and *Arabidopsis thaliana* . Plant Mol Biol Reporter 11: 338–344.
10. Bloom RA, Mullin BC, Tate RL (1989) DNA restriction patterns and DNA-DNA solution hybridisation studies of *Frankia* isolates from *Myrica pensylvanica* (bayberry). Appl Environ Microbiol 55: 2155–2160.
11. De Bruijn FJ (1992) Use of repetitive (Repetitive extragenic palindromic and enterobacterial repetitive intergenic consensus) sequences and the polymerase chain reaction to fingerprint the genomes of *Rhizobium meliloti* isolates and other soil bacteria. Appl Environ Microbiol 58: 2180–2187.
12. Cheung WY, Hubert N, Landry BS (1993) A simple and rapid DNA microextraction method for plant animal and insect suitable for RAPD and other PCR analyses. PCR Methods Applic 3: 69–70.
13. Chomzynski P, Sacchi N (1987) Single step method of RNA isolation by acid guanidium thiocyanate-phenol-chloroform extraction. Anal Biochem 162: 156–159.
14. Clarke BC, Moran LB, Appels R (1989) DNA analysis in wheat breeding. Genome 32: 334.
15. Collins GG, Symons RH (1992) Extraction of nuclear DNA from grape vine leaves by a modified procedure. Plant Mol Biol Reporter 10: 233–235.
16. Constantinidou HA, Hirano SS, Baker LS, Upper CD (1990) Atmospheric dispersal of ice-nucleation-active bacteria: The role of rain. Phytopathol 80: 934–937.
17. Cresswell N, Saunders VA, Wellington EMH (1991) Detection and quantification of *Streptomyces violaceolatus* plasmid DNA in soil. Lett Appl Microbiol 13:193–197.
18. Dellaporta SL, Wood J, Hicks JB (1983) A plant DNA minipreparation: Version II. Plant Mol Biol Reporter 1: 19–21
19. Do N, Adams RP (1991) A simple technique for removing plant polysaccharide contaminants from DNA. BioTechniques 10: 162–166.
20. Donegan K, Matyac C, Seidler R, Porteous A (1991) Evaluation of methods for sampling, recovery and enumeration of bacteria applied to the phylloplane. Appl Environ Microbiol 57: 51–56.
21. Doyle JJ, Doyle JL (1987) A Rapid DNA isolation procedure for small quantities of fresh leaf tissue. Phytochem Bull 19: 11–15.
22. Edwards K, Johnstone C, Thompson C (1991) A simple and rapid method for the preparation of plant genomic DNA for PCR analysis. Nucl Acid Res 19:1349.
23. Ercolani GL (1991) Distribution of epiphytic bacterai on olive leaves and the influence of leaf age and sampling time. Microbial Ecol 21: 35–48.

24. Fang G, Hammar S, Grumet R (1992) A quick and inexpensive method for removing polysaccharides from plant genomic DNA. BioTechniques 13: 52–55.

25. Fokkema NJ, Schippers B (1986) Phyllosphere versus rhizosphere as environments for saprophytic colonisation. In: Fokkema NJ, van den Heuvel J (eds) Microbiology of the Phyllosphere, pp. 137–159. Cambridge University Press.

26. Guillemaut P, Maréchal-Drouard L (1992) Isolation of plant DNA: A fast, inexpensive and reliable method. Plant Mol Biol Reporter 10: 60–65.

27. Hattori J, Gottlob-McHugh SG, Johnson DA (1987) The isolation of high-molecular weight DNA from plants. Anal Biochem 165: 70–74.

28. Hirano SS , Upper CD (1989) Diel variation in population size and ice nucleation activity of Pseudomonas syringae on snap bean leaves. Appl Environ Microbiol 55: 623–630.

29. Hirano SS, Upper CD (1986) Temporal, spatial and genetic variabilty of leaf-associated bacterial populations. In: Fokkema NJ, van den Heuvel J (eds) Microbiology of the Phyllosphere, pp. 235–251. Cambridge University Press.

30. Hirano SS, Nordheim EV, Arny DC, Upper CD (1982). Lognormal distribution of epiphytic bacterial populations on leaf surfaces. Appl Environ Microbiol 44: 695–700.

31. Hwang I, Farrand SK (1994) A novel gene tag for identifying microorganisms released into the environment. Appl Environ Microbiol 60: 913–920.

32. Innis MA, Gelfand DH, Sninsky JJ, White TJ (1990) In: PCR Protocols A Guide to Methods and Applications. Academic Press.

33. Kanazawa A, Tsutsumi N (1992) Extraction of restrictable DNA from plants of the genus Nelumbo. Plant Mol Biol Reporter 10: 316–318.

34. Lane DJ, Pace B, Olsen GJ, Stahl DA, Sagin ML, Pace NR (1985) Rapid determination of 16S ribosomal RNA sequences for phylogenetic analysis. Proc Nat Acad Sci USA 82: 6955–6959.

35. Larkin JC, Hunsperger JP, Culley D, Rubinstein I, Silflow CD (1989) The organisation and expression of the maize ribosomal protein gene family. Genes Devel 3: 500–509.

36. Lazarus CM, Baulcombe DC, Martienssen RA (1985) Alpha-amylase genes of wheat are two multigene families which are differentially expressed. Plant Mol Biol 5: 8–24.

37. Leben C, Whitmoyer RE (1979) Adherence of bacteria to leaves. Can J Microbial 25: 896–901.

38. Lindemann JHA, Constantinidou HA, Barchet WR, Upper CD (1982) Plants as sources of airborne bacteria, including ice nucleaction activity. Appl Environ Microbiol 56: 3375–3381.

39. Lindow SE, Arny DC, Upper CD (1978) Distribution of ice-nucleation active bacteria on plants in nature. Appl Environ Microbiol 36: 831–8.

40. Lindow SE (1993) Novel method for identifying bacterail mutants with reduced epiphytic fitness. Appl Environ Microbiol 59: 1586–1592.

41. Loomis WD (1974) Overcoming problems of phenolics and quinones in the isolation of plant enzymes and organelles. Meth Enzymol 31: 528–545.

42. Mazurier SI, Wernars K (1992) Typing of Listeria strains by random amplification of polymorphic DNA. Res Microbiol 143: 499–505.

43. Menard C, Brousseau R, Mouton C (1992) Application of polymerase chain reaction with arbitrary primer (AP-PCR) to strain identification of Porphyromonas (Bacteroides) gingivalis. FEMS Microbiol Lett 95: 163–168.

44. Mullis K, Faloona F (1987). In: Wu R (ed) Methods in Enzymology, 155, p.335. Academic Press, New York and London.

45. Murphy G, Kavanagh TA (1988) Speeding up the sequencing of double-stranded DNA. Nucl Acids Res 16:5198.

46. Muyzer G, de Waal EC, Uitterlinden AG (1993) Profiling of complex microbial populations by denaturing gel electrophoresis analysis of polymerase chain reaction-amplified genes coding for 16S rRNA. Appl Environ Microbiol 59: 695–700.

47. McMillin DE, Muldrow LL (1992) Typing of toxic strains of *Clostridium-Difficile* using DNA fingerprints generated with arbitary polymerase chain reaction primers. FEMS Microbiol Lett 92: 5–10.
48. McPherson MJ, Quirke P, Taylor GR (1991) PCR: A Practical Approach. IRL Press.
49. Oard JH, Dronovalli S (1992) Rapid isolation of rice and maize DNA for analysis by random primer PCR. Plant Mol Biol Reporter 10: 236–241.
50. Paterson AH, Brubaker CL, Wendel JF (1993) A rapid method for extraction of cotton (*Gossypium* spp.) genomic DNA siutable for RFLP or PCR analysis. Plant Mol Biol Reporter 11: 122–127.
51. Pedgley DE (1991) Aerobiology: The atmosphere as a source and sink for microbes. In: Andrews JH, Hirano SS (eds) Microbial Ecology of Leaves, pp. 43–59. Springer Verlag.
52. Rand K H, Houck H (1990) Taq Polymerase Contains Bacterial DNA of Unknown Origin. Mol Cell Probes 4:445–450.
53. Rether B, Delmas G, Laouedj A (1993) Isolation of polysaccharide-free DNA from plants. Plant Mol Biol Reporter 11: 333–337.
54. Rodriguez E, Healey PL, Mehta I (1984) Biology and Chemistry of Plant Trichomes. Plenum Press, New York and London.
55. Rogers HJ, Parkes HC (1994) (unpublished observations).
56. Sambrook J, Fritsch EF, Maniatis T (1989) Molecular Cloning: A Laboratory Manual, 2nd ed. Cold Spring Harbor Laboratory, CHS, New York.
57. Schneider B, Ahrens U, Kirkpatrick BC, Seemuller E (1993) Classification of plant-pathogenic mycoplasma-like organisms using restriction-site analysis of PCR-amplified 16S rDNA. J Gen Microbiol 139: 519–527.
58. Sørensen B, Falk, ES, Wisloff-Nilsen E, Bjorvatu B, Kristiansen BE (1985) Multivariate analysis of Neisseria DNA restriction endonuclease patterns. J Gen Microbiol 131: 3099–3104.
59. Tai TH, Tanksley SD (1990) A rapid and inexpensive method for the isolation of total DNA from dehydrated plant tissue. Plant Mol Biol Reporter 8: 297–303.
60. Thompson IP, Bailey MJ, Ellis RJ, Lilley AK, McCormack PJ, Purdy KJ, Rainey PR (1995) Short-term community dynamics in the phyllosphere microbiology of field-grown sugar beet. FEMS Microbiol Ecol in press.
61. Thompson IP, Bailey MJ, Fenlon JS, Fermor TR, Lilley AK, Lynch JM, McCormack PJ, McQuilken M, Purdy KJ, Rainey PB, Whipps JM (1993) Quantitative and qualitative seasonal changes in the microbial community from the phllosphere of sugar beet (*Beta vulgaris*). Plant and Soil 150: 177–191.
62. Trevors JT, van Elsas JD (1989) A review of selected methods in environmental microbial genetics. Can J Microbiol 35: 895–902.
63. Tsai Y-L, Olson BH (1992) Rapid method for the separation of bacterial DNA from humic substances in sediments for polymerase chain reaction. Appl Environ Microbiol 58: 2292–2295.
64. Wang H, Qi M, Cutler J (1993) A simple method of preparing plant samples for PCR. Nucl Acids Res 21: 4153–4154.
65. Zhu H, Qu F, Zhu L-H (1993) Isolation of genomic DNAs from plants, fungi and bacteria using benzyl-chloride. Nucl Acids Res 21: 5279–5280.

Molecular Microbial Ecology Manual **1.5.1**: 1-17, 1995.

Direct and simultaneous extraction of DNA and RNA from soil

SONJA SELENSKA-POBELL

Department of Genetics, University of Bayreuth, D-95440 Bayreuth, Germany

Introduction

The direct extraction of DNA and RNA from soil is important for molecular ecological studies of the terrestrial environment. The structure and activities of natural bacterial communities can be better understood by analyses of their nucleic acids when the latter are recovered from soil samples by direct lysis instead of isolating them from bacterial cells, which have been separated from soil. One reason for this is that bacteria often are sticking strongly to soil particles and only part of them can be extracted [18, 30, 37]. On the other hand, most of the environmental bacteria, even when they are well extracted from the soil particles, fail to be cultured on laboratory media [6, 10, 44, 45]. In addition, the natural population densities of many bacteria could be rather low [22, 24].

It was believed that the problems of ineffective extraction, non-culturability and low densities of bacteria in soil could be overcome by polymerase chain reaction (PCR) amplification of total DNA, recovered by direct lysis of the cells present in the tested samples. However, PCR analyses are based on detection of intact DNA sequences rather then intact cells and the possibility of positive PCR amplifications from dead cells or even from naked DNA molecules exists [10, 16]. Potential templates for PCR amplification products can also be representatives of the indigenous bacterial communities, which harbour the DNA sequence of interest. It is necessary to consider the latter possibility, because it was demonstrated in many studies that genetic information can be exchanged between bacteria in soil by conjugation [8, 26, 36], transduction [11, 46] or/and transformation [12, 14, 28]. It was supposed that the identification problems of the targets of PCR amplifications can be solved by direct extraction and analysis of mRNA and rRNA [1, 6, 7, 23, 24, 34, 43] in addition to DNA from the environmental samples. In general, the presence of rRNA of the target bacteria in soil samples can be used as indirect evidence for their natural activity. On the other hand, by analysing the mRNA of particular genes, including marker genes, studies of the natural gene expression can be performed.

During the last years many rather effective methods for recovery of DNA [2, 3, 5, 18, 22, 25, 30, 32, 33, 35, 37, 39–42] or RNA [7, 15, 20, 43] from soil have been published. However, it is well known that bacteria form clusters in the soil environment [17, 37]. Hence, two samples from the same piece of soil can have significant differences in their densities of the investigated bacteria. This can be a reason for recovery of quite different amounts of DNA or RNA in these samples, which can lead to misinterpretations of the natural state of the studied organisms. In this respect, simultaneous extraction and analyses of DNA and RNA from the same soil sample is important. Such a technique [34] will be presented in this chapter.

In addition, the technique presented here is also recommended to monitor the fate of deliberately or accidentally released genetically manipulated bacteria and study their impact on natural bacterial communities. It has been shown that released bacteria could be present in natural environments in low densities and often they could become undetectable [9, 21, 22, 32–35, 38] for analyses by conventional micro-biological methods. In this respect, obviously, the application of direct molecular techniques for studying the stability and activity of the bacteria newly introduced to the soil environment is preferable.

Experimental approach

The direct isolation of DNA from soil was pioneered by Ogram *et al.* [18]. The resulting DNA contained large amounts of brown colored humic material and was considered to be too impure for molecular experiments. Steffan *et al.* [37] improved the quality of the DNA recovered by introducing cleaning steps such as polyvinylpolypyrrolidone treatments, CsCl-ethidium bromide density gradient centrifugation and hydroxyapatite column chromatography, in addition to the extensive phenol extraction of the original Ogram *et al.* procedure. These steps increased the complexity of the procedure and caused large losses of DNA. According to the authors [37], only 10% of a radioactively labelled DNA sample, introduced to the soil as a control, was recovered by their procedure.

Since that time many new DNA extraction techniques, rather effective, fast and significantly simplified were published [2, 3, 5, 10, 16, 20–22, 25, 27, 30, 32, 33, 35, 39–42]. All these techniques include detergent treatment and in most of them steps of intensive mechanical disruption of the cells are involved [16, 18, 22, 30, 35, 39]. These steps are considered to improve the effectivity of bacterial and DNA extraction from the soil samples. However, it was shown that these steps cause fragmentation of the DNA molecules and the yield of high-molecular weight DNA, recovered by such procedures, is significantly reduced. This may negatively influence further PCR analysis of the recovered soil DNA, because it was shown that the tendency toward formation of hybrid chimeric products is increased if the PCR is performed on fragmented DNA instead of on high molecular weight DNA [13].

In many techniques the mechanical disintegration of the cells is combined with some other treatments such as: hot phenol [25] or cold phenol [35, 39] extractions, freeze-thaw cycles [3, 41], freeze-boil cycles [22], microwave heating [22], etc. By such drastic treatments not only the DNA but also other soil compounds) humic acids, for example are extensively extracted, and this has undesirable effects on the properties of the DNA recovered [18, 30, 31, 37, 39]. For this reason, additional intensive cleaning steps are required for molecular analyses of the DNA extracted. Different column chromatography purifications were reported to improve the purity of the extracted soil DNA [3, 5, 35, 37, 41, 42]. These steps are rapid, easy to handle and the resulting DNA is pure enough for PCR amplifications. However, during gel filtration procedures, significant losses of large DNA fragments occur, because most of the fragments larger than 20 kb remain irreversibly bound to the column material. For this reason, DNA purified by column exchange chromatography is not relevant for Southern hybridization and restriction fragments length polymorphism (RFLP) analyses. Simple and fast techniques [3, 5, 35, 37, 41, 42] are preferable for PCR detection of a particular short gene sequence in extensive field experiments. However, the risk of chimera formation, because of the small sizes of the template DNA, should be considered in these tests [13].

In many cases it is important to identify the organisms carrying the sequences which are targeted by PCR amplification. This can be done on the basis of a specific hybridization pattern, which is characteristic for the organism of interest. In general, to obtain reliable restriction patterns, large DNA fragments (not smaller than 25 kbp) should be analysed in Southern hybridization experiments [29]. For such analyses, the DNA should be recovered from soil in a gentle way. Several direct DNA extraction procedures from soil do not use drastic disintegration steps [25, 32–34, 40], but rather base lysis of soil microorganisms on the use of SDS [25], a combination of SDS and heat treatments [32–34], or sarcosyl and heat treatments [40]. Soil and cellular debris are discarded by low speed centrifugation of the samples [25, 32–34, 40]. In the method of Selenska and Klingmüller [32–34] the nucleic acids, which are present in the crude lysate, are precipited with PEG-6000. This step not only concentrates DNA and RNA molecules in the soils lysate, but also results in partial purification, because many brown-colored substances do not precipitate by treatment with PEG-6000 and are discarded by centrifugation. In contrast, the crude lysate in the technique of Thiem *et al.* [40] is concentrated by reverse dialysis against PEG-6000 and in the procedure of Porteous *et al.* [25] by isopropanol precipitation. Further purification of the DNA in the latter case [25] is performed by gel electrophoresis and otherwise [32–34, 40] by isopycnic CsCl centrifugation. The method described in [34] recovers not only DNA but also RNA from soil. This method has been developed on the basis of that for direct DNA extraction from soil [32, 33]. The DNA resulting from these procedures [25, 32–34, 40] is characterized by its large size (above 25 kb), and that recovered by the techniques of Selenska and Klingmüller [32–34] has been shown to be pure enough for Southern/RFLP analysis and PCR amplifications. The digestibility of the DNA, recovered by Porteous *et al.* [25] and Thiem *et al.* [40] was not discussed and only PCR amplification was shown to be successful [25, 40]. Concentration of the total crude soil lysates [25, 40] could also concentrate impurities together with the DNA. In the method of Thiem *et al.* [40], a second CsCl density gradient purification could increase the purity of the DNA. In the other technique [25], where an isopropanol precipitation is applied, copurification and contamination of DNA with undesirable soil compounds may be more problematic. Additional purification of the DNA obtained by these two techniques may be needed, since it was noticed in many studies [30, 31, 41, 42] that restriction endonucleases are more sensitive to soil compounds than Taq polymerases. In case of positive restriction endonuclease treatments, however, the two techniques [25, 40] can be recommended for RFLP analysis of total soil DNA, because they recover high molecular weight DNA and are fast and simple.

The performance of Southern-RFLP analyses using large DNA fragments recovered from soil is important not only because of the general requirement to digest high molecular DNA for obtaining reliable restriction patterns [29]. Here, it is important to stress that large DNA fragments released from dead cells normally do not persist for long in natural environments. Naked DNA fragments larger than 9 kb are subject to relatively rapid degradation in soil conditions [19, 28], and smaller DNA fragments are preferentially adsorbed in different soils [19]. This can be explained by the fact that in soil shorter DNA fragments have better chances to be protected from degradation [19]. Hence, the large-sized DNA fraction (above 20 kb) in total DNA recovered from soil represents material derived from intact cells rather than from the shorter extracellular DNA fragments in soil.

The subject of this chapter is the procedure for direct simultaneous recovery of DNA and RNA from soil [34]. The DNA recovered by this method, which is actually a simplification of the Ogram *et al.* protocol, is pure enough for molecular analyses, despite the fact that the phenol extraction steps of the original protocol are omitted. The reason for omitting these was the finding of other authors [30, 31, 37] that phenol extractions purify together with the DNA some other compounds from soil, which have a strong inhibitory effect on the functioning of restriction endonucleases. In this respect, it is important to stress that the inclusion of one phenol treatment step before the CsCl gradient step, makes the DNA recovered by the method presented below irreversibly brown colored and undigestable by most restriction endonucleases. In addition, the yield of the recovered soil DNA is reduced, because of DNA losses unavoidable in phenol treatments.

The DNA extracted by the method described was analysed by Southern hybridization, where a specific restriction pattern of the inoculant bacteria, consisting of specific (34, 21 and 9.6 kb) fragments [4, 32], was obtained in long-term inoculation experiments. On the basis of this, we concluded that the hybridization signals were evidence for the presence of intact inoculant cells.

For full analysis of soil samples, not only the presence, but also the natural activity of the target bacteria can be analysed e.g. by analyses of mRNA and rRNA. rRNA molecules have very compact secondary structure and could be rather stable in soil. It was postulated by Nannipieri *et al.* [17] that RNA and DNA have the same probability of becoming stabilized in soil once released after cell lysis. Considering the smaller size of rRNA molecules, the chances of their protection are higher in comparison to those of the larger DNA molecules. Normally, dying or dormant bacteria have strongly reduced amounts of mRNA and rRNA. However, in case of accidental release of a large population of physiologically active bacteria, one can expect that a large number of them will rapidly die and at least some part of their rRNA can be adsorbed and stabilized on colloidal soil particles. Naked rRNA molecules could hence be also targets for reverse transcriptase dependent (RT)-PCR amplifications. Since RNA molecules can be subject to fast degradation by RNases, prescriptions for work free of RNases should be strictly followed in the procedures presented below.

Procedures

Extraction of DNA and RNA from soil samples

The method described in the following is very gentle and is recommended for recovery of large fragments of DNA as well as mRNA and rRNA from the same soil sample [34]. The recovered high-molecular DNA is pure and can be used for hybridization analysis [4, 32, 33] using Southern blotting. The recovered DNA is also an excellent template for PCR amplifications [4, 34], because the risk of forming chimeric PCR products, connected with the fragmentation of the DNA [13] does not exist. The recovered mRNA and rRNA are also pure enough to be used for assessing the activity of soil bacteria in natural environments by oligonucleotide probes [7] as well as in northern hybridization experiments and RT-PCR amplifications [15, 34].

Steps in the procedure

The glassware for this procedure has to be filled with 0.1% diethyl-pyrocarbonate (DEPC) containing double-distilled water, allowed to stand for 2 h at 37 °C, after which the liquid is discarded. It is heated afterwards to 100 °C for 15 min, then rinsed with sterile double distilled water and autoclaved.

Plasticware has to be filled with chloroform for 5 to 10 min (under a hood). The chloroform should be collected for the next treatments and the plasticware should be rinsed with pretreated double-distilled water and autoclaved afterwards. Instead of autoclaving, overnight baking at 280 °C is also effective against RNases. In addition, all solutions and distilled water should be pretreated as presented below (see Solutions) for inhibition of RNases.

1. DEPC-treatment of the samples for inactivation of soil RNase activity:
    Add 15 ml sterile double distilled water, containing 1% DEPC, to 10 g of soil. Shake the sample for 20 min at room temperature. Centrifuge the sample at 2,800 × g and 10 °C.
    Wash the pellet twice with 10 ml sterile double distilled water, pretreated with 0.1% DEPC, by shaking (15 min) and centrifugation as described above.
2. Direct lysis of the cells present in the soil sample and extraction of the released DNA and RNA:
    Add 25 ml of extraction buffer (0.12 M Na_2HPO_4, pretreated with 0.1% DEPC), and 5 ml of a 5% SDS solution and shake the samples gently in a water bath for 1 h at 70 °C.
    Centrifuge the samples for 20 min at 2,800 × g and 10 °C. Collect supernatants and keep these at 4 °C.
    Extract the DNA and RNA from the pellet two more times with 30 ml of the same extraction buffer by shaking for 20 min at 70 °C. Add the resulting two supernatants to that of the first extraction and store the mixture for at least one hour at 4 °C.

Centrifuge the mixtures for 30 min at 8,000 × g and 4 °C. By this step the SDS, which can negatively influence the purification and the properties of the extracted nucleic acids, is separated from the rest of the supernatant.

3. Concentration, purification and separation of the extracted DNA and RNA:

Add 5 M NaCl (1/10 volume of supernatant) to the cleared supernatant fluid and mix well.

Precipitate DNA and RNA overnight at 4 °C by adding PEG-6000 to a final concentration of 15%.

Centrifuge the mixture at 5,000 × g and 4 °C for 20 min.

Add 9.5 g CsCl to exactly 8.5 ml of the nucleic acid solution in a tube. Close the tube with parafilm and dissolve the CsCl crystals by gentle inversion of the tube. Then mix the solution carefully with 1 ml of a stock solution (5 mg/ml) of ethidium bromide.

Discard the dark red coloured soil and cellular debris by centrifugation for 15 min at 5,000 × g and 4 °C.

Fill a 12 ml quick-seal tube with the cleared ethidium bromide coloured sample (the refraction index should be read in a refractometer and adjusted, if necessary, to 1.388).

Close the quick-seal tubes and run the gradient at 36,000 r.p.m. in a Ti 50 rotor of the (Beckman) ultracentrifuge for 40 h. By this step, separation of DNA from proteins (staying on top of the CsCl gradient) and humic acids (forming a diffuse band below the DNA band) is achieved. In addition, mRNA as well as small and large ribosomal subunit RNA is concentrated in the resulting compact and stable pellet of the gradient on the bottom of the centrifuge tube.

4. Final purification of soil DNA and RNA: After stopping the ultracentrifuge, the resulting DNA band should be removed from the gradient with a syringe [29] and the RNA pellet should be immediately washed and purified.

4.1. Final purification of the soil DNA

Collect the DNA band (approximately 1 ml) from the gradient tube with a syringe (20 G 1$_{1/2}$) and perform a dialysis for 2 h at 4 °C against 1000 volumes of TE buffer, pH 8, according to Sambrook *et al.* [29] to reduce the concentration of CsCl in the DNA sample. (After the dialysis is set up, immediately start purification of the RNA pellet as described in 5.)

After two hours of dialysis, transfer the DNA sample to two Eppendorf tubes and extract the ethidium bromide with Tris-HCl saturated phenol. (Note: do not use other reagents for the extraction of the ethidium bromide, e.g. isopropanol, because this is the only phenol-purification step of the DNA and it is important to be performed at this stage.) This step does not spoil the properties of the DNA, because the compounds which phenol copurifies from soil together with DNA in other methods [18, 37, 41, 42] have already been removed by the density gradient centrifugation. Phenolization at the end of the procedure, as described, significantly improves the purity of the recovered DNA.

Perform one additional extraction of the resulting water phase with chloroform to extract traces of phenol, which can negatively influence the restriction endonuclease activity in the consequent experiments.

Transfer the water phase, containing the DNA, to a fresh dialysis tube and continue the dialysis overnight with three changes of the TE buffer at 4 °C. The resulting DNA has to be stored for future analyses at 4 °C.

5. Final purification of the RNA pellet of the CsCl density gradient:
Carefully remove the ethidium bromide containing liquid from the tube. Cut the tube 1.5 cm from the bottom with a scalpel. Wash the red pellet several times (at least five) with 1 ml of 80% ice-cold ethanol until the washing solution does not turn pink anymore.

Resuspend the pellet in 100 µl TE buffer, pH 7.5, containing 0.1% SDS. Transfer the mixture into an Eppendorf tube. For completely dissolving the yellow-brown particles of the pellet, freeze the sample in liquid nitrogen and melt it at 45 °C. If necessary, repeat the freeze-thaw step once again.

Add 150 µl TE buffer pH 7.5 and mix by inversion of the tube.

Add 670 µl of 30% PEG 6000 solution and 345 µl of phosphate (PP) buffer, pH 7.6. Gently mix the liquid by inversion of the tube and put it on ice for 10 min; continue mixing the sample from time to time.

Spin in a centrifuge (4 °C for 10 min at 6,000 × g to separate the two liquid phases, the lower of which contains the RNA. Collect the lower fraction, add four volumes of sterile double-distilled water and precipitate the RNA with 0.1% CTAB and 0.01% NaCl (final concentrations) for 2 h at 4 °C.

Spin at 8,000 × g and 4 °C and wash the resulting pellet with 70%

ice-cold ethanol containing 200 mM NaOAc.

Resuspend the pellet in 25 µl sterile double-distilled water and store in 5 µl aliquots at minus 80 °C.

6. Additional purification of RNA for PCR amplifications by spun-column gel filtration: In some cases the rRNA after step 5 is not pure enough for performing RT-PCR amplifications. In these cases the following cleaning procedure is recommended:

Pack a sterile blue 1 ml pipet with a glass wool plug and fill it with 1 ml of Sephadex G-75 equilibrated overnight in sterile double-distilled water.

Treat the resulting column 3 to 5 times by addition of 200 µl sterile double-distilled water and centrifuge in a swing-bucket centrifuge at 1,400 × g for 5 min (for details see [15, 29].

Take the 25 µl of the RNA recovered at the last stage of the purification procedure described in 5., mix with 25 µl cold double-distilled water and add the RNA to the preconditioned column of Sephadex G-75.

Collect the purified RNA in a microcentrifuge tube fitted over the bottom of the tip by centrifugation at 1,400 × g for 1 min.

Freeze the resulting purified RNA immediately at minus 80 °C and keep it for further analysis.

Solutions

Before beginning the extraction procedures at least 2 l of double-distilled water should be pretreated with 0.1% DEPC. The DEPC should be dissolved by intense mixing, and then the water should be incubated for at least 2 h (better overnight) at 37 °C. After this, it should be boiled for 15 min to inactivate the DEPC, which can negatively influence the properties of the extracted RNA.

Other solutions should be treated in the way described above, The stock solutions of Tris- and MOPS-buffers should be prepared in fresh RNase free flasks, using DEPC pretreated double-distilled water.

- Sterile double-distilled water, containing 1% DEPC.
- 0.12 M Na_2HPO_4, pH 8.0.
- 5% SDS
- 5 M NaCl

- 50% PEG 6000
- 30% PEG 6000
- 1 M Tris-HCl buffer, pH 8 – stock solution
- 1 M Tris-HCl buffer, pH 7.5 – stock solution
- 0.5 M EDTA, pH 8 – stock solution
- TE buffer, pH 8 (10 mM Tris-HCl pH 8 and 1 mM EDTA, pH 8)
- TE buffer, pH 7.5 (10 mM Tris-HCl, pH 7.5 and 1 mM EDTA, pH 8)
- PP buffer: 5.93 g KH_2PO_4 and 30 g K_2HPO_4, disolved in 100 ml sterile double-distilled water, pH adjusted to 7.6
- 2% CTAB (cetyltrimethyl ammonium bromide)
- 1% NaCl
- 2 M NaAc, pH 4

Analyses of the extracted soil DNA and RNA:

1. Analysis of DNA

1.1 Electrophoretic analysis
Mix five µl of the extracted DNA with one µl DNA sample loading mixture and load into the well of a 0.7% agarose gel. Perform electrophoresis for one hour in a small gel-electrophoresis chamber, with a distance of 15 cm between the electrodes, at 40 V. Stain the gel in a solution of TE buffer, containing 1 µg/ml ethidium bromide for 10 min, wash it for 5 min in water and photograph it under 304 nm UV light (see Fig. 1).

1.2 Spectrophotometrical analysis
The adsorbance of the DNA sample should be measured in 1 ml cuvettes at 230, 260 and 280 nm. An absorbance value of 1 at A_{260} is assumed to be equal to a concentration of 50 µg DNA per ml. The A_{260}/A_{280} and the A_{260}/A_{230} ratios can be used as measures for DNA purity [32, 33, 37].

1.3 Digestion of soil DNA with restriction endonucleases
*Eco*R I, *Hin*d III, *Bam*H I or other enzymes can be tested. *Eco*R I as the cheapest endonuclease is suggested to be used, because the amounts of enzyme necessary for complete digestion are relatively large.

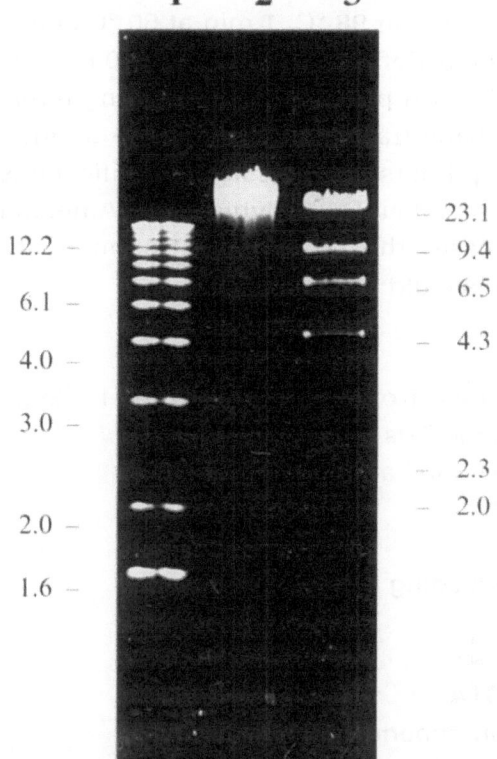

Figure 1. Electrophoretical analysis of total soil DNA recovered by the direct lysis procedure. 1) 1 kilobase DNA ladder (GIBCO BRL); 2) DNA recovered from a noninoculated soil sample of the experimental field of the University of Bayreuth, Germany; 3) λ-DNA, digested with *Hin*d III endonuclease.

Take 35 µl (approximately 1.5 µg) of soil DNA; add 4 µl reaction buffer, as suggested by the manufacturer, and 1 µl *Eco*RI (50 U). Incubate overnight at 37 °C.

1.4 PCR amplification

For successful amplifications the use of the AmpliTaq DNA polymerase, Stoffel fragment (Perkin Elmer Cetus) is preferable. The recommended conditions are as follows: 10 mM KCl, 10 mM Tris-HCl, pH 8.3, 200 µM of each dNTP, 0.4 µM of the two chosen primers, 5 mM $MgCl_2$, 0.2 µg of the recovered total DNA and 2.5 units of the enzyme. The volume of the samples can be 50 or 100 µl, and the following cycling profile is successful, when *npt* I and *npt* II gene sequences are amplified [4, 34]: initial melting of DNA for 5 min at 95 °C; fifty cycles – 1 min at 98 °C, 1 min at 50 °C and 1.5 min at 72 °C. Extension of the reaction for an additional 10 min at 72 °C increases the amount of the PCR product. The annealing temperature and the length of the polymerization step should be optimized for different templates and primers. Successful amplifications can be also obtained, by using other Taq polymerases (Amersham, Boehringer, BRL). In these cases the final concentration of the $MgCl_2$ in the reaction mixture should be 3 mM.

Solutions:
– TBE buffer for electrophoresis (according to [29])
    – 0.5 ×: 0.045 M Tris
        – 0.045 M Boric acid
        – 1 mM EDTA
        – pH 8.3
– DNA-sample-loading buffer:
    – 4 M urea
    – 50% sucrose
    – 50 mM EDTA
    – 0.25% bromophenolblue
    – 0.25% xylenecyanol

2. Analyses of soil RNA

2.1 Electrophoretic analyses
The analysis of the recovered total RNA should be performed in a 1.2% formaldehyde containing agarose gel. (Preparation of the gel should be done under the hood, because of the poisonous evaporation of the formaldehyde).

Preparation of the gel
Add 1.34 g agarose to 70 ml of DEPC treated double-distilled water. Melt the agarose by boiling. Take the agarose solution under the hood and add 20 ml formaldehyde solution (12.3 M) and 22 ml 5 × formaldehyde buffer. Cover the agarose solution and mix carefully. Keep the agarose very well covered at 65 °C.

Preparation of the RNA sample for electrophoresis
(Keep the solution of RNA and other solutions on ice bath).
Add to an Eppendorf tube:
- 4.5 µl RNA
- 2.0 µl 5 × formaldehyde buffer
- 3.5 µl formaldehyde
- 10 µl formamide

Mix well and denature the RNA for 15 min at 65 °C. Put the sample immediately on ice and centrifuge briefly at 4 °C. Add 2 ml formaldehyde gel-loading buffer.

Running conditions
- Prerun the gel for 5 min at 70 V.
- Load the RNA samples after their denaturation
- Run the electrophoresis for 4 h at 100 V.

Staining the gel
The gel has to be colored in a 0.1 M NH_4OAc solution, containing 0.5 µg/ml ethidium bromide for 45 min. For taking a picture, the gel has to be washed for at least one hour in sterile double-distilled water, pretreated with 0.1% DEPC (see Fig. 2).

Solutions

All solutions for RNA analysis must be prepared with sterile double-distilled water, pretreated with 0.1% DEPC.
- Formaldehyde gel-loading buffer:
    - 50% glycerol
    - 1 mM EDTA
    - 0.25% bromophenol blue
    - 0.25% Xylene cyanol
- Formaldehyde buffer 5 ×:
    - 0.1 M MOPS buffer, pH 7.0
    - 40 mM NaOAc
    - 5 mM EDTA, pH 8.0

    Prepare the buffer as follows: Prepare 800 ml 50 mM NaOAc, pH 7.0 and treat it with 0.1% DEPC. Add 20.6 g MOPS, 10 ml 0.5 M EDTA, pH 8.0 and adjust the volume to 1l with double-distilled sterile water.

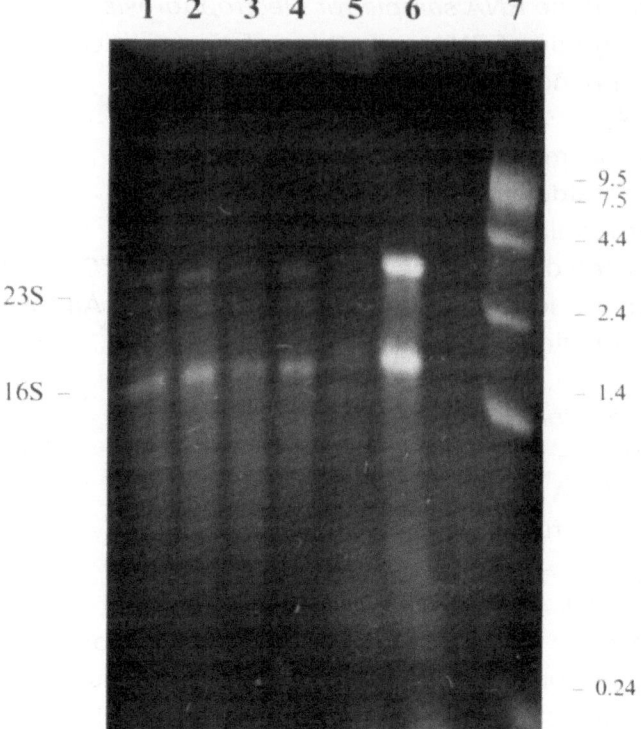

Figure 2. Electrophoretical analysis of total soil RNA, recovered by the direct lysis procedure. 1) and 2) total soil RNA recovered from two parallel soil samples inoculated with 3×10^7 cells per g soil of *E. agglomerans*; 3) and 4) total soil RNA recovered from two parallel noninoculated soil samples; 5) total bacterial RNA isolated from 5 ml of *E. agglomerans culture, containing* 1×10^3 cells per ml; 6) total bacterial RNA isolated from 5 ml of *E. agglomerans* overnight culture (1×10^9 cells per ml); 7) RNA kilobase ladder (GIBCO BRL).

  − 0.5 M EDTA, pH 8.0, pretreated with 0.1% DEPC

Recovered RNA can also be analysed in Northern blotting or Hybri-slot hybridizations by using standard procedures [29], as well as in RT-PCR amplification experiments (see Chapter 2.7.5).

References

1. Amann RI, Zarda B, Stahl DA, Schleifer KH (1992) Identification of individual procariotic cells by using enzyme-labelled rRNA-targeted oligonucleotide probes. Appl Environ Microbiol 58: 3007–3011.
2. Dijkmans RA, Jagers A, Kreps S, Collard JM, Mergeay M (1993) Rapid method for purification of soil DNA for hybridization and PCR analysis. Microb Releases 2: 29–34.
3. Erb RW, Wagner-Döbler I (1993) Detection of polychlorinated biphenyl degradation

genes in polluted sediments by direct DNA extraction and polymerase chain reaction. Appl Environ Microbiol 59: 4065–4073.

4. Evguenieva-Hackenberg E, Selenska-Pobell S, Klingmüller W (1994) Persistance and stability of genetically manipulated derivatives of *Enterobacter agglomerans* in soil microcosms. FEMS Microbiol Ecology 15: 179–192.

5. Flemming CA, Leung KT, Lee H, Trevors JT, Greer CW (1994) Survival of *lux-lac*-marked biosurfactant-producing *Pseudomonas aeruginosa* UG 2L in soil monitored by nonselective plating and PCR. Appl Environ Microbiol 60: 1606–1613.

6. Giovannoni SJ, Britschgi TB, Moyer CL, Field G (1990) Genetic diversity in Sargasso Sea bacterioplankton. Nature 345: 60–63.

7. Hahn D, Kester R, Starrenburg MJC, Akkermans ADL (1990) Extraction of ribosomal RNA from soil for detection of *Frankia* with oligonucleotide probes. Arch Microbiol 154: 329–335.

8. Henschke RB, Schmidt FRJ (1990) Plasmid mobilization from genetically engineered bacteria to members of the indigenous soil microflora *in situ*. Current Microbiol 20: 105–110.

9. Herrick JB, Madsen EL, Batt CL, Ghirose WC (1993) Polymerase chain reaction amplification of naphtalene catabolic and 16S rRNA gene sequences from indigenous sediment bacteria. Appl Environ Microbiol 59: 687–694.

10. Josephson KL, Gerba CP, Pepper TL (1993) Polymerase chain reaction detection of nonviable bacterial pathogens. Appl Environ Microbiol 59: 3513–3515.

11. Kidambi SP, Ripp S, Miller RV (1994) Evidence for phage-mediated gene transfer among *Pseudomonas aeruginosa* strains on the phyloplane. Appl Environ Microbiol 60: 496–500.

12. Khanna M, Stotzky G (1992) Transformation of *Bacilis subtilis* by DNA bound on montmorillonite and effect of DNase on the transforming ability of bound DNA. Appl Environ Microbiol 58: 1930–1939.

13. Kopczynski ED, Bateson MM, Ward DD (1994) Recognition of chimeric small subunit ribosomal DNAs composed of genes from uncultivated microorganisms. Appl Environ Microbiol 60: 746–748.

14. Lorenz MG, Wackernagel W (1990) Natural genetic transformation of *Pseudomonas stutzeri* by sand adsorbed DNA. Arch Microbiol 154: 380–385.

15. Moran AM, Torsvik VL, Torsvik T, Hodson RF (1993) Direct extraction and purification of RNA for ecological studies. Appl Environ Microbiol 59: 915–918.

16. Moré MI, Herrick JB, Silva MC, Ghiorse WC, Madsen EL (1994) Quantitative cell lysis of indigenous microorganisms and rapid extraction of microbiol DNA from sediment. Appl Environ Microbiol 60: 1572–1580.

17. Nannipieri P, Ciardi C, Badalucco L, Casella S (1986) A method to determinate soil DNA and RNA. Soil Biol and Biochem 18: 275–281.

18. Ogram A, Sayler GS, Barkay TJ (1987) DNA extraction and purification from sediments. J Microbiol Meth 7: 57–66.

19. Ogram AV, Mathot ML, Harsh JB, Boyle J, Pettigrew CA, JR (1994) Effects of DNA polymer length on its adsorption to soils. Appl Environ Microbiol 60: 393–396.

20. Ogunseitan OA, Deldago IL, Tsai YL, Olson BH (1991) Effect of 2-Hydroxybenzoate on the maintenance of naphtalene degrading *Pseudomonas* in seeded and unseeded soil. Appl Environ Microbiol 57: 2873–2879.

21. Pillai SD, Josephson, Bailey RL, Gerba CP (1991) Rapid method for processing soil samples for polymerase chain reaction amplification of specific gene sequences. Appl Environ Microbiol 57: 2283–2286.

22. Picard C, Ponsonnet C, Paget E, Nesme X, Simonet P (1992) Detection and enumeration of bacteria in soil by direct DNA extraction and polymerase chain reaction. Appl Environ Microbiol 58: 2717–2722.

23. Pichard SL, Paul JH (1992) Detection of gene expression in genetically engineered microorganisms and natural phytoplankton populations in the marine environment by mRNA analysis. Appl Environ Microbiol 57: 1721–1727.

24. Pichard SL, Paul JH (1993) Gene expression per gene dose, a special measure of gene expression in aquatic microcosms. Appl Environ Microbiol 59: 451–457.
25. Porteous LA, Armstrong JL (1993) A simple mini-method to extract DNA directly from soil for use with polymerase chain reaction amplification. Curr Microbiol 27: 115–118.
26. Richaume A, Angle JC, Sadowsky M (1989) Influence of soil variables on *in situ* plasmid transfer from *Escherichia coli* to *Rhizobium fredii*. Appl Environ Microbiol 5: 1730–1734.
27. Recorbet G, Picard C, Normand P, Simonet P (1993) Kinetics of the persistance of chromosomal DNA from genetically engineered *Escherichia coli* introduced into soil. Appl Environ Microbiol 59: 4289–4294.
28. Romanowski G, Lorenz M, Sayler G, Wackernagel W (1992) Persistence of free plasmid in soil monitored by various methods, including a transformation assay. Appl Environ Microbiol 58: 3012–3019.
29. Sambrook J, Fritsch EF, Maniatis T (1989) Molecular Cloning: A laboratory Manual 2nd Ed. New York: Cold Spring Harbor Laboratory Press.
30. Sayler GS, Fleming J, Applegate B, Werner C, Nikbakht K (1989) Microbial community analysis using environmental nucleic acid extracts. In: Hattori T, Ishida Y, Maruyama Y, Morita Y, Ushida A (eds) Recent Advances in Microbial Ecology, Proceedings of the 5th International Symposium on Microbial Ecology, pp. 658–662.
31. Sayler GS, Layton AC (1990) Environmental application of nucleic acid hybridization. Ann Rev Microbiol 44: 625–648.
32. Selenska S, Klingmüller W (1991) DNA recovery and direct detection of Tn5 sequences from soil. Lett Appl Microbiol 13: 21–24.
33. Selenska S, Klingmüller W (1991) Direct detection of *nif*-gene sequences of *Enterobacter agglomerans* in soil. FEMS Microbiol Lett 80: 243–246.
34. Selenska S, Klingmüller W (1992) Direct recovery and molecular analysis of DNA and RNA from soil. Microb Releases 1: 41–46.
35. Smalla K, Cresswell N, Mendonca-Hagler LC, Wolters A, Van Elsas JD (1993) Rapid DNA extraction protocol from soil for polymerase chain reaction-mediated amplification. J Appl Bacteriol 74: 78–85.
36. Smit E, Van Elsas JD, Van Veen JA, De Vos WM (1991) Detection of plasmid transfer from *Pseudomonas fluorescens* to indigenous bacteria in soil using bacteriophage φ-R2F for donor counterselection. Appl Environ Microbiol 57: 3482–3488.
37. Steffan RJ, Goksoyr J, Bej AK, Atlas RM (1988) Recovery of DNA from soils and sediments. Appl Environ Microbiol 54: 2908–2915.
38. Steffan RJ, Atlas RM (1988) DNA amplification to enhance detection of genetically engineered bacteria in environmental samples. Appl Environ Microbiol 54: 2185–2191.
39. Tebbe CC, Vahjen W (1993) Interference of humic acids and DNA extracted directly from soil in detection and transformation of recombinant DNA from bacteria and a yeast. Appl Environ Microbiol 59: 2675–2665.
40. Thiem SM, Krumme ML, Smith RL, Tiedje JM (1994) Use of molecular techniques to evaluate the survival of a microorganism injected into an aquifer. Appl Environ Microbiol 60: 1059-1067.
41. Tsai YL, Olson BH (1991) Rapid method for direct extraction of DNA from soil and sediments. Appl Environ Microbiol 57: 1070–1074.
42. Tsai YL, Olson BH (1992) Rapid method for separation of bacterial DNA from humic substances in sediments for polymerase chain reaction. Appl Environ Microbiol 58: 2292–2295.
43. Tsai Y, Park MJ, Olson BH (1991) Rapid method for direct extraction of mRNA from seeded soils. Appl Environ Microbiol 57: 765–768.
44. Van Kuppeveld FJM, Van der Logt JTM, Angulo AF, Van Zoest MJ, Quint WGV, Niesters HGM, Galama JMD, Melchers WJG (1992) Genus- and species-specific identification of mycoplasmas by 16S rRNA amplification. Appl Environ Microbiol 58: 2606–2615.

45. Ward DM, Weller R, Bateson MM (1990) 16S rRNA sequences reveal numerous unculturated microorganisms in a natural community. Nature 345: 63–65.
46. Zeph LR, Stotzky G (1989) Use of botinylated DNA probe to detect bacteria transduced by bacteriophage P_1 in soil. Appl Environ Microbiol 55: 661–665.

Molecular Microbial Ecology Manual **1.6.2**: 1–11, 1995
© 1995 *Kluwer Academic Publishers.*

Extraction of ribosomal RNA from microbial cultures

ERKO STACKEBRANDT and NAOMI WARD

*DSM-German Collection of Microorganisms and Cell Cultures GmbH, Mascheroder Weg 1b,
38124 Braunschweig, Germany*

Introduction

The methods outlined in this chapter for the purification of large ribosomal ribonucleic acids (16S and 23S rRNA), are designed for use on pure cultures. Isolation of rRNA from mixed cultures will be covered in Chapter 1.5. Various purification methods have been published in recent years, depending on the use of different RNA species in systematics and phylogenetic studies. The following protocols are relatively simple to apply and have proved satisfactory for most cell types.

The majority of the nucleic acid within a bacterial cell is RNA. Since the number of ribosomes in a single bacterial cell is about 10^4, the 16S and 23S rRNA account for most of the RNA, while the 5S rRNA, transfer RNA (tRNA) and messenger RNA (mRNA) account for lesser amounts. Transfer RNAs and 5S rRNAs have been analyzed in recent years to determine the complexity of bacterial populations in natural samples. This topic will be covered in Chapter 3.3.7. 16S and 23S rRNAs, either as purified rRNA species or as bulk rRNA, have been used extensively in rDNA cistron similarity studies and all three rRNA species have been analyzed for phylogenetic studies.

General comments

In contrast to the isolation of DNA, which in most cases does not require special precautions, as DNases can be easily inactivated (see Chapter 1.6.1.), RNases are ubiquitous and more difficult to inactivate. It must be the prime goal, during the whole process of isolation of rRNA, to limit wherever possible the action of any endogenous nucleases, and to eliminate any introduction of nucleases from external sources. The following are some guidelines to increase the likelihood of purifying undegraded rRNA species. Extensive protocols have been published by Johnson [4,5].

In order to remove nucleases, all glassware should be either oven-baked at least 150 °C for several hours, autoclaved or rinsed with alcohol before use. Use either disposable plastic ware, or autoclave material in the presence of detergents. Commercial products are available for removing RNase from glassware and plastic ware. One of the most simple means of preventing the introduction of nucleases is to wear plastic gloves while isolating RNA, and to avoid touching the inside of tubes, beakers and flasks.

Solutions should be prepared from fresh, double-distilled water, and the whole isolation procedure should be carried out whenever possible on ice or at 4 °C. Water can be treated by adding 0.2% diethylpyrocarbonate (DEPC) to flasks of distilled water. After two or more hours with occasional shaking to disperse DEPC droplets, the water can be used after autoclaving for making up the solutions and buffers indicated below.

RNA can be stored as a dry pellet at 4 °C, or resuspended at 2.0 mg/ml in distilled water and stored at —20 °C. For long-term storage, the RNA is best stored as an ethanol slurry at —70 °C.

Standard solutions

- NH$_4$Ac: 7.5 M, pH 6. Sterilize by filtration through a 0.22 µm nitrocellulose or cellulose acetate membrane.
- Electrophoresis buffer: 0.02 M sodium acetate, 1 mM EDTA, 0.2% (w/v) SDS, 0.04 M Tris-HCl, pH 7.2. Autoclave.
- Lysis buffer: 0.02 M NaAc, 1 mM EDTA, 0.04 M Tris-HCl, pH 7.2. Autoclave.
- Pellet-dissolving buffer: 20 mM Tris-HCl, pH 7.5, 10 mM vanadyl ribonucleoside complexes, and 100 µg/ml proteinase K. Autoclave before addition of proteinase K.
- Ribosome buffer: 0.01 M MgCl$_2$, 0.5 M NH$_4$Cl, 0.01 M Tris-HCl, pH 7.4. Autoclave.
- Sodium acetate: 3 M, pH 5. Sterilize by filtration through a 0.22 µm membrane.
- Sodium dodecyl sulphate (SDS): 20% w/v.
- Standard saline citrate (SSC): Prepare a 20 × SSC stock solution (3.0 M NaCl, 0.3 M trisodiumcitrate, pH 7.2), add 0.2% DEPC and autoclave. Prepare appropriate dilutions using treated water.
- Standard saline citrate-HEPES (SSC-HEPES): Prepare the SSC in the usual manner (see above) with an additional 10 mM HEPES (pH 7.0), add 0.2 % DEPC and autoclave.
- Subunit buffer: 0.05 M KCl, 1 mM MgCl$_2$, 5 mM Tris-HCl, pH 7.4. Autoclave.
- TE buffer: 10 mM Tris-HCl, pH 7.5, 1 mM EDTA. Autoclave.
- TBE buffer: 10 × TBE = 0.89 M Tris-base, 0.89 M boric acid, 0.02 M EDTA, pH 8.3.

- Water-saturated phenol: Let distilled phenol drop directly into bidistilled water. Two phases will appear, once the water is saturated with phenol, with water at the top and phenol at the bottom. Mix vigorously and keep in a light-proof glass bottle in the dark at 4-10 °C. If phenol shows a reddish colour, it is oxidized and should be replaced by a fresh solution.

Procedures

Lysis of bacterial cells [11]

After harvesting, store bacterial pellets as soon as possible at −70 °C until required. For use, quickly thaw a pellet of 2–5 g (wet weight) and resuspend in ice-cold lysis buffer, ca. 1 ml buffer per gram cells. Transfer the suspension to a French pressure cell (Aminco), pre-cooled to 4 °C. There are alternative physical disruption procedures such as homogenization using glass beads or ultrasonication. In each case the method of choice depends on the chemical compo-sition of the cell wall and several options should be tested.

Steps in the procedure

1. Add a few drops of 20% (w/v) SDS. Since most Gram-negative cells, and also some Gram-positive cells, will start lysing im-mediately, force the suspension through the pressure cell at 20,000 psi (= 138 MPa).
2. Let the lysate drip into 10 ml ice-cold distilled and water-saturated phenol in a chilled, nuclease-free (baked at 200 ° overnight), 15 ml Corex tube.
3. Deproteinize the lysate by successive phenol extractions: add an equal volume of phenol, shake or vortex, centrifuge 20 min at 12,000 × g (Sorvall SS34), and transfer the upper (aqueous) phase with a pasteur pipette to a clean 15 ml Corex tube; the protein interface and phenolic phase are discarded. The number of phenol extractions depends on the thickness of the interphase layer. Usually it will be removed after 2–4 extraction steps.
4. Take the upper phase, precipitate nucleic acids by the addition of 2.5 volumes of ethanol, and store the suspension for at least 1 h at −20 °C. Centrifuge for 20 min at 12,000 × g (Sorvall SS34).

5. Remove the last traces of phenol by resuspending the pellet in 1–2 ml lysis buffer, reprecipitate the nucleic acid as above and store at 4 °C.
6. If the nucleic acids are to be subjected to preparative gel electrophoresis, centrifuge the suspension for 30 min at 12,000 × g, and redissolve the nucleic acid pellet in 1 ml cold electrophoresis buffer, shortly before application to the gel.

Isolation of crude 16S and 23S rRNA [6]

These variations of the Kirby procedure [6] are for the isolation of 16S and 23S rRNA, although the preparations also contain fragmented mRNA. The RNA preparations are essentially free of DNA and of small RNAs (tRNA and 5S rRNA).

Steps in the procedure
1. Centrifuge the ethanol precipitate of total nucleic acids (from Step 5 above) for 20 min at 4 °C, 12,000 × g, and remove excess ethanol from the precipitate by placing the tube, opening down, on a paper tissue. Resuspend the final pellet in 1 ml of TE buffer.
2. Determine the concentration of the nucleic acid by measuring the absorbance at 260 nm of an appropriately diluted aliquot. Assuming that a 50 µg/ml RNA solution has an A_{260} of 1, adjust the concentration of the nucleic acid solution to ca. 5 mg/ml (either by diluting with TE buffer or resuspending in a smaller volume after reprecipitation with ethanol).
3. Add an equal volume of 3.0 M NaCl, mix, and allow the high-molecular-weight RNA to precipitate overnight at 4 °C (this is most conveniently done in a 1.5 ml microcentrifuge tube).
4. Collect the precipitate by centrifugation for 5–10 min at 4 °C in a microcentrifuge. Resuspend the pellet of high-molecular-weight RNA in 1 ml TE buffer.
5. Determine the concentration of RNA spectrophotometrically, as above.

Alternative procedure
1. Overlay the ethanol precipitate (Step 5 above) with 5 ml ice-cold 3 M sodium acetate, pH 5.0.

2. Resuspend the precipitate twice (10 × 5 s each time) with a DEPC-treated Ultra-Turrax homogenizer (IKA, Jahnke & Kunkel, Stauffen i. Breisgau, Germany). Chill the solution for 1 min between the homogenizing steps. The RNA stays as a precipitate while DNA goes into solution.
3. Centrifuge and discard the DNA-containing supernatant.
4. Repeat this treatment until the precipitate appears opalescent.
5. Collect the RNA by centrifugation (5–10 min at 4 °C at 12,000 × g). Dry the pellet lightly under a low vacuum and resuspend in 500 µl distilled water.

Isolation of 70S ribosomes [12]

This procedure is necessary when ribosomal RNA species can not be recovered as intact molecules, but are "nicked" to result in the formation of two fragments. This has been described as occurring in the 23S rRNA of several organisms e.g., *Gemmobacter aquatilis*, *Rhodobacter sphaeroides*, *Agrobacterium tumefaciens* and a few other members of the α-subclass of proteobacteria [10]. Since the two 23S rRNA fragments have a size similar to that of the 16S rRNA, gel electrophoresis does not separate them satisfactorily.

Steps in the procedure
1. Resuspend 7–8 g wet weight cells with 50 g glass beads (diameter 0.1 mm) and 20 ml ribosome buffer in a glass-stopped bottle.
2. Shake 2 × 30 s in a cell homogenizer (e.g., Braun, Melsungen). Check disruption of cells by microscopy and repeat procedure until cells are lysed.
3. Allow the glass beads to settle, remove the supernatant and shake the glass beads with fresh ribosome buffer. Combine the supernatants.
4. Centrifuge for 40 min at 4 °C at 45,000 × g to pellet cell debris.
5. Centrifuge the supernatant by ultracentrifugation for 2 h at 4 °C at 150,000 × g.
6. Dissolve the pellet in 25 ml ribosome buffer and underlay with 10 ml of a 30% (w/v) sucrose solution (in ribosome buffer).
7. Centrifuge for 4 h at 4 °C at 80,000 × g.
8. Dissolve the pellet in 2–4 ml subunit buffer.

9. Estimate yield (1 optical density unit at 260 nm = 60 µg ribosome ml^{-1}) [13]

Isolation of total rRNA from isolated ribosomes [12]

Steps in the procedure
1. Dissolve each ribosome pellet from Step 7 above in 1.0 ml of the pellet-dissolving buffer. Pool and incubate at 50 °C for 1–2 h.
2. Add ¼ volume phenol-chloroform (1:1), shake for 20 min and centrifuge at 17,000 × g for 10 min. Transfer the upper aqueous phase to a 50 ml screw-cap centrifuge tube.
3. For the isolation of large rRNA species, add ½ volume of 7.5 M ammonium acetate to the aqueous layer, mix well and place on ice for 20 min. Centrifuge at 17,000 × g for 10 min to pellet RNA. Repeat the ammonium acetate precipitation step.
4. Alternatively, to isolate total RNA, add 1/10 volume of 20 × SSC, mix, add 2 volumes of 95% (v/v) ethanol, mix, cool at –20 °C for 15 min and centrifuge at 17,000 × g for 15 min.
5. Dissolve the RNA in 10 ml of 1 × SSC-1 mM HEPES buffer pH 7.0, and reprecipitate by adding 2 volumes of 95% ethanol and centrifuging.
6. Dissolve the partially dried RNA pellet in 5–10 ml of 1 × SSC-1 mM HEPES (pH 7.0), add SDS to a final concentration of 0.5 %, and store at –20 °C.

Isolation of ribosomal subunits by sucrose gradient ultracentrifugation [12]

Steps in the procedure
1. Prepare a 17–25% (w/v) sucrose gradient in subunit buffer with an appropriate apparatus.
2. Overlay the solution of 33 ml in a centrifuge tube with 200 to 1000 µl ribosome solution (depending on the yield).
3. Centrifuge in a vertical rotor for 1.5 h at 4 °C at 80,000 × g.
4. Fractionate each tube by recording the optical density (260 nm) of the tube contents, starting with the concentrated sucrose solution, and collect fractions of about 0.75 ml.
5. Combine the fractions that appear to contain each subunit, using

stringent criteria for selection by selecting only those fractions that are well separated from adjacent peaks.
6. Add 1/10 volume 3 M sodium acetate, pH 6.0, and 3 volumes of ethanol and precipitate subunits for 12 h at –20 °C.

Isolation of rRNA from subunits [12]

Steps in the procedure
1. Concentrate the precipitated subunits by centrifugation for 20 min at 4 °C at 9,000 × g.
2. Dissolve the pellet in 1 ml bidistilled water.
3. Phenolize (add an equal volume of water-buffered phenol) and centrifuge in a bench-top centrifuge (e.g., Eppendorf) for 10 min.
4. Extract the upper aqueous phase with 2 volumes ether, remove the ether and repeat the extraction step 5 times.
5. Add 1/10 volume 3 M sodium acetate, pH 6.0, and 3 volumes ethanol and store for 4 h at – 20 °C.
6. Centrifuge at 12,000 × g for 20 min, dry pellet, dissolve it in 200 µl water and reprecipitate.
7. Dissolve pellet in 100 µl water and determine the yield spectrophotometrically.
    While the 16S rRNA is the only RNA species in the small subunit, the large one contains 5S and 23S rRNA. The two species need to be separated and purified by gel electrophoresis as outlined below.

Analytical identification of rRNA species

The presence of 16S and 23S rRNA in preparations can be checked quickly by conventional agarose gel electrophoresis.

Steps in the procedure
1. Prepare a 1% (w/v) agarose gel in 1 × TBE, containing 0.5 µg/ml ethidium bromide.
2. Combine a small aliquot (e.g., 20 µl) of the RNA preparation with tracking dye (e.g., 0.25% bromophenol blue, 30% glycerol in water) and load onto the gel, together with an appropriate molecular weight marker (kb ladder, BRL) and/or a standard of commercially available rRNA species.

3. Electrophorese at 100 V for 1–2 h in 1 × TBE electrophoresis buffer. Check the presence of rRNA by transillumination of the gel with UV light of wavelength 254 nm. The two large rRNA species should appear as distinct bands. If only a bright blur can be seen, the gel is overloaded, try diluting the aliquot before loading.

Isolation of defined rRNA species

Preparative gel electrophoresis
For purification of 16S and 23S rRNA, a 2.8% (w/v) polyacrylamide gel has been used routinely [1]. For separation of 5S rRNA, a 10% polyacrylamide gel is used. Gels are poured as slabs of 20 cm × 20 cm × 3 mm.

Steps in the procedure
1. For one gel, dissolve 4.2 g acrylamide and 0.21 g N,N'-bis-methyleneacrylamide in 150 ml electrophoresis buffer. Degas the solution under vacuum, add 125 µl tetra methylethylenediamine (TEMED) and 250 µl 50% (w/v) ammonium persulphate solution (stored at −20 °C in small aliquots), and quickly pour the gel between two glass plates, inserting a comb with two or three large pockets (ca. 5 cm × 0.5 cm × 0.3 cm). Allow the gel to polymerise for at least one hour before use.
2. Pre-electrophorese at 80 mA (ca. 160 V), for 30–60 min.
3. Dissolve the nucleic acid sample in 1 ml electrophoresis buffer, made 10% (v/v) in glycerol, and add a small drop (ca. 10 µl) of a saturated solution of bromophenol blue. Carefully layer the sample(s) into the well(s). Electrophorese initially at 40 mA, until the bromophenol blue marker has migrated into the gel (about 20 min), and subsequently at 80 mA (about 160 V constant voltage), for 6–16 h.
4. After electrophoresis, remove the gel from the glass plates, and place it on polyethene film wrap. To visualise the nucleic acid bands, place the gel on a fluorescent (at 260 nm) thin-layer plate (Merck), and illuminate from above with UV light of wavelength 260 nm. Nucleic acids are visible as dark bands. Cut out the rRNA bands with a scalpel, wrap in parafilm and store at 4 °C. The presence of SDS in the gel prevents the degradation of nucleic

acids by nucleases. RNA species are electro-eluted from the gel piece using a Bio-trap (Schleicher & Schuell) type electroelution unit. Alternatively, 5S rRNA can be eluted as published by Maxam et al. [9].

Isolation of mRNA

Messenger RNA (mRNA) is relatively easy to isolate from mammalian cells, due to poly (A) chains carried at the 3' termini of the molecules which serve as a target for purification by affinity chromotography on oligo (dT) cellulose. The recovery of mRNA from prokaryotic cells is considerably more difficult, as this structure is absent. Most studies involving the use of prokaryotic mRNA (e.g. ribosome binding site and initiation of translation studies) have been carried out using mRNA synthesized *in vitro* [2, 3].

Labelling of RNA

Most studies of bacterial RNA, such as ribosomal RNA/DNA cistron similarity studies and phylogenetic investigations, have required the use of labelled molecules, either as hybridization probes or for sequence analysis. In early studies, nucleic acids were radiolabelled in vivo by growing the organism in the presence of ^{3}H-, ^{14}C-, or ^{32}P-labelled precursors. In vivo labelling of nucleic acids depends upon the medium required for culturing the organisms, and also upon the nucleic acid precursors the organisms need to take up from the medium [4]. As label incorporation is often inefficient, and as it is difficult to predict the specific activities of the labelled nucleic acids, the use of in vivo labelled nucleic acids in nucleic acid hybridization studies has decreased. Ribosomal RNA for sequence analysis has been labelled most efficiently with ^{32}P when labelled orthophosphate was added to a dephosphorylized growth medium. However, the health risk caused by high contaminant radiation of medium, glass ware, French press and other equipment has stopped the use of this powerful labelling method. Current in vitro labelling procedures reflect state-of-the-art technology and commercial kits are available for the 5' and 3' end-labelling of RNA. Label can also be introduced into RNA by in vitro transcription or by iodisation. Advantages of these procedures include less manipulation of radioactive components, less radioactive waste material, efficient incorporation of the labelled isotope, and predictable levels of specific activity.

A substitute for radioactive labelling has been the incorporation of biotin- or digoxygenin-containing nucleotides into RNA molecules. Biotin-labelled nucleic acids are most usually detected by a reaction with a streptavidin-AP (alkaline phosphatase) conjugate [7,8], followed by a colour reaction with a chromogenic substrate such as X-phosphate and NBT. Chemiluminescent detection of biotin-labelled nucleic acids is also possible with commercial kits such as FLASH (Stratagene) and Gene-Lite (BioRad). Similarly, digoxygenin-labelled nucleic acids can be detected by an enzyme-linked immunoassay using an antidioxygenin-AP conjugate, together with substrates for a chromogenic or chemiluminescent reaction (e.g., the DIG system from Boehringer Mannheim). These non-isotopic techniques are more fully covered in Chapter 2.3.

References

1. Bishop HDL, Claybrook JR, Spiegelman S (1967) Electrophoretic separation of viral nucleic acids on polyacrylamide gel. J Mol Biol 26: 373–387.
2. Burgess RR (1969) A new method for the large scale purification of *Escherichia coli* deoxyribonucleic acid-dependent ribonucleic acid polymerase. J Biol Chem 244: 6160–6167.
3. Vendrisak JJ, Burgess RR (1975) A new method for the large scale purification of wheat germ DNA-dependent RNA polymerase II. Biochemistry 14: 4639–4645.
4. Johnson JL (1985) DNA reassociation and RNA hybridization. In: Gottschalk G (ed) Methods in Microbiology, Vol. 18, pp. 33–74. Academic Press, London.
5. Johnson JL (1992) Isolation and purification of nucleic acids. In: Stackebrandt E, Goodfellow M (eds) Nucleic Acid Techniques in Bacterial Systematics, pp. 1–19. Wiley, Chicester.
6. Kirby KS (1968) Isolation of nucleic acids with phenolic solvents. Meth Enzymol 12: 87–99.
7. Langer PR, Waldrop AA, Ward DC (1981) Enzymatic synthesis of biotin-labeled polynucleotides: Novel nucleic acid affinity probes. Proc Natl Acad Sci USA 78: 6633–6637.
8. Leary JJ, Brigati DJ, Ward DC (1983) Rapid and sensitive colorimetric method for visualising biotin-labeled DNA probes hybridised to DNA or RNA immobilised on nitrocellulose: Bio-blots. Proc Natl Acad Sci USA 80: 4045–4049.
9. Maxam AM, Donis-Keller H, Gilbert W (1977) Mapping adenines, guanines and pyrimidines in RNA. Nucl Acids Res 4: 2527–2543.
10. Rothe R, Fischer A, Hirsch P, Sittig M, Stackebrandt E (1986) The phylogenetic position of the budding bacteria *Blastobacter aggregatus* and *Gemmobacter aquatilis* gen. nov., spec. nov. Arch Microbiol 147: 92–99.

11. Stackebrandt E, Ludwig W, Fox GE (1985) 16S ribosomal RNA oligonucleotide cataloguing. In: Gottschalk G (ed) Methods in Microbiology, Vol. 18, pp. 75–107. London, Academic Press.
12. Traub P, Mizushima S, Lowry CV, Nomura N (1971) Reconstitution of ribosomes from subribosomal components. In: Grossman L, Moldave K (eds) Methods in Enzymology, Vol. XX, pp. 391–407. C. Academic Press, London.
13. Wobus U (1980) Isolierung, Fraktionierung und Hybridisierung von Nukleinsäuren. Verlag Chemie, Weinheim.

Molecular Microbial Ecology Manual **2.2.1**: 1–34, 1995
© 1995 *Kluwer Academic Publishers.*

Preparation of radioactive probes

MARTIN CUNNINGHAM

Research and Development Department, Amersham International plc, Amersham, Bucks, UK

Introduction

Labelled nucleic acid probes are widely used for the detection of homologous sequences in solution, immobilised on membranes, or *in situ* in cells or tissues. Although an increasing proportion of this work is carried out using non-radioactively labelled probes (Chapter 2.3.1) and/or by target amplification using the polymerase chain reaction, a very significant proportion is still reliant on the use of radioactive hybridization probes [5]. Such probes provide a high degree of reliability coupled with maximum sensitivity and allow accurate quantification of target levels. A variety of detection methods can be used. For example, for solution assays, samples can be counted directly by β-scintillation, on membranes autoradiography by exposure to X-ray film is generally used while, for *in situ* applications, slides are coated with autoradiographic emulsion to allow the most sensitive detection of signal. Quantification of membrane-based signal can be achieved by use of a phosphorimager, by densitometry of autoradiographic images or by scintillation counting of excised bands, while *in situ* images can be quantified in exposed emulsions by counting grains or using an image analyzer.

Radioactive labels and nucleotides

The radioisotopes ^{32}P and ^{35}S have most commonly been used as labels for nucleic acid probes, and, more recently, ^{33}P has become available as a label with properties intermediate between those of ^{35}S and ^{32}P [8]. ^{3}H and ^{125}I have also been used, to a much lesser extent, as labels for specific applications. For example, ^{3}H can be used in *in situ* hybridization to give high resolution but at the expense of low sensitivity and long exposure times. The features and applications of the most important radiolabels are summarized in Table 1. They are available in the form of radiolabelled nucleoside triphosphates which can be incorporated into nucleic acids by

Table 1. Properties of radioisotopes used in labelling nucleic acid probes

Isotope	^{32}P	^{33}P	^{35}S	^{125}I	^{3}H
Energy of) emission (MeV)	1.71	0.249	0.167	0.035[a]	0.0018
Half-life (days)	14.3	25.4	87.4	60.0	12.4 years
Resolution[b] (μm)	20–30	15–20	10–15	1–10	0.5–1.0
Main applications	Membrane hybridization Macroscale *in situ*	*In situ* localization at cellular level High target membrane hybridization	*In situ* localization at cellular level High target membrane hybridization	*In situ* sub-cellular localization	*In situ* sub-cellular localization
Features	High sensitivity Rapid detection with X-ray film Low resolution	Rapid exposure for *in situ* Good resolution for membranes	Short exposure, medium resolution for *in situ* Good resolution for membranes	High sensitivity, short exposure, good resolution for *in situ*	High resolution for *in situ*

[a] Auger electrons
[b] Scatter around point of source

a variety of enzymatic methods [6].

^{32}P is the most commonly used isotope for membrane applications. The high energy of the β-particles, together with the high specific activity of the available nucleotides, allows it to be used for applications requiring the highest degree of sensitivity such as the detection of single copy genes in complex mammalian genomes, where there may be less than 1 pg of a specific gene in a 1 μg sample of DNA. The lower energy of ^{35}S makes it more appropriate for applications where sensitivity is less of an issue but where a higher degree of image resolution is required, for example *in situ* hybridization. ^{33}P is also being increasingly used in such applications. It gives similar results to ^{35}S, while its shorter half-life allows easier disposal of waste materials and, in some instances, lower backgrounds may be obtained. A membrane-based application that is appropriate for ^{35}S-labelled probes is in screening phage plaques or bacterial colonies.

Nucleoside triphosphates are available at a variety of specific activities containing ^{32}P, ^{33}P or ^{35}S at either the α- or γ-phosphate position. α-labelled nucleotides are used in most nucleic acid labelling procedures since they are incorporated as monophosphates with release of the unlabelled β-

and γ-phosphates. γ-labelled nucleotides are used in a reaction catalyzed by the enzyme polynucleotide kinase in which the terminal phosphate of the nucleotide is transferred to the 5'-end of the nucleic acid. All four deoxynucleoside triphosphates (dATP, dCTP, dGTP and TTP) and all four ribonucleoside triphosphates (ATP, CTP, GTP and UTP) are available labelled at the α-phosphate with either ^{32}P or ^{35}S for the preparation of DNA and RNA probes. ^{35}S-labelled nucleotides are thionucleotide analogues in which a sulphur atom replaces a nonbridging oxygen atom in one of the phosphate groups. The presence of this altered group can, in some cases, affect the kinetics of enzyme-mediated labelling reactions. Many of these radiolabelled nucleotides have recently become available in a stabilised formulation which allows them to be stored in a refrigerator, so avoiding the necessity of carrying out freeze-thaw cycles, and which contains an intense red dye which significantly improves visualisation of the nucleotide during use (Redivue range of nucleotides, Amersham International, plc).

^{32}P-labelled probes can be stored at –20 °C for several days, although prolonged storage can lead to substantial probe degradation. High specific activity ^{32}P-labelled probe should be stored for no more than three days. Longer storage is possible with ^{33}P- and ^{35}S-labelled probes, up to approximately 2 weeks and 1 month respectively.

Types of probe

There are two main categories of nucleic acid probe, long probes and oligonucleotide probes. Long probes are usually derived from cloned sequences and range in length from approximately 100 base pairs (bp) to several kilobase pairs (kb). The entire clone, including plasmid or bacteriophage vector sequence, is often labelled, but in many cases it is preferable to avoid possible cross-hybridization between vector and target by labelling only the cloned insert. In this case the insert is excised from the vector by restriction endonuclease digestion and separated from it by agarose gel electrophoresis. The insert can then be labelled either after further purification or in the presence of agarose, in which case low melting point agarose should be used so that the labelling mix remains liquid at room temperature. Depending on application, the label can be introduced more or less uniformly throughout the length of the molecule or specifically at its ends. Uniform labelling is the most common approach for long probes as label density, and hence probe specific activity and final detection sensitivity, is higher.

Oligonucleotide probes are typically between 10 and 100 bases in length and are readily prepared using automated synthesizers. It is generally simpler to end-label oligonucleotides than to introduce a uniform label. As the molecules are short, label density remains acceptably high, and probes can be used at a high molar concentration during hybridization to maximize sensitivity. Methods are available for introducing a labelled group at either the 3'- or the 5'-end of the molecule.

Procedure

Table 2 summarises the properties of the probe labelling methods to be described in this section. The three most commonly used methods of uniform labelling are random primer labelling, nick translation and phage RNA polymerase-based labelling. The first two methods generate DNA probes which are used widely in membrane hybridizations and so will be described primarily for use with ^{32}P-labelled nucleotides, while RNA polymerase-based methods produce RNA probes that have proved to be of particular use for *in situ* hybridization and so will be primarily described for use with ^{35}S-labelled nucleotides. End labelling methods will be described primarily for oligonucleotides. 5'-end labelling occurs with low efficiency with thionucleotides and so will be described for use with ^{32}P-labelled nucleotides. Oligonucleotide probes 3'-end labelled with ^{35}S are finding increasing use in *in situ* hybridization, and this method will be described chiefly for ^{35}S-labelled nucleotides. Radiolabelled nucleotides can also be incorporated during the polymerase chain reaction (PCR) either to facilitate detection of amplified sequence or to enable it to be used as a probe. This approach, together with related methods for generating RNA probes, will also be briefly discussed. Finally, methods will be given for determination of incorporation efficiency and for removal of unincorporated nucleotides from the labelled nucleic acid probe. In practice, probe purification is not usually necessary for most membrane applications, although it may sometimes be found useful to remove unincorporated nucleotides to reduce background in filter hybridizations for probes with specific activities > 10^9 dpm/µg or when the labelling reaction yields an incorporation of less than 40–50%.

Table 2. Properties of major nucleic acid radiolabelling methods

	Random primer	Nick Translation	Phage polymerase	3′-end labelling	5′-end labelling
Enzyme	Klenow polymerase	DNA polymerase I	SP6/T7/T3 RNA polymerase	Terminal transferase	Polynucleotide kinase
Template	ssDNA (and denatured dsDNA)	dsDNA	dsDNA	ssDNA preferred, oligos	DNA, RNA or oligos with dephosphorylated 5′-end
Amount of template	25 ng	50–1000 ng	1–2 µg	10 pmols ends	10 pmols ends
Reaction time	5 min – overnight	1–3 h	1 h	1–2 h	30 min
Efficiency of incorporation	> 60%	> 50%	> 70%	Varies with molar ratio label: ends	Varies with molar ratio label:ends
Nature of probe	DNA	DNA	RNA	oligo/DNA	oligo/DNA
Competing strand	yes	yes	no	no	no
Amount of probe	40–50 ng	50–1000 ng	150–250 ng	e.g. 3.3 µg of 1 Kb duplex; 65 ng of 20-mer oligo	e.g. 3.3 µg of 1 Kb duplex; 65 ng of 20-mer oligo
Potential specific activity of probe (^{32}P dpm/µg)	5×10^9	$5 \times 10^8 – 5 \times 10^9$	$1–2 \times 10^9$	1×10^9	up to 1×10^9
Insert specific	no	no	yes	–	–
Requirement for subcloning	no	no	yes	no	no

Values quoted are for commonly used protocols

1. Random primer labelling

This method uses oligonucleotides of random sequence, originally hexamers, to prime synthesis along the length of a single-stranded DNA template [9, 10]. The primers anneal to short stretches of complementary DNA on the template and, in the presence of an appropriate DNA polymerase enzyme and deoxynucleoside triphosphates (dNTPs), a new complementary DNA strand is synthesized. If a radioactive nucleotide is present, this will be incorporated into the new strand. Figure 1 gives a diagrammatic outline of the reaction.

The most commonly used enzyme is the Klenow fragment of *E. coli* DNA polymerase I [12]. This retains the polymerase activity of the whole enzyme, which synthesizes the new strand in a 5'-3' direction, but lacks the 5'-3' exonuclease activity which, if present, would remove incorporated nucleotides in the same direction starting with the 5'-end of the primer. Other polymerases are sometimes used including bacteriophage T7 DNA polymerase and an exonuclease-free derivative of the Klenow fragment which also lacks the 3'-5' exonuclease activity of DNA polymerase I and has been observed to give somewhat higher levels of incorporation (B. Harvey, unpublished observations). Primers of different length, particularly nonamer primers, are also frequently used.

The random primer labelling method has a number of useful properties. It can be used to label small quantities of DNA, as little as 25 ng, and high probe specific activities are achievable, up to 5×10^9

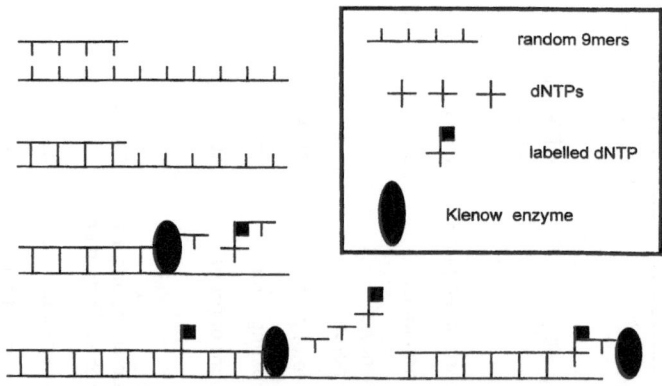

Figure 1. Principles of random prime labelling.

dpm/µg using high specific activity [α-^{32}P]dNTPs (3000–6000 Ci/mmol). The reaction can be carried out very rapidly, in 5–10 min at 37 °C, or over a longer period, for example overnight, at room temperature. The ability to label small amounts of DNA is useful when labelling insert DNA which can only readily be purified in µg quantities. Impure DNA, for example in low melting point agarose, can also be labelled efficiently.

Careful choice of appropriate nucleotide concentrations and ratio of primer to template concentrations is important in order to synthesize probe of an optimal length for hybridization. The protocol below should give acceptable results over a range of template concentrations. Commercially available systems containing pretested reagents to ensure optimal labelling have also proved popular (for example, Megaprime System, Amersham International, plc) and dried pre-aliquotted reactions requiring only the addition of label and denatured template (for example, Rediprime System, Amersham International, plc) have significant advantages in terms of throughput, convenience and reproducibility.

Steps in the procedure
1. Prepare an appropriate nucleotide mix from the 300 µM stocks by mixing 3.3 µl of each of three dNTP stocks, excluding that to be used as label. Alternatively, if the same labelled nucleotide is to be used frequently, a stock 5 × nucleotide mix can be prepared containing each dNTP at 100 µM except that to be used as label. For example, if [α-^{32}P]dCTP is to be used, the 5 × nucleotide mix contains 100 µM dATP, 100 µM dGTP and 100 µM TTP.
2. Dissolve the DNA to be labelled in either distilled water or TE buffer to a concentration of 2–25 µg/ml. Place all solutions, except for the enzyme, at room temperature. Leave the enzyme in a freezer until required, and return immediately after use.
3. Place 25 ng of the DNA to be labelled into a microcentrifuge tube and to it add 5 µl of primers and the appropriate volume of water to give a final volume of 50 µl in the labelling reaction. Denature by heating to 95–100 °C for 5 min in a boiling water bath. Spin the tube briefly in a microcentrifuge to bring the contents to the bottom of the tube.
4. Keeping the tube at room temperature, add the following in the order given:

10 × labelling buffer	5 µl
5 × nucleotide mix	10 µl
[α-^{32}P]dNTP (3000 Ci/mmol, 10 mCi/ml)	5 µl
Klenow enzyme	2 units

5. Mix gently by pipetting up and down, and cap the tube. Spin for a few seconds in a microcentrifuge to bring the contents to the bottom of the tube. Avoid vigorous mixing as this can cause loss of enzyme activity.
6. Incubate at 37 °C for 10 min.
7. Terminate the reaction by the addition of 5 µl of 0.2 M EDTA.
8. A small aliquot (1–2 µl) can be removed for determination of incorporation as described in 'Determination of incorporation' below.
9. For use in hybridization, the DNA should be denatured by heating to 95–100 °C for 5 min, then chilled on ice.

Calculation of probe specific activity
During random primer labelling there is net synthesis of DNA while the initial substrate remains unlabelled. Both can participate in the subsequent hybridization, so:

Probe yield = ng template DNA + ng DNA synthesized.

Average molecular weight of a nucleotide in DNA is 350, so for a labelled nucleotide of specific activity $X \times 10^3$ Ci/mmol:

$$\text{ng DNA synthesized} = \frac{\mu\text{Ci incorporated} \times 0.35 \times 4}{X}$$

Note that a multiplication factor of 4 is included as there are four nucleotides, only one of which is labelled. This assumes equal abundance of each nucleotide.

Once the probe yield has been calculated, the specific activity can be determined:

$$\text{specific activity (dpm/µg)} = \frac{\text{total activity incorporated (dpm)}}{\text{probe yield (µg)}}$$

Specific example
Assume 70% incorporation of labelled nucleotide at 3000 Ci/mmol with 25 ng template DNA.

Amount of labelled nucleotide incorporated = 50 µCi × 0.7 = 35 µCi.

$$\text{ng DNA synthesized} = \frac{35 \times 0.35 \times 4}{3} = 16.3 \text{ ng}$$

Total DNA = 25 + 16.3 = 41.3 ng = 0.041 µg.

As 1 µCi = 2.2 × 10^6 dpm, total activity incorporated = 35 × 2.2 × 10^6 dpm = 7.7 × 10^{7}dpm.

$$\text{So probe specific activity} = \frac{7.7 \times 10^7}{0.041} = 1.9 \times 10^9 \text{ dpm/µg.}$$

Notes

2. DNA in most restriction enzyme buffers can also be used.
3a. When labelling DNA in low melting point agarose, first place the tube containing the stock DNA in a boiling water bath for 30 s to melt the agarose before removing the required volume. The volume of low melting point agarose DNA should not exceed 25 µl in a 50 µl reaction.
3b. More than 25 ng may be labelled in a 50 µl reaction but the highest specific activity is obtained if the reaction is scaled up appropriately.
6. Purified DNA can be labelled to high specific activity in 10 min at 37 °C but, if desired, can be labelled for up to 1 h at this temperature. When labelling DNA in low melting point agarose, longer incubations of 15–30 min at 37 °C are required for optimum labelling. Longer incubations of up to 60 min are required when nucleotide analogues such as [^{35}S]dNTPαS are used. Reactions can also be routinely left to proceed overnight at room temperature with any radioisotope.
8a. With purified DNA template, the reaction should give > 60% incorporation of label, equivalent to a specific activity of > 1.7 × 10^9 dpm/µg.

Solutions

– 10 × labelling buffer (600 mM Tris-HCl, pH 7.8, 100 mM MgCl$_2$, 100 mM 2-mercaptoethanol)
– TE buffer, pH 8.0 (10 mM Tris-HCl, pH 8.0, 1 mM EDTA)
– 300 µM dATP in TE buffer
– 300 µM dCTP in TE buffer
– 300 µM dGTP in TE buffer
– 300 µM TTP in TE buffer
– [α-^{32}P]dNTP at 3000 Ci/mmol, 10 mCi/ml
– DNA polymerase I (Klenow fragment)
– Random nonamer primers dissolved at a concentration of 30

absorbance units/ml in TE buffer containing nuclease-free bovine serum albumin (BSA) at 4 mg/ml.
- 0.2 M EDTA, pH 8.0

2. Nick translation

The nick translation reaction [15, 17] was developed several years before the random primer approach and has to a large extent been superseded by it. However, it is still a widely used procedure and is particularly suitable for labelling large quantities of DNA. The reaction uses two enzymes, deoxyribonuclease I (DNAse I) and DNA polymerase I, both from *E. coli*. Single strand nicks are introduced into a double-stranded DNA template by the action of DNAse I then, starting from the nicks, the 5'-3' exonuclease activity of DNA polymerase I progressively removes nucleotides from the exposed 5'-end while the polymerase activity replaces them in the same direction from the exposed 3'-end, incorporating radioactive nucleotides present in the reaction mix. The position of the nick therefore moves along the DNA in a 5'-3' direction, hence the term nick translation.

Compared to random primer labelling, nick translation has the advantages that an initial denaturation step for double stranded template is unnecessary and the probe concentration for hybridization is more readily determined as there is no net synthesis of DNA during the reaction. The reaction is often used to label relatively large (µg) amounts of DNA, although specific activities approximately 10-fold lower than with the standard random primer labelling reaction detailed above are generally obtained. However, it is possible to adapt the reaction to label as little as 25-50 ng DNA giving specific activities greater than 10^9 dpm/µg within 30 min. Because of the presence of DNAse I and the 5'-3' exonuclease activity of DNA polymerase I, it is necessary to control the reaction time and temperature carefully to avoid removal of incorporated nucleotides. This is generally achieved by incubating the reaction at 15–16 °C for a maximum of 2–3 h, or less when using lower amounts of template. Probe size is determined largely by the DNAse I to template ratio and, again, commercially available kits are useful as a source of pre-optimised reagents (for example, Nick Translation Kit, Amersham International, plc).

Steps in the procedure
1. Prepare an appropriate nucleotide mix from the 300 μM stocks by mixing 3.3 μl of each of three dNTP stocks, excluding that to be used as label. Alternatively, if the same labelled nucleotide is to be used frequently, a stock 5 × nucleotide mix can be prepared containing each dNTP at 100 μM except that to be used as label. For example, if [α-^{32}P]dCTP is to be used, the 5 × nucleotide mix contains 100 μM dATP, 100 μM dGTP and 100 μM TTP.
2. Dissolve the DNA to be labelled in either distilled water or TE buffer to a concentration of 5–50 μg/ml. Place all solutions, except for the enzyme mix, at room temperature. Leave the enzyme mix in a freezer until required and return immediately after use.
3. Add the following in the order given to a microcentrifuge tube on ice:

DNA	50–500 ng
5 × nucleotide mix (dATP, dGTP, TTP)	10 μl
Water to a final reaction volume of 50 μl	
10 × labelling buffer	5 μl
[α-^{32}P]dNTP (3000 Ci/mmol, 10 mCi/ml)	10 μl
Enzyme mix	5 μl

4. Mix gently by pipetting up and down, and cap the tube. Spin for a few s in a microcentrifuge to bring the contents to the bottom of the tube. Avoid vigorous mixing as this can cause loss of enzyme activity.
5. Incubate at 15 °C for 60 min.
6. Terminate the reaction by the addition of 5 μl of 0.2 M EDTA.
7. A small aliquot (1–2 μl) can be removed for determination of incorporation as described in 'Determination of incorporation' below.
8. For use in hybridization, the DNA probe should be denatured by heating to 95–100 °C for 5 min, then chilled on ice.

Calculation of probe specific activity
During nick translation, nucleotides are excised and replaced and there is usually no net synthesis of DNA. Thus the probe specific activity is calculated simply as:

$$\text{specific activity (dpm/µg)} = \frac{\text{total activity incorporated (dpm)}}{\text{amount of template DNA added (µg)}}$$

Thus, 50% incorporation in the above reaction would give a specific activity of 2×10^8 dpm/µg with 500 ng template DNA and 2×10^9 dpm/µg with 50 ng (as 1 µCi = 2.2×10^6 dpm/µg).

Notes

2. Any double stranded DNA can be used as template for the nick translation reaction. DNA solutions which are too dilute to be used should be concentrated by ethanol precipitation and redissolved in an appropriate volume of water of TE buffer.
3. This protocol may be used to label up to 2 µg of DNA but the volume of $[\alpha\text{-}^{32}P]dNTP$ should be increased to 20 µl (66 pmol). For lower amounts of DNA (50–100 ng), 5 µl labelled nucleotide will be adequate. Lower specific activity nucleotides (for example, 400 Ci/mmol) can be used, particularly with the higher amounts of DNA, but the rate of reaction will be lower.
5. Shorter reaction times (30 min) can be used for lower amounts of DNA (50–100 ng), while longer times (up to 3 hours) can be used with higher amounts (250–500 ng) or if using nucleotide analogues such as $[^{35}S]dATP\alpha S$. Careful control of temperature is necessary to avoid the generation of 'snap back' regions in the labelled probe
6. With purified DNA template, the reaction should give > 50% incorporation of label.

Solutions

- 10 × labelling buffer (600 mM Tris-HCl, pH 7.8, 100 mM $MgCl_2$, 100 mM 2-mercaptoethanol)
- TE buffer, pH 8.0 (10 mM Tris-HCl, pH 8.0, 1 mM EDTA)
- 300 µM dATP in TE buffer
- 300 µM dCTP in TE buffer
- 300 µM dGTP in TE buffer
- 300 µM TTP in TE buffer
- $[\alpha\text{-}^{32}P]dNTP$ at 3000 Ci/mmol, 10 mCi/ml
- Enzyme mix (0.006 units/ml DNaseI and 500 units/ml DNA polymerase I)
- 0.2 M EDTA, pH 8.0.

3. Labelling with bacteriophage RNA polymerases

This approach takes advantage of the high degree of specificity for their own promoters shown by a number of bacteriophage RNA polymerases, to allow the generation of high specific activity RNA probes [7, 13]. By cloning a DNA sequence downstream of an RNA polymerase promoter in a suitable vector, it is possible to use the polymerase to synthesize large amounts of RNA transcript from the insert in the presence of the ribonucleoside triphosphates ATP, CTP, GTP and UTP. By replacing one of the nucleotides with a radiolabelled equivalent, it is possible to synthesize smaller amounts of highly labelled transcript.

The most frequently used polymerases are from the bacteriophages SP6, T7 and T3, and a variety of vectors are available incorporating one or more of their promoters next to multiple cloning sites for a variety of restriction enzymes. Transcription of vector sequence can be avoided by cutting the vector with a restriction enzyme just downstream of the insert, so that a run-off transcript is produced. Frequently the cloning site is flanked by two different promoters in opposite orientation so that, by choice of polymerase, it is possible to produce a transcript of either strand of the insert (see Fig. 2). This can provide a valuable negative control, for example for *in situ* hybridization to cytoplasmic mRNA, where the transcript from one strand (the non-coding strand) is complementary to the mRNA and will therefore hybridize, while probe from the other strand is identical to the mRNA and will not hybridize.

A relatively high concentration of template (1–2 µg) is used in standard labelling reactions but, as this is intact vector rather than purified insert, there is usually little difficulty in obtaining adequate quantities. Transcript, and hence probe, size will vary with the size of the insert, usually between 100 bases and several kilobases. To avoid a high proportion of prematurely-terminated transcripts, the chemical concentration of the radioactively labelled nucleotide should be at least equal to that of the K_m of the enzyme for that nucleotide, approximately 12 µM for most nucleotides. UTP and CTP are most frequently used as the labelling nucleotide because GTP is involved in the transcription initiation step and ATP has a higher K_m with some enzymes. The non-radioactively labelled nucleotides are,

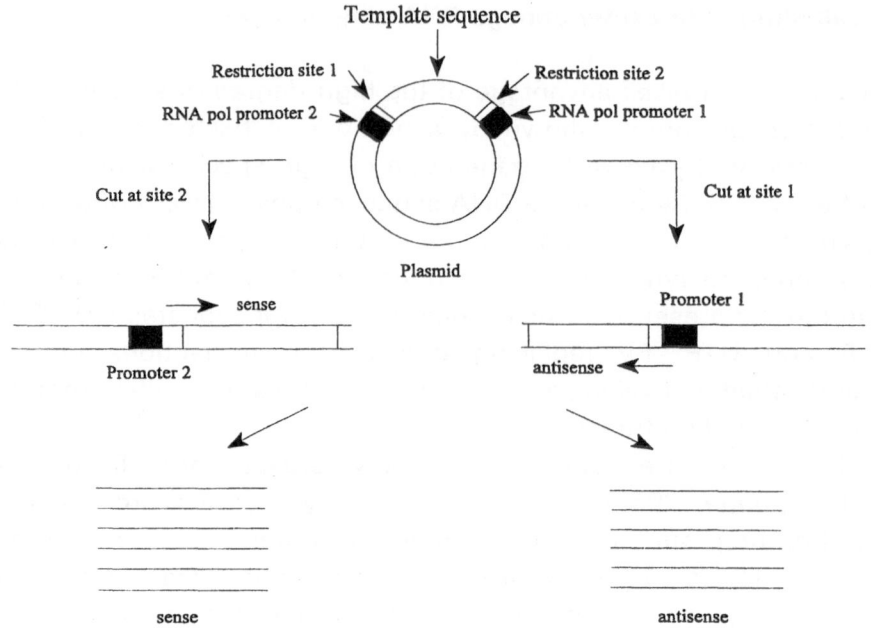

Figure 2. RNA polymerase-based probe labelling.

as with other uniform labelling reactions, present in excess. Under these conditions, it is possible to obtain up to 80% incorporation of labelled nucleotide within 1 h at 37–40 °C, generating probes of approximately 1–2×10^9 dpm/µg with ^{32}P-labelled nucleotide and 5×10^8 dpm/µg with ^{35}S-labelled nucleotide. The method has the advantage that strand-specific probes free of vector sequence can be readily made, although it is necessary initially to clone the insert into a suitable vector.

RNA probes have a number of properties that distinguish them from DNA probes. They are single-stranded and hence do not require denaturation before hybridization, and will not re-anneal to themselves in solution during hybridization. In the presence of formamide, RNA-RNA and RNA-DNA hybrids are more stable than DNA-DNA hybrids, and it is often found that the optimum ratio of signal to background for RNA probe hybridizations is obtained at a higher temperature than with an equivalent DNA probe. In general, RNA probes are less frequently used than DNA probes for membrane hybridizations, but are often used in particular applications such as RNase mapping and *in situ* hybridization. For the latter application the method is most commonly used with ^{35}S-labelled nucleotides, and increasingly with ^{33}P-labelled nucleotides. Accordingly, the following protocol describes the incorporation of [35S]UTPαS. Probe can be used directly in a hybridization following the transcription stage. However, lower backgrounds may result if the initial DNA template is removed, and for *in situ* hybridization, optimal probe size is 200–400 bases so, for longer transcripts it is advisable to carry out an alkaline hydrolysis before hybridization. Protocols for both of these stages are therefore also given below.

Steps in the procedure

In vitro transcription
1. Dissolve the DNA in either distilled water or TE buffer at a concentration of approximately 500 µg/ml. Place all solutions, except for the enzyme, on ice to thaw. Leave the enzyme in a freezer until required and return immediately after use.
2. Add the following in the order given to a microcentrifuge tube at room temperature.

Transcription buffer	4	µl
DTT solution	1	µl
HPRI	20	units
ATP	0.5	µl
CTP	0.5	µl
GTP	0.5	µl
Linearized template DNA	1	µg
[35S]UTPαS (> 1000 Ci/mmol, 20 mCi/ml)	10	µl
RNA polymerase (SP6, T7 or T3)	20	units

3. Mix gently by pipetting up and down, and cap the tube. Spin for a few s in a microcentrifuge to bring the contents to the bottom of the tube. Avoid vigorous mixing as this can cause loss of enzyme activity.
4. Incubate for a minimum of 1 h. For SP6 polymerase incubate at 40 °C and for T7 and T3 incubate at 37 °C.
5. If the reaction is to be used in a hybridization with no further processing, it may be terminated by the addition of 2 µl 0.2 M EDTA and a small aliquot (1–2 µl) can be removed for determination of incorporation as described in 'Determination of incorporation'

below. Denaturation is not required for use in hybridization as the probe is single stranded.

Template removal

6. If the DNA template is to be removed before hybridization, proceed from 4. above. Do not add EDTA as it will inhibit the action of DNAse I.
7. Add 10 units of RNase-free DNase I to the labelling mix from 4. above.
8. Incubate at 37 °C for 10 min.
9. If the reaction is to be used in a hybridization at this stage, it may be terminated by the addition of 2 μl 0.2 M EDTA and a small aliquot (1–2 μl) can be removed for determination of incorporation as described in 'Determination of incorporation' below. Denaturation is not required for use in hybridization as the probe is single stranded.

Alkaline hydrolysis

10. To the reaction mix from 8. above add 20 μl 0.4 M $NaHCO_3$, 20 μl 0.6 M Na_2CO_3 and 60 μl sterile water. Mix gently.
11. Incubate at 60 °C for a time based on the transcript length and the probe size required. This can be determined from t(min) =

$$\frac{L_0 - L_f}{k \, L_0 \, L_f}$$

where t = incubation time in minutes, L_0 = primary transcript length (kbases), L_f = average probe length required (kbases) and the rate constant k = 0.11 cuts/kbase/minute.

12. Add 1.3 μl glacial acetic acid, 20 μl 3 M sodium acetate solution, 2 μl yeast tRNA solution and 500 μl ethanol. Precipitate the RNA at −20 °C for at least 2 h.
13. Pellet the RNA in a microcentrifuge at 13,000 rpm for 15 min. Remove supernatant.
14. Rinse the pellet in 70% (v/v) ethanol at −20 °C. Centrifuge for 5 min at 13,000 rpm. Remove the supernatant.
15. Redissolve the RNA probe in sterile water (or sterile 0.1 M dithiothreitol for ^{35}S-labelled probes) at a concentration of 10–20 × that required in the hybridization procedure. Store the RNA

probe at –20 °C. Denaturation is not required for use in hybridization as the probe is single stranded.

Calculation of probe specific activity
As unlabelled vector DNA does not participate in the hybridization,
the specific activity of the probe depends solely on the specific activity of the labelled nucleotide, while probe yield can be calculated
from the % incorporation of label. The specific activity of the probe
is independent of labelling efficiency.

Specific example
Assume 70% incorporation of 200 µCi of [^{35}S]UTPαS at > 1000
Ci/mmol.

Calculation of probe yield
Amount of labelled nucleotide incorporated is 200 × 0.7 = 140 µCi.
This is equivalent to 0.14 nmol (1000 Ci = 1 mmol), or 0.14 × 350 = 49
ng of incorporated label (molecular weight of a nucleoside
monophosphate in DNA is approximately 350). With a single labelled
nucleotide species this is equivalent to 49 × 4 = 196 ng of probe. Note
that the amount of probe is significantly less than the amount of
whole vector present in the reaction.

Calculation of specific activity
1 mCi = 2.2 × 10^9 dpm, so the specific activity of the [^{35}S]UTPαS can
be expressed as

$$\frac{1000 \times 2.2 \times 10^9}{350} = 6.3 \times 10^9 \text{ dpm/µg.}$$

Therefore, the specific activity of the probe = 6.3 × 10^9/4 = 1.6 × 10^9
dpm/µg. If unlabelled UTP is added to the reaction, the specific
activity of the probe will be reduced, although the yield may be improved.

Notes
2a. The reaction must be set up at room temperature to avoid precipitation of the
    template DNA by spermidine in the transcription buffer.
2b. There are several formulations of [^{35}S]UTPαS available which are appropriate for
    use with phage polymerase labelling reactions. However with some high specific

activity formulations, it is difficult to achieve the required nucleotide concentration of 12.5 µM, based on the apparent Km of the enzyme for UTP, to ensure a predominance of full length transcripts. As transcripts may therefore be of variable length, it is then more difficult to predict optimum conditions for alkali digestion. This concentration is more readily achieved with nucleotides of lower specific activity (for example, 400 or 800 Ci/mmol), but the specific activity of the probe is correspondingly lower. The formulation suggested (1000 Ci/mmol, 20 mCi/ml) is at a sufficiently high concentration to allow 12.5 µM to be achieved in the reaction whilst also maximising specific activity. Higher specific activities for *in situ* hybridization are achievable with [α-^{33}P]UTP, which is available at 1000–3000 Ci/mmol, 20 mCi/ml (Amersham code BF1002), but the transcript length is likely to be reduced due to low nucleotide concentration using 10 µl per 20 µl reaction. As there are currently no alternative formulations, the hydrolysis time must be reoptimized when using this nucleotide. As a general rule, the hydrolysis time can be calculated for the theoretical full length transcript and then halved. Several formulations of [α-^{32}P]UTP are also available. Again, it is not possible to achieve 12.5 µM in the labelling reaction with the highest specific activities (3000 Ci/mmol, 10 mCi/ml, Amersham code PB10203), but lower specific activity alternatives (for example, 800 Ci/mmol, 20 mCi/ml, Amersham code PB20383) produce probes of adequate specific activity (1.3×10^9 dpm/µg) for most *in situ* and membrane hybridizations.

2c. A similar reaction can be set up with the same concentration of labelled CTP with unlabelled ATP, GTP and UTP. ATP and GTP are not generally recommended as labels.

2d. Plasmids containing an insert to be used as the template for RNA probe production should be linearized with an appropriate restriction enzyme. The restriction site should ideally be as close as possible to the end of the insert sequence, to avoid production of labelled plasmid sequences which may cross-hybridize with target sequences.

4. Nucleotide analogues such as [^{35}S]UTPαS are incorporated less efficiently than normal nucleotides. With [α-^{32}P]UTP and [α-^{33}P]UTP label it is possible to use 4–10 units of polymerase. Under these conditions, a reaction time of 1 hour is adequate for all nucleotides.

5a. With purified DNA template, the reaction should give > 70% incorporation of label. For many membrane hybridization applications it will be found that satisfactory results will be obtained using probe at this stage. For *in situ* hybridization however, template removal and alkaline hydrolysis is recommended.

5b. RNA probe at this stage should not be denatured before hybridization as this will also render the unlabelled template single stranded. This is generally present in excess over labelled probe and will also hybridize with target.

7. If a DNAse is used that is not known to be RNase-free, then an additional 20 units HPRI should be added.

8. This stage is to reduce the size of the probe by alkaline hydrolysis for use in *in situ* hybridization. The optimum probe size is between 200 and 400 bases but should be determined empirically for each application.

12. The precipitation can be aided by the addition of 2 µl of 10 mg/ml yeast tRNA.

15. For longer-term storage, RNA probes can be stored at –70 °C.

Solutions

- TE buffer (10 mM Tris-HCl, pH 8.0, 1 mM EDTA).
- Transcription buffer (200 mM Tris-HCl, pH 7.5, 30 mM $MgCl_2$, 10 mM spermidine, 0.05% (w/v) BSA).
- 20 mM ATP
- 20 mM GTP
- 20 mM CTP
- Human placental ribonuclease inhibitor (HPRI) (for example, Amersham code, E2310Y)
- 0.2 M dithiothreitol (DTT), freshly prepared (30.8 mg DTT in 1 ml sterile water).
- Linearized DNA template containing either an SP6, T7 or T3 RNA polymerase promoter upstream of the sequence to be transcribed.
- [^{35}S]UTPαS, > 1000 Ci/mmol, 20 mCi/ml (Amersham code, SJ603).
- RNA polymerase as appropriate for promoter and probe sequence required (SP6, Amersham code E2520Y; T7, code E0707Y; T3, United States Biochemicals, code 70051).
- 0.2 M EDTA, pH 8.0
- RNase-free DNase I (for example, United States Biochemicals, code 14367) freshly diluted to 10 units/80 µl in sterile water.
- 0.4 M $NaHCO_3$ (3.66 g in 100 ml water). Sterilize by autoclaving.
- 0.6 M Na_2CO_3 (6.36 g in 100 ml water). Sterilize by autoclaving.
- Sterile water
- Glacial acetic acid
- 3 M sodium acetate (24.6 g anhydrous sodium acetate dissolved in 90 ml water. pH adjusted to 5.2 by the addition of glacial acetic acid. Made up to 100 ml and sterilized by autoclaving).
- 10 mg/ml yeast tRNA (for example, Sigma code R8759)
- Ethanol

4. 3'-end labelling

In the presence of deoxynucleoside triphosphates, the enzyme terminal deoxynucleotidyl transferase (TdT) will introduce a series of nucleotides at the 3′-end of DNA molecules in a manner that is not dependent on the presence of a template strand [3]. By including only a single nucleotide species, a 3′-homopolymer tail can be synthesized, the length of which can be controlled by a number of factors such as reaction time, divalent cation species and the molar ratio of nucleotides to 3′-ends available for labelling. If a radiolabelled dNTP is used, the addition of multiple residues will increase the specific activity of the probe, and hence maximise the sensitivity of detection. However, in some cases, the presence of too many additional nucleotide residues can reduce the sequence specificity of the probe during hybridization. The use of a 2′,3′-dideoxynucleotide, available as [α-^{32}P]ddATP (Amersham, codes PB10233 and PB10235), restricts the addition to a single residue as there is no 3′-hydroxyl group available for formation of a phosphodiester bond to a further nucleotide. However, for some applications, it may be found necessary to introduce multiple labelled residues and, in these instances, the length of the tail can be controlled by the molar ratio of template and nucleotide to avoid problems of hybridization specificity.

For preparation of oligonucleotide probes for *in situ* hybridization it is possible to use either [α-^{33}P]dATP (Amersham, code BF1001) or [^{35}S]dATPαS which is available in a formulation specifically designed for use in 3′-end labelling reactions (Amersham, code SJ1334). The protocol below details the use of [^{35}S]dATPαS for labelling an oligonucleotide. The reaction uses 50 pmols labelled nucleotide and 10 pmols oligonucleotide, so that only a short chain of labelled residues will be introduced. The 3′-end labelling reaction will also label double stranded DNA, with more efficient incorporation reported for molecules with protruding 3′-ends. The equation below demonstrates how to calculate the concentration of 3′-ends present in samples containing double stranded DNA of defined length:

$$\text{picomols of 3′-ends/µg DNA} = \frac{2 \times 10^6}{660 \times \text{length (base pairs)}}$$

If the sample is a restriction digest of DNA containing a variable set of lengths, the amount of 3′-ends can be determined by multiplying the number of picomols of 3′-ends for the undigested sample by twice the number of restriction enzyme digestion sites.

Steps in the procedure

1. Place all solutions except for the enzyme on ice to thaw. Leave the enzyme in a freezer until required and return immediately after use.
2. Add the following in the order given to a microcentrifuge tube on ice:

Cacodylate buffer	5 µl
Oligonucleotide	10 pmols
Water to a final reaction volume of 50 µl	
[^{35}S]dATPαS (Amersham, code SJ1334, > 1000 Ci/mmol, 10 mCi/ml)	5 µl
Terminal deoxynucleotidyl transferase	10 units

3. Mix gently by pipetting up and down, and cap the tube. Spin for a few seconds in a microcentrifuge to bring the contents to the bottom of the tube. Avoid vigorous mixing as this can cause loss of enzyme activity.
4. Incubate for 1–2 h at 37 °C.
5. Terminate the reaction by the addition of 5 µl of 0.2 M EDTA.
6. A small aliquot (1–2 µl) can be removed for determination of incorporation as described in 'Determination of incorporation' below.
7. If a single stranded oligonucleotide is labelled then the probe can be used directly in a hybridization without denaturation. If the substrate is double stranded, then the probe should be denatured by heating to 95–100 °C for 5 min then chilled on ice. For longer term storage the probe can be kept at –20 °C.

Calculation of specific activity

$$\text{Specific activity (dpm/µg)} = \frac{\text{total activity incorporated (dpm)}}{\text{amount of substrate added (µg)}}$$

An incorporation of 50% in the above reaction would give the following specific activity:

Total activity incorporated = $0.5 \times 50 \times 2.2 \times 10^6 = 5.5 \times 10^7$ dpm

Amount of substrate added = 50 ng = 0.05 µg (for a 15-mer oligonucleotide).

Specific activity = 1.1×10^9 dpm/µg.

Notes

2a. It is recommended that the probe to be labelled is dissolved or diluted in sterile distilled water.

2b. Calculation of oligonucleotide concentration from absorbance (A) varies according to the base composition. The molar extinction coefficient (ε) at 260 nm (pH 8.0) for a given oligonucleotide can be obtained by summing the contribution of each nucleotide: A = 15200, C = 7050, G = 12010 and T = 8400. Concentration (mol/l) = A_{260}/ε. The weight of oligonucleotide corresponding to 10 pmol is dependent on the length of the sequence. As an approximate estimate, for each additional residue the amount required to give 10 pmol increases by 3.3 ng. Thus, for example, 50 ng of a 15-mer oligonucleotide is equivalent to 10 pmol, while 166 ng of a 50-mer is required.

2c. The reaction is illustrated for use with a formulation of [^{35}S]dATPαS that has been optimised for efficient incorporation in 3'-end labelling reactions. Other formulations may contain levels of dithiothreitol that inhibit the reaction. ^{32}P- and ^{33}P-labelled nucleotides do not contain dithiothreitol and may be used successfully in this reaction.

5. A typical incorporation obtainable using an oligonucleotide in the reaction detailed above would be approximately 50% labelled nucleotide.

Solutions

- Cacodylate buffer (1.4 M sodium cacodylate, pH 7.2, 10 mM cobalt (II) chloride, 1 mM dithiothreitol). *Warning:* cacodylate is an arsenic-containing compound which is highly toxic by contact with skin or if swallowed. It may be a carcinogen and with danger of cumulative effects. Cobalt (II) chloride is also harmful if swallowed, inhaled or absorbed through the skin and may cause eye and skin irritation. Manufacturers' safety data sheets should be consulted for these compounds and appropriate safe handling procedures followed.
- [^{35}S]dATPαS, 1000 Ci/mmol, 10 mCi/ml (Amersham, code SJ1334)
- Terminal deoxynucleotidyl transferase
- 0.2 M EDTA, pH 8.0

5. 5'-end labelling

Unlike other enzymes previously discussed, polynucleotide kinase (PNK) does not catalyze the incorporation of a nucleotide but the transfer of a phosphate group from the terminal γ-position of a ribonucleoside triphosphate (most commonly ATP) to the 5'-end of the terminal nucleotide of a nucleic acid molecule which may be either DNA, RNA or a chemically synthesized oligonucleotide [16]. This can be achieved in one of two ways. If the 5'-end contains a 3'-hydroxyl group, as is the case with chemically synthesized oligonucleotides or with DNA that has been dephosphorylated with alkaline phosphatase [19], a direct transfer of phosphate can take place (the forward reaction). With [γ-^{32}P]ATP or [γ-^{33}P]ATP as donor, probe can be labelled with a high degree of efficiency. Transfer of a phosphorothioate group from [^{35}S]ATPγS occurs with lower efficiency and it is necessary to use a higher enzyme concentration and longer reaction times for effective labelling. If the 5'-end contains a phosphate group, then PNK can be used to catalyze an exchange reaction [2] which occurs, in the presence of an excess of ADP, with a lower efficiency. The 5'-phosphate of the nucleic acid is transferred to the ADP converting it to ATP, then the 5'-end is rephosphorylated using a nucleoside triphosphate donor, again usually ATP, as in the forward reaction.

Unlike 3'-end labelling, it is not possible to incorporate multiple labels by 5'-end labelling and so for probe-based hybridization applications ^{32}P remains the most frequently used label, although both ^{35}S- and, increasingly, ^{33}P-labelled probes are used in *in situ* hybridization. 5'-end labelled DNA and RNA have been traditionally used in chemical sequencing methods. An advantage of 5'-end labelling is that oligonucleotides labelled in this way can, unlike 3'-end labelled oligonucleotides, be used as primers, for example in sequencing, PCR and probe capture experiments.

The following protocol can be employed for labelling both oligonucleotides and long DNA or RNA using the forward reaction. It includes details of the 5'-dephosphorylation of DNA with calf intestinal alkaline phosphatase (CIAP). If an oligonucleotide is to be labelled, dephosphorylation can be omitted and the labelling started from step 6. This is followed by an abbreviated protocol for the exchange reaction which can be used for long DNA and RNA molecules with phosphorylated 5'-ends.

A. Forward reaction

Steps in the procedure

1. Dissolve the DNA to be labelled at a concentration of 10 pmols 5′-ends in 5–35 µl 10 mM Tris-HCl, pH 8.0. Place all solutions, except for the enzymes, on ice to thaw. Leave the enzymes in a freezer until required, and return immediately after use.

2. Add the following in the order given to a microcentrifuge tube on ice:

DNA/RNA/oligonucleotide	10 pmol ends
10 × CIAP buffer	5 µl
Water to final reaction volume of 50 µl	
Calf intestinal alkaline phosphatase	0.05 units

3. Mix gently by pipetting up and down, and cap the tube. Spin for a few s in a microcentrifuge to bring the contents to the bottom of the tube. Avoid vigorous mixing as this can cause loss of enzyme activity.

4. Incubate at 37 °C for 30 min for duplex DNA or at 55 °C for 30 min for RNA.

5. Briefly centrifuge as in 3. above, then add an equal volume of buffer-saturated phenol. Extract twice with phenol and twice with chloroform-isoamyl alcohol or ether. Add a one tenth volume of 0.5 M NaCl followed by 2 volumes of cold absolute ethanol for DNA or 3 volumes for RNA. Precipitate the nucleic acid at −20 °C overnight or at −80 °C for 30 min.

6. Resuspend the pellet in 10 µl TE buffer and add the following in the order given to a microcentrifuge tube on ice:

DNA/RNA	10 µl
10 × PNK buffer	5 µl
Water to a final reaction volume of 50 µl	
[γ-^{32}P]ATP, 3000 Ci/mmol, 10 mCi/ml (Amersham, code PB10168)	20 µl
T4 polynucleotide kinase	10 units

7. Mix gently by pipetting up and down, and cap the tube. Spin for a few s in a microcentrifuge to bring the contents to the bottom of the tube. Avoid vigorous mixing as this can cause loss of enzyme activity.

8. Incubate at 37 °C for 30-60 min.
9. Terminate the reaction by the addition of 5 µl 0.2 M EDTA
10. A small aliquot (1–2 µl) can be removed for determination of incorporation as described in 'Determination of incorporation' below.
11. If a single stranded oligonucleotide is labelled then the probe can be used directly in a hybridization without denaturation. If the substrate is double stranded, then the probe should be denatured by heating to 95–100 °C for 5 min then chilled on ice. For longer term storage the probe can be kept at –20 °C.

Calculation of probe specific activity
Specific activity of 5'-end labelled probes can be calculated as for 3'-end labelled probes described in section 4. above.

Notes
2. Calculation of oligonucleotide concentration from absorbance (A) varies according to the base composition. The molar extinction coefficient (ε) at 260 nm (pH 8.0) for a given oligonucleotide can be obtained by summing the contribution of each nucleotide: A = 15200, C = 7050, G = 12010 and T = 8400. Concentration (mol/litre) = A_{260}/ε. The weight of oligonucleotide corresponding to 10 pmol is dependent on the length of the sequence. As an approximate estimate, for each additional residue the amount required to give 10 pmol increases by 3.3 ng. Thus, for example, 50 ng of a 15-mer oligonucleotide is equivalent to 10 pmol, while 166 ng of a 50-mer is required.

6a. The reaction contains 10 mol 5'-ends and 67 pmols [γ-^{32}P]ATP in order to maximise the specific activity of the probe. As the label is in molar excess, the % incorporation will be low and removal of unincorporated label is advisable. This can be achieved by precipitation [19], by gel filtration chromatography as described in 'Removal of unincorporated label' below or alternatively, for oligonucleotides, by thin layer chromatography or polyacrylamide gel electrophoresis in the presence of 7 M urea [21]. This has the additional benefit of separating labelled and unlabelled oligonucleotide based on the different mobilities of 5'-phosphorylated and 5'-hydroxyl oligonucleotides, although it will be found that, for most hybridization applications, this further purification will not be necessary. A more economical use of labelled nucleotide can be achieved by the addition of a lower amount of nucleotide (20–50 µCi), although a lower proportion of labelled oligonucleotides may result.

6b. Both the forward and exchange reactions are most efficient with single-stranded molecules or double-stranded molecules with protruding 5'-ends. If the DNA molecules have blunt or recessed 5'-ends, then a short incubation at 70 °C followed by rapid chilling on ice prior to labelling may improve efficiency.

Solutions

- 10 × CIAP buffer (0.5 M Tris-HCl, pH 9.0, 10 mM $MgCl_2$, 10 mM $ZnCl_2$, 10 mM spermidine)
- Calf intestinal alkaline phosphatase (CIAP)
- 10 mM Tris-HCl, pH 8.0
- Phenol saturated with TE buffer
- 24:1 (v/v) chloroform:isoamyl alcohol, or ether
- 5 M NaCl
- Ice-cold absolute ethanol
- TE buffer (10 mM Tris-HCl, pH 8.0, 1 mM EDTA)
- 10 × kinase buffer (0.5 M Tris-HCl, pH 7.6, 0.1 M $MgCl_2$, 50 mM dithiothreitol, 1 mM spermidine)
- T4 polynucleotide kinase
- [γ-^{32}P]ATP (3000 Ci/mmol, 10 mCi/ml, Amersham code PB 10168)
- 0.2 M EDTA, pH 8.0

B. Exchange reaction

Steps in the procedure

1. Dissolve the DNA to be labelled at a concentration of 10 pmols 5′-ends in 5–20 µl 10 mM Tris-HCl, pH 8.0. Place all solutions except for the enzyme on ice to thaw. Leave the enzyme in a freezer until required, and return immediately after use.
2. Add the following in the order given to a microcentrifuge tube on ice:

DNA/RNA	10 pmol ends
10 × exchange buffer	5 µl
water to a final reaction volume of 50 µl	
[γ-^{32}P]ATP, 3000 Ci/mmol, 10 mCi/ml	
(Amersham, code PB 10168)	20 µl
T4 polynucleotide kinase	10 units

3. Continue as in steps 7–11 for the forward reaction

Calculation of probe specific activity
As for the forward reaction.

Solutions

- 10 × exchange buffer (0.5 M imidazole-HCl, pH 6.6, 0.1 M MgCl$_2$, 50 mM dithiothreitol, 3 mM ADP, 1 mM spermidine)
- T4 polynucleotide kinase
- [γ-^{32}P]ATP, 3000 Ci/mmol, 10 mCi/ml (Amersham, code PB 10168).

6. PCR-based labelling

The polymerase chain reaction (PCR) allows the amplification of a specific sequence from a heterogeneous population of sequences [18]. A pair of primers, complementary to opposite strands of a DNA template and separated by the sequence to be amplified, are annealed to the DNA sample of interest which has been previously heat-denatured. Synthesis of complementary DNA is then catalyzed by a thermostable DNA polymerase such as Taq polymerase. Following a further heat denaturation stage in which the DNA is again made single-stranded, primers are re-annealed and the synthesis repeated. By carrying out a number of cycles of annealing, synthesis and denaturation, it is possible to produce an enormous amplification of the sequence of interest (approximately 1 million-fold after 20 cycles).

As the specific amplification product can usually be visualised by ethidium bromide staining following gel electrophoresis, the polymerase chain reaction can be used to circumvent a membrane hybridization. Alternatively, a membrane hybridization may be carried out with amplified target so that a lower level of sensitivity is required, an approach amenable to the use of non-radioactively labelled hybridization probes. However, it is also frequently useful to produce a radiolabelled amplification product either as a means of quantifying yield from the PCR or for preparing vector-free probe, possibly from a limited source of template (for example, genomic DNA). A number of approaches are possible. For example, a 5'-end labelled primer can be used if a low specific activity probe is required. Alternatively, a labelled nucleotide can be added either together with the inactive nucleotide mix at the beginning of the PCR or separately for the final cycle(s). The specific activity of the product can be adjusted by altering the ratio of labelled to unlabelled nucleotide. Strand-specific probe can be produced by carrying out asymmetric PCR [11] which uses a limiting concentration of one of the primers. Amplified cDNA probe can also be produced by carrying out an initial reverse transcriptase reaction before the PCR [23]. As optimal PCR conditions are dependent on a number of factors such as template or primer-pair, and because different labelling approaches are possible, a detailed protocol will not be given here. As a rough guide, if using radioactivity to quantify PCR product, a radioactive concentration of 50 µCi/ml of [α-^{32}P]dCTP in a PCR mix containing each cold nucleotide at 200 µM is appropriate. If the PCR product is to be used as a probe, significantly higher amounts of label will be required when incorporating label during the reaction, although care must be taken if reducing any of the cold nucleotide concentrations to maximize specific activity, as this will adversely affect the yield of PCR product.

An alternative, and often more convenient approach, is to carry out a normal cold PCR and then to label the product by a conventional method such as random primer labelling, following purification of the amplified product and removal of the cold nucleotides from the PCR reaction. A number of purification methods, involving precipitation or column purification can be used [19], and a variety of products are commercially available. This approach allows only as much PCR product as necessary to be labelled and can be more economical than labelling during PCR.

The combination of chemical oligonucleotide synthesis and PCR has provided approaches to RNA polymerase-based probe synthesis which avoid both the requirement for subcloning into specific vectors and the limitations on the 3'-end sequence dependent on the presence of an appropriate restriction site for run-off transcription. Phage promoter sequences can be incorporated into synthetic oligonucleotide sequences upstream of the sequence to be transcribed. The promoter sequence can then be rendered double stranded by annealing with a complementary oligonucleotide and *in vitro* transcription initiated [4, 14]. Recently, polymerase promoter sequences have been incorporated into PCR primers, allowing amplification of relatively long sequences prior to transcription. Microgram amounts of PCR products have been generated with T7 or T3 promoters at both ends using as template either plasmid DNA [1], cDNA [22], or a set of self-priming oligonucleotide primers [20]. These products have then been used as templates for T7 or T3 polymerase transcription to generate radiolabelled or non-radiolabelled probes which have been used successfully in both membrane and *in situ* hybridization. This approach offers particular advantages for *in situ* hybridization because probes can be designed to have the optimum size, so avoiding the requirement for alkali denaturation, and they are unlikely to be contaminated either with cross-hybridizing plasmid sequences or with bacterial ribonucleases.

Determination of incorporation

When establishing a new labelling method or using a new preparation of DNA or RNA, it is often advisable to check the efficiency of the labelling reaction by determining the percentage incorporation of label. One of the most commonly used methods is precipitation of nucleic acid by trichloroacetic acid (TCA) followed by capture of the precipitate on a glass fibre or similar filter [19]. However, TCA is highly corrosive, the method requires vacuum filtration and is inconvenient when processing a large number of samples. The method given below utilises absorption of nucleic acid to DEAE paper which avoids the use of corrosive materials, can be used with both long and oligonucleotide probes, and is suitable for either multiple or single samples.

Steps in the procedure
The method as described is suitable for multiple samples but 2.4 cm discs or individual squares of DEAE paper can be used for single samples. Labelled DNA is absorbed to the paper while nucleotides are washed off.
1. Mark an appropriate grid in pencil on a sheet of DE-81 paper (Whatman). For each sample to be analysed four 1 cm × 1 cm squares will be required, two for determining total label in duplicate and two for label incorporated into nucleic acid. The

aliquots for determining incorporated counts should be grouped together on one sheet and those for total counts on a separate sheet. A row of blank squares surrounding the squares for incorporated counts should be included to avoid damage during the washing stages.

2. Dilute an aliquot of labelling reaction into 0.2 M EDTA to give between 10^4 and 10^6 dpm in the final samples counted. Usually a 2 µl aliquot diluted into either 18 or 198 µl 0.2 M EDTA is adequate.

3. Carefully spot 2 µl aliquots of diluted reaction mix onto four squares on the gridded DE-81 paper sheets resting on a double layer of 3 MM paper (Whatman). Allow the sheets to dry for approximately 10 min.

4. Wash the sheet to be used for determining incorporated counts for 2 × 10 min in 2 × SSC (approximately 200 ml for a 10 cm × 5 cm sheet). Briefly rinse in distilled water, then in absolute ethanol. Allow to dry for 10–15 min.

5. Cut out all the squares and count by liquid scintillation. If ^{32}P has been used it is possible to count directly without scintillant (Cerenkov counting).

6. Calculate mean values for incorporated and total counts for each sample and determine percentage incorporation as:

$$\% \text{ incorporation} = \frac{\text{incorporated counts} \times 100}{\text{total counts}}$$

7. If desired, the efficiency of washing of the filters can be checked by carrying out a negative control labelling reaction without substrate or without enzyme and processing as above.

Notes

3. Samples should be applied slowly and carefully to avoid excessive spread.
4. DE-81 paper is fragile in aqueous solution and handling once wet should be avoided. Liquid can be readily poured off into a designated sink without handling the filter itself. The filter can be handled again carefully once in ethanol.

Solutions

- 2 × SSC (0.3 M NaCl, 30 mM sodium citrate)
- 0.2 M EDTA, pH 8.0
- Absolute ethanol

Removal of unincorporated label

For many purposes it will be found unnecessary to remove unincorporated label, especially for incorporation levels of greater than 40–50%. However, if required, label can be removed from the labelled nucleic acid by a variety of methods, including ethanol precipitation and gel filtration chromatography. When using ^{32}P label, these methods can be used as a rough and ready guide to incorporation efficiency by monitoring incorporated label with a β-monitor. More precise measurements can be obtained by β-scintillation counting, which can also be used for ^{33}P, ^{35}S and ^{3}H labels. However, such measurements assume a complete separation of incorporated and unincorporated label, which is not always achieved. More accurate determinations of incorporation efficiency are obtained by following the procedure in the previous section. The following protocols give procedures for both ethanol precipitation and gel filtration chromatography. For short oligonucleotide probes (< 20 bases, approximately), ethanol precipitation can give poor recovery and alternative methods such as precipitation with cetylpyridinium bromide, purification by gel electrophoresis, TLC, HPLC or size exclusion or reverse phase chromatography are recommended [19, 21]. The protocol given below for gel filtration chromatography should be appropriate for most oligonucleotides.

Steps in the procedure

1. Ethanol precipitation. The method is for a labelling reaction of 50 μl. For other volumes, the amounts of reagents should be scaled up or down accordingly.
1. To a 50 μl labelling reaction, add the following:

0.2 M EDTA, pH 8.0	20 μl
5 M NaCl or 3 M sodium acetate, pH 7.0	10 μl
carrier DNA	20 μl
ice-cold absolute ethanol	400 μl

2. Leave to precipitate at –80 °C for 30 min or at –20 °C overnight.
3. Spin in a microcentrifuge for 15 min. Carefully remove the supernatant and dispose of in a suitable manner.
4. Wash the pellet in ice-cold absolute ethanol, centrifuge for 5 min,

pour off or aspirate the supernatant, and dry the remaining pellet under vacuum.

5. Resuspend the labelled nucleic acid pellet in an appropriate volume of TE buffer for use as a probe.

6. If desired, a small aliquot of the solution can be removed for scintillation counting to calculate the specific activity of the probe. With ^{32}P, it is frequently sufficient to check the remaining activity with a β-monitor.

Notes

1a. Carrier DNA is optional, but improves the recovery of probe at low concentrations, for example during random primer labelling. Alternative carriers are total RNA, for example from yeast, or glycogen. As little as 1 μg may be used effectively.

1b. For larger volumes it is adequate to use 3 volumes of ice-cold ethanol, to enable precipitation to be carried out in a single microcentrifuge tube.

Solutions

- 0.2 M EDTA, pH 8.0
- 5 M NaCl or 3 M sodium acetate, pH 7.0
- Carrier DNA (calf thymus or herring sperm DNA at 1 mg/ml)
- Ice-cold absolute ethanol
- TE buffer, pH 8.0 (10 mM Tris-HCl, pH 8.0, 1 mM EDTA)

2. Gel filtration spun column chromatography

1. Plug the bottom of a 1 ml disposable syringe with a small amount of sterile glass wool, preferably siliconized to prevent adsorption of nucleic acid.

2. In the syringe, prepare a column of Sephadex G-50 equilibrated with TE buffer for long probes or Sephadex G-25 for long or oligo probes, by the addition of Sephadex suspension.

3. Place the syringe in a 10 ml conical tube, into which a decapped 1.5 ml microcentrifuge tube has been inserted. Centrifuge at 1,600 *g* for 4 min.

4. Continue to add the Sephadex suspension until the packed volume is 0.9 ml.

5. Add 100 μl TE buffer and recentrifuge as in step 3.

6. Repeat step 5. The volume collected should now be

approximately 100 µl. If the volume is greater than 150 µl, repeat step 5 again.

7. Apply the labelling reaction to the column in a total volume of 100 µl. If the reaction volume is less than 100 µl, it may be diluted in TE buffer.

8. Recentrifuge as above. The unincorporated nucleotides should remain in the syringe, while the labelled probe is eluted in approximately 100 µl.

9. If desired, a small aliquot of the eluate can be removed for scintillation counting to calculate the specific activity of the probe. With ^{32}P, it is frequently sufficient to check the remaining activity with a β-monitor.

Solutions

– Sephadex G-25 or G-50 (Pharmacia) equilibrated in TE buffer
– TE buffer, pH 8.0.

References

1. Bales KR, Hannon K, Smith II CK, Santerre RF (1993) Single-stranded RNA probes generated from PCR-derived DNA templates. Mol Cell Probes 7: 269–275.
2. Berkner KL, Folk WR (1977) Polynucleotide kinase exchange reaction EcoRI cleavage and methylation of DNAs containing modified pyrimidines in the recognition sequence. J Biol Chem 252: 3176–3184.
3. Bollum FJ (1974) Terminal deoxynucleotidyl transferase In: Boyer PD (ed) The Enzymes, Vol 10, pp. 145–171. New York: Academic Press.
4. Brysch W, Hagendorff G, Schlingensiepen K-H (1988) RNA probes, transcribed from synthetic DNA, for *in situ* hybridization. Nucl Acids Res 15: 2333.
5. Cunningham MW, Harris DW, Mundy CR (1990) *In vitro* labelling A. Nucleic acids In Slater RJ (ed) Radioisotopes in Biology: A Practical Approach, Oxford: Oxford University Press, pp. 138–191.
6. Cunningham MW, Simmonds AC (1993) Detection systems In Harper DR (ed) Virology Labfax, Oxford: Bios Scientific Publishers, pp. 123–149.
7. Durrant I, Cunningham MW (1994) Synthesis of riboprobes In Hames BD, Higgins S (eds) Gene Probes: A Practical Approach, Vol 1, Oxford: Oxford University Press, pp. 189–210.
8. Evans MR, Read CA (1992) ^{32}P, ^{33}P and ^{35}S: selecting a label for nucleic acid analysis. Nature 358: 520–521.
9. Feinberg AP, Vogelstein B (1983) A technique for radiolabelling DNA restriction endonuclease fragments to high specific activity. Anal Biochem 132: 6–13.
10. Feinberg AP, Vogelstein B (1984) Addendum: A technique for radiolabelling DNA restriction endonuclease fragments to high specific activity. Anal Biochem 137: 266–267.
11. Gyllensten UB, Erlich HA (1988) Generation of single-stranded DNA by the polymerase

chain reaction and its application to direct sequencing of the HLA-DQA locus. Proc Natl Acad Sci USA 85: 7652–7656.

12. Klenow H, Henningsen I (1970) Selective elimination of the exonuclease activity of the deoxyribonucleic acid polymerase from *Escherichia coli B* by limited proteolysis. Proc Natl Acad Sci USA 65: 168–175.

13. Melton D, Krieg PA, Rebagliati MR, Maniatis T, Zinn K, Green MR (1984) Efficient *in vitro* synthesis of biologically active RNA and RNA hybridization probes from plasmids containing a bacteriophage SP6 promoter. Nucl Acids Res 12: 7035–7056.

14. Milligan JF, Groebe DR, Witherell GW, Uhlenbeck OC (1987) Oligoribonucleotide synthesis using T7 RNA polymerase and synthetic DNA templates. Nucl Acids Res 15: 8783-8798.

15. Mundy CR, Cunningham MW, Read CA (1991) Nucleic acid labelling and detection In Brown TA (ed) Essential Molecular Biology: A Practical Approach, Vol 2. Oxford: Oxford University Press, pp. 57–109.

16. Richardson CC (1981) Bacteriophage T4 polynucleotide kinase In Boyer PD (ed) The Enzymes. Vol 14, New York: Acad Press, pp. 299–314.

17. Rigby PWJ, Dieckmann M, Rhodes C, Berg P (1977) Labelling deoxyribonucleic acid to high specific activity *in vitro* by nick translation with DNA polymerase I. J Mol Biol 113: 237–251.

18. Saiki RK, Scharf S, Faloona F, Mullis KB, Horn GT, Erlich HA, Arnheim N (1985) Enzymatic purification of β-globin genomic sequences and restriction site analysis for diagnosis of sickle cell anemia. Science 230: 1350–1354.

19. Sambrook J, Fritsch EF, Maniatis T (1989) Molecular Cloning: A Laboratory Manual (2nd edition). New York: Cold Spring Harbor Laboratory Press.

20. Sitzmann JH, Le Motte PK (1993) Rapid and efficient generation of PCR-derived riboprobe templates for *in situ* hybridization histochemistry. 41: 773–776.

21. Thein SL, Ehsani A, Wallace RB (1993) The use of synthetic oligonucleotides as specific hybridization probes in the diagnosis of genetic disorders In Davies KE (ed) Human Genetic Disease Analysis: A Practical Approach (2nd edition), Oxford: Oxford University Press pp. 21–33.

22. Urrutia R, McNiven MA, Kachar B (1993) Synthesis of RNA probes by the direct *in vitro* transcription of PCR-generated DNA templates. J Biochem Biophys Methods 26: 113–120.

23. Veres G, Gibbs RA, Scherer SE, Caskey CT (1987) The molecular basis of the sparse fur mouse mutation. Science 237: 415–417.

Molecular Microbial Ecology Manual **2.3.1**: 1-28, 1995.
© 1995 *Kluwer Academic Publishers.*

Detection of nucleic acids by chemiluminescence

MARTIN CUNNINGHAM, BRONWEN HARVEY and
MARTIN HARRIS
Research and Development Department, Amersham International plc, Amersham, Bucks, UK

Introduction

Non-radioactive methods for the detection of nucleic acids by hybrid-ization on blots have found increasing acceptance in recent years. This has been made possible by the development of detection chemistries based on chemiluminescence, which can provide greatly improved sensitivity over earlier colorimetric techniques.

Two sensitive chemiluminescent reactions have been most frequently used with both nucleic acid blots and protein blots, namely enhanced chemiluminescence catalyzed by the enzyme horseradish peroxidase (HRP) and dioxetane-based chemiluminescence catalyzed by alkaline phosphatase (AP). These two reactions have quite distinct properties which, when used in combination with appropriate probe labelling methods, make each especially suitable for particular blotting applications. These are based on i) the kinetics of light output of the two reactions, ii) the relative sensitivity achievable with each system, and iii) the approaches available for probe labelling with either detection method.

The reactions can be used in microbial ecology for the detection and identification of specific microbial nucleic acid sequences. AP-catalyzed chemiluminescence is useful for detecting minute amounts of sequence, at levels which might otherwise require the use of radioactivity (see Applications of chemiluminescent systems). In contrast, HRP-catalyzed enhanced chemiluminescence is particularly appropriate for samples containing moderate to high levels of detectable nucleic acid, for example to detect a relatively prevalent species or in samples where the target nucleic acid has been amplified by PCR. HRP is also a less common contaminating enzyme than AP, so lower backgrounds are likely to be obtained in mixed or impure biological samples when using enhanced chemiluminescence.

The chemiluminescent reactions

1. HRP-catalyzed enhanced chemiluminescence

The enhanced chemiluminescence reaction involves the oxidation of luminol in the presence of hydrogen peroxide in a reaction catalyzed by horseradish peroxidase. Although the oxidative degradation of luminol and other cyclic diacylhydrazides, with the associated emission of light, has been known for a number of years [6], the output of light from this unenhanced reaction is limited. Light output can, however, be stimulated more than 1000-fold by the addition of chemical enhancers, generally derivatives of phenol such as p-iodophenol, which increase both intensity and duration [10].

The general principle of the enhanced chemiluminescence reaction is illustrated in Fig. 1. The conversion of hydrogen peroxide to water causes oxidation of HRP to an activated form which returns to the resting state following gain of electrons from either luminol (unenhanced reaction) or from the enhancer (enhanced reaction). The enhancer radicals so produced may then react with luminol to form luminol radicals. In the subsequent light emitting pathway, the luminol radical is converted to luminol endoperoxide which then degrades to form nitrogen and an electronically excited 3-aminophthalate which, on falling back to the ground state, emits light.

In practice, the peroxide is supplied as a peracid salt. This salt and the enhancer/luminol mixture must be kept separate until immediately before use to avoid gradual chemical degradation of the substrates. Individually, the two substrate formulations can be stored at 4 °C for at least 3 months. The maximum emission of light in the enhanced chemiluminescence

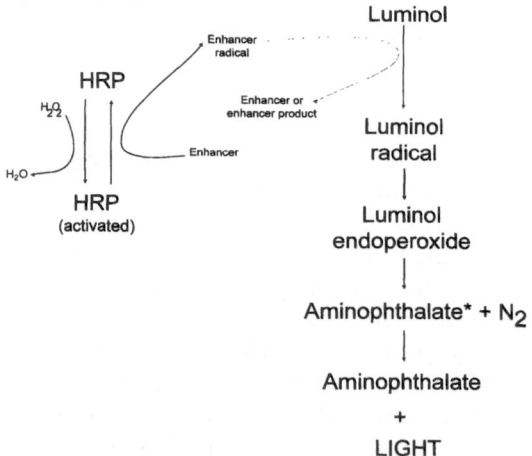

Figure 1. The mechanism of HRP-catalyzed enhanced chemiluminescence.

reaction is at 420-430 nm, appropriate for film sensitive to blue light. In blotting applications, good levels of sensitivity can be obtained with low background.

Light output from the HRP-catalyzed enhanced chemiluminescence reaction occurs rapidly with no detectable lag phase. There is a peak of light output at approximately 5–10 min, followed by a decay with a half-life of approximately 1 h. This characteristic decay is thought to be due to a gradual loss of enzyme activity resulting from free radical damage.

2. AP-catalyzed dioxetane chemiluminescence

Chemiluminescence from the alkaline phosphatase catalyzed decomposition of a substituted dioxetane provides the basis for high sensitivity detection methods. 1,2-dioxetanes are cyclic peroxides which undergo decomposition into two carbonyl containing fragments. One of these fragments is initially formed in an electronically excited state, undergoing relaxation to its ground state with the emission of light [4]. A phenyl-phosphate-substituted dioxetane containing a single adamantyl group (4-(phenyl-3-phosphate)-4-methoxyspiro[1,2-dioxetane-3,2'-adamantane]) has proved to be of particular value in a variety of assay formats [7]. It is highly stable but, in the presence of calf intestinal alkaline phosphatase, undergoes rapid conversion to the luminescent form via an unstable aryloxide dioxetane. The reaction is illustrated in Fig. 2. Such enzyme-triggerable dioxetanes are available from Tropix, Inc. and from Lumigen, Inc.

The maximum light output from this reaction is at 470-480 nm. Under some conditions, for example in solution assays, the light output can be greatly increased by the inclusion of surfactant or polymers which contain fluorophores such as fluorescein that can be activated by energy transfer from the luminescent intermediate. Using fluorescent micelles formed from cetyltrimethylammonium bromide (CTAB) and 5-(N)-tetradecanoyl-aminofluorescein, chemiluminescence efficiency can be increased at least 400-fold [8], with maximum light output now at 520-530 nm. It has more recently been found that certain blotting membranes, including nylon membranes, can themselves enhance light output.

Dioxetane-based chemiluminescence allows particularly high levels of sensitivity to be obtained. Low background can also be achieved, although it is necessary to ensure that a number of the reagents used during detection, in particular the antibody conjugate and membrane blocking agent, are optimised for this purpose. Additionally, solutions can easily become contaminated with bacterial alkaline phosphatases which can cause extensive spotting on the final results on film. It is therefore necessary to use sterile solutions and clean apparatus whenever possible.

In contrast to the HRP-catalysed chemiluminescence reaction, light output from a standard dioxetane luminescence reaction is initially slow,

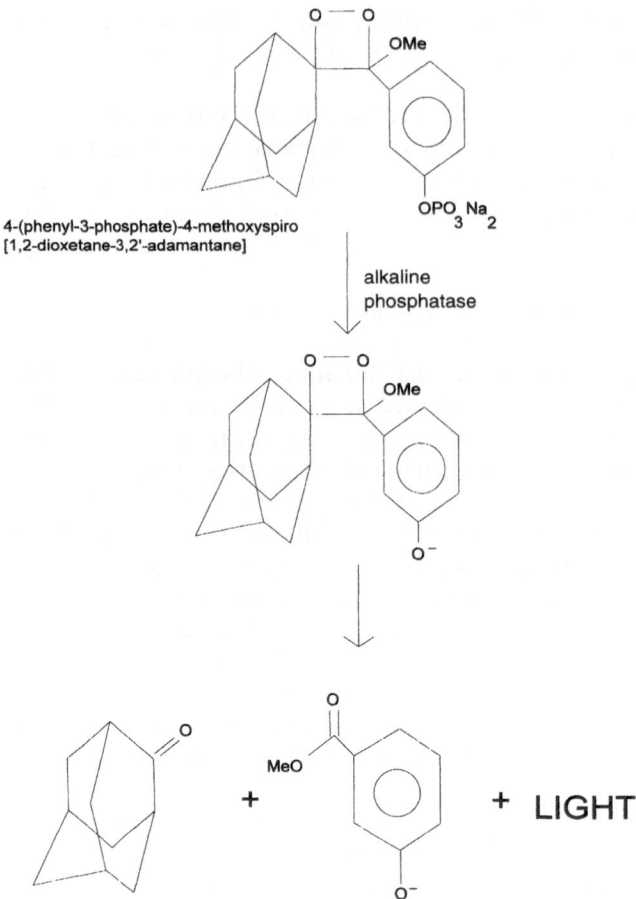

4-(phenyl-3-phosphate)-4-methoxyspiro [1,2-dioxetane-3,2'-adamantane]

OPO$_3$Na$_2$

alkaline phosphatase

OMe

O⁻

O

MeO

+

+ LIGHT

O⁻

Figure 2. The mechanism of alkaline phosphatase-mediated dioxetane chemiluminescence.

reaching maximum output only after several hours. This lag phase is due to the relatively slow accumulation of aryloxide intermediate which itself decays with a finite half-life. However, light output can continue at a high level for several days with only a slow decline. This allows accumulation of maximum signal during exposure to film, with sensitivity generally being limited by the level of background. The prolonged light output also gives the opportunity for multiple film exposures if required. More recently, substituted dioxetane variants have become available which have significantly shorter lag periods, allowing signal to be gained more rapidly (for example, CDP STAR from Tropix, Inc.).

Procedures

Probe labelling with HRP

1. The ECL™ direct system with enhanced chemiluminescent detection

It is possible to take further advantage of the rapid kinetics of light generation in the enhanced chemiluminescence reaction by coupling this detection with a labelling approach in which the nucleic acid hybridization probe is itself labelled directly with HRP. This allows the detection of hybrids to be carried out more rapidly than with other non-radioactive systems that require incubation with an antibody conjugate prior to the chemiluminescent reaction.

The direct labelling approach is made possible by the use of a chemical modification procedure [5]. The enzyme is modified by parabenzoquinone and polyethyleneimine which render it positively charged, thus allowing it to bind to single stranded, negatively charged nucleic acid such as denatured DNA. The modified enzyme is then covalently linked to the nucleic acid through amine groups on the bases using glutaraldehyde. This reaction is illustrated in Fig. 3.

This approach is used in the ECL direct system (Amersham International). It is rapid (20 min) and can be scaled up or down to label between 100 ng and 1.5 µg nucleic acid per reaction. Either DNA or RNA can be labelled and it is not necessary to remove unincorporated

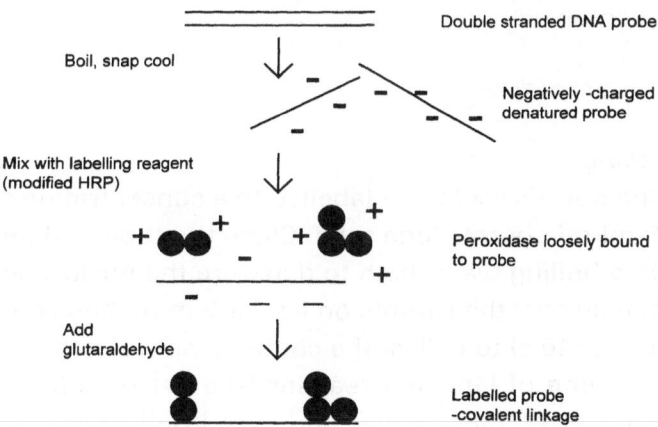

Figure 3. The mechanism of direct probe labelling with HRP.

MMEM-2.3.1/5

modified enzyme before hybridization. Although hybridization at elevated temperatures such as 65 °C would denature the HRP, it is possible to carry out hybridization at 42 °C in a denaturing buffer containing 6M urea. This is equivalent to incubation at 42 °C in 50% formamide or 65 °C in a non-denaturing buffer. The ECL direct system uses a 'Gold' hybridization buffer which also contains a novel rate enhancer which maximizes sensitivity and minimizes backgrounds. Stringency control is achieved by varying salt concentration either during hybridization, during the stringency washes or both. A probe concentration of 10 ng/ml is used routinely with overnight hybridization although, for low sensitivity applications such as the analysis of PCR products or plaque and colony screens, a reduced concentration of 2–5 ng/ml with short (1–2 h) hybridization times may be used.

The kinetics of light production by enhanced chemiluminescence have a further advantage for reprobing. The HRP bound to the probe is progressively inactivated during the detection step so that bands due to the presence of this probe will not be visible after a subsequent hybridization and detection. It is therefore possible to reprobe membranes that have been detected using ECL direct without removal of the previous probe. The majority of enzyme activity will have decayed 3–4 h after detection, although it is recommended to check the filter by adding fresh detection reagent and exposing to film to ensure no detectable signal remains before commencing reprobing.

The following protocol describes the direct labelling of probes with HRP together with appropriate hybridization conditions. A procedure for detection of the signal using enhanced chemiluminescence is also given.

Steps in the procedure

Probe labelling
1. Dilute the nucleic acid to be labelled to a concentration of 10 ng/μl in a 1.5 ml microcentrifuge tube. Close the tube and incubate for 5 min in a boiling water bath to denature the nucleic acid.
2. Immediately cool the sample on ice for 5 min, then centrifuge the tube briefly (10 s) to collect the contents at the bottom.
3. Add a volume of labelling reagent (see list of solutions below) equal to that of the sample and mix briefly by taking up the solution in the tip of a micropipette and gently expelling.

4. Add a volume of glutaraldehyde solution (see solutions below) equal to that of the labelling reagent. Again mix gently by pipette.
5. Incubate at 37 °C for 10 min. The probe is now labelled with HRP and can be stored on ice for up to 15 min before use.

Hybridization
6. Prepare the hybridization buffer as follows. Take the required volume of 'Gold' hybridization buffer (see solutions below) and add solid sodium chloride to a final concentration which is most suitable for effective probe hybridization. This may vary for different probes although 0.5 M NaCl has been found to give satisfactory results in most cases. Add blocking agent (see solutions below) to a final concentration of 5% (w/v). Rapidly mix the blocking agent into the buffer to avoid the formation of clumps. Continue to mix thoroughly using a magnetic stirrer at 42 °C for 30–60 min. A few undissolved particles of blocking agent will not affect the hybridization.
7. Place the blot containing the target nucleic acid sequence onto the surface of the fully prepared hybridization buffer prewarmed to 42 °C. A volume of buffer equivalent to 0.25 ml/cm^2 of membrane is satisfactory for most applications. Allow the blot to saturate fully before submerging in the buffer. Prehybridize at 42 °C for 15–16 min in a shaking water bath with gentle agitation.
8. Add the labelled probe from step 5 above to the prehybridization buffer at a concentration of either 10 ng/ml or 2–5 ng/ml depending on application (see above). Avoid placing the probe directly on the membrane and mix gently. Continue incubation with gentle agitation for between 2 h and overnight at 42 °C.

Stringency washes
9. Preheat enough primary wash buffer to 42 °C to allow two washes of 2 ml buffer/cm^2 membrane.
10. Carefully transfer the blot from the hybridization container to half of the prewarmed primary wash buffer in a clean container. Incubate at 42 °C in a shaking water bath for 20 min. Discard the wash buffer, replace with the remaining buffer and incubate with agitation for a further 20 min at 42 °C.
11. Discard the wash buffer, place the blot in a fresh container, and add an excess of secondary wash buffer (at least 2 ml buffer/cm^2

membrane). Incubate for 5 min at room temperature with gentle agitation. Discard the wash buffer and repeat this step with fresh secondary wash buffer. The blot may be temporarily stored at this stage if required.

Detection

12. Combine equal volumes of detection reagents 1 and 2 (see solutions below) so that there is sufficient for 0.125 ml combined reagents/cm^2 membrane.

13. Using forceps, carefully remove the blot from the final secondary wash buffer. Drain off the excess wash buffer by touching the bottom edge of the blot onto a paper towel. Place the blot, sample side uppermost, onto a sheet of SaranWrap on the bench.

14. Cover the blot evenly with the freshly mixed detection reagents and leave for 1 min. From this stage it is advisable to work as quickly as possible to minimise the delay between addition of substrates and exposure to film

15. Using forceps to hold the blot, drain off the excess detection reagents by touching the bottom edge of the blot onto a paper towel. Place the blot nucleic acid side down onto a sheet of SaranWrap in a film cassette.

16. Fold over the SaranWrap to completely wrap the blots, but ensure that only a single layer covers the sample side. Turn the blots so that the sample side is uppermost.

17. In a darkroom with a red safelight, place a piece of blue-light sensitive film on top of the blot, shut the cassette and expose at room temperature for the required length of time (please see notes). Develop film under normal conditions.

Notes

1a. Between 10 μl (100 ng) and 150 μl (1.5 μg) probe may be labelled in each tube. Labelling efficiency will be reduced if the volume of nucleic acid exceeds 150 μl per tube.

1b. Complete denaturation of double stranded probes is essential for maximal labelling. Extending the length of this step should be considered when using heating blocks as these can be less efficient at denaturation than boiling waterbaths. The denaturation step is not strictly necessary for single stranded probes although, in general, denaturation of RNA is advisable to avoid secondary structure.

1c. Since labelling is achieved in part through an ionic interaction between the negatively charged nucleic acid and the positively charged labelling reagent, the concentration of salt in the sample should not exceed 10 mM.

3. It is important to use equivalent volumes of nucleic acid and labelling reagent for maximum labelling efficiency. Accurate assessment of probe concentration (10 ng/μl) is essential to achieve maximum sensitivity.

5. Probes once labelled are single-stranded and may be used in any membrane hybridization application. If they are to be stored for more than 15 min, sterile glycerol should be added to a final concentration of 50% (v/v) and the probe stored at —20 °C until required. Probes stored for up to 12 months have been used in hybridization with no appreciable loss of sensitivity.

7a. Blots greater than 100 cm^2 should be pre-wet in 5 × SSC before addition to the hybridization buffer. The highest sensitivity detection is obtained with charged nylon hybridization membranes such as Hybond-N + (Amersham International). However, for many applications, satisfactory results can be obtained with uncharged nylon (e.g. Hybond-N) or nitrocellulose (e.g. Hybond-ECL).

7b. For hybridization carried out in boxes a volume of buffer equivalent to 0.25 ml/cm^2 membrane is recommended. If the box is significantly larger than the blot, then the volume used should correspond to the area of the bottom of the box. Several blots may be hybridized together if adequate circulation of the buffer is maintained so that the blots can move freely. Increased backgrounds will otherwise result.

7c. Hybridizations may also be carried out in bags or tubes. In these cases a buffer volume of 0.125 ml/cm^2 may be sufficient. When using tubes, it is important to exclude any air bubbles trapped between the hybridization tube and the membrane as this can lead to patches of high background following detection Nylon mesh may be used if there is significant overlap of the blot, to ensure good access of probe.

8. 10 ng/ml final probe concentration and overnight hybridization will give maximum sensitivity and is necessary for high sensitivity work such as the detection of single copy genes on genomic Southerns. However, for low sensitivity applications such as many plaque or colony screens or for detection of PCR products, it may be possible to use a lower probe concentration (2–5 ng/ml) in conjunction with a 2 h hybridization.

10. Stringency is controlled in the primary washes by the concentration of SSC. The primary wash recipe given below contains 5 × SSC but higher stringency and hence more specific hybridization signal can be obtained with lower concentrations of SSC such as 0.1 × SSC. Generally this wash will contain 6 M urea and is performed at 42 °C. However, it is possible to achieve similar results using primary wash without urea with incubation at 55 °C. In this case, the total wash time should not exceed 20 min in order to conserve peroxidase activity.

11a. The secondary washes are included to ensure removal of SDS which may interfere with the enhanced chemiluminescence reaction.

11b. Blots may be stored for up to 24 h at 4 °C, wetted in secondary wash buffer and wrapped in SaranWrap. After storage, the blots should be rinsed briefly in fresh secondary wash buffer just before detection.

12. The detection reagents should ideally be kept separate until immediately before detection to avoid gradual chemical decomposition of the substrates. However, if not used immediately it is possible to store the prepared solution on ice for up to 30 min.

17. For initial experiments an exposure of 30–60 min is recommended for low target applications. An alternative approach is to carry out a short 1–2 min initial exposure and use this to judge the length of the subsequent exposure. For plaque and colony screens in contrast, it is important to use the minimum film exposure time which enables positive colonies/plaques to be unambiguously determined in order to avoid false positives.

Solutions and reagents

Components for labelling, hybridization and enhanced chemiluminescent detection are available in kit format as the ECL direct nucleic acid labelling and detection system (Amersham International, plc). Detailed protocols are given with each kit. The following components (plus controls) are included in the system.

- Labelling reagent, comprising HRP crosslinked to PEI.
- Glutaraldehyde solution (1.5%(v/v) in water). Although glutaraldehyde is classified as harmful, in this kit it is present at a concentration at which it is not so classed. However, like all chemicals, it should be handled within the principles of good laboratory practice.
- 'Gold' hybridization buffer. In addition to 6 M urea, this buffer contains a detergent, a pH stabilizing compound, and a novel rate enhancement system.
- Blocking agent.
- ECL detection reagent 1. This contains specially purified luminol and ECL enhancer in a borate buffer.
- ECL detection reagent 2. This contains a peracid salt in borate buffer.

Additional solutions and reagents

- Sodium chloride (analytical grade)
- 20 × SSC (0.3 M trisodium citrate, 3.0 M sodium chloride, pH 7.0)
- Primary wash buffer with urea (360 g urea, 4 g sodium dodecyl sulphate (SDS), 25 ml 20 × SSC. Make up to 1 l)
- Primary wash buffer without urea (4 g sodium dodecyl sulphate,

25 ml 20 × SSC. Make up to 1 l)
- Secondary wash buffer (2 × SSC)
- Blue-sensitive autoradiography film (e.g. Hyperfilm™-ECL)
- Saran Wrap™ (Dow Chemical Company).

Probe labelling with fluorescein hapten

An alternative method for labelling probes non-radioactively is to introduce a small, relatively inert molecule such as a hapten or biotin. These molecules are usually coupled to a nucleoside triphosphate such as dUTP which is then incorporated into a nucleic acid probe by an enzyme-mediated reaction such as random prime labelling for long probes or terminal transferase-catalyzed 3′-end labelling for oligonucleotides. Following hybridization, detection takes place using an antibody (or, in the case of biotin, streptavidin) linked to an enzyme such as alkaline phosphatase or horseradish peroxidase which is used to catalyze the detection reaction. A variety of haptens have been used, most commonly digoxygenin and fluorescein. This approach involves a longer detection protocol than the direct enzyme labelling methodology because of the additional steps involved in antibody binding and detection. However, it can be more readily adapted to high sensitivity detection with dioxetane substrates. It can also be used more readily with oligonucleotide probes, in this case coupled to HRP-catalyzed enhanced chemiluminescence detection to take advantage of the rapid light output and low backgrounds of this reaction.

These two applications are described in the following protocols using fluorescein as the hapten. Systems are commercially available for labelling with a variety of haptens, for example fluorescein-11-dUTP (Amersham International, UK), digoxygenin-12-dUTP (Boehringer Mannheim, Germany) and biotin-11-dUTP (Enzo Diagnostics, Inc., USA). Fluorescein is used here as the hapten because it has several features of particular use in nucleic acid systems. These can be listed as follows.

a) It is stable during hybridizations even at elevated temperatures.
b) Fluorescein nucleotide conjugates are accepted by DNA polymerases for incorporation into probes.

c) High affinity antibodies can be prepared against the fluorescein moiety.

d) The resulting probes have a low affinity for hybridization membranes when suitable blocking agents are used so that there is no necessity for probe purification.

e) There is little chance of sample interference from endogenous agents that would be recognised by the antibody, as may occur with biotin, for example in *in situ* hybridization.

f) A final, significant advantage of fluorescein as a hapten which is not shared by, for example, biotin or digoxygenin, is that its natural fluorescence can be exploited in a rapid (20 min) transilluminator-based assay to monitor the success of a probe labelling reaction before embarking on the hybridization stage. This is analogous to the rapid check on incorporation that can be carried out with ^{32}P-labelled probes using beta-monitors. A protocol for performing this assay is given below.

1. Random prime labelling with fluorescein-11-dUTP and dioxetane detection

This approach is used in the Fluorescein Gene Images system (Amersham International). Fluorescein is incorporated into a probe by a random prime labelling reaction [2] catalyzed by exonuclease-free Klenow in the presence of the modified nucleotide fluorescein-11-dUTP. Exonuclease-free Klenow is particularly efficient at incorporation of fluorescein-11-dUTP and there can be considerable net synthesis of probe during the labelling reaction. For example, starting with 50 ng probe, it is possible to synthesize a further 300 ng probe in a 1 h reaction. The labelled probe can be used directly in the hybridization (following denaturation) without probe purification, while the stringency of the hybridization itself can be controlled by either salt concentration or temperature during the hybridization or the stringency washes. Recommended probe concentrations are the same as for the ECL direct system.

Both the hybridization stage and the subsequent membrane blocking step prior to addition of antibody conjugate use a blocking agent which, if supplied as a solid, can be difficult to dissolve fully, leading to variable levels of background. However, this component is available as a concentrated solution, thus helping to ensure consistent, low backgrounds. Backgrounds have been further minimized by careful preparation of the antibody conjugate used with the system. This has involved optimization of the methods used for preparation of the conjugate, and the use of a monoclonal anti-fluorescein antibody which gives a high signal to noise in this system.

Steps in the procedure

Preparation of labelled probe
This protocol allows the labelling of 50 ng of template DNA. The standard reaction can be used to label 25 ng–2 µg of template DNA. Synthesis of labelled probe is most efficient at 50 ng of template, although net synthesis tends to increase with the amount of template.
1. Dilute the DNA to be labelled to a concentration of 2–25 ng/µl in either distilled water or TE buffer (10 mM Tris-HCl, 1 mM EDTA, pH 8.0).
2. Place the nucleotide mix (see list of solutions below), primers (see solutions below) and water in an ice bath to thaw. Leave the enzyme (see solutions below) at —20 °C until required, and return it to the freezer immediately after use.
3. Denature the DNA sample by heating for 5 min in a boiling water bath, then chill on ice. It is advisable to denature in a volume of at least 20 µl.
4. To a 1.5 ml microcentrifuge tube, placed in an ice bath, add the appropriate volume of each reagent in the following order:
   Water to a final reaction volume of 50 µl

Nucleotide mix	10 µl
Primers	5 µl
Denatured DNA	50 ng
Enzyme solution (exo-free Klenow) 5 units/µl	1 µl
5. Mix gently by pipetting up and down and cap the tube. Spin briefly in a microcentrifuge to collect the content at the bottom of the tube.
6. Incubate the reaction mix at 37 °C for 1 h.
7. If any of the probe is to be stored, rather than being used in a hy-

bridization immediately, then the reaction should be terminated by the addition of EDTA to a final concentration of 20 mM. Probes can then be stored in a freezer at −20 °C in the dark for at least 6 months. Do not use a frost-free freezer.

8. The yield of probe after labelling 25 or 50 ng of template will typically be between 6–8 ng/µl. With higher template levels, while the amount of labelled template produced will increase, the overall reaction will be less efficient.

   At this stage, the rapid labelling assay (see protocol below) can be used as a check to ensure that the probe has been successfully labelled.

Hybridization and stringency washes

The hybridization and wash conditions given in the following protocol are appropriate for a majority of probes, allowing detection of single copy mammalian genes without significant cross-hybridization to non-homologous sequences. However, if these conditions are found to be insufficiently stringent for particular probes, then hybridized filters can be washed in 0.2 or 0.1 × SSC at 60 °C. Alternatively, stringency can be increased by raising the hybridization or wash temperature to 65 °C. Such alterations may lead to some decrease in specific signal, although an overall improvement in signal-to-noise should result.

9. Prepare the hybridization buffer as follows:

   5 × SSC

   0.1% (w/v) SDS

   5% (w/v) dextran sulphate

   20-fold dilution of liquid block (see list of solutions below).

   100 µg/ml denatured sheared heterologous DNA (optional).

   Combine all the components and make up to the required volume. Gentle heating with continuous stirring will be required to completely dissolve the dextran sulphate. The liquid block has been shown to contain low enough ribonuclease activity to be used in the hybridization buffer for Northern blots. However, if required, liquid block can be autoclaved, on a liquid cycle, but must be sub-aliquoted into a suitable autoclavable container prior to sterilisation.

10. Preheat the required volume of hybridization buffer (0.3 ml/cm^2

for small blots and 0.125 ml/cm^2 or less for large blots) to 60 °C, then place the blots in the buffer and prehybridize for at least 30 min at 60 °C with constant, gentle agitation.

11. Remove the required amount of probe to a clean micro-centrifuge tube. If the volume is less than 20 μl, make up to this volume with water or TE buffer. Denature the probe by boiling for 5 min and snap cool on ice.

12. Centrifuge the denatured probe briefly, then add to the prehybridization buffer, (avoiding placing it directly on the membrane), and mix gently.

13. Hybridize overnight at 60 °C with gentle agitation.

14. Prepare stringency wash solutions of 1 × SSC, 0.1% (w/v) SDS and 0.5 × SSC, 0.1% (w/v) SDS and preheat both solutions to 60°C. Carefully transfer the blots to the 1 × SSC, 0.1% (w/v) SDS and wash for 15 min at 60 °C with gentle agitation. Carry out a further wash in preheated 0.5 × SSC, 0.1% (w/v) SDS at 60 °C for 15 min. These wash solutions should be used in excess volumes of approximately 2–5 ml/cm^2 of membrane.

Blocking, antibody incubation and washes
The following steps are performed at room temperature and all the incubations require constant agitation of the blots. Recipes are given in the list of solutions below.

15. Following the hybridization washes, briefly rinse the blots in an excess (2 ml/cm^2) of buffer A at room temperature.

16. With gentle agitation, incubate blots for 1 h at room temperature in approximately 0.75-1.0 ml/cm^2 of a 1 in 10 dilution of liquid blocking agent in buffer A.

17. Briefly rinse blots in buffer A as in 15.

18. Dilute the anti-fluorescein- AP conjugate 5,000-fold in freshly-prepared 0.5% (w/v) bovine serum albumin in buffer A. Incubate the blots in diluted conjugate (0.3 ml/cm^2 of membrane) with gentle agitation at room temperature for 1 h.

19. Remove unbound conjugate by washing for 3 × 10 min in 0.3% (v/v) Tween-20 in diluent buffer at room temperature with agitation. An excess volume is again used (2-5 ml/cm^2).

20. Rinse blots in buffer A as in 15.

Signal generation and detection
Please read through this whole section before proceeding.
If possible, wear powder-free gloves. The dioxetane detection reagent should be applied in a fume cupboard.

21. After completion of the antibody incubation stage and subsequent washes, drain off any excess wash buffer from the blots (by touching the corner of the blot against the box used for washing the blots or other convenient clean surface) and place them on a sheet of clear plastic. The bags supplied in the detection module can be used for this purpose as follows:
    a) Cut a section from the bag large enough to cover the blot(s) with a small border of at least 1 cm.
    b) Cut along the sides of the bag, leaving one side uncut. Open out the bag and after draining off the blots, arrange them sample side up on one half of the opened bag. Do not allow the blots to dry out.
22. With the blots in a fume cupboard and the plastic opened to expose the blot surface, hold the dioxetane spray applicator (see list of solutions and reagents below) 2–3 cm above the blot and apply the reagent. Approximately one spray per 14 cm^2 should be sufficient, e.g. a 50 cm^2 blot will require 4 sprays.
23. After spraying, fold the plastic over the top of the blots and immediately spread the reagent evenly over the blot(s). This can be done either by rolling a 5 ml pipette over the surface or wiping the surface with a gloved hand.
24. For best results, the blots can now be transferred to a fresh bag before being placed in the film cassette. Ensure the outside of the bag is dry before exposing to film. Place the bag containing the blots (sample side up) in a film cassette and take to a darkroom.
25. Switch off the lights and place a sheet of autoradiography film on top of the blots. Close the cassette and expose for 1–2 h for high target/low sensitivity applications. Longer exposures (4–16 h) should be used for higher sensitivity applications. Remove film and develop.

Notes
1. If the DNA solution is too dilute to be used directly, it should be concentrated by ethanol precipitation followed by redissolution in an appropriate volume of water

or TE buffer. If an intact plasmid probe is to be used it may be necessary to correct for the proportion of plasmid sequence present during the hybridization.

3. Closed circular double-stranded DNA can be linearized to avoid more rapid renaturation following this step, although this is not usually necessary.

4. A volume of DNA solution up to 25 µl may be used for fragments in low melting point agarose.

6. The temperature of incubation and the reaction time can be chosen for convenience as the reaction reaches a plateau and does not decline significantly overnight at room temperature. However, the rate of reaction does depend to some extent on DNA purity so, for samples of lower purity such as miniprep DNA or DNA in agarose, a longer incubation period may be required. With purified DNA samples, an increase in probe yield can be obtained with longer reaction times (up to 4 h at 37 °C).

7. Labelled probe can be used directly in a hybridization, following denaturation, without removal of unincorporated nucleotide. For long term storage, it is advisable to keep probes in the dark as the fluorescein molecule is light-sensitive.

9. The hybridization buffer can be stored in suitable aliquots at −20 °C for at least 12 months. The addition of sheared, denatured heterologous DNA to the hybridization buffer can reduce non-specific hybridization although some reduction in sensitivity may also be observed.

10. Hybridization can be carried out in either boxes, bags or tubes provided there is sufficient buffer to allow adequate access of probe to the blot. With larger blots (greater than 50 cm^2), or if using minimal hybridization volumes in bags or tubes, the blots should be pre-wetted in 5 × SSC. It is also possible to hybridize several blots in the same solution, again providing there is adequate volume and circulation of buffer. A maximum of two blots should be hybridized together in a single bag and, when using more than one blot in a tube, it is important that the blots do not overlap one another. High backgrounds can result if there is insufficient hybridization buffer or if a blot is not totally immersed. Agitation improves access of buffer to the blot, particularly if several blots are to be hybridized in the same solution.

11. Recommended probe concentrations are 10–20 ng/ml for low target applications (for a 50 ng template reaction, this is approximately 1–2 µl of reaction product per ml of hybridization buffer) and 1–5 ng/ml for high target applications.

12. Alternatively, a small aliquot of the hybridization buffer may be removed and mixed with the probe before returning the mixture to the bulk of the buffer. For high target applications it is possible to use shorter hybridization times. Some loss of sensitivity will result, but if necessary such loss can be offset by the use of higher probe concentrations.

14. During the second wash the SSC concentration can be varied in the range 0.1–1 × SSC to achieve the desired stringency [3]. The temperature of these washes may also be increased to achieve the desired stringency. Several blots can be washed in the same solution provided that they can move freely.

15. Buffer A is an autoclaved solution of 100 mM Tris-HCl, 300 mM NaCl (pH 7.5). Once opened after autoclaving, unused buffer A should be discarded at the end

of that working day to avoid contamination with exogenous alkaline phosphatases.

16. Diluted block can be stored frozen (in aliquots) for several weeks and is easily thawed when required. However, it is recommended that no aliquot is subjected to repeated thawing and re-freezing. Also, it may be convenient to separate any remaining undiluted liquid block into appropriate aliquots for re-freezing.

18. Diluted conjugate should be used immediately. Loss of sensitivity occurs with diluted conjugate that has been stored or been through a freeze-thaw cycle. Several blots can be incubated together but it is important that there is free access of solution to the blot. With larger blots, it is possible to carry out this stage in hybridization bags or tubes to help minimize volume. For a 20 × 20 cm blot, a volume of 50 ml can be used. It may be found that the dilution of conjugate can be optimised further. Greater dilutions will give lower signal but may reduce background, while more concentrated conjugate will give the opposite effect. Bovine serum albumin fraction V (for example, Sigma Product no. A-2153) should be used in the buffer.

21. This step can either be carried out directly in the fume cupboard or else the blots can be arranged while working on the bench and the plastic temporarily closed over the top of the blots while they are taken to the fume cupboard. A sheet of SaranWrap™ can be used in place of the detection bags, although we have found bags to give the most satisfactory results.

24. Changing the bag in this way will avoid the transfer of excess liquid into the cassette. Alternatively, squeezing out all excess liquid from the original bag may suffice. If using SaranWrap, ensure no liquid can escape.

25. If required, expose a second film for an appropriate length of time. Depending on the application, the high sensitivity of this system may mean that overnight exposures of the blot to film result in too dark an image. By the second day, however, the light output will have stabilised at a significantly higher level than that emitted during the first few hours after application of the detection reagent, allowing significantly shorter optimum exposure times. Once the light output has stabilised, it is easier to judge the optimum exposure time.

Solutions and reagents

Components for labelling and dioxetane chemiluminescent detection are available in kit format as the Fluorescein Gene Images labelling and detection system (Amersham International plc, UK). The following components (plus controls) are included in the system.

- Nucleotide mix (5 × stock solution of fluorescein-11-dUTP (Fl-dUTP), dATP, dCTP, dGTP and dTTP in Tris-HCl, pH 7.8, 2-mercaptoethanol and $MgCl_2$). This mixture has been optimized to give both a high yield of probe and efficient substitution with fluorescein.

- Primers (random nonamers)
- Enzyme solution (5 units/µl exonuclease free Klenow)
- Water
- Liquid block (for use in the recommended hybridization buffer and for membrane blocking prior to addition of antibody conjugate)
- Anti-fluorescein alkaline phosphatase conjugate (5000 × stock)
- Dioxetane detection reagent in a spray bottle
- Detection bags for use during application of the detection reagent and subsequent exposure to film

Additional solutions and reagents

- 20 × SSC (3 M NaCl, 0.3 M sodium citrate, pH 7.0)
- Hybridization buffer (5 × SSC, 1 in 20 dilution liquid block, 0.1% (w/v) sodium dodecyl sulphate (SDS), 5% (w/v) dextran sulphate) +/–100 µg/ml sheared, denatured heterologous DNA (e.g. herring sperm DNA)
- TE buffer (10 mM Tris-HCl, pH 8.0, 1 mM EDTA)
- 0.2 M or 0.5 M EDTA pH 8.0
- 10% or 20% (w/v) SDS
- Buffer A (100 mM Tris-HCl, pH 7.5, 300 mM NaCl). Autoclave in 500 ml aliquots in 1 litre bottles for 15 min at 105 kPa (15 psi). Once opened after autoclaving, do not use for more than one day.
- Dextran sulphate (molecular weight 500,000, for example, Sigma Product no. D-6001)
- Bovine serum albumin (BSA) Fraction V (for example, Sigma Product no. A-2153, BDH Product no. 44155)
- Tween™ 20 (Polyoxyethylene sorbitan monolaurate; for example, Sigma Product no. P-1379, BDH Product no. 66368).
- Autoradiography film such as Hyperfilm-MP (Amersham International plc, UK)

2. 3′-end labelling of oligonucleotides with fluorescein-11-dUTP, and enhanced chemiluminescent detection

In this reaction, the enzyme terminal transferase is used to introduce a short tail of fluorescein nucleotides to the 3′-end of an oligonucleotide [1] followed by enhanced chemiluminescent detection of the labelled hybrids. This approach is used in the ECL 3'-oligolabelling and detection system for hybridizations on membranes.

Steps in the procedures

Oligonucleotide labelling
This protocol allows the labelling of single-stranded oligonucleotides bearing a 3'-hydroxyl group. Solutions are described in the list of solutions and reagents below.
1. Place the required tubes from the labelling components, excluding the enzyme, on ice to thaw.
2. To a 1.5 ml conical polypropylene tube on ice, add the labelling reaction components in the following order:

Oligonucleotide (100×10^{-12} moles)	x µl
Fluorescein-11-dUTP	10 µl
Cacodylate buffer	16 µl
Water	y µl
Terminal transferase	16 µl
Total	160 µl

The volumes corresponding to x and y should be adjusted so that the total reaction volume is 160 µl.
3. Mix gently by pipetting up and down in the pipette tip.
4. Incubate the reaction mixture at 37 °C for 60–90 min.
5. If desired, the reaction can be checked using the rapid labelling assay (see protocol below).
6. Store labelled probe on ice for immediate use or place at —20 °C in the dark for long term storage.

Hybridization
Hybridization temperature is dependent on the particular probe sequence to be used. Ideally, stringency is controlled by the post-hybridization washes so that, for many systems, a hybridization temperature of 42 °C will suffice. Generally, the hybridization temperature should be 5–10 °C below the Tm (melting temperature) of the oligonucleotide [9] (the presence of the 3'-tail does not significantly alter the Tm value).
7. The recommended hybridization buffer consists of the following components
$5 \times$ SSC

0.1% (w/v) hybridization buffer component (see solutions below)

0.02% (w/v) SDS

20-fold dilution of liquid block (see solutions below)

Combine all the components, make up to the required volume and mix well by stirring.

8. Place 8 blot(s) into the hybridization buffer and prehybridize at the required temperature, for a minimum of 30 min, in a shaking water bath.

9. Add the labelled oligonucleotide probe to the buffer used for the prehybridization step at a final concentration of 5–10 ng/ml.

10. Hybridize at the required temperature for 1–2 h in a shaking water bath.

Washing the membrane

The stringency of the hybridization is controlled at this stage. The presence of the 3'-tail does not affect the stringency conditions that would be required for an unlabelled sequence. Stringency can be altered by a combination of temperature and SSC concentration during the wash steps.

11. Remove the blot(s) from the hybridization solution and place in a clear container. Cover with an excess of $5 \times$ SSC, 0.1% (w/v) SDS.

12. Incubate at room temperature for 5 min with constant agitation.

13. Replace the wash buffer with fresh $5 \times$ SSC, 0.1% (w/v) SDS and incubate for a further 5 min.

14. Discard the wash solution. Place blot(s) in a clean container and cover with an excess of appropriate pre-warmed stringency wash buffer. For many systems a suitable wash buffer would contain $1 \times$ SSC, 0.1% (w/v) SDS.

15. Incubate at the desired temperature (typically 42–50 °C) for 15 min in a shaking water bath.

16. Replace the stringency wash buffer with fresh solution (as in 14) and incubate for a further 15 min at the same temperature as in step 15.

Membrane blocking, antibody incubations and washes

The following steps are performed at room temperature and all the incubations require constant agitation of the blot(s).

17. Place the filters in a clean container and rinse with buffer 1 for 1 min.
18. Discard buffer 1 and replace with block solution (a 20-fold dilution of liquid block in buffer 1). Incubate for at least 30 min.
19. Rinse blot(s) briefly (1 min) in buffer 1.
20. Dilute the anti-fluorescein HRP conjugate 1000-fold in buffer 2 containing 0.5% (w/v) bovine serum albumin (fraction V).
21. Incubate the blot(s) in the diluted antibody conjugate solution for 30 min.
22. Place filters in a clean container and rinse with an excess of buffer 2 for 5 min.
23. Repeat step 22 a further three times to ensure complete removal of non-specifically bound antibody.

Signal generation and detection
This section is carried out as for ECL direct detection steps 12–17. The exposure time will vary depending on the level of target material. Typical exposure times will vary from 1–10 min for high target applications to 10–60 min for low target applications. With high target systems, it is possible to perform multiple exposures.

Notes
1. Place the enzyme on ice immediately prior to use. Return the enzyme to a freezer at −20 °C as soon as the required amount has been removed.
2. It is recommended that the oligonucleotide is dissolved (or diluted) in sterile distilled water. The weight corresponding to 100×10^{-12} moles of an oligonucleotide is dependent on the length of the sequence:

Number of bases	Amount (ng)
15	500
20	660
30	1000
50	1660

If alternative levels of probe are to be labelled, then the volumes of the reactants and the final reaction volume should be altered in the same ratio as the change in the amount of probe. Reactions ranging from $25–250 \times 10^{-12}$ moles have been undertaken successfully.
5. Purification of the probe after labelling is not necessary.
7. For convenience it is recommended that the buffer is made up in a large volume and stored in suitable aliquots at −20 °C. Under these conditions, the buffer is stable for at least 3 months.
8. The volume of buffer should be equivalent to 0.25 ml/cm² of membrane. This may be reduced to 0.125 ml/cm² for large blots hybridized in plastic bags or hybrid-

izations in ovens. It is possible to alter the volume of buffer depending on the size of the container and the number of blots to be hybridized. It is essential that the blots should move freely within the buffer.

9. Avoid placing the probe directly on to the blot. Alternatively, a small aliquot of the buffer may be removed and mixed with the probe before returning the mixture to the bulk of the hybridization buffer.
10. Hybridization time can be increased up to 17 h without any significant change in sensitivity.
11. Use the wash buffers at a volume equivalent to 2 ml/cm^2 of membrane. The blot(s) should move freely. If the wash solution shows any precipitation during storage at room temperature it should be warmed slightly before use, until all material has dissolved.
14. The SSC concentration can be varied in the range 0.1–1 × SSC to achieve the desired stringency (the SDS concentration should remain constant at 0.1% (w/v)). At a temperature 3–5 °C below the Tm of the probe, these conditions should allow the discrimination of perfectly matched sequences from those containing mismatches [9].
15. The wash temperature can be as high as 65 °C, although when combined with changes in the SSC concentration it may not be necessary to go higher than 50 °C to achieve the desired stringency.
17. Use 2 ml of buffer for each cm^2 of membrane.
18. The amount of diluted liquid block solution prepared should be equivalent to 0.25 ml/cm^2 of membrane.
19. Volume as in 17.
20. The amount of diluted antibody-conjugate solution prepared should be equivalent to 0.25 ml/cm^2 of membrane.
22. Volume as in 17.
23. It is possible to store the blots in the final wash buffer for up to 30 minutes prior to proceeding to the detection stage. If longer term storage is required, remove the filter from the wash buffer, wrap in SaranWrap™ and store in a refrigerator at 2–8 °C. Do not allow the filter to dry out.

Solutions and reagents

Components for labelling and enhanced chemiluminescent detection are available in kit format as the ECL 3'-oligolabelling and detection system (Amersham International plc, UK). The following components (plus controls) are included in the system.

− Liquid block (for use in hybridization buffer and membrane blocking solution)
− Anti-fluorescein HRP conjugate (1000 × stock)
− Hybridization buffer component
− Fluorescein-11-dUTP

- Terminal transferase (2 units/µl)
- Cacodylate buffer (10× concentrated buffer, pH 7.2). (*Warning: Cacodylate is a highly toxic compound; please follow all safety warnings and precautions*)
- Detection reagent 1
- Detection reagent 2

Additional solutions and reagents

- 20 × SSC (0.3 M sodium citrate, 3 M NaCl)
- Buffer 1 (0.15 M NaCl, 0.1 M Tris, pH 7.5)
- Buffer 2 (0.4 M NaCl, 0.1 M Tris, pH 7.5)
- 20% (w/v) SDS stock solution
- TE buffer (10 mM Tris-HCl, pH 8.0, 1 mM EDTA)
- Bovine serum albumin (BSA) fraction V (e.g. Sigma (A-2153))

3. Monitoring incorporation using the rapid labelling assay

The efficiency of incorporation of fluorescein-11-dUTP can be estimated using the following rapid labelling assay. The basic protocol shown is for use with random prime labelled long probes, but modifications to the protocol for use with oligonucleotides are detailed in the notes. If sample identification is required, a pencil should be used to mark the membrane since some inks can interfere with the assay.

Steps in the procedure
1. Prepare 1/5, 1/25, 1/50, 1/100, 1/250, 1/500, and 1/1000 dilutions of the 5× nucleotide mix from the Fluorescein Gene Images system in TE buffer.
2. Dot out 5 µl of labelled probe and 5 µl of the 1/5 dilution of nucleotide mix (negative control) on to a strip of Hybond-N+, placed on a non-absorbent backing. Allow the liquid to absorb, (but do *not* allow to dry) and wash the strip with gentle agitation in excess pre-heated 2 × SSC at 60 °C for 15 min.
3. Prepare a reference strip by dotting 5 µl of each nucleotide mix dilution, except the 1/5, on to a separate strip of Hybond-N+. This reference strip can be re-used and can be stored, wrapped in SaranWrap, for several weeks at −20 °C in the dark.

4. Place both the reference and the washed strips on a piece of Whatman 3 MM paper lightly moistened with TE buffer and take to the darkroom. Visualize both strips (sample side down) on a UV transilluminator. Optimum contrast is obtained using a short wavelength (254 nm) transilluminator. The labelled probe should be visible as a fluorescent spot with an intensity between the 1/25 and 1/250 dilutions on the reference strip. Such a result indicates that the probe has successfully incorporated the fluorescein label. The closer the intensity of the labelled DNA is to the lower dilution (1/25), the more efficient is the labelling reaction. The washed negative control should retain little or no fluorescence indicating that the fluorescence of the probe is due only to incorporated fluorescein and not unincorporated Fl-dUTP. If significant fluorescein remains in the negative control, wash the filter for a further 15 min. This is possible only if the strip has not dried.
5. For a hard copy of the results the strips can be photographed. A Kodak Wratten No 9 filter (or a similar yellow filter) should be used in conjunction with Polaroid 667 black and white film.

Notes
1. For oligonucleotide probes prepared using the ECL 3′-oligolabelling system the recommended dilutions are 1/16, 1/100, 1/250, 1/500, 1/1000 and 1/2000 of the fluorescein-11-dUTP (Fl-dUTP) in TE buffer.
2a. When using the 3′-oligolabelling system, dot out 5 µl of each labelled oligonucleotide and 5 µl of the 1/16 dilution of Fl-dUTP (negative control).
2b. Whatman DE81 paper can also be used as an alternative to Hybond-N+ and is recommended for use with oligonucleotides. In this case $2 \times$ SSC, 0.1% (w/v) SDS should be used to wash the strip. A 1 cm edge should be allowed around the edge of the filter to avoid damage during washing. Following the wash stage, rinse the strip successively with water and then ethanol before handling (DE81 paper is fragile in aqueous solutions).
3. If using DE81 paper, the reference strip can be wetted (after applying the dots) successively in water and then ethanol to reduce distortion of the filter.
4a. An efficiently labelled oligonucleotide should be visible as a fluorescent spot with an intensity between the 1/500 and 1/2000 dilutions.
4b. When using DE81 paper dry Whatman 3 MM paper should be used to transport the strips to the darkroom.

Solutions and reagents

- 20 × SSC (3 M NaCl, 0.3 M sodium citrate). Dilute as required.
- 10% (w/v) sodium dodecyl sulphate (SDS) stock solution. Dilute as required.
- Nylon membrane (e.g. Hybond-N+, Amersham International)
- DE 81 paper (Whatman)
- 3 MM paper (Whatman)

Applications of chemiluminescent labelling and detection systems

The combination of rapid probe labelling with HRP, a short detection protocol and the fast kinetics of the enhanced chemiluminescence reaction together with its simplicity of reprobing, make the ECL direct system ideally suited for applications with multiple samples requiring rapid throughput. This is especially true for applications requiring only low sensitivity for which a rapid hybridization may also be possible. The high resolution of chemiluminescent signal compared to that from ^{32}P, together with its low background, is also

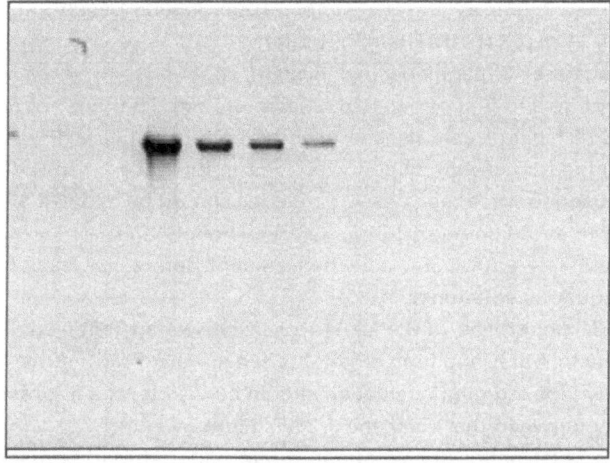

Figure 4. Single copy gene detection with Fluorescein Gene Images. Human placental DNA, digested with the restriction enzyme *Hin*d III, was probed with a fluorescein-labelled 1.5 kb insert against N-*ras* proto-oncogene. Detection was by dioxetane-based chemiluminescence, and filters were exposed to autoradiography film for 4 h following addition of detection reagent. Samples loaded were 1000, 500, 250 and 125 ng DNA, representing 500, 250, 125 and 60 fg complementary sequence.

a significant advantage in screening applications where there may be multiple bands or high density colonies or plaques.

Provided that probe and target concentrations are accurately known and a high efficiency of transfer is obtained during blotting, the sensitivity of the ECL direct system is generally adequate for the detection of single copy genes of higher organisms. For example, a gel loading of 1 µg is required to ensure an appropriate level of target when using a 1 kb probe (equivalent to approximately 500 fg target in human DNA). As a rough guide, experiments requiring exposures of 24 h or less when using ^{32}P-nucleotides should allow an acceptable result to be obtained. However, the greater sensitivity and prolonged signal of dioxetane chemiluminescence allows lower gel loadings to be used and provides a greater degree of robustness in the detection of low levels of nucleic acid. Using this system a sensitivity of at least 100 fg can be achieved in genomic hybridization, using a 0.2 µg loading of human genomic DNA and a 1 kb probe. Figure 4 shows the application of this system to a Southern blot. The lowest loading represents the detection of approximately 60 fg of complementary sequence. It is also possible to use the Fluorescein Gene Images system in many Northern blot applications. Again, as a rough guide, exposures of less than 48 h with ^{32}P-nucleotides should allow acceptable results to be obtained. The system is, of course, also compatible with lower sensitivity applications.

In general, oligonucleotide probes are used for lower sensitivity applications and the combination of fluorescein-labelling and enhanced chemiluminescent detection allows as little as 20×10^{-18} moles of homologous target to be detected within a 1 h exposure.

References

1. Bollum FJ (1974) Terminal deoxynucleotidyl transferase. In: Boyer PD (ed) The Enzymes, Vol 10, pp. 145–171. New York: Academic Press.
2. Feinberg AP, Vogelstein B (1983) A technique for radiolabelling DNA restriction endonuclease fragments to high specific activity. Anal Biochem 132: 6–13.
3. Meinkoth J, Wahl G (1984) Hybridization of nucleic acids immobilized on solid supports. Anal Biochem 138: 267–284.
4. Reguero M, Bernardi F, Bottoni A, Olivucci M, Robb M (1991) Chemiluminescent decomposition of 1,2-dioxetanes: an MC-SCF/MP2 study with VB analysis. J Am Chem Soc 113: 1566–1571.

5. Renz M, Kurz C (1984) A colorimetric method for DNA hybridization. Nucl Acids Res 12: 3435–3444.
6. Roswell DF, White EH (1978) The chemiluminescence of luminol and related hydrazides. Meth Enzymol 57: 409-423.
7. Schaap AP (1988) Chemical and enzymatic triggering of 1,2-dioxetanes. Photochem Photobiol 47: 50.
8. Schaap AP, Akhavan-Tafti H, Romano LJ (1989) Chemiluminescent substrates for alkaline phosphatase: application to ultrasensitive enzyme-linked immunoassays and DNA probes. Clin Chem 35: 1863–1864.
9. Wallace RB, Miyada CG (1990) Oligonucleotide probes for the screening of recombinant DNA libraries. Meth Enzymol 152: 432–442.
10. Whitehead TP, Thorpe GHG, Carter TJN, Groucutt C, Kricka LJ (1983) Enhanced chemiluminescence procedure for sensitive determination of peroxidase-labelled conjugates in immunoassay. Nature 305: 158-159.

Molecular Microbial Ecology Manual **2.6.1**: 1 12. 1995
ⓒ 1995 *Kluwer Academic Publishers.*

Detection of microbial DNA sequences by colony hybridization

PENNY R. HIRSCH

Rothamsted Experimental Station, Harpenden, Herts AL5 2JQ, UK

Introduction

Colony hybridization, the technique of lysis *in situ* of bacterial colonies on filters followed by denaturation and binding of the cellular DNA, allowing hybridization to DNA or RNA probes, was first reported in 1975 [3]. It was developed initially for screening *Escherichia coli* transformants in cloning experiments, to identify colonies containing a specific gene. Subsequently many variations of the method have been generated, appropriate for different applications, including analysis of environmental isolates. The main benefit of *in situ* colony hybridization is that isolates containing a target sequence can be identified from amongst hundreds or thousands of colonies and recovered for further analysis from either the original plate or a replicate.

The original method [3] relied on the ability of nitrocellulose to bind single-stranded DNA non-covalently, an attachment that can be made permanent by baking. Bacterial cells grown up on nitrocellulose membrane filters laid over nutrient agar were subjected to alkaline lysis, releasing cellular DNA which was simultaneously denatured. After neutralizing the filter (nitrocellulose is unstable in alkaline conditions) and washing to remove loosely-bound cellular debris, the filters were incubated with DNA probes to locate the position of colonies containing homologous sequences. Subsequent modifications of the method have included transferring the colonies from agar to filters by blotting (colony lifts), and using high wet-strength cellulose paper filters [2]. More recently, nylon membranes specifically developed for binding DNA, RNA and protein have become available and both neutral and positively-charged nylon have been used for colony hybridization.

Before the development of non-radioactive probing systems, most workers used ^{32}P-labelled DNA or RNA probes. The use of biotinylated DNA probes for colony hybridization was reported in 1986 [4] and there are now a variety of alternative non-radioactive systems for labelling and detection. Manufacturers and suppliers of kits for making DNA probes may recommend particular membrane filters (usually supplied by the same company) which have been verified for use with their probing/detection system. However, our experience is that alternative filters of an appropriate type can be used successfully.

There are two main applications for colony hybridization in microbial ecology, and these affect the experimental approach. The first, analogous to the original application of colony hybridization [3], is to monitor cells containing a particular genetic marker in environmental samples. In this case, there is usually a probe available with 100% homology to the target sequence, so DNA hybridization and washing unbound probe from the filters can be carried out under stringent conditions. For example, in our laboratory, a Tn5 DNA probe is used for monitoring survival of a Tn5-marked rhizobial inoculant in field soil. If the genetic marker is not anticipated to have moved from the cells being monitored, then optimal conditions for lysis will be known. However, when mixed populations are screened, it should be recognised that some colonies may not be lysed efficiently.

The second approach is to use a more general probe to enumerate all bacteria with a particular property in an environmental sample, such as the presence of catabolic plasmids [9]. An example from our laboratory is using a probe containing the structural gene for nitrogenase (*nif*) from *Klebsiella* as a probe to identify nitrogen-fixing bacteria in soil samples plated on semi-selective agar. This requires an efficient general lysis method and hybridization conditions that are less stringent to allow for imperfect homology to the probe.

Experimental approach

Preparation of plates

The choice of media and whether any preliminary screening is appropriate will be dictated by the nature of the research. Since the process of cell lysis and DNA denaturation is destructive, the next consideration is how to maintain viable replicates of the colonies which are being screened. With some bacteria, young colonies can be lifted off a plate on a filter leaving sufficient cells to regrow, but this is not always possible. Therefore for any bacteria where the capacity for regrowth is uncertain, including mixed environmental isolates, plates should be replicated and grown up before making a colony lift from one replica plate. When mixed cultures are grown up on filter membranes overlying agar plates, the colony debris tends bind DNA probes non-specifically, resulting in false positives, especially with nylon membranes. However this method can be useful with some types of bacteria, using nitrocellulose membranes: after colony growth, the membrane can be replicated to agar, or to a second membrane filter (which can be stored frozen) before lysis.

Choice of filter membrane

Each type of filter has distinct properties and the choice may depend upon the type of probe to be used, as well as personal preference and budget. If many plates are to be processed, the availability of the membrane of choice in pre-cut circles is a factor to consider, 82 mm diameter membranes being the appropriate size for standard plastic Petri dishes which have an internal diameter of 90 mm.

In our experience nitrocellulose filter membranes consistently give the best results with the lowest background:signal ratio, but they are fragile and also unstable in alkaline conditions. However the supported nitrocellulose membranes which are now available overcome these problems. The advantages of nylon membranes, apart from their strength, is that they bind the released DNA covalently and have a higher binding capacity than nitrocellulose, so that they can be reprobed many times without loss of signal. A disadvantage that we have observed is that they appear to give higher background noise than nitrocellulose, especially with non-radioactive probes, particularly positively-charged nylon membranes. This may be due in part to the greater efficiency with which both DNA and protein are bound. Filter papers are relatively cheap, have a high DNA binding capacity, and can be reprobed several times, but in our experience removal of bound probe is inefficient and they cannot be used with non-radioactive probes where an antibody-binding step is required to detect bound probe, because of non-specific binding to the paper.

Filters need not be sterile unless they are to be used to make a replicate for storage, or the plate from which they are taken is to be kept. However, they must never be touched by ungloved hands, as fingerprints tend to produce clear and indelible images!

It is difficult to provide an exhaustive list of different membranes that are available, and some companies supply membranes manufactured by another company, under a different name. The following list may provide some guidance, and lists the membrane which we have used in our laboratory over the past 12 years:

Nitrocellulose
    Schleicher & Schüll BA 83, BA 85
Supported Nitrocellulose
    Schleicher & Schüll BA-S 83, BA-S 85
    Gelman Biotrace NT
    Sartorius NC mit Gewebe
High wet-strength cellulose paper
    Whatman 541
Nylon
    New England Nuclear GeneScreen
    Pall Biodyne A

Positively-charged nylon
   Gelman Biotrace RP
   Schleicher & Schüll Nytran-plus
   Biorad Zeta-Probe
   Boehringer Mannheim Nylon Membrane – Positively Charged

Processing the colonies

The colony blot or lift is done by carefully placing the filter membrane over the agar surface in the Petri dish. It is then peeled off the agar and laid with the bacteria facing upwards, on a stack of absorbent paper saturated with alkaline solution to lyse cells and denature the DNA. Nitrocellulose filters must be treated simultaneously with high salt to facilitate DNA binding, but this is unnecessary for nylon membranes, where it may actually reduce the efficiency of DNA binding [8]. Some workers recommend a pre-treatment with detergent (SDS) before alkaline lysis but we have not found this to be necessary, although we expose our filters to alkali for longer than is usually recommended. Subjecting the filters to steam for a few minutes is also reported to improve lysis [6] but we have not tested this method. For some Gram positive bacteria it may be necessary to treat the filters with lysozyme prior to alkaline denaturation [5, 10]. The filters are then transferred to a fresh stack of paper saturated with neutralization buffer. Finally they are removed separately with forceps and rinsed by passing through a dish containing SDS to remove loose cellular debris, blotted with absorbent paper to remove surface moisture and loose debris, then baked at 80 °C. Alternatively, attachment to nylon filters can be facilitated by exposure to UV light [8]. Some protocols include a rinse in ethanol to precipitate the DNA and improve binding, but we have not found this to be necessary.

 After fixing the DNA by baking or UV treatment, the filters can be stored in dry conditions for many years but normally they are processed immediately so that the replicate colonies remain viable for further testing. Treatment of the filters with protease is sometimes recommended to remove proteins that can interfere with probe hybridization and detection. A method utilizing both lysozyme and proteinase K, followed by phenol chloroform, and isoamyl alcohol treatments, is reported to greatly reduce the background on paper filters [12], and a similar approach has been described for nitrocellulose filters prior to hybridization with biotinylated probes [4].

Probing the filter membranes

Before pre-hybridizing the filters, we wash them in SDS for 30 min to remove any remaining cellular debris, and blot them dry before placing in a bag with pre-hybridization buffer. If there are many filters to be screened it is an advantage to place them in a stack of single membranes, each layer separated by a plastic mesh. This avoids the problems of transfer of any remaining loose cellular material between filters. The stack can then be sealed into a plastic bag or tube with the pre-hybridization solution. We usually use buffer containing 50% formamide so that the incubations can be done at 42 °C. Several different pre-hybridization and hybridization buffers are recommended by the various commercial suppliers of membranes and kits for probe labelling and detection. We have tried several different systems but have not been able to detect significant differences. After pre-hybridization it is advisable to discard the solution and add fresh buffer containing the probe, even when following a protocol where the same buffer is used in both steps. This reduces background, as cellular debris can be released during the pre-hybridization step.

Both DNA and RNA can be used to probe colony lifts. When an unknown population of environmental isolates is being screened using a cloned gene on a plasmid there is often homology to vector sequences and this can mask any homology specific to the target gene. Therefore it can be an advantage to use a purified fragment. For example, we have found homology between soil isolates and sequences in pBR322, pAT153 and pUC vectors and similar observations are reported in [9].

The labelling system for the probe is a matter of personal choice and convenience. DNA labelled with ^{32}P can be very sensitive but has a short half-life so that probes cannot be reused many times. Non-radioactive labelling is often very stable so that one batch of probe can be made and fresh aliquots removed for use for over 1 year, in addition to re-using the probe in hybridization solution for subsequent experiments. A drawback is the reduced sensitivity of colorimetric detection systems, compared with radioactive methods. Also, pigments released from some bacteria can interfere with the interpretation of colorimetric results. These problems of low sensitivity are largely overcome by using chemiluminescent detection. However, because the detection of probe bound to the target DNA relies on antibody binding, there may be significant background where cellular proteins remain on the filter. Non-specific binding also prevents use of the cheaper but efficient paper filters in these systems.

The washing step is the major determinant of hybridization stringency, *via* temperature and salt concentration. It is important to rinse off excess formamide-containing hybridization buffer in relatively non-stringent conditions before raising the temperature and lowering the salt concentration.

Autoradiographic methods for detecting hybridization to radioactive probes are well documented. Colorimetric and chemiluminescent detection of non-radioactive probes depends upon the system used. However, because of the background due to non-specific binding of probe or detection system to colony debris on the filter, it is essential to include both positive and negative controls, preferably on each filter but certainly in each batch tested.

Procedures

The methods described in detail below appear to work with Gram negative eubacteria, and we have tested them with various enteric bacteria, rhizobia and pseudomonads. For Gram-positive bacteria other methods have been developed to provide more vigorous cellular lysis. For example, actinomycete spores are germinated on filters, treated with lysozyme, then lysed and denatured in boiling NaOH with SDS [5]. Staphylococcal colonies can be lysed by steaming in alkali [7].

Preparation of plates and materials

The agar media and selection procedure will depend upon the requirements of each experiment. If it is impractical to incorporate positive and negative controls on each plate (for example, when plating out environmental isolates), control cultures can be grown up separately and a small amount of bacterial cells are transferred to each filter using a sterile toothpick or loop, after the colony lift has been completed. We make a cross or write a number using the positive control at a point on the edge of the filter which will also act as an orientation marker. The negative control, in a different-shaped mark, is placed next to it.

Ensure that replicates of the plates to be blotted have been made, or that the colonies are at a stage where regrowth is possible. Sterilize membrane filters if necessary.

Cut 8 sheets of absorbent paper (Whatman 3MM or equivalent) to fit each shallow dish or tray large enough to accommodate all the filters (the number of trays needed will depend upon the number of steps). For each step, add sufficient of the appropriate solution to saturate the paper, without any surface liquid. Some sheets of absorbent paper for drying the filters should also be available.

Colony transfer, lysis, and binding cellular DNA to filters

1. The colony blot or lift is done by carefully placing the filter membrane over the Petri dish, holding both sides with blunt-ended forceps, placing one side on the edge of the plate then carefully lowering the rest of the membrane to exclude air

bubbles. The filter can be smoothed lightly with a glass spreader to ensure that contact has been made, then peeled off. Unless the colonies have been subcultured and arranged in an orderly array, it is absolutely essential to mark the filter for subsequent orientation with respect to the plate. This includes marking the side which is in contact with the bacteria. We use a soft pencil to write on the contact side of the filter. It is also possible to make small "nicks" on the edge of the membrane, or puncture it with a sharp needle, to produce asymmetric orientation marks. If necessary the positive and negative controls can be added at this stage.

2. Place filters on paper saturated with the following solutions, carefully eliminating air bubbles whilst lowering onto the surface, bacterial side upwards.

3. First stage (optional) 10% SDS 10 min, then transfer to dry paper to remove excess liquid.

4. Immediately transfer to alkaline lysis solution: for nitrocellulose membranes use Southern Denaturation Solution [11]: NaOH, 0.5 M; NaCl, 1.5 M. For paper or nylon, use 0.5 M NaOH. Leave 30 min then transfer to dry paper to remove excess alkaline solution.

5. Immediately transfer to neutralization buffer. For nitrocellulose filters, use Southern Neutralization Buffer [11]: 0.5 M Tris-Cl pH 7.0, 3 M NaCl, or a modification 0.5 M Tris-Cl pH 7.2 – pH 7.4, 1.5 M NaCl. For paper and nylon filters use 3 M Na acetate pH 5.5 or 1 M Tris pH 7.2–7.4. Leave for 30 min, then transfer to dry paper.

6. Repeat step 5.

7. Lift off filters singly and pass through wash solution with gentle agitation: $2 \times$ SSC, 0.1% SDS, then place on dry paper and carefully blot upper surface. Some loose cellular debris may be removed in this step. The filters should be left to air-dry for about 10 min.

8. Place the air-dried filters between two sheets of absorbent paper, fold the edges to seal loosely, then bake at 80 °C for 2 h. Nylon membranes may alternatively be subjected to UV light but since sources differ, each must be assessed to establish the dose required.

9. Store in a desiccator or sealed in a plastic bag, between two sheets of absorbent paper, unless processed further immediately.

Hybridization to probes and detection

This stage depends to a certain extent upon the nature of both the probe and the membrane used, and the method below is one we have used for nitrocellulose, nylon, and paper filters, using ^{32}P-labelled DNA probes. Generally, a small amount of DNA labelled to a high specific activity gives lower background than more DNA with a lower specific activity. For a rough guide, 500 ng DNA at about 10^8 dpm per µg specific activity is appropriate for up to 10 filters in 20–40 ml hybridization buffer. Although hybridization rates are reputed to be slower in formamide than in aqueous salt solutions, we prefer the lower temperatures which formamide makes possible, and we leave our hybridizations overnight.

1. Prewash filters in $2 \times$ SSC, 0.1% SDS at 68 °C for 1 h. Blot between two sheets of absorbent paper to remove any loosened cellular debris.
2. Arrange in a single layer or in several layers separated by plastic mesh in a plastic bag or glass tube and add pre-hybridization buffer. The volume depends upon the number of filters and the container, but usually 5 ml per filter is sufficient. Remove air bubbles from bags before sealing or cap tubes carefully. Incubate bags shaking or tubes rolling at 42 °C for at least 4 h.
3. Discard the pre-hybridization buffer and add fresh hybridization buffer (2–5 ml per filter) and ^{32}P-labelled DNA probe previously denatured by boiling for 5 min then placing immediately on ice. If the probe volume is a significant proportion of the total hybridization volume (i.e. more than 5%), adjust with salt and formamide (add 1 volume $20 \times$ SSPE and 2 volumes formamide) before addition. Incubate as in (2) at 42 °C overnight.
4. Decant hybridization solution and retain – it can be reused if stored frozen. Wash formamide from filters in $2 \times$ SSC, 0.1% SDS at room temperature for 5 min. Transfer the filters to an open dish and repeat the wash with agitation (i.e. shaking) with fresh $2 \times$ SSC. Stringency is applied in subsequent washes which are usually performed in dishes in a shaking incubator.
5. *Stringent wash conditions* for probes with 100% homology to the target sequence: $2 \times$ SSC, 0.1% SDS 68 °C 10 min; $1 \times$ SSC, 1% SDS 68 °C 10 min; $0.1 \times$ SSC, 0.1% SDS 68 °C 15 min; repeat final wash once more.

Less stringent wash to retain probe hybridization with imperfect homology: 2 × SSC, 0.1% SDS 65 °C 10 min, repeated 3 more times.

6. Blot filters dry, cover in Saran Wrap (Dow Chemical Company) or equivalent plastic film and set up for autoradiography by standard methods. We use cassettes with calcium tungstate intensifying screens and store at −75 °C before developing the film.

7. The first exposure is usually made after 1–2 h to assess whether an overnight exposure is necessary. (If the orientation markings on the membrane filters become damaged, the colonies usually stain pink after 15 min in ethidium bromide solution, 0.5 µg per ml).

Reprobing filters

If the filters are to be reprobed, they should not be allowed to dry out during the preceding stages. The probe can be stripped by several methods but we never find 100% removal and therefore auto-radiography is necessary to determine the efficiency of the stripping step. Even if the specific activity of the probe has decayed beyond detection, the probe should be stripped to avoid "sandwich" effects where previously-attached probe hybridizes to the new probe. We have had better results in stripping nitrocellulose and paper filters than nylon.

Nitrocellulose and paper
1. Wet filters in 6 × SSC.
2. Transfer to 1 volume Southern Denaturation Solution diluted with 4 volumes H_2O. This 1 in 5 dilution provides 0.1 M NaOH. Agitate gently by shaking for 5 min.
3. Rinse in 6 × SSC.
4. Transfer to Southern Neutralization Buffer, agitate gently for 10 min.
5. Rinse in 6 × SSC, then transfer to absorbent paper to blot dry.
6. Repeat pre-hybridization and hybridization steps.

Nylon

For nylon filter membranes, follow the manufacturers recommendations. It is particularly important not to let these dry out after the first hybridization.

Methods with which we have had some success are:

1. Incubate in 10 mM Na phosphate pH 6.5, 50% formamide, 65 °C, 1 h.
2. Rinse in 2 × SSC, 0.1% SDS, at room temperature for 10 min.
3. Repeat pre-hybridization and hybridization steps.
   Or, alternatively:
1. Boil in 0.1% SDS for 1 h.
2. Repeat pre-hybridization and hybridization steps.

Solutions

- 2 × SSC
  - 0.3 M NaCl
  - 0.03 M Na citrate
- 20 × SSPE
  - 3.6 M NaCl
  - 0.2 M Na phosphate pH 8.3
  - 0.02 M EDTA
- Pre-hybridization and hybridization buffer
  - 50% formamide
  - 5 × Denhardt's solution
  - 5 × SSPE
  - 0.1% SDS
  - 100 µg per ml denatured salmon sperm DNA

Formamide should be fresh molecular biology grade, or reagent grade purified by recrystallization or treatment with BioRad AG 501-X8 mixed bed ion-exchange resin. It can be stored at –20 °C after purification for up to 1 year.

- Denhardt's solution
  - 1% Ficoll
  - 1% PVP
  - 1% BSA (Pentax fraction V)

Dissolve in H_2O then filter sterilize, store at –20 °C.

- denatured salmon sperm DNA
  Dissolve in H_2O at 10 mg per ml then shear by ultrasonication until viscosity is noticeably reduced and average fragment size (determined by gel electrophoresis) is less than 50 kb. To sonicate, keep solution on ice, and adjust sonicator to the maximum power setting that gives cavitation but not foaming (usually c. 100 W). Sonicate for 20 s then rest for 20 s, repeat 5 times. Extract with phenol and chloroform, boil each aliquot to be added to hybridization solution for 5 min and plunge onto ice before adding to hybridization buffer.
- $6 \times$ SSC
  - 0.9 M NaCl
  - 0.09 M Na citrate

Non-radioactive probes

There are several different systems available, and the manufacturers instructions should be followed. We have used digoxigenin-labelled probes prepared using the DIG System (known as the "Genius" system in the USA) from Boehringer Mannheim, with nitrocellulose membranes (Schleicher and Schüll BA85, 0.45 µm pore) and charge-modified nylon (Gelman Biotrace RP) as well as the positively charged nylon membrane supplied by Boehringer Mannheim. Using our standard laboratory method for colony lifting and processing, and the hybridization conditions recommended by Boehringer Mannheim [1], we obtained satisfactory results with all three membranes with the colorimetric detection kit. Nitrocellulose filters gave the strongest signal and lowest background. However, when these filters were reprobed with the chemiluminescent detection system, nitrocellulose gave no signal although on the nylon membranes the detection of positive colonies was improved. Boehringer Mannheim warn that stripping off the colour prior to chemiluminescent detection damages nitrocellulose membranes and also that these membranes quench chemiluminescence [1]. However, we have obtained satisfactory signals with nitrocellulose where chemiluminescent detection was used in the first instance, although the signals were weaker than those obtained with nylon.

References

1. Anonymous (1993) The DIG System User's Guide for Filter Hybridization (The Genius System User's Guide for Filter Hybridization). Mannheim: Boehringer Mannheim GmbH Biochemica.
2. Gergen JP, Stern RH, Wensink PC (1979) Filter replicas and permanent collections of recombinant DNA plasmids. Nucl Acids Res 7: 2115–2136.
3. Grundstein M, Hogness DS (1975) Colony hybridization: a method for the isolation of cloned DNAs that contain a specific gene. Proc Natl Acad Sci USA 72: 3961–3965.
4. Haas MJ, Flemming DJ (1986) Use of biotinylated DNA probes in colony hybridization. Nucl Acids Res 14: 3976.
5. Hopwood DA, Bibb MJ, Chater KF, Kieser T, Bruton CJ, Kieser HM, Lydiate DJ, Smith CP, Ward JM, Schrempf H (1985) Genetic Manipulation of *Streptomyces* – A Laboratory Manual. Norwich: John Innes Foundation.
6. Maas R (1983) An improved colony hybridization method with significantly increased sensitivity for detection of single genes. Plasmid 10: 296–298.
7. Notermans S, Heuvelman KJ, Wernars K (1988) Synthetic enterotoxin B DNA probes for detection of enterotoxic *Staphylococcus aureus* strains. Appl Environ Microbiol 54: 531–533.
8. Reed KC, Mann DA (1985) Rapid transfer of DNA from agarose gels to nylon membranes. Nucl Acids Res 13: 7207–7221.
9. Sayler GS, Shields MS, Tedford ET, Breen A, Hooper SW, Sirotkin KM, Davies JW (1985) Application of DNA-DNA colony hybridization to the detection of catabolic genotypes in environmental samples. Appl Environ Microbiol 49: 1295–1303.
10. Schrempf H (1982) Plasmid loss and changes within the chromosomal DNA of *Streptomyces reticuli*. J Bacteriol 151: 701–707.
11. Southern EM (1975) Detection of specific sequences among DNA fragments separated by gel electrophoresis. J Mol Biol 98: 503–507.
12. Taub F, Thompson EB (1982) An improved method for preparing large arrays of bacterial colonies containing plasmids for hybridization: *in situ* purification and stable binding of DNA on paper filters. Anal Biochem 126: 222–230.

Molecular Microbial Ecology Manual **2.7.2**: 1–10, 1995

Polymerase chain reaction (PCR) analysis of soil microbial DNA

JAN DIRK VAN ELSAS and ANNEKE WOLTERS
IPO-DLO, P.O. Box 9060, 6700 GW Wageningen, The Netherlands

Introduction

Microbial communities as well as specific microorganisms and genes in soil can be detected and analysed by using methods based on nucleic acids [4, 8, 20, 21]. Analysis can be performed directly on colonies obtained from the environment by plating, using colony hybridization [14], or via Northern, Southern, dot or slot blot procedures unlashed on either microbial cells (dot/slot blots) or DNA extracted from cells or directly from the environment [15, 16]. In addition, environmental DNA can be characterized in terms of its internal reannealing [20] or hybridization behaviour [11], and these criteria have been taken as being characteristic for the complexity of microbial communities. However, a major problem often encountered with hybridization analysis directly on environmental DNA extracts is the lack of sensitivity, limiting the analysis to populations of cells or genes which occur in high numbers in the environmental samples [23].

A major step forward in the study of microorganisms in the environment via their DNA and/or RNA has been the development of polymerase chain reaction (PCR) amplification [3, 5, 10, 15, 16, 18, 22]. PCR is based on repeated cyclic enzymatic extension of primers at two opposite ends of a DNA template, resulting in the generation of numerous copies of this template. The amplification cycle, composed of template denaturing, primer annealing and extension steps, is achieved by concerted changes in reaction temperature, most easily performed in a programmable thermal cycler. Due to the high denaturing temperature (often 94 °C), DNA polymerases used in PCR have to be thermostable, like for instance the frequently used *Thermus aquaticus Taq* polymerase.

DNA extracted from the environment can be directly used as a template for PCR amplification, whereas RNA can be is reverse-transcribed into cDNA using reverse transcriptase. Multiplication of target sequences can then ensue to facilitate subsequent detection or identification via restriction (RFLP-PCR) or hybridization to a specific probe, or cloning, sequencing and analysis [6]. Using PCR, DNA sequence information can even be obtained from non-culturable organisms or dwarf forms which are known to abound in soil but traditionally form an enigmatic group there. For instance, based on PCR amplification of 16S ribosomal sequences, Liesack and Stackebrandt [6] recently described a novel group of bacteria (Planctomycetales), which had previously been unknown. The application of temperature or denaturing gradient gel electrophoresis (TGGE and DGGE, respectively) for separation of PCR products produced with primers of conserved regions of the 16S ribosomal RNA gene may further provide a quick entry into studies on species diversity in sediments or soils [7]. PCR also allows for the direct detection of specific gene sequences in bacterial populations in soil, as shown recently with *npt*II and Tn*5* sequences [16].

The specificity of the PCR reaction is dictated by a variety of factors which affect the fidelity of the reaction. For instance, the specificity of the primers used, the cycling regime and reaction components and their concentrations used, and the presence of additives (adjuvants) are crucial in determining the success of amplification [3, 5]. The difficulty when working with environmental, and in particular soil-derived DNA, is to fine-tune the reaction conditions to the quality of the template DNA obtained.

This chapter gives an outline of procedures used in our laboratory for PCR amplification for detection of specific gene sequences in soil. The protocols have been successfully used by us for several objectives, such as the specific detection of *Mycobacterium chlorophenolicum* using a specific region of its 16S ribosomal RNA gene sequence [1], the detection of genetically modified bacteria in soil [15, 23] and the detection of *npt*II and Tn*5* sequences in soil [16]. PCR protocols of soil DNA developed in other laboratories may differ in their approach to circumvent the problem of inhibition of the thermostable DNA polymerase, and we will not include these different approaches. Also, we will not extensively address the principles and details of the polymerase chain reaction, for which the reader is referred to Erlich [3] and Innis *et al.* [5].

Experimental approach

In order to successfully perform PCR amplification of a specific target in soil DNA, adequate template DNA has to be obtained. This manual contains several protocols which can achieve this goal (see e.g. Chapters 1.3.1, 1.3.3, 1.3.4 and 1.5.1), and the choice of the extraction method will depend on the nature of the sample material as well as the specific aims of the experiment. In our laboratory, the direct DNA extraction protocol described in Chapter 1.3.3 has been found to work well with a variety of soils, including sandy, clayey and high organic matter ones. In all cases,

DNA of adequate purity and reasonably high molecular weight was obtained, which was suitable for subsequent PCR amplification analysis.

PCR amplification of soil DNA and analysis of PCR products may proceed along the following steps: 1) Primer design, 2) *In vitro* testing and optimization of PCR protocol using pure target DNA, 3) Testing and optimization of PCR protocol with soil DNA extracts, 4) Routine use of PCR for detection in soil DNA samples, and 5) Analysis, cloning and identification of PCR products.

A first condition for successful application of PCR to soil DNA is the availability of suitable primers for specific amplification of the target in the soil DNA extract. Such primers, which often span a region 0.2 to 1 kb in size, are designed using criteria such as a G+C% around 50%, absence of substantial complementarity between primers or secondary structure within them, and no annealing to other regions of the target or flanking sequences. Since the 3' primer ends are crucial for adequate priming, the presence of purine bases in the last two positions has been advocated. Current computer programs such as PRIMER (Whitehead Institute, MIT, Cambridge) or OLIGO (Natural Biosciences, Plymouth) apply some or all of these criteria and are helpful for quick unambiguous primer design. In addition, they provide information about the best primer annealing temperature to be used in the PCR (see below).

Before application to soil DNA extracts, it is primordial that the amplification of any new primer/target system is optimized *in vitro*, i.e. optimal PCR conditions as to melting, primer annealing and extension temperatures, duration of each step per cycle and number of cycles, concentration of reagents, in particular $MgCl_2$, and use of additives which enhance the fidelity of the PCR reaction, should be sought. Often, it is convenient to start optimization using 25–40 times a standard cycle of 94 °C for melting (1 min), 55 °C for primer annealing (1 min) and 72 °C (1–2 min depending on the length of the target sequence) for extension, at a Mg^{2+} concentration of 2.5 mM in dependance of the enzyme used (for ampliTaq DNA polymerase Stoffel fragment, 3.75 mM is used in our laboratory). The primer annealing temperature will depend on the G+C% of both primers. Even though the computer program used for design indicates an optimal annealing temperature, it is often useful to experimentally test it. Starting with standard conditions as to cycling and reaction conditions, PCR may be fine-tuned with regards to fidelity and sensitivity, by varying cycling parameters such as the number of cycles, temperature and duration of each step, the Mg^{2+} concentration and the use of additives (adjuvants) such as formamide, T4 gene 32 protein [19], bovine serum albumin or formamide [2]. As an example, initial results with new PCR systems are sometimes plagued by the appearance of aspecific products (incorrect size) or smears which do not hybridize to a template-specific probe, or even by the total absence of any product. In the former case, this suggests PCR conditions were too permissive, e.g. due to a too low initial annealing temperature, resulting in priming at wrong sites, whereas in the latter case the initial annealing temperature might have been too high or the Mg^{2+} concentration suboptimal.

To improve PCR specificity, several strategies have been devised which are of relevance for amplifications of soil DNA. The use of a "hot start" of the PCR reaction, which is based on completing the PCR reaction mix at denaturing temperature, is strongly recommended, since it avoids initial mispriming by only allowing extension to occur with all reaction components mixed at denaturing temperature. In practice, the hot start can be achieved by preparing PCR reaction mixes lacking one key component (e.g. target DNA or enzyme) at room temperature, and then adding the missing key component at denaturing temperature. Booster PCR, which to enhance specificity runs a few initial cycles at low primer concentrations followed by the remainder of cycles with standard primer concentrations, has been successfully applied to soil DNA [9]. Touch-down PCR attempts to limit initial priming to high-fidelity (primer-target) annealing by using restrictive (high) initial primer annealing temperatures, which in the following cycles are progressively lowered to allow for more efficient subsequent amplification cycles. Finally, nested PCR enhances specificity by allowing a first round of amplification with "external" primers to be followed by a second one with internal primers, using fresh enzyme and reaction components and PCR product of the first amplification round as the template. Further details on these potential improvements can be found in the Journal "Amplifications", freely available with Perkin-Elmer/Cetus. See also Erlich [3] and Innis *et al.* [5].

An *in vitro* optimized amplification protocol can be tested on soil DNA, however it may show to work poorly due to inhibitions from soil. Hence, PCR amplification of target sequences in soil-derived DNA requires fair purity of the DNA preparation. The degree of purity also determines the maximum volume of DNA extract allowable in the PCR reaction mix (of 50 µl) without causing inhibition of the *Taq* DNA polymerase, thereby affecting the sensitivity of detection via PCR. We have routinely applied 1 to maximally 3 µl of DNA extract prepared according to the direct extraction protocol (Chapter 1.3.3) to 50 µl PCR reaction volumes, and found these concentrations allowed for consistent amplification of the target DNA. When working with soil DNA, it is primordial to check for the activity of the DNA polymerase in reaction mixes with soil extracts, by adding pure target DNA. In addition, possible contamination of samples by target DNA should also be carefully controlled in appropriate negative controls without added template. PCR products should be assayed by ethidium bromide visualization after electrophoresis on agarose gels [13], and it is often necessary to control the identity of the amplification products by blotting and hybridization to a specific internal probe or by direct restriction analysis. For the latter purpose, it is useful to have internal restriction sites present in the PCR target region [23].

Procedures and cautionary notes for setting up PCR facilities

Since DNA amplification via PCR is extremely sensitive, care should be taken not to contaminate the sample material or new PCR reaction mixes with target DNA or PCR products resulting from previous reactions, since this might lead to false positives. Aerosols which may form when samples containing target DNA or PCR product are handled, are notorious sources of contamination. Therefore, sample preparation and processing, setting up of the PCR reaction mixes, and in particular handling of the PCR products all have to be performed with extreme care. To avoid the occurrence of false positive results, we find it adequate to have separate sample preparation, PCR and product analysis rooms, with separate equipment, including pipettes. In addition, all glass- and plasticware used in the PCR room should be exclusively handled there. Further, PCR reagents should only be prepared in the PCR room, and divided in aliquots, which are then stored at −20 °C.

A notorious problem when using PCR amplification systems based on conserved regions of the small subunit (16S) ribosomal gene, is the presence of these bacterial sequences as contaminants in many commercially available enzymes. To avoid amplification of this instead of target DNA, resulting in false results when amplifying environmental DNA with conserved 16S primers, enzyme solutions as well as PCR reaction mixes should be treated so as to remove any 16S ribosomal gene sequences. Several strategies have been developed for this, e.g. treatment with psoralen, UV irradiation or treatment with DNase [12]. We commonly treat the PCR reaction mixes and the *Taq* polymerase with DNase I, as outlined below.

In the description of the PCR protocol below, we will only address points 4 and 5 raised above since the other points are too comprehensive. Aspects are treated extensively in PCR manuals [3, 5], as well as the literature [e.g. 9, 12, 23].

Treatment of (Taq) DNA polymerase with DNase I to remove contaminating bacterial DNA

1. Add 10 µl DNase I (10 U/µl) to PCR reaction mix (without *Taq* polymerase and primers) for 10 reactions.

2. Add 0.1 μl DNase I to 5 μl *Taq* DNA polymerase Stoffel fragment (10 U/μl).
3. Incubate tubes for 30 min at 37 °C, then inactivate DNase I by heating at 98 °C for 10 min. The reaction mix as well as enzyme can be used directly for PCR.

PCR for detection of target DNA in soil DNA samples

Steps in the procedure

Used enzyme: AmpliTaq DNA Polymerase, Stoffel fragment (Perkin Elmer/Cetus; Cat. no. N808–0038). In our laboratory, this enzyme was shown to be least inhibited by soil impurities. However, several other enzymes may work as well on soil-derived DNA.

1. Prepare master mix of reagents for any number of 50 μl reactions. Each mixture contains (final amounts in 50 μl final reaction volume):

Sterile deionized water	23.45 μl	
Stoffel buffer*10×	5 μl	(final concentration 1×)
Each dNTP (1 mM) mix –5×	10 μl	(final concentration 200 μM)
$MgCl_2$ 25 mM	7.5 μl	(final concentration 3.75 mM)
Primer 1 10 μM	1 μl	(final concentration 0.2 μM)
Primer 2 10 μM	1 μl	(final concentration 0.2 μM)
Formamide**	0.5 μl	(final concentration 1%)
T4 gene 32 protein (5 mg/ml)*	0.05 μl	(final conc. 0.25 μg/50 μl)
Taq DNA polymerase Stoffel fragment (10 U/μl)	0.5 μl	(final conc. 0.1 U/μl)

Total volume 49 μl. Cover with 2 drops of heavy mineral oil. An aliquot of 1 μl soil DNA extract is added through the oil to the reaction mixture containing all other components in the "hot start" procedure (see 3) prior to starting thermal cycling.

Note
* Stoffel buffer: commercially-available buffer (Perkin Elmer/Cetus) for Stoffel fragment.
** Formamide (Merck, Darmstadt) is added to enhance specific primer annealing [2]. Its effect is shown in Fig. 2. T4 gene 32 protein (Boehringer, Mannheim) is added since it enhances the stability of single-stranded DNA, facilitating primer annealing [19]. We found a 6-fold lower concentration than that used by Tebbe and Vahjen [19] to be optimal (Fig. 1, lanes 11–13).

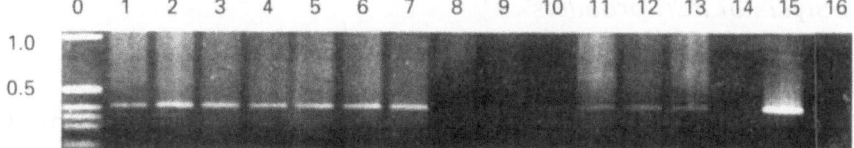

Figure 1. Detection in soils of *Pseudomonas fluorescens* carrying inserted marker genes (chr::KTG, product size 0.4 kb; 17) by using PCR applied on total soil DNA extracts. Effect of added T4 gene 32 protein on PCR efficiency. Lanes: 0, 1-kb ladder; 1–7, DNA obtained from Flevo silt loam; 8–14, DNA obtained from Löss; 15, positive control; 16, negative control. Amount of T4 gene 32 protein added per PCR reaction for lanes: 1, 8: 1.5 μg; 2, 9: 1.25 μg; 3, 10: 1 μg; 4, 11: 0.75 μg; 5, 12: 0.5 μg; 6, 13: 0.25 μg, 7, 14: no addition.

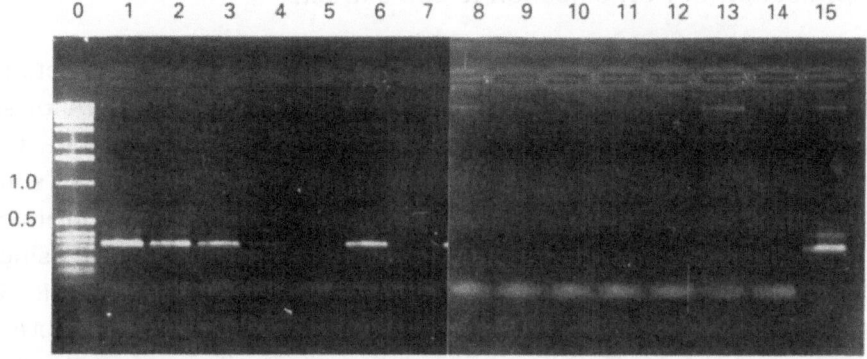

Figure 2. Effect of formamide addition on the efficiency of PCR amplification of part of the *cry*IVB gene (product size 0.33 kb) in a silt loam soil DNA extract. Lanes 1–5, amplification with 1% formamide added to reaction mix; 6–10, without formamide. Lanes: 0: 1 kb ladder (Gibco-BRL Life technologies, Gaithersburg, MD, USA), 1 and 8, 2 and 9, 3 and 10, 4 and 11, 5 and 12: soil DNA extracts with, respectively undiluted, 10-fold, 100-fold, 1000-fold and infinitely diluted target DNA, 6, 15: positive control (*P. fluorescens (chr::KTG)* DNA), 7, 14: negative control (water).

2. Prepare positive and negative controls, as indicated in the foregoing.

3. "Hot start" procedure: preheat the PCR mixture at 95 °C for 2 min prior to adding enzyme or target DNA, in order to denature any bonds adventitiously or erroneously formed between primers and target.

4. Run PCR in a programmable thermal cycler, for 25 to 40 cycles. Often, the machine can be set so as to run the amplification cycles overnight.

5. After thermal cycling, perform a final primer extension at 72 °C (e.g. 10 min), and keep reaction mixtures at 4 °C until analysis.

Analysis of PCR products

6. Analyse PCR products on 0.8–1.2% agarose gel [13].
7. Blot gel to a nylon or nitrocellulose membrane for hybridization analysis [13] using a specific internal probe.
8. (Alternatively) restrict PCR product with adequate restriction enzyme to confirm identity [13].
9. (Eventually) clone PCR product using a PCR cloning vector (e.g. pCR, Invitrogen), and determine sequence [13].

Application of PCR detection to soil extracts

PCR detection of specific gene sequences in soil DNA extracts has been routinely applied in our laboratory. With target DNA of sufficient purity, PCR was shown to be efficient to detect down to 10^2 to 10^3 cells with low copy numbers of target sequences per g of soil in several different soils tested [1, 15]. For instance, introduced genetically modified pseudomonads have been detected using a PCR amplification system based on the *pat* marker on plasmid RP4 [15] at a limit of detection of 10^3 added cells per g soil. Further, *Mycobacterium chlorophenolicum* was detected in soil at a detection limit in the order 10^2 cells per g soil [1]. Similar limits of detection have been reported by Tsai and Olson [22], who detected *Escherichia coli* cells in soil by using a 7-copy 16S ribosomal RNA sequence as a target, and by Picard *et al.* [9], who detected *Agrobacterium tumefaciens* cells in soil with an optimized PCR protocol based on a region of the *vir* gene as a target [9]. Furthermore, the application of two rounds of amplification reportedly permitted the detection of transposon Tn*5* labelled *Rhizobium leguminosarum* cells at levels of 1–10 cells per g soil [10].

Figure 1 shows an example of the detection of cells of *Pseudomonas fluorescens* carrying a unique marker gene cassette (KTG; [17]) as a chromosomal marker in two soils, Flevo silt loam and Löss silt loam, by specific PCR followed by analysis of the PCR product on agarose gel. PCR analysis of total DNA extracts also enabled us to detect naturally occurring *npt*II as well as *Inc* determinant sequences in soil microbial communities [16].

Acknowledgements

This work was supported by grants provided by the EC-BIOTECH (BIO2–92–0491) and EC-Environment programmes. The Netherlands Integrated Soil Research Programme also provided financial support. We thank the collaborators in the Soil Biotechnology Laboratory for help with several experiments.

References

1. Briglia M, Eggen R, De Vos WM, Van Elsas JD (1994) Detection in soil of *Mycobacterium chlorophenolicum* using a PCR system based on 16S ribosomal sequences and a specific gene probe. Appl Environ Microbiol, submitted.
2. Bruce KD, Hiorns WD, Hobman JL, Osborn AM, Strike P, Ritchie DA (1992) Amplification of DNA from native populations of soil bacteria by using the polymerase chain reaction. Appl Environ Microbiol 58: 3413–3416.
3. Erlich HA (1989) PCR Technology. Principles and Applications for DNA Amplification. New York/London: Stockton Press.
4. Holben WE, Jansson JK, Chelm BK, Tiedje JM (1988) DNA probe method for the detection of specific microorganisms in the soil bacterial community. Appl Environ Microbiol 54: 703–711.
5. Innis MA, Gelfand DH, Sninsky JJ, White TJ (1990) PCR Protocols. San Diego: Academic Press.
6. Liesack W, Stackebrandt E (1992) Occurrence of novel groups of the domain bacteria as revealed by analysis of genetic material isolated from an Australian terrestrial environment. J Bacteriol 174: 5072–5078.
7. Muyzer G, De Waal EC, Uitterlinden AG (1993) Profiling of complex microbial populations by denaturing gradient gel electrophoresis analysis of polymerase chain reaction-amplified genes coding for 16S rRNA. Appl Environ Microbiol 59: 695–700.
8. Ogram A, Sayler GS, Barkay TJ (1987) DNA extraction and purification from sediments. J Microbiol Meths 7: 57–66.
9. Picard C, Ponsonnet C, Paget E, Nesme X, Simonet P (1992) Detection and enumeration of bacteria in soil by direct DNA extraction and polymerase chain reaction. Appl Environ Microbiol 58: 2712–2722.
10. Pillai SD, Josephson KL, Bailey RL, Gerba CP, Pepper IL (1991) Rapid method for processing soil samples for polymerase chain reaction amplification of specific gene sequences. Appl Environ Microbiol 57: 2283–2286.
11. Ritz K, Griffiths BS (1994) Potential application of a community hybridization technique for assessing changes in the population structure of soil microbial communities. Soil Biol Biochem 26: 963–971.
12. Rochelle PA, Will JAK, Fry JC, Jenkins JS, Parkes RJ, Weightman AJ (1994) Extraction and amplification of 16S rRNA genes from deep marine sediments and seawater to assess bacterial community diversity. In: Trevors JT, Van Elsas JD (eds) Nucleic Acids in the Environment; Methods and Applications. Heidelberg: Springer Verlag, in press.
13. Sambrook J, Fritsch EF, Maniatis T (1989) Molecular Cloning. A Laboratory Manual. Second Edition. New York: Cold Spring Harbor.
14. Sayler GS, Shields MS, Tedford ET, Breen A, Hooper SW, Sirotkin KM, Davis JW (1985) Application of DNA-DNA colony hybridization to the detection of catabolic genotypes in environmental samples. Appl Environ Microbiol 49: 1295–1303.
15. Smalla K, Cresswell N, Mendonca-Hagler LC, Wolters AC, Van Elsas JD (1993) Rapid

DNA extraction protocol from soil for polymerase chain reaction-mediated amplification. J Appl Bacteriol 74: 78-85.

16. Smalla K, Van Overbeek LS, Pukall R, Van Elsas JD (1993) Prevalence of *npt*II and Tn5 in kanamycin-resistant bacteria from different environments. FEMS Microbiol Ecol 13: 47-58.

17. Smit E, Van Elsas JD (1992) Conjugal gene transfer in the soil environment: new approaches and developments. In: Gauthier M (ed) Gene Transfer and Environment, pp. 79-94. Heidelberg: Springer-Verlag.

18. Steffan RJ, Goksoyr J, Bej AK, Atlas RM (1988) Recovery of DNA from soils and sediments. Appl Environ Microbiol 54: 2908-2915.

19. Tebbe CC, Vahjen W (1993) Interference of humic acids and DNA extracted directly from soil in detection and transformation of recombinant DNA from bacteria and a yeast. Appl Environ Microbiol 59: 2657-2665.

20. Torsvik V (1994) Extraction, purification and analysis of DNA from soil bacteria. In: Trevors JT, Van Elsas JD (eds) Nucleic Acids in the Environment: Methods and Applications. Heidelberg: Springer Verlag, in press.

21. Trevors JT, Van Elsas JD (1989) A review of selected methods in environmental microbial genetics. Can J Microbiol 35: 895-902.

22. Tsai Y-L, Olson BH (1992) Detection of low numbers of bacterial cells in soils and sediments by polymerase chain reaction. Appl Environ Microbiol 58: 754-757.

23. Van Elsas JD, Van Overbeek LS, Fouchier R (1991) A specific marker, *pat*, for studying the fate of introduced bacteria and their DNA in soil using a combination of detection techniques. Plant Soil 138: 49-60.

Molecular Microbial Ecology Manual **2.7.5**: 1-14, 1995.

Detection of mRNA and rRNA via reverse transcription and PCR in soil

SONJA SELENSKA-POBELL

Department of Genetics, University of Bayreuth, D-95 440 Bayreuth, Germany

Introduction

Only an estimated 20% of naturally occurring bacteria have been isolated from their habitats and can be grown and studied in laboratory conditions [38, 39]. The reason for this is, that for the majority of the microorganisms in the environment it is still impossible to mimic the natural conditions required for their proliferation, and they can not be handled under laboratory conditions [6, 38]. In addition, many microorganisms are very strongly bound to sediment or soil particles and thus their extraction by conventional methods is not effective [23].

Molecular biological methods offer new opportunities for characterizing structures of complex microbial communities. By amplification of 16S rRNA genes using the polymerase chain reaction (PCR) directly in natural samples and consequent hybridization with oligonucleotides complementary to specific regions of these genes, the presence of particular members of the natural bacterial communities can be determined [2, 15, 36, 38]. Detection and monitoring of many bacteria deliberately introduced into the environment can also be performed by PCR amplifications of their particular intrinsic [7, 8, 15, 22, 29] or especially inserted marker sequences [10, 30, 31, 33] in total DNA recovered from the fields where they have been released.

However, for a better understanding of the behaviour and function of natural microorganisms and of the fate of released bacteria in different environments not only their presence, but also their activity should be studied. Microbial gene expression in the environment can be studied on the level of proteins – by measuring specific enzyme activities [14, 18] or by analyzing protein synthesis [20] as well as on the level of mRNA – by direct analysis of the gene transcripts [16, 24, 25, 31, 35]. The latter is preferable, because the cellular content of many enzymes is regulated not only on the transcriptional – but also on the translational level [34]. By mRNA analysis the transcription of particular bacterial genes in the environmental samples can be studied. However, for the identification of microorganisms from natural communities carrying the gene of interest, additional analyses should be performed. The unique structure of 16S rRNA, which is characterized by the presence of highly conserved regions and ones which display sequence variability at the genus and species levels, makes these molecules an excellent target for developing identification procedures for bacteria. By using oligonucleotides complementary to specific rRNA regions as hybridization probes, *in situ* analysis of the complex natural microbial communities can be performed [1, 3, 4, 11, 17].

On the other hand, it is well known that the variations in ribosome content are intimately linked to growth regulation of bacteria [20]. In metabolically active bacteria rRNAs are naturally present in high copy numbers, which can be up to 10,000 molecules per cell [37]. In addition, in contrast to the mRNA molecules, which are in general rather labile [16], rRNA molecules are much more stable, because of their specific, compact secondary structures and their isolation from the environmental samples can be performed with high yields [13, 19, 21]. For these reasons, rRNA molecules can be used not only for identification of bacteria [13, 17, 21], but also for the measurement of their natural activity [19].

However, in natural conditions many bacteria may be in a dormant state, in which the expression of many of their genes can be strongly inhibited [31, 34] and the amount of their ribosomes reduced [6, 21]. The latter can make the *in situ* identification of such microorganisms by oligonucleotide rRNA probing very difficult. In these cases, reverse transcriptase dependent (RT) – PCR amplification of total RNA recovered by direct lysis from the environmental samples can be very useful [13, 21, 23, 31, 37].

In this chapter the RT-PCR amplification procedures of mRNA and rRNA for the detection of bacteria in soil environmental samples will be presented.

Procedures

I. Reverse transcriptase linked (RT-PCR) amplification of mRNA recovered from soil by direct lysis

The mRNA template for the experiment described can be prepared by the method of Selenska-Pobell (see Ch. 1.5.1., and ref. 31), or by the method published by Tsai *et al*. [35]. The latter is preferable for mRNA analyses as this method is very fast and the danger of losing mRNA, because of the short half life and lability of many classes of these molecules, is minimized. However, an additional purification on the mRNA recovered by the method of Tsai *et al.* [35] is necessary for successful RT-PCR amplification. This purification can be performed by the liquid phase procedure as described in Ch. 1.5.1. of this manual.

An example of the detection of a 550 bp fragment of the Tn5 *npt* II-gene transcript in total soil RNA will be presented. The detection of this gene and it's natural expression is important because Tn5 may be present in many soils and sediments as a result of the wide application of this marker in laboratory [32] and field studies [12, 26, 27, 33, 40] during the last decade. In general, the introduction in natural fields of Tn5 is undesirable, because of the existing probability for transfer of the latter into eventual recipients in the environment and to cause unpredictable changes in them [9]. The introduction of new bacterial inoculants into fields contaminated with Tn5 should be also avoided. For this reason, soils where bacterial releases are planned should be tested for the presence of Tn5 or other movable genetic elements.

Steps in the procedure

1. *Transcription of the nptII mRNA by reverse transcriptase reaction:*
   1.1 Compose a 30 µl reaction sample as follows:
      - 10 µl double distilled sterile water
      - 4 µl of 10 mM $MnCl_2$ solution
      - 3 µl of 10×rTth reverse transcriptase buffer
      - 1 µl of a 10 mM solution of dATP
      - 1 µl of a 10 mM solution of dCTP
      - 1 µl of a 10 mM solution of dGTP
      - 1 µl of a 10 mM solution of dATP
      - 3 µl (2 µM) of the downstream primer (see materials and solutions)
      - 2 µl total soil RNA (approximately 5 µg)
      - 4 µl of rTth DNA polymerase (10 U)
   1.2 Overlay the mixture with 50 µl of mineral oil, close the tubes and incubate in a thermal cycler at 70 °C for 15 min.
   1.3 Place the tube on ice and keep it there for the next step.

2. PCR amplification

Two procedures for performing of this step will be presented. The first procedure is preferable for amplifications of very well puri-fied RNA. However, the total RNA recovered from some soils samples remains yellow-brown and contaminated with humic compounds even after several cycles of spun-column filtration. Then the second one of the procedures presented below is recommended for PCR amplifications. This procedure was shown to be successful for the amplification of the *npt* II-gene even in rather contaminated samples of total soil RNA [31]. A simple liquid phase purification of total soil RNA as described in Ch. 1.5.1. is usually enough for successful RT-PCR experiments involving PCR procedure 2.

Steps of the PCR procedure # 1:

2.1.1 Prepare 80 µl of a reaction mixture consisting of:
   –   60 µl of sterile double distilled water
   –   8 µl of the 10× chelating buffer 1
   –   10 µl of a 25 mM $MgCl_2$ solution
   –   2 µl (1 µM) of the upstream *npt* II primer (see materials and solutions)
2.1.2 Combine the resulting 80 µl mixture with a 20 µl aliquote of the reverse transcription tube, mix and centrifuge the tube for 20 seconds in a microcentrifuge.
2.1.3 Add 50 µl of mineral oil and perform the PCR amplification under the following conditions:
   –   Melt the mRNA/cDNA complex by incubation for 2 min at 95 °C.
   –   Perform 45 cycles with the parameters: 1 min at 95 °C and 1 min at 60 °C.
   –   Extend the reaction by incubation for 10 minutes at 72 °C.
   –   Stop the reaction by placing the tube on ice. The resulting sample should be kept at minus 20 °C for further analyses.

Steps of the PCR procedure # 2:

2.2.1 Prepare 80 µl of a mixture consisting of:
- 49 µl sterile double-distilled water
- 8 µl at a 25 mH HgCl$_2$ solution of the 10× chelating buffer 2
- 20 µl MgCl$_2$ solution
- 1 µl (10 U) of Ampli Taq DNA polymerase (Stoffel fragment)

2.2.2 Combine the resulting mixture with the rest of 10 µl of the transcription reaction sample, mix well and centrifuge for 20 sec in a microcentrifuge.

2.2.3 Add 50 µl of mineral oil and incubate in a thermal cycler at the following conditions:
- Melt the mRNA/cDNA complex for 3 min at 95 °C.
- Perform 50 cycles as follows: 1 min at 98 °C, 1 min at 55 °C and 2 min at 72 °C.
- Extend the reaction for 10 min at 72 °C.
- Stop the reaction by placing the tube on ice. The resulting samples should be kept at minus 20 °C until needed.

Analysis of the npt II mRNA amplification products

1. *Electrophoretical analysis:*
   Take 10 µl of the PCR reaction mixture and mix with 2 µl of a DNA-sample loading buffer. Load the sample into a well of a 2% agarose gel and perform an electrophoresis at 2 V/cm. After finishing the electrophoresis stain the gel for 20 min with 1 µg/ml water solution of ethidium bromide and take a picture under UV light.

2. *Confirmation the identity of the RT-PCR products:*
   Total soil or sediment RNA samples are very complex targets because they consist of molecules which originate from many different microorganisms. For this reason, the determination only of the size of the PCR-amplification products is not enough to confirm the presence of the gene of interest in the samples investigated. For discrimination of specific from unspecific PCR-fragments additional analyses of the RT-PCR products, should be performed (see below).

2.1. Hybridization with a 930 bp Pst I fragment of the npt II-gene:
The identity of the products obtained after RT-PCR amplification of the *npt* II in total soil RNA samples can be confirmed in hybridization experiments. A 930 bp *Pst* I fragment of the *npt* II gene [5] can be used as hybridization probe (see Fig. 1). Both PhotoGene (Gibco-BRL) or DIG (Boehringer) nonradioactive techniques can be used according to the standard procedures described by the manufacturer (see chapters 2.1 and 2.4)

2.2. Restriction analysis of the resulting PCR products:
This analysis is extremely easy and can be used to confirm the identity of the RT-PCR amplified sequences in a few hours instead of days, as in the case of the hybridization experiment presented above. The method is preferable in cases when the exact sequences of the amplified products is known and/or if

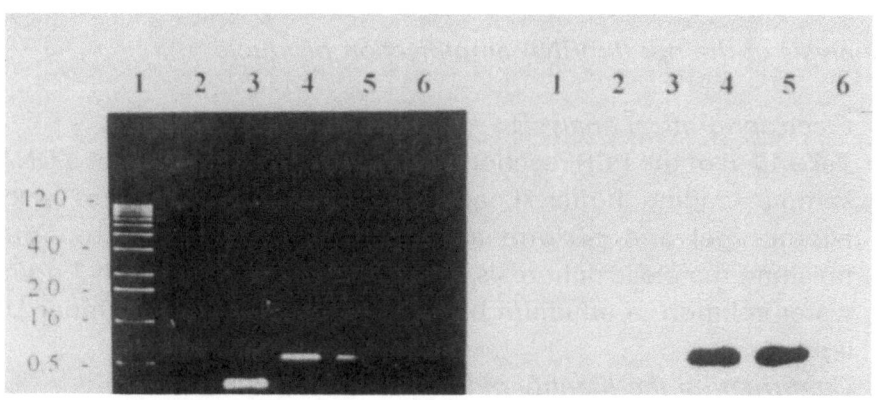

Figure 1. A) RT-PCR amplification of soil RNA: 1 - kilobase ladder (GIBCO-BRL); 2 - amplification in total RNA, recovered from noninoculated soil; 3 - amplification of the positive control mRNA (0.5 µg), supplied by Perkin Elmer Cetus, in a sample containing 5 µg of RNA, recovered from noninoculated soil; 4 - amplification in total bacterial RNA of *Enterobacter agglomerans* (0.5 µg); 5 - amplification in soil RNA (5 µg) recovered from soil 14 days after inoculation with *E. agglomerans.*

appropriate sites for restriction endonucleases exist in the PCR products.

2.2.1 For this analysis first separate the products of the RT-PCR amplification electrophoretically in a 1% low melting Sea Plaque GTG (FMC-Biozym) agarose. Visualize the resulting DNA bands by stain the gel with 1 µg/ml aquaeous solution of ethidium bromide and than excise the 550 bp fragment under UV light.

2.2.2 Perform a digestion of the DNA fragment in the gel as follows:
   - Melt the gel slice by incubation for 10 min at 68 °C.
   - Take 16 µl of the melted gel, put in a new microtube and continue the incubation at 37 °C. Add to the liquid gel 2 µl of buffer 6 (Gibco-BRL), 1 µl BSA and 1 µl (10 U) Sph I restriction endonuclease.
   - Mix well by pipetting and incubate for at least 2 hours at 37 °C.

2.2.3 Add 2 µl DNA-sample loading buffer to the sample, mix by pipetting and load the resulting mixture in a well of a 3% NuSieve 3:1 (FMC-Byozym) agarose gel.

2.3.4 Perform an electrophoresis at 2 V/cm for 2–3 hours, stain the gel with ethidium bromide solution and analyse under UV light. As a result of the digestion of the 550 bp fragment of the npt II-gene, two DNA fragments with sizes of 250 bp and 300 bp can be visualized (see Fig. 2).

II Reverse transcriptase linked (RT-PCR amplification of rRNA recovered by direct lysis from soil

Steps in the procedure:

1. Transcription of the 16S rRNA by reverse transcriptase reaction
   1.1 Compose a 20 µl reaction sample as follows:
      - 8 µl double distilled sterile water
      - 3 µl of a 10 mM solution of $MnCl_2$
      - 2 µl of rTth reverse transcriptase buffer
      - 0.5 µl of a 10 mM solution of dATP

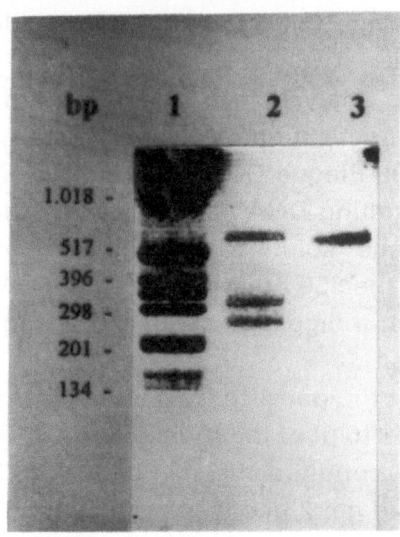

Figure 2. Restriction analysis of the RT-PCR amplified 550 bp sequence of the *npt* II gene:
1 - kilobase ladder; 2 - *Sph* I in gel digested RT-PCR product; 3 - nondigested RT-PCR product.

- 0.5 µl of a 10 mM solution of dCTP
- 0.5 µl of a 10 mM solution of dGTP
- 0.5 µl of a 10 mM solution of dTTP
- 2 µl (1 µM) of the downstream 16S rRNA primer described below (see materials)
- 2 µl total soil RNA (approximately 5 µg)
- 1 µl or rTth DNA polymerase (5 U).

1.2 Overlay the mixture with 50 µl of mineral oil, close the tube and incubate in a thermal cycler for 15 min at 70 °C.

1.3 Stop the reaction by placing the tube on ice.

2. *PCR amplification*

2.1 Prepare the reaction mixture for the PCR amplification step:
- 56 µl of sterile double distilled water
- 8 µl of the 10× chelating buffer 1
- 12 µl of a 25 mM solution of $MgCl_2$
- 2 µl (1 µM) of the upstream primer (see materials and solutions)

–   1 μl of rTth DNA polymerase (5 U)
2.2 Add the resulting 80 μl mixture to the reverse transcription tube (1.3) above) mix well and spin the mixture down in a microcentrifuge.
2.3 Perform 45 cycles in a thermal cycler with the following parameters:
–   1 min at 94 °C
–   1 min at 55 °C
–   3 min at 72 °C.
2.4 Extend the reaction by incubation for 15 min at 72 °C.
2.5 Stop the reaction by placing the tube on ice.

Analysis of the rRNA amplification products

1. *Electrophoretical analysis:*
   Perform an electrophoretical analysis of a 10 μl aliquot of the RT-PCR products from step 2.5 above in 2% agarose.
2. *Purification of the PCR product and hybridization with specific 16S rRNA oligonucleotides:*
   2.1 Perform a preparative gel electrophoresis of the PCR products from step 2.5 above in 1% Sea Plaque agarose (FMC-Biozym). Stain the gel in a water solution of ethidium bromide (1 μg/ml), excise the DNA band containing the amplified rRNA sequence with a size of approximately 1,500 bp. Elute and precipitate the DNA fragments as described by Sambrook *et al.* [28].
   Resuspend the eluted PCR amplified material in 10 to 30 μl of double distilled water, dependent on the efficiency of the amplification reactions.
   2.2 Denaturate 10 μl of the purified PCR product for 15 min at 65 °C with 30 μl of denaturation solution (see below). Cool on ice and load the sample on nylon membrane by using the Hybri-slot apparatus (Gibco-BRL). Fix the DNA to the nylon membrane by cross linking with UV light and bake in vacuum for 20 min at 80 °C.
   2.3 Perform hybridization with a digoxigenin labelled oligonucleotide representing the universal eubacterial probe [2, 21] (see Materials and solutions and chapters 2.1.1, 2.3.2 and 3.3.6).

Materials and solutions:

In the RT-PCR-amplification experiments the GeneAmp Thermostable rTth DNA polymerase from PERKIN ELMER CETUS should be used.

For the DNA PCR AmpliTaq DNA Polymerase, Stoffel fragment can be also used.

Structure of the downstream primer for the amplification of the *npt* II mRNA:

5′-GCAGCTGTGCTCGACGTTGTCACTGAAG-3′.

This oligonucleotide is complementary to the antisense DNA strand and it is situated 177 nucleotides after the beginning of the open-reading frame of the *npt* II gene [5, 31].

Structure of the upstream primer for the amplification of the *npt* II mRNA:

5′-CAAGAAGGCGATAGAAGGCGATG-3′.

This oligonucleotide is complementary to the sence DNA strand and it is situated six nucleotides before the two stop codons at the end of the gene [5, 31].

Structure of the downstream primer for the amplification of the 16S rRNA:

5′-AGAGTTTGATYMTGGCTCAG-3′.

This primer corresponds to the positions 8-27, according to the 16S rDNA nomenclature of *Escherichia coli* [15].

Structure of the upstream primer for the amplification of the 16S rRNA:

5′-AGAAAGGAGGTGATCC-3′

This primer corresponds to the positions 1540–1528, according to the 16S rDNA nomenclature of *Escherichia coli* [15].

Structure of the 16S rRNA universal eubacterial oligonucleotide probe for hybri-slot hybridizations [2, 21]:

5′-GCTGCCTCCCGTAGGAGT-3′

This oligonucleotide is complementary to all bacterial 16S rRNA so far sequenced, and can be used as a positive control by testing the RT-PCR amplifications in total soil RNA.

10× rTth reverse transcriptase buffer:
  100 mMTris-HCl pH 8.3
  900 mM KCl

10 mM dATP
10 mM dCTP
10 mM dGTP
10 mM dTTP

Chelating buffer 1.
   50% glycerol
   100 mM Tris-HCl, pH 8.3
   1 M KCl
   7.5 mM EGTA
   0.5% Tween 20

Chelating buffer 2:
   50% glycerol
   100 mM Tris-HCl pH 8.3
   7.5 mM EGTA
   0.5% Tween 20

   10 mM $MnCl_2$
   25 mM $MgCl_2$
   BSA – Albumin from bovine serum – 20 mg/ml Boerhringer Mannheim.

DNA-sample loading buffer:
   4 M Urea
   50% sucrose
   50 mM EDTA
   0.25% bromphenolblue
   0.25% xylencyanol

MOPS buffer pH 7.5:
   20 mM MOPS
   5 mM Natrium acetat
   1 mM EDTA

Denaturating solution:
   1.2 ml MOPS buffer
   6 ml Formamid
   1.95 ml Formaldehyd

References

1. Amann RI, Krumholz L, Stahl DA Fluorescent oligonucleotide probing of whole cells for determinative, phylogenetic and environmental studies in microbiology (1990) J Bact 172: 672–770.
2. Amann RI, Springer N, Ludwig W, Schleifer K-H, Peterson N (1991) Identification *in situ* and phylogeny of uncultured bacterial endosymbionts. Nature 351: 161–164.
3. Amann RI, Stromley J, Devereux R, Key R, Stahl DA (1992) Molecular and microscopic identification of sulfate-reducing bacteria in multispecies biofilms. Appl Environ Microbiol 58: 614–623.
4. Amann RI, Zarda B, Stahl DA, Schleifer K-H (1992) Identification of individual procariotic cells by using enzyme-labelled rRNA-targeted oligonucleotide probes. Appl Environ Microbiol 58: 3007–3011.
5. Beck E, Ludwig G, Auerswald EA, Reiss B, Schaller H (1982) Nucleotide sequence and exact localisation of the neomycin phosphotransferase gene from transposon Tn5. Gene 19: 327–336.
6. Colwell RR, Brayton PR, Grimes DJ, Rosak DB, Huq SA, Palmer LM (1985) Viable but non-culturable *Vibrio cholerae* and related pathogens in the environment: implications for release of genetically engineered microorganisms. Bio/Technology 3: 817–820.
7. DeLong EF, Wickham GS, Pace NR (1989) Phylogenetic strains: Ribosomal RNA-based probes for the identification of single cells. Science 243: 1360–1363.
8. Erb RW, Wagner-Döbler I (1993) Detection of polychlorinated biphenyl degradation genes in poluted sediments by direct DNA extraction and polymerase chain reaction. Appl Environ Microbiol 59: 4065–4073.
9. Evguenieva-Hackenberg E, Selenska-Pobell S, Klingmüller W (1994) Persistance and stability of genetically manipulated derivatives of *Enterobacter agglomerans* in soil microcosms. FEMS Microbiol Ecol 15: 179–192.
10. Flemming CA, Leung KT, Lee H, Trevors JT, Greer CW (1994) Survival of lux-lac-marked biosurfactant-producing *Pseudomonas aeruginosa* UG 2L in soil monitored by nonselective plating and PCR. Appl Environ Microbiol 60: 1606–1613.
11. Hahn D, Amann RI, Ludwig W, Akkermans ADL, Schleifer K-H (1992) Detection of microorganisms in soil after *in situ* hybridization with rRNA-targeted, fluorescently labelled oligonucleotides. J Gen Microbiol 138: 879–887.
12. Frederickson JK, Bezdicek DF, Brockman FJ, Li SW (1988) Enumeration of Tn5 mutant bacteria in soil by using a most-probable-number-DNA hybridization procedure and antibiotic resistance. Appl Environ Microbiol 54: 446–453.
13. Hahn D, Kester R, Starrenburg MJC, Akkermans ADL (1990) Extraction of ribosomal RNA from soil for detection of *Frankia* with oligonucleotide probes. Arch Microbiol 154: 329–335.
14. Heitzer A, Webb OF, Thonnard JE, Sayler GS (1992) Specific and quantitative assessment of naphthalene and salicylate bioavailability by using a bioluminescent catabolic reporter bacterium. Appl Environ Microbiol 58: 1839–1846.
15. Herrick JB, Madsen EL, Batt CA, Ghiorse WC (1993) Polymerase chain reaction amplification af naphthalene catabolic and 16S rRNA gene sequences from indigenous sediment bacteria. Appl Environ Microbiol 59: 687–694.
16. Jeffrey WH, Nazaret S, Von Haven R (1994) Improved method for recovery of mRNA from aquatic samples and its application to detection of *mer* expression. Appl Environ Microbiol 60: 1814–1821.
17. Kane MD, Poulsen LK, Stahl DA (1993) Monitoring the enrichment and isolation of sulfate-reducing bacteria by using oligonucleotide hybridization probes designed from environmentally derived 16S rRNA sequences. Appl Environ Microbiol 59: 682–686.
18. King RJ, Short K, Seidler RJ (1991) Assay for detection and enumeration of genetically engineered microorganisms which is based on the activity of a deregulated 2,4-dichlorophenoxyacetate monooxygenase. Appl Environ Microbiol 57: 1790–1792.

19. Kramer JG, Singleton FL (1993) Measurement of rRNA variations in natural communities of microorganisms on the southeastern U.S. continental shelf. Appl Environ Microbiol 59: 2430–2436.
20. Matin A (1990) Molecular analysis of the starvation stress in *Escherichia coli*. FEMS Microb Ecol 74: 185–196.
21. Moran AM, Torsvik VL, Torsvik T, Hodson RF (1993) Direct extraction and purification of rRNA for ecological studies. Appl Environ Microbiol 59: 915–918.
22. More MI, Herrick JB, Silva MC, Ghiorse WC, Madsen EL (1994) Quantitative cell lysis of indigenous microorganisms and rapid extraction of microbial DNA from sediment. Appl Environ Microbiol 60: 1572–1580.
23. Muyzer G, De Waal EC, Uitterlinden AG (1993) Profiling of complex microbial populations by denaturing gradient gel electrophoresis analysis of polymerase chain reaction-amplified genes coding for 16S rRNA. Appl Environ Microbiol 59: 695–700.
24. Pichard SL, Paul JH (1991) Detection of gene expression in genetically engineered microorganisms and natural phytoplankton populations in the marine environment. Appl Environ Microbiol 57: 1721–1727.
25. Pichard SL, Paul JH (1993) Gene expression per gene dose, a special measure of gene expression in aquatic microcosms. Appl Environ Microbiol 59: 451–457.
26. Pillai SD, Pepper IL (1991) Transposon Tn5 as an identifiable marker in *Rhizobia*: Survival and genetic stability of Tn5 mutant bean *Rhizobia* under temperature stressed conditions in desert soils. Microb Ecol 21: 21–33.
27. Recorbet G, Givaudan A, Steinberg C, Bally R, Normand P, Faurie G (1992) Tn5 to assess soil fate of genetically marked bacteria: screening for aminoglycoside-resistance advantage and labelling specifity. FEMS Microbiol Ecol 12: 97–105.
28. Sambrook J. Fritsch EF, Maniatis T (1989) Molecular Cloning: A Laboratory Manual 2nd adn New York: Cold Spring Harbor Laboratory Press, Cold Spring Harbor.
29. Selenska S, Klingmüller W (1991) Direct detection of *nif*-gene sequences of *Enterobacter agglomerans* in soil. FEMS Microbiol Lett 80: 243-246.
30. Selenska S, Klingmüller W (1991) DNA recovery and direct detection of Tn5 sequences in soil. Lett Appl Microbiol 13: 21–24.
31. Selenska S, Klingmüller W (1992) Direct recovery and molecular analysis of DNA and RNA from soil. Microbial releases 1: 41–46.
32. Simon R (1984) High frequency mobilization of Gram-negative bacterial replicons by the in vitro constructed Tn5-*Mob* transposon. Mol Gen Genet 196: 413–420.
33. Smalla K. Van Overbreek LS, Pukall R, Van Elsas JD (1993) Prevalence of *npt* II and Tn5 in kanamycin-resistant bacteria from different environments. FEMS Microbial Ecology 13: 47–58.
34. Stanier RY, Ingraham JL, Wheelis ML, Painteer PR (1986) The microbial world, 5th ed. Prentice Hall Inc., Englewood Cliffs, NJ.
35. Tsai Y, Park MJ, Olson BH (1991) Rapid method for direct extraction of mRNA from seeded soils. Appl Environ Microbiol 57: 765–768.
36. Tsai Y, Olson BH (1992) Detection of low numbers of bacterial cells in soils and sediments by polymerase chain reaction. Appl Environ Microbiol 58: 754–757.
37. Van Kuppeveld FJM, Van der Logt JTM, Angulo AF, Van Zoest MJ, Quint WGV, Niesters HGM, Galama JMD, Melchers WJG (1992) Genus- and species-specific identification of Mycoplasmas by 16S rRNA amplification. Appl Environ Microbiol 58: 2606–2615.
38. Ward DM, Weller R, Bateson MM (1990) 16S rRNA sequences reveal numerous uncultured microorganisms in a natural community. Nature 345: 63–65.
39. Wayne LG, Brenner DJ, Colwell RR, Grimont PAD, Kandler O, Krichevsky MI, Moore LH, Moore WEC, Murray RGE, Stackebrandt E, Starr MP, Truper HG (1987) Report of the ad hoc committee on reconciliation of approaches to bacterial systematics Int J Syst Bacteriol 37: 463–464.

40. Wilson M, Lindow SE (1993) Release of recombinant microorganisms. Annu Rev Microbiol 4: 913–944.

Molecular Microbial Ecology Manual **3.1.1**: 1–17, 1995

Partial and complete 16S rDNA sequences, their use in generation of 16S rDNA phylogenetic trees and their implications in molecular ecological studies

ERKO STACKEBRANDT and FRED A. RAINEY

DSM – German Collection of Microorganisms and Cell Cultures GmbH, Mascheroder Weg 1b, 38124 Braunschweig, Germany

Introduction

When the first reports of 16S rRNA sequence analysis were published about 20 years ago, the potential of ribosomal RNA species for studies in evolution, phylogeny, systematics and identification [47] and ecology [28, 29] were far from being fully acknowledged. What started with short sequences from a limited range of pro- and eukaryotic species turned out to revolutionize our understanding of (i) the origin of life on the planet Earth, (ii) the development of unknown, ancient, ancestral forms into main lineages leading to extant organisms such as Bacteria, Archaea and Eukaryotes, (iii) the relatedness among recently evolved species, and (iv) the richness of microbial forms in the environment. The results obtained have not only resusciated interest in the work of systematists, but have stimulated the cooperative work of researchers from different disciplines. Both, microbial taxonomy and microbial ecology have revived from Sleeping Beauty-types, somewhat boring fields into exciting disciplines in which objectivity has replaced decades of subjective judgement of facts. Molecular biology, using techniques such as PCR amplification, cloning, sequencing, hybridization and probing has finally been fully accepted by these disciplines. Organisms known to exist as symbionts and parasites were suddenly available for further characterization, and the extent of living but uncultured microorganisms became obvious without, however, allowing an estimation of the order of magnitude with which these dormant forms exceed the culturable organisms. In the past, uncultured strains have been observed in natural samples solely by light and electron microscopy. The determination of phylogenetic relationships of some these obligately inter- and intracellular organisms has been one of the scientific highlights in the eighties. Using the cataloguing approach, the task was easy when the symbiont occurred abundantly as an almost pure culture, as in the case of *Prochloron* species within their ascidian (Didemnid) hosts [38]. The reverse transcriptase technique of 16S rRNA sequencing was used to determine the phylogeny of the extremely slow growing pathogenic organism *Mycobacterium leprae* [39], but the main breakthrough occurred recently when the 16S rDNA of small numbers of cells could be amplified by the PCR approach and sequenced, either directly or following a cloning step. Examples are symbionts of shipworms [8], *Holospora* [1], the nematode lysing *Pasteuria penetrans* [Stackebrandt, Dorsch and Paschke, unpublished], the giant fish bacterium [2], *Buchnera* species which are symbions of aphids [43] or the methanogenic symbionts of anaerobic ciliates [11].

Despite this progress, the majority of prokaryotic organisms in the environment, and the role they play in the the function of an ecosystem, remained unidentified. As the vast majority of organisms cannot be cultured, the number of existing prokaryotic species is a guess, ranging from several thousands to at least as many as there are insect species. Today, microbiologists are at least able to show convincingly that the 3300 validly described species (excluding cyanobacteria), are only a minute fraction of the number, which is probably greater than a million. In the context of conservation of biodiversity, gene-pool preservation and the role of microorganisms in concert with more highly evolved organisms for maintaining the stability of an ecosystem, the basis for a discussion about the importance of microorganisms will be strengthened when the estimated number of species of Bacteria, Archaea, fungi, yeasts, algae and protozoa is scientifically more sound than previously determined, and when the role of the as yet uncultured organisms as partners equal in importance to higher evolved organisms becomes determinable.

This communication will not cover the methodologies of nucleic acid extraction, cloning and sequence analysis. These topics will be covered elsewhere. We assume that the reader is informed about

1. the methodologies applied to determine the presence of genetic variety in an environmental sample; the basic steps of these techniques have been published [17, 24, 26, 46].
2. the results of molecular studies on the composition of the microflora in different environments, such as soil [23, 24, 42], groundwater [10], marine environment ([4, 6, 16–18], bioleach environment [19, 20], hot springs [3, 45], antarctic site [5], invertebrates [1, 8, 9, 11, 27, 43], and human origin [32, 33].
3. and the limitation of these studies to actually (i) quantify the composition of taxa determined [12, 30], and (ii) infer the physiological role the organisms detected play in the environment.

We would rather focus interest on the importance of correct handling of sequence information. It is a well acknowledged fact that the generation of sequence data is the easy part of this kind of ecological study. Sequence alignment, selection and limitation of sequence regions, choice of reference strains and the importance of the data base are parameters that in the end will determine the quality of any environmental study.

The semantides: The search for optimal markers for phylogenetic studies

The potential of macromolecule primary structure comparison for the determination of the phylogenetic structure of prokaryotic species was glimpsed as far back as the 1960s, when the spectrum of evolutionary rates of change within a molecular sequence was recognized [50]. Based on only a few protein sequences, such as cytochrome c and the hemoglobins, it was stated that biological molecules fall into three categories of information. The most prominent ones are the *semantophoric molecules* or *semantides*, which are genes or transcripts thereof (primary: DNA; secondary: RNA; and tertiary: proteins). Genotypic information, i.e. sequences of these molecules are the historical record of evolution, and the determination of the primary structure provided a powerful approach to measure evolutionary relationships, not only between the molecules sequenced but ideally reflecting the evolution of the organisms from whch the sequence was generated. *Episemantic molecules* are synthesized under the control of tertiary semantides, such as ATP, carotenoids, and chemotaxonomic markers. These molecules were not considered useful for deriving evolutionary conclusions because enzymes with different primary structures can lead to the synthesis of identical or similar episemantic molecules in different organisms as long as the active sites of proteins with enzymatic activity are similar. Episemantic molecules are of no phylogenetic value, but they are extremely useful in the characterization of taxa defined by semantides. *Asemantic molecules*, such as vitamins, phosphate ion, oxygen, viruses, do not express any of the information that this organism contains, and consequently are of no phylogenetic value but can eventually be used in the general description of a strain.

As it is presently impossible to use the entire primary structure of DNA, the choice can be made between protein gene sequences, amino acid sequences and ribosomal RNA species. The number and nature of sequence differences among proteins and genes coding for rRNA and proteins reflect their phylogenies, and consequently allow the recognition of pairs or groups of molecules which had been derived from common ancestors, as well as the ordering of the sequenced organisms according to their evolutionary history. In order to exactly measure the time elapsed since the divergence of the most recent ancestor, changes must be selectively neutral and independent of the overlying phenotype and the selection pressure dictated by its function. From this follows that the most reliable molecular chronometers would be represented by a functionless part of the genome that evolves randomly. Considering that there are only a limited number of possible character states at particular sequence position (the four nucleotides), and neutral changes become fixed very rapidly, only very recent evolutionary events could be measured. The reality, however, is different as all sequenced genes (and rRNA species) constitute a chimeric structure of regions with varying degree of conservation. The rate and nature of substitution differ at individual sequence positions depending on their importance for the function of the molecule or the corresponding gene product, but are not subject to neutral changes over long revolutionary changes. Rather, multiple changes are highly likely to accumulate at variable regions; these parts will not reflect the accurate number of mutational events but will obscure the history of the molecules by simulating "false" identities. As molecules are composed of functionally important and functionless regions, the degree of sequence dissimilarity can not be directly correlated with elapsed time. As pointed out below, the information on conserved and variable regions in small subunit RNA is perfectly suited to analyse environmental nucleic acids.

The larger rRNA species fulfill all prerequisites for a suitable phylogenetic marker

Today, the larger rRNA species can be considered the most useful molecular chronometers as they meet all those characteristics that define a phylogenetic marker: they are ubiquitously distributed, hence they have the potential to cover the complete range of taxa. They derived from a common ancestor, hence they are homologous. The function remained similar over billions of years, and as function is determined by sequence and inversely exacts selective pressure on the sequence, a comparable mode of evolution for the most distantly related organisms can only expected for functionally equivalent molecules [47]. Ribosomal RNA genes appear to be genetically stable as they are housekeeping molecules and genetic crossing-over in ribosomal sequences only between highly related organisms has been discussed to occur [40]. 16S and 23S rDNA are reasonably large molecules; this factor is important as the number of detectable phylogenetic levels are increased with the number of independently evolving positions or regions. The primary structures of these marker molecules contain independently evolving domains, allowing the relevance of local sequence identities or differences for phylogenetic conclusions to be tested. Different regions within a rRNA vary at different rates: over a range changing so slowly that the region remains constant in composition over the full expanse of evolutionary time, to so rapidly changing that differences in composition can often be seen in a given region between two strains of the same species.

Several proteins that have been used as marker for phylogeny, such as cytochromes, ferredoxins and azurins [37], ATPase and tuf factor [36] and heat shock proteins [44] are suitable semantides, but not as easy to handle as ribosomal RNAs. The reasons are (i) the degeneracy of the genetic code prevents the formation of conserved regions necessary for the development of PCR amplification primers that could cover the full range of prokaryotic diversity; material to be sequenced must be obtained following time consuming cloning steps, (ii) most proteins used in the past are not ubiquitously distributed, which restricts the analysis to the establishment of only partial phylogenetic trees, and (iii) the occurence of protein families which sometimes makes it difficult to decided whether proteins are truly homologous, or paralogous.

Higher order structures of rRNA sequences facilitate alignment

The conserved role of ribosomal RNA during ribosome assembley and translation has led to the assumption that the conserved function is reflected by structure similarity. This was later shown by demonstrating coordinated nucleotide changes at homologous positions in 16S rRNA sequences from phylogenetically diverse origins. Higher order structures features are visualized at the level of secondary structure by the presence of helices formed by intramolecular interactions of inverse complementary sequence stretches [21, 48]. Although native rRNA interacts in the ribosomes in their tertiary or quarternary structures and long range intramolecular interactions are rarely known, the secondary structure allows a first glance at those regions which might form stable base pairs. 16S rRNA and 23S rRNA molecules contain about 50 and 100 higher order structure elements, respectively. The distribution of helices and loops, e.g. 5' and 3' helix halves of double stranded regions as well as internal or terminal single stranded loops are conserved features of rRNA, although the degree of variation in certains regions may differ from taxon to taxon. These regions (see below, Table 1) can be regarded as homologous elements which can facilitates alignment even if their primary structures differ.

The quality of phylogenetic trees derived from sequence data depends, among other factors outlined below, significantly from the quality of the sequence alignment. It is the goal to arrange homologous residues which are derived from a common position within the ancestral sequence in columns. Given the high content of invariant and conserved positions or regions along the sequence of rRNAs, aligning of these regions is a straight forward procedure. However, in contrast to most proteins, rRNAs differ significantly in length. Although automatic alignment programs are available and useful for proteins sequences, most researchers that analyse rRNAs still align sequences by hand using a number of prealigned reference sequences as a reference. The reason for this is the presence of large deletions and insertions that are frequently found. Insertion usually do not extent 10 to 15 nucleotides but more than 100 nucleotides have been found in the 16S rDNA of thermophilic members of the *Clostridium/Bacillus* subphylum of Bacteria (Fig. 1), and the 23S rRNA of actinomyctes [34], while certain members of *Proteobacteria* lack a stretch of about 80 bases. Within the variable and highly variable regions

Table 1. Degree of similarity between different hypervariable regions of 16S rDNAs

a. Streptomyces (S.) ambofaciens and representatives from different main lines of descent of the domain Bacteria.

Organisms	61–108	170–227	Region 433–490	1009–1044	1433–1465	total	16S rDNA G+C content
S. ambofaciens/S. lividans	100	**83.3**	*100*	100	**93.9**	98.8	59.1/58.8
S. ambofaciens/A. globiformis	68.4	**62.1**	*97.8*	94.4	**53.1**	90.3	59.1/57.0
S. ambofaciens/B. subtilis	55.6	**50**	*72.3*	64.7	**56.2**	80.8	59.1/55.1
S. ambofaciens/E. coli	60.5	**54.5**	*76.1*	69.7	**50.0**	78.4	59.1/54.5
S. ambofaciens/Bt. vulgaris	43.2	**49.2**	*67.4*	58.3	**31.2**	71.6	59.1/51.2
S. ambofaciens/P. staleyi	40.5	**45.5**	*60.0*	48.5	**34.4**	72.2	59.1/54.4
S. ambofaciens/T. maritima	65.7	**52.6**	*76.6*	60.0	**37.5**	78.2	59.1/63.6

b. Escherichia (E) coli and representatives from different main lines of descent of the domain Bacteria.

	61–108	170–227	Region 433–490	1009–1044	1433–1465	total
E. coli/S. ambofaciens	60.5	**54.5**	*76.1*	**50.0**	69.7	78.4
E. coli/S. lividans	60.5	**54.5**	*76.1*	**56.9**	69.7	78.4
E. coli/A. globiformis	62.5	**63.6**	*73.9*	**52.8**	70.7	76.9
E. coli/B. subtilis	59.5	**57.7**	*53.8*	**5.3**	62.5	77.8
E. coli/Bt. vulgaris	60.9	**50.0**	*54.2*	**34.4**	62.5	71.3
E. coli/P. staleyi	43.2	**57.1**	*54.0*	**44.4**	39.4	72.3
E. coli/T. maritima	48.6	**60.0**	*51.6*	**50.0**	57.6	74.2

Abbreviations: S. – Streptomyces; A. – Arthrobacter; B. – Bacillus; E. – Escherichia; Bt. – Bacteroides; P. – Pirellula; T. – Thermotoga
The sequence of Methanosarcina barkeri, a representative of the domain Archaea, was used as a outside reference.
Columns in bold indicate highest variation

D.austra	AGUCGAGCGGGUAACGGAGGUCGGUCAUCGGAGGUCAGAAGUCAGAUUAAGAAGCAAGUGGC--CUGA--GGCAGUUGCGACGAAAAGAUCUGGCAUCCGAUUUCCGACGGCCGACUUCCGUUGCCAGCGCGG--ACGGGUGA
Sph.aur	AGUCGAGCGAAC--GG-------------------------------------ACGAGAAG--CUUG--CUACUGU----------------------------------GAU-----GUGGCGG--ACGGGUGA
L.planta	AGUCGAACGAAC--UCU----------------------GGUAUUGAUUGGUG--CUUG--CAUCAUGAUUUACA----------------------UUUGA--GUGAGUGGCGA--ACUGGGUGA
M.pneun	AGUCGAUCGAAA-------------------------------------GUA--GUAA--UAC-------------------------------UUUUAGAAGCGA--ACGGGUGA
Stm.amb	AGUCGAACGAUG-------------------------AACCAC--UUCG--GUGGGG-----------------------AUUUGUGGCGA--ACGGGUGA
M.lep	AGUCGAACGGAA-A------------------------GGGCUCUAAAAA--AUCU--UUUUUAGAGA----------------UAC--UCGAGUGGCGA--ACGGGUGA
Mic.lac	AGUCGAACGGUGAA-----------------------GGCGGAG--CUUG--CUCUGCU---------------------GG--AUCAGUGGCGA--ACGGGUGA
Arb.gl	AGUCGAACGAUG-A------------------------UCCGGUG--CUUG--CACCGGG---------------------G--AUUAGUGGCGA--ACGGGUGA
C.butyri	AGUCGAGCGAUG-------------------------AAGCUCC--UUCG--GGAGUGG-------------------AUUAGCGGCGA--ACGGGUGA
B.subtil	AGUCGAGCGGACA--GGU-------------------GGGAG--CUUG--CUCCC---------------------GAU--GUUGAGCGGCGG--ACGGGUGA

Abbreviations: D. austra- *Desulfotomaculum australicum*; Sph.aur-*Staphylococcus aureus*;

L.planta-*Lactobacillus plantarum*; M.pneun-*Mycoplasma pneumoniae*; Stm.amb-*Streptomyces ambofaciens*;

M.lep-*Mycobacterium leprae*; Mic.lac-*Microbacterium lacticum*; Arb.gl-*Arthrobacter globiformis*;

C.butyri-*Clostridium butyricum*; B.subtil-*Bacillus subtilis*

Figure 1. Some examples for length variation in the hypervariable region around position 100 of the 16S rDNA/RNA sequences of Gram-positive members of the Domain Bacteria.

which show the highest degree of length variation it is often difficult or even impossible to recognise homology from the primary structure. The alignment of these regions can in many cases be improved by taking into account the predicted higher order structure. Nevertheless, because of the significant degree of uncertainty to to homologize these regions they are usually omitted from the analysis of taxa above the genus rank.

The beginner, who is not familiar with secondary order structure elements may find it helpful to use certain universal oligonucleotides [49] for prealignment of sequences. A simple search for these short stretches facilitates the arrangement of sequences which will lead to the alignment of flanking region to the nearest variable regions.

Phylogenetic trees

The phylogenetic relationships of organisms, expressed in similarity or dissimilarity values, can be visualised in graphs. These consist of internal and terminal points (nodes) connected by edges (branches). The terminal nods represent the molecules of the analysed organisms while an internal nod represents a common stage in the evolution of these molecules. In unrooted trees only the interrelationships of the organisms are visualised and emphasis is placed on the relationship of neighboring taxa, whereas in rooted trees the position of the common ancestor is indicated, hence the order at which the organisms evolved is displayed. The distances between two nodes (organisms) are measured by the sum of edges between the nodes (Fig. 2). Radial trees are suitable for the depiction of relationships when only a few organisms are analyzed. In dendrograms organisms are arranged in a fork-like fashion. This mode is usually choosen when

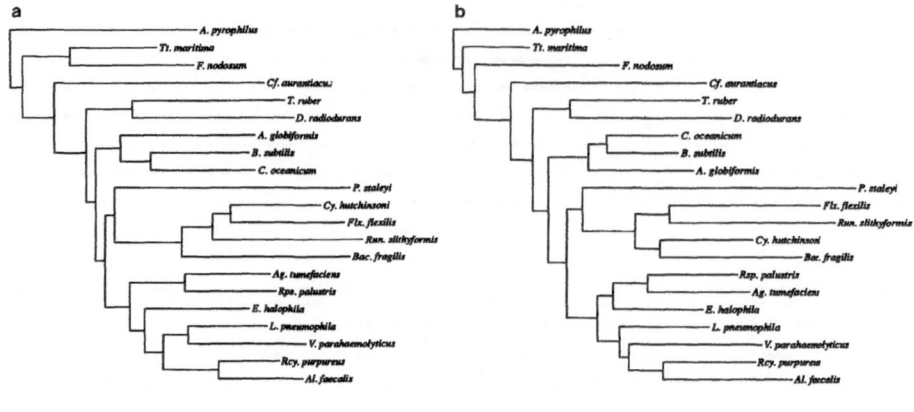

Figure 2. Comparison of phylogenetic dendrograms based on different information content of the 16S rDNA sequences: ability of partial 16S rDNA sequences to assign organisms to the genus level by a. Analysis based on almost complete sequences, b. Analysis based on about 500 nucleotides from the 5' terminus. Bar represents 10% sequence divergence (Dendrograms kindly provided by Naomi Ward, DSM). A., *Aquifex*; Tt., *Thermotoga*; F., *Fervidobacterium*; Cf., *Chloroflexus*; T., *Thermus*; D., *Deinococcus*; A. *Arthrobacter*.; B., *Bacillus*; C., *Clostridium*; P., *Planctomyces*; Cy., *Cytophaga*; Flx., *Flexibacter*; Run., *Runella*, Bac., *Bacteroides*; Ag., *Agrobacterium*; Rps., *Rhodopseudomonas*; E., *Ectothiorhodospira*; L., *Legionella*; V., *Vibrio*; Rcy., *Rhodocyclus*; Al., *Alcaligenes*.

analysing many organisms as in principle there are no limitations to the number of organisms to be depicted graphically. Only the horizontal components of connecting lines are summed to read the distances. In contrast to the tree mode, branches are not separated by an angle but by vertical lines of constant length which can be considered common nodes of edges that have been "elongated" to allow graphical representation (Fig. 2).

Treeing methods

Of the three major types of tree inferring approaches the distance methods [13, 14] and maximum parsimony [15] are the most widely used one. The maximum likelihood method, seeking for an explicit model of evolution by analysing the sequences on a site by site basis, is computational intensive and less suitable for PC, and has received little attention to date. Maximum parsimony methods seek for the most parsimonious trees among all possible tree topologies by determining the sum of changes which must have occurred to give the sequences in the alignment. In this procedure the number of necessary base changes along the edges connecting the terminal nodes is determined for the individual alignment positions. The tree with the the minimal overall number of changes needed is the most parsimonious one, hence the one that reflects evolution most closely. It is a disadvantage of parsimony methods, however, to often misplace individuals or groups of organisms if the rate of evolution differs significantly in different lineages.

For the application of distance methods, such as Neighbor Joining [35] or the algorithm of De Soete [7] a matrix of pairwise dissimilarity values is calculated from the sequences that are selected in a sequence alignment. In order to compensate for multiple substitutions at the same site these dissimilarity values are usually transformed into phylogenetic distances. The Jukes and Cantor transformation [22] is one widely used algorithm. It must be kept in mind that the number of possible characters at a particular site is limited and that only contemporary sequences are available. As a consequence, the true extent of exchanges, namely evolutionary events, cannot be fully recognised and are underestimated. The number of unrecognized evolutionary events increase with the dissimilarity values.

Distance values are the basis for phylogenetic trees which are commonly reconstructed applying additive treeing methods. These distance methods use dissimilarity values while information about individual sequence positions is not considered. The branch length between any to pairs of organisms are adjusted to match the binary dissimilarity values for these organisms most closely. Due to differences in the evolutionary rate of genes coding for ribosomal RNAs distances calculated from real sequence data usually are not ultrametric; distances between organisms A and B to a third related organism C are in most cases different and, as more organisms are included in the analysis, compromises have to be made concerning the position of internal nodes and the lengths of branch. The best tree topology is considered the one that requires minimal deviations from the values of the underlying matrix. Laboratories that are rarely involved in sequence analysis of ribosomal RNA may not be willing to become too deeply involved in data analysis. In this case the Ribosomal Data Base Project (RDP) offers reliable analysis free of charge.

Factors influencing the topologies of trees

All available treeing methods are based on assumptions, such as that, that individual sites evolve independently from each other. However, this may not be true with real data. Furthermore, due to the enormous number of possible tree topologies and the expense in computing time needed to generate them programs usually do not perform exhaustive tests of all possible tree topologies. The optimal tree may therefore not be found. In order to save computing time most programs add and treat the data according to their order in an alignment which introduces a bias. This can be eliminated by changing the order of entries randomly and to perform several runs.

The novice in the field of microbial phylogeny may assume that complete sequences provide the maximum of available phylogenetic information accessible from a given sequence. In principle this is true but it does imply that the inclusion of all positions gives optimal results. The choice of regions to select depends on the phylogenetic level of relatedness. Absolutely invariant and highly conserved positions indicate homology and are a valuable help for aligning sequences; however, these residues confer little phylogenetic information. On the other hand the highly variable regions which are subjected to multiple changes at a given site and for which homology is difficult to determine should be excluded from the analysis because they will underestimate remote relationships. The closer the degree of relatedness the higher will be the number of conserved nucleotides and differences will mainly be found in the variable regions. For highly related organisms the hypervariable regions are the only regions that may show some differences. Whether these regions reflect phylogeny is still under discussion. The influence of the hypervariable positions on tree topologies can be tested by performing several analyses while successively removing these positions or regions.

The statistical significance of the order of particular subtrees in a phylogenetic tree can be tested by resampling methods such as the bootstrap method. This approach randomly resamples alignment positions with the result that some of them are included more often than others while some are not included at all. The procedure is usually repeated between 100 to 1000 times with alternatively truncated or rearranged data sets. The higher the fraction of runs of recomputation in which the taxa defined by the branching points appear as a monophyletic subtree, the higher is the significance of the individual branching points.

The root of a subtree can be determined by including homologous sequences from phylogenetically moderately related organisms. This "outgroup" reference could be represented by a single sequence but it is preferrable to include entries from different phyla in order to avoid artefacts which may be due to differences in the evolutionary rate between the outgroup and the group under investigatin. It should be kept in mind that the addition of any new homologous sequence to an existing data set influences the topology of the tree. This is caused by the fact that each new sequence (which is different from those in the data set) gives rise to a new set of dissimilarity values that causes the algorithm to make adjustments to branch length in order to match the binary values most closely (see above). The inclusion of this additional information may improve the tree locally or even globally but may also have a negative influence on the stability of clusters. Lineages represented by a single organism often can not be stably positioned in phylogenetic trees. The addition of related sequences usually stabilises the branching point of the respective lineage. The addition of incomplete or incorrect sequences may reduce the stability of subtrees of related sequences and consequently the influence of sequences on the topology need to be checked carefully.

Another potential source of treeing artefacts is compositional bias. For the majority of prokaryotes the G+C contents of rRNAs vary within a narrow margin between 50 and 55 mol% but extreme values may be as low as 40% and as high as 64 mol%. The values are usually higher in molecules from thermophilic bacteria but not exclusively so. As the purine content is rather stable among bacterial rRNAs, transversion analysis is able to compensate for this bias [31].

The restricted information of partial rdna sequences to reflect phylogenetic relationships

From the above statements it is obvious that phylogenetic positions of taxa should be based on analysis of complete sequences. The existing 16S rDNA and 16S rRNA database contains only more or less complete sequences and gaps are mainly caused by regions not unambiguously determinable and stretches located upstream and downstream the 5' and upstream the 3' PCR primers, respectively. Sequence analysis of rDNAs has been very much facilitated that is can almost be considered a routine technique. The number of automated DNA sequencers is increasing and the amount of time and costs involved for the analysis of full sequences is only insignificantly higher than that for partial analysis.

The restriction of sequence analysis to selected portions of 16S rDNA that contain a mixture of conserved and variable regions has frequently been used in the past. The following example on the genus *Vibrio* should demonstrate some of the problems encountered by using partial 16S rDNA sequences for phylogenetic studies (Fig. 3). Based on full sequences, *V. fischeri* and *V. salmonicida* represent the most deeply branching pair of organisms, while *V. hollisae* branches between *V. proteolyticus* and relatives and *V. vulnificus*. When the analysis is restricted to the 3' terminal 900 nucleotides, *V. hollisae* appears as the deepest branch, while *V. fischeri* and *V. salmonicida* clusters with *V. anguillarum* and *V. ordalii*. Yet another topology is found when analysing the first 450 nucleotides. Although the branching pattern resembles more closely that based on the complete sequences, significant deviation can be detected in the branching points of almost all species. The variation of tree topologies is due to the significant differences in the similarity values determined for certain regions of the sequence from the same set of organisms (see also Table 1). The position of hypervariable regions within the 16S rDNA primary structure differs from taxon to taxon and need to be determined individually for their use in determination of relatedness [41].

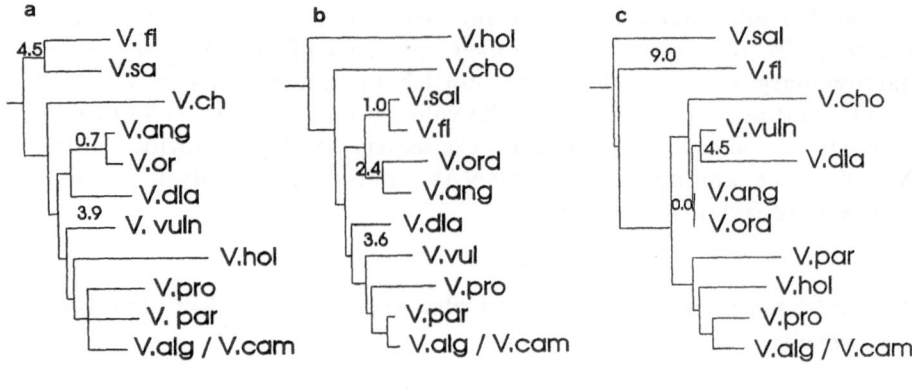

Figure 3. Comparison of phylogenetic dendrograms based on different information content of the 16S rDNA sequences: inability of partial 16S rDNA sequences to determine the intrageneric position of species as exemplified on the genus *Vibrio*. a. Analysis based on almost complete sequences, b. Analysis based on a region covering positions 501 through 1450, c. Analysis based on about 500 nucleotides from the 5′ terminus. Bar represents 2.5% sequence divergence.

Partial sequences and the recovery of the main lines of descent

While the above examples demonstrate the inability of partial sequences to reliably unravel the intrageneric phylogeny, these stretches are helpful to rapidly allocate isolates and clone sequences to higher taxa. The two phylogenetic dendrograms shown in Fig. 2 demonstrate the ability to differentiate the main lines of descent within the domain Bacteria based on the analysis of both, full 16S rDNA/rRNA sequence data and partial sequence comprised of the first 500 bases (5′ → 3′). So, by determination of, for example, the sequence of the first 500 bases of an environmental 16S rDNA clone, it can be assigned to a main line of descent. This approach can be useful in determination of both the degree of diversity in a particular clone library, and in the design of oligonucleotide probes used for screening clone libraries for abundance of probe-positive clones. In this context it should be noted that in order to assess biodiversity at a global scale it appears timely to discuss a strategy which involves sequence analysis of the same region of the 16S rDNA and of a minimum of nucleotides. As most investigations have already included the 5′ terminus of this molecule it is recommended to sequence the 5′ terminal 300 to 500 nucleotides.

Determination of the phylogenetic position of a clone partial 16s rDNA sequence

Having obtained the partial sequence of a 16S rDNA clone, the next step is the determination of the identity, or the source, of this fragment. The identification of the microorganism or group of microorganisms from which this clone sequence originated leads to the assumption that that microbe was present in the environment for which the microbial diversity is being investigated.

The first step in the process of clone identification is alignment of the sequence data to a set of sequences in an alignment. Such an alignment would contain sequences of representatives of the main lines of descent within the Bacteria (see Fig. 2, right dendrogram). During manual alignment, some idea as to the identity of the clone sequence can be gained from considering the main line to which it best aligns.

Once the sequence has been roughly aligned, a similarity matrix can be constructed. In most cases the main phylogenetic group to which the clone sequence shows highest similarity determined. A phylogenetic tree containing the main line representatives and the unidentified clone can also be made to give some idea as to the phylogenetic position of a clone which may have given similarity values in a range equal to more than one main group. The similarity values shown in Table 2 demonstrate the use of this approach in determination of the main group to which a clone sequence shows highest similarity. However, it should be stressed that in those cases

Table 2. 16S rDNA similarity values obtained for environmental clone sequences and their phylogenetically closest culturable neighbors. The analysis is based on the 5' terminal 500 nucleotides

	CLONES				
	H7	H101	H3	H79	Mc100
Agrobacterium tumefaciens	**78.5**	70.2	76.8	74.0	60.8
Rhodopseudomonas palustris	**77.9**	71.3	74.6	70.2	61.7
Rhodocyclus purpureus	75.7	**85.6**	73.5	73.5	61.9
Alcaligenes faecalis	67.4	**81.2**	76.8	69.6	61.9
Escherichia coli	71.3	71.3	**82.9**	69.6	56.9
Vibrio parahaemolyticus	73.5	75.1	76.2	66.9	62.4
Desulfobacter postgatii	73.5	65.7	74.0	75.7	60.2
Pirellula staleyi	67.4	68.0	72.4	66.9	**74.6**
Arthrobacter globiformis	71.3	68.5	70.2	**86.7**	62.4
Bacillus subtilis	70.2	67.4	70.7	75.1	65.2
Clostridium oceanicum	74.0	71.8	68.5	71.8	63.5
Bacteroides fragilis	63.0	64.1	65.7	72.4	59.1
Cloroflexus aurantiacus	66.3	69.6	65.2	74.6	61.9
Thermus ruber	73.5	72.4	70.7	72.4	64.6

Source: H, intestinal tract of holothurian, Great Barrier Reef (Naomi Ward, unpublished); MC, soil sample from Mount Coot-tha, Brisbane, Australia [24]

were the branching point is close to the root of a phylum the position may be changed from one phylum to another when the full sequence becomes available and/or a different selection of reference organisms is used.

When the main group to which the unidentified clone sequence belongs has been determined, the sequence can then be compared to those available for that group. Such a comparison can lead to placement of the clone sequence at one of various taxonomic levels from family down to species. It is at this stage that a detailed knowledge of the group within which the clone falls is necessary, since failure to include sequences of all representatives could cause one to make the erroneous assumption that the clone represented an as yet unknown or unsequenced taxon.

The lack of sequence data currently available for many genera and species is currently a hindrance to the accurate determination of clone sequence identity. More than 2000 complete or nearly complete sequences are available for comparison and the majority of species can now phylogenetically be placed into one of the phyla and kingdoms, respectively. For several genera, containing about 370 species (about 12% of total), no sequence data is available, and so the discovery of new lines of descent within the main groups must be viewed cautiously. When a sequence is found to cluster within a group that is considered to represent a particular genus, but is equidistantly related to all species for which sequence data has been determined, the researcher should consider whether other species of this genus for which sequence data is currently not available exist, before coming to the conclusion that the clone sequence represents a new, as yet uncultivated, species. The same is true for other taxa higher than genera.

The occurrence of chimeric clones, generated by the PCR reaction in the presence of a high fraction of degraded DNA [25] must also be checked carefully. Their undetected presence among the sequences may give the impression of unique prokaryotes in the environment. In order to avoid a tedious analysis by deriving the secondary structure of each clone sequence the ribosomal data base project (RDP) offers a check for chimera.

Only for the identification of very closely related clones, for which the only differences in 16S rDNA sequences are concentrated in one or two hypervariable regions, the restriction of sequence analysis to these stretches is an appropriate approach. However, as outlined above, analysis should be based solely on comparison of similarity values and one should refrain from attempts to indicate phylogenetic relationships by depicting phylogenetic trees.

References

1. Amann RI, Springer N, Ludwig W, Görtz H-D, Schleifer K-H (1991) Identification *in situ* phylogeny of uncultured bacterial endosymbionts. Nature 351: 161–164.

2. Angert ER, Clements KD, Pace NR (1993) The largest bacterium. Nature 362: 239–241.
3. Barns SM, Fundyga RE, Jeffries MW, Pace NR (1994) Remarkable archaeal diversity detected in a Yellowstone National Park hot spring environment. Proc Natl Acad Sci USA 91: 1609–1613.
4. Britschgi T, Giovannoni SJ (1991) Phylogenetic analysis of a natural marine bacterioplankton population by rRNA gene cloning and sequencing. Appl Environ Microbiol 57: 1707–1713.
5. DeLong EF, Wu KY, Prezelin BB, Jovine RVM (1994) High abundance of Archaea in Antarctic marine picoplancton. Nature 371: 695–697.
6. DeLong EF (1992) Archaea in coastal marine environments. Proc Natl Acad Sci USA 89: 5685–5689.
7. De Soete G (1983) A least square algorithm for fitting additive trees to proximity data. Psychometrika 48: 621–626.
8. Distel DL, DeLong EF, Waterbury JB (1991) Phylogenetic characterisation and *in situ* localisation of the bacterial symbiont of shipworms (Teredinidae: Bivalvia) by using 16S rRNA sequence analysis and oligodeoxynucleotide probe hybridisation. Appl Environ Microbiol 57: 2376–2382.
9. Distel DL, Wood AP (1992) Characterization of the gill symbiont of *Thyasira flexuosa* (*Thyasiridae*: Bivalvia) by use of polymerase chain reaction and 16S rRNA sequence analysis. J Bacteriol 174: 6317–6320.
10. Ekendahl S, Arlinger J, Stahl F, Pedersen K (1994) Characterization of attached bacterial populations in deep granitic groundwater from the Stripa research mine by 16S rRNA gene sequencing and scanning electron microscopy. J Gen Microbiol 140: 1575–1583.
11. Embley TM, Finlay BJ (1994) The use of rRNA sequences to unravel the relationships between anaerobic cilaites and their methanogen endosymbionts. Microbiology 140: 225–235.
12. Farelly V, Rainey FA, Stackebrandt E (1994) Effect of genome size and *rrn* gene copy number on PCR amplification of 16S rRNA genes from a mixture of bacterial species. Appl Environ Microbiol, submitted.
13. Felsenstein J (1978) Numerical methods for inferring evolutionary trees. Q Rev Biol 57: 379–404.
14. Fitch WM, Margoliash E (1967) Construction of phylogenetic trees. Science 155: 279–284.
15. Fitch WM (1971) Toward defining the course of evolution: minimum change for a specified tree topology. Syst Zool 20: 406–416.
16. Fuhrman JA, McCallum K, Davis AA (1992) Novel major archaebacterial group from marine plankton. Nature 356: 148–149.
17. Fuhrman JA, McCallum K, Davis AA (1993) Phylogenetic diversity of subsurface marine microbial communities from the Atlantic and Pacific oceans. Appl Environ Microbiol 59: 1294–1302.
18. Giovannoni SJ, Britschgi TB, Moyer CL, Field KG (1990) Genetic diversity in Sargasso Sea bacterioplankton. Nature 345: 60–63.
19. Goebel BM, Stackebrandt E (1994) Cultural and phylogenetic analysis of mixed microbial populations found in natural and commercial bioleach environments. Appl Environ Microbiol 60: 1614–1621.
20. Goebel BM, Stackebrandt E (1994) The biotechnological importance of molecular biodiversity studies for metal bioleaching. In: Priest F, Fewson CA (eds) Identification of Bacteria: Current Status – Future Prospects. New York: Plenum Press, in press.
21. Gutell RR, Weiser B, Woese CR, Noller HF (1985) Comparative anatomy of 16S-like ribosomal RNA. Progr Nucl Res Molec Biol 32: 155–216.
22. Jukes TH, Cantor CR (1969) Evolution of protein molecules. In: Munro HN (ed) Mammalian Protein Metabolism, pp. 21–132. New York: Academic Press.
23. Liesack W, Stackebrandt E (1992) Unculturable microbes detected by molecular sequences and probes. Biodiv Conserv 1: 250–262.

24. Liesack W, Stackebrandt E (1992) Occurrence of novel groups of the domain *Bacteria* as revealed by analysis of genetic material isolated from an Australian terrestrial environment. J Bacteriol 174: 5072–5078.

25. Liesack W, Weyland H, Stackebrandt E (1991) Potential risks of gene amplification by PCR as determined by 16S rRNA analysis of a mixed-culture of obligately barophilic bacteria. Microb Ecol 21: 188–201.

26. Muyzer G, De Waal E, Uitterlinden AG (1993) Profiling of complex microbial populations by denaturating gradient gel electrophoresis analysis of polymerase chain reaction-amplified genes coding for 16S rRNA. Appl Environ Microbiol 59: 695–700.

27. O'Neill SL, Giordano R, Colbert AME, Karr TL, Robertson HM (1992) 16S rRNA phylogenetic analysis of the bacterial endosymbionts associated with cytoplasmic incompatability in insects. Proc Natl Acad Sci USA 89: 2699–2702.

28. Olsen GJ, Lane DJ, Giovannoni SJ, Pace NR (1986) Microbial ecology and evolution: a ribosomal RNA approach. Ann Rev Microbiol 40: 337–355.

29. Pace NR, Stahl DA, Lane DJ, Olsen GJ (1985) The analysis of natural microbial communities by ribosomal RNA sequences. Microb Ecol 9: 1–56.

30. Rainey FA, Ward N, Sly LI, Stackebrandt E (1994) Dependence on the taxon composition of clone libraries for PCR amplified, naturally occurring 16S rDNA, on the primer pair and the cloning system used. Experientia 50: 796–797.

31. Rainey FA, Ward NL, Morgan HW, Stackebrandt E (1993) A phylogenetic analysis of anaerobic thermophilic bacteria: an aid for their reclassification. J Bacteriol 175: 4772–4779.

32. Relman DA, Loutit JS, Schmidt TM, Falkow S, Tomkins LS (1990) The agent of bacillary angiomatosis. New Engl J Med 323: 1573–1580.

33. Relman DA, Schmidt TM, MacDermott RP, Falkow S (1992) Identification of the uncultured bacillus of Whipple's disease. New Engl J Med 327: 293–301.

34. Roller C, Ludwig W, Schleifer K-H (1992) Gram-positive bacteria with a high G+C content are characterized by a common insertion within their 23S rRNA genes. J Gen Microbiol 138: 1167–1175.

35. Saitou N, Nei M (1987) The neighbor-joining method: a new method for reconstructing phylogenetic trees. Mol Biol Evol 4: 406–425.

36. Schleifer KH, Ludwig W (1989) Phylogenetic relationships among bacteria. In: Fernholm B, Bremer K, Jörnwall H (eds) The Hierarchy of Life, pp. 103–117. Amsterdam: Elsevier.

37. Schwartz RM, Dayhoff, MO (1978) Origins of prokaryotes, mitochondria and chloroplasts. Science 199: 395–403.

38. Seewaldt E, Stackebrandt E (1982) Partial sequence of 16S rRNA and the phylogeny of *Prochloron*. Nature 295: 618–620.

39. Smida J, Kazda J, Stackebrandt E (1988) Molecular-genetic evidence for the relationship of *Mycobacterium leprae* to slow growing mycobacteria. J Leprosy 56: 449–454.

40. Sneath PHA (1993) Evidence from *Aeromonas* for genetic crossing-over in ribosomal sequences. Int J Syst Bacteriol 43: 626–629.

41. Stackebrandt E, Liesack W, Witt D (1992) Ribosomal RNA and ribosomal DNA sequence analysis. Gene 115: 255–260.

42. Stackebrandt E, Liesack W, Goebel BM (1993) Bacterial diversity in a soil sample from a subtropical Australian environment as determined by 16S rDNA analysis. FASEB J 7: 232–236.

43. Unterman BM, Bauman P, McLean DL (1989) Pea aphid symbiont relationships established by analysis of 16S rRNAs. J Bacteriol 171: 2970–2974.

44. Viale AM, Arakaki AK, Soncini FC, Ferreyra RG (1994) Evolutionary relationships among eubacterial groups as inferred from GroEL (Chaperonin) sequence comparisons. Int J Syst Bacteriol 44: 527–533.

45. Ward DM, Weller R, Bateson MM (1990) 16S rRNA sequences reveal numerous uncultured inhabitants in a natural community. Nature 345: 63–65.

46. Ward DM, Bateson MM, Weller R, Ruff-Roberts, AL (1992) Ribosomal analysis of microorganisms as they occur on nature. Adv Microbiol Ecol 12: 219–286.
47. Woese CR (1987) Bacterial evolution. Microbiol Rev 51: 221–271.
48. Woese CR, Gutell R, Gupta R, Noller HF (1983) Detailed analysis of the higher-order structure of 16S ribosomal ribonucleic acids. Microbiol Rev 47: 621–669.
49. Woese CR, Stackebrandt E, Macke TJ, Fox GE (1985) A phylogenetic definition of the major eubacterial taxa. Syst Appl Microbiol 6: 43–151.
50. Zuckerkandl E, Pauling L (1965) Molecules as documents of evolutionary history. J Theor Biol 8: 357–366.

Molecular Microbial Ecology Manual **3.3.1**: 1–11, 1995
© 1995 *Kluwer Academic Publishers.*

Amplification of ribosomal RNA sequences

RICHARD DEVEREUX[1] and STEPHANIE G. WILLIS[2]

[1] *Microbial Ecology and Biotechnology, United States Environmental Protection Agency, 1 Sabine Island Dr., Gulf Breeze, Fl 32561, U.S.A.;* [2] *Institute for the Study of Earth, Oceans, and Space, University of New Hampshire, Durham, NH 03824, U.S.A.*

Introduction

Emergence of the major outlines of bacterial phylogenetic relationships marks a defining advancement in the recent history of microbiology. This was made possible through the work of Woese and his colleagues who pioneered the comparison of rRNA sequences [30]. The ability to measure bacterial phylogenetic relationships, and those among microscopic eukaryotes, has had a direct impact on systematics that extends well into microbial ecology. While the driving force has been basic phylogenetic research, application of molecular techniques based on rRNA sequence comparisons to determinative and environmental microbiology is a blossoming area [9,26].

An advantage of rRNA sequence comparison is the generation of an increasingly expanding data set against which newly determined sequences may be compared [3,21]. The number of available rRNA sequences is over two thousand [16]. For the most part, the rRNA sequence data has been compiled from well established and thoroughly characterized species obtained from culture collections. As recognized by Pace et al. [22], the data set serves not only in the comparison of rRNA sequences obtained from pure culture isolates, but also for those obtained by cloning genes from genetic material extracted from the microbiota in an environmental sample.

As divergence of the primary lines of bacterial descent occurred during early biotic history, highly conserved molecular chronometers are best suited to the task of bacterial phylogenetics [30]. An integral element of the protein synthesizing apparatus, the basic components and mechanism of which are present in all primary kingdoms, rRNAs are among the most highly conserved cellular molecules. Yet, rRNAs also contain sufficient sequence variability so that relationships between closely and distantly related groups can be determined.

Abundance of rRNAs in actively growing cells made them readily obtainable in the purified form needed for earlier methodologies which determined their sequences directly. At first, rRNAs (primarily 16S rRNAs) were compared by oligonucleotide cataloguing [30]. This entailed the digestion of purified rRNA with a ribonuclease to generate fragments that were then electrophoretically separated and individually sequenced. Comparison of oligonucleotide catalogs and contiguous sequences determined from cloned rRNA genes led to the identification of highly conserved nucleotide tracts in dispersed regions of the 16S rRNA. The conserved tracts, serving as priming sites, made it possible to rapidly determine nearly complete sequences using rRNA as the template for reverse transcriptase in dideoxynucleotide terminating sequencing reactions [14,15]. Highly conserved nucleotide tracts have also now been identified in the 23S rRNA [14].

Although determination of rRNA sequences with reverse transcriptase is straightforward, use of rRNA templates fosters occasional ambiguities. This is generally attributed to degradation of the template or to stable secondary structures that cause multiple bands to appear across several or all lanes of the sequencing gel autoradiogram. In some instances, multiple bands might be the result of microheterogeneities in sequence between rRNA operons [20]. Often, the overall quality of sequencing gel autoradiograms is greater with reactions performed with DNA templates making them faster and easier to read with fewer ambiguities.

The most recent advance in obtaining rRNA sequence information was heralded with the advent of PCR [24]. In any PCR advanced knowledge of sequence at the distal and proximal ends of the sequence to be amplified is necessary. The conserved nature and availability of a large number of rRNA sequences has enabled the design of primers that amplify nearly full length 16S rRNA. As the template is the rRNA gene, the amplification target or product is referred to as rDNA. Medlin et al. [18] first described amplification of 16S-like rDNA from algae, fungi, and protozoa, and reports using 16S rDNA of bacteria and other eukaryotes soon followed [7,8,27,28]. The small amounts of DNA needed as template also enable amplification from minute amounts of material when seeking phylogenetic information from a few available cells or even eukaryotic cellular organelles. The amplifications provide a simple, rapid approach (without the need to screen a large genomic library) to obtain rDNA that may either be sequenced directly [4,5,7,9], or cloned and sorted through a recombinant library. In pure culture studies, more effort is expended in the generation of the plasmid templates than for the rRNA templates used with reverse transcriptase. However, as mentioned above, the resulting autoradiograms are easier to interpret.

Experimental approach

Precautionary measures

The great sensitivity of PCR makes the possibility of amplifying contaminating DNA a real concern. Care should be taken to avoid the introduction of unwanted DNA into PCR reagents or reaction mixtures. This is particularly true when performing amplification of rDNA; ribosomal RNA genes are ubiquitous, and the primers used in rDNA amplifications are often designed for broad specificity.

There are two types of contaminant DNA of which to be wary. First, and perhaps most serious, is DNA carried over from previous amplifications of homologous targets. Second is exogenous DNA from other sources such as genomic DNA preparations, or microbial growth in reagents. Contamination can be effectively eliminated with careful laboratory practice. Guidelines, provided by Kwok [12] and Kwok and Higuchi [13], include physically separating PCR preparations from amplification products, aliquoting reagents, briefly spinning tubes to avoid splashes when they are opened, using positive displacement pipettes (Ranin, Woburn, MA), changing gloves frequently, autoclaving solutions, and carefully selecting positive and negative controls. We have found aerosol resistant tips a convenient alternative to positive displacement pipettes. The investigator should become familiar with these guidelines and adhere to them as closely as possible.

Template

The first section of this manual describes procedures for the isolation of microbial nucleic acids from pure cultures and environmental samples. DNA templates prepared by these methods should be suited to PCR amplification. However, particularly when purified from the environment, highly degraded or sheared DNA can lead to the formation of chimeric rDNA PCR products. During the annealing cycle, amplified rDNA from different templates may reassociate leading to heterogeneous products in proceeding cycles [17]. It is also possible to proceed with amplification beginning with an rRNA template. In the first step of the process a cDNA is generated through reverse transcription [19].

The sensitivity of PCR also makes it possible to amplify rDNA directly from small amounts of cells [10,25], a portion of a lyophilized pellet [29], or with DNA obtained using rapid nucleic acid purification protocols [9].

Primers

The goal in designing primers is obtaining a pair that will effectively hybridize to the target DNA and achieve the desired amplification. Beyond

Table 1. PCR primers useful for the amplification of nearly full length 16S-like rDNA

Primer[a]	Sequence[b]	Priming site[c]	Designed specificity	Reference/comments[d]
fD1	ccgaattcgtcgacaacAGAGTTTGATCCTGGCTCAG	8–27	Most eubacteria	[27] for, *EcoRI*, *SalI*
fD2	ccgaattcgtcgacaacAGAGTTTGATCATGGCTCAG	8–27	Enterics and relatives	[27] for, *EcoRI*, *SalI*
fD3	ccgaattcgtcgacaacAGAGTTTGATCCTGGCTTAG	8–27	*Borrelia* spirochetes	[27] for, *EcoRI*, *SalI*
fD4	ccgaattcgtcgacaacAGAATTTGATCTTGGTTCAG	8–27	Chlamydiae	[27] for, *EcoRI*, *SalI*
rD1	cccgggatccaagcttAAGGAGGTGATCCAGCC	1541–1525	Many eubacteria	[27] rev, *HindIII*, *BamHI*, *XmaI*
rP1	cccgggatccaagcttACGGTTACCTTGTTACGACTT	1512–1492	Enterics (and most eubacteria)	[27] rev, *HindIII*, *BamHI*, *XmaI*
rP2	cccgggatccaagcttACGGCTACCTTGTTACGACTT	1512–1492	Most eubacteria	[27] rev, *HindIII*, *BamHI*, *XmaI*
rP3	cccgggatccaagcttACGGATACCTTGTTACGACTT	1512–1492	Fusobacteria (and most eubacteria)	[27] rev, *HindIII*, *BamHI*, *XmaI*
5'	gcgggatccGAGTTTGATCCTGGCTCAG	9–27	most eubacteria	[17] for, *BamHI*
3'	cgcggatccAGAAAGGAGGTGATCCAGCC	1542–1525	most eubacteria	[17] rev, *BamHI*
pA	AGAGTTTGATCCTGGCTCAG	8–27	eubacteria	[6,7] for
pH	AAGGAGGTGATCCAGCCGCA	1541–1522	eubacteria	[7] rev
reverse	GGTTACCTTGTTACGACTT	1510–1492	eubacteria	[6] rev
Primer A	ccgaattcgtcgacAACCTGGTTGATCCTGCCAGT	1–21	eukaryotic 16S-like rRNA	[18] for, *EcoRI*, *SalI*
Primer B	cccgggatccaagcttTGATCCTTCTGCAGGTTCACCTAC	1795–1772	eukaryotic 16S-like rRNA	[17] rev, *SmaI*, *BamHI*, *HindIII*, *PstI*

Table 1. Continued

Primer[a]	Sequence[b]	Priming site[c]	Designed specificity	Reference/comments[d]
Primer A	ccgtcgacgagctAGAGTTTGATCMTGGCTCAG	8–27	eubacterial	[9] for *SalI*, *SacI*
Primer B	cccgggtaccaagcttAAGGAGGTGATCCANCCRCA	1541–1518	eubacterial	[9] rev, *SmaI*, *KpnI*, *HindIII*
forward	CTCCGGTTGATCCTGCC	7–23	used with a hyperthermophilic methanogen	[2] for
reverse	GGAGGTGATCCAGCCG	1539–1524	used with a hyperthermohilic methanogen	[2] rev
27f	AGAGTTTGATCMTGGCTCAG	9–27	most eubacteria	[14] for, adapted from a sequencing primer
1492r	TACGGYTACCTTGTTACGACTT	1513–1492	most eubacteria, archaebacteria	[14] rev, sequencing primer adaptable to PCR
1525r	AAGGAGGTGWTCCARCC	1541–1525	most eubacteria, archaebacteria	[14] rev, sequencing primer adaptable to PCR

[a] Some primers were not specifically named by the authors. Note the names "Primer A" and "Primer B" were used by different authors for different primers.

[b] Sequences written in the 5' to 3'direction. Lower case indicates linker sequence that contains restriction endonuclease recognition sites, upper case indicates region hybridizing with the rDNA. M= A and C; R= G and A; Y= C and T; W= A and T; N= C,A,T, and G; positions where the synthesized primer contains equimolar amounts of more than one nucleotide.

[c] Priming sites indicated by *E. coli* 16S rRNA numbering for bacterial and archaeal specific primers, and by the numbering of the *Saccharomyces cerevisiae* 16S-like rRNA for the eukaryotic primers.

[d] For; primer is homologous to the rRNA sequence with extension occurring to the 3' end of the rRNA, rev; primer is complementary to the rRNA with extension occurring to the 5' end of the rRNA. Restriction enzymes cutting within the linker are given, the *PstI* site in eukaryotic Primer B is located in the region hybridizing with the rDNA.

MMEM-3.3.1/5

that, consideration of a few variables during the design process can enhance the efficiency and yield of the reaction. The finer points of primer design have been detailed by Innis and Gelfand [11] and by Saiki [23]. Briefly, the primers should lack palindromic sequences and complementarity, particularly at the 3' ends, which may lead to the formation of "primer dimers". The calculated T_m of the primers should be closely matched and in the range of 55 to 72 °C.

Primers that amplify nearly full length 16S-like rDNA are listed in Table 1. The primers generally target the same regions of the 16S rDNA with minor variations in length or sequence used by different authors. Restriction endonuclease sites engineered at the 5' ends are recognized by enzymes that rarely cut within a 16S rDNA. A primer pair, one forward and one reverse, selected for the broadest specificity among the organisms of interest (i.e., Bacteria, Archaea, or Eukarya) should enable amplification of most 16S-like rDNAs. It may be necessary to try several different combinations or design an additional primer. Selection of primers can often be guided by comparison of sequences in a data base, or by what has proven successful in previous amplifications from closely related strains. Additional primers of broad specificity that will amplify shorter lengths can be adapted from those used for rRNA sequencing [14]. It is also possible to design primers that can be used to selectively amplify rDNA of a phylogenetically defined group [1].

Reaction conditions

The "standard" PCR conditions [11] will amplify most target sequences and are a good starting point from which to optimize amplifications when necessary. In general, a reaction component at too high a concentration can promote misincorporation of nucleotides and non-specific priming. A less than optimal concentration can reduce the yield of the reaction. The products obtained from a reaction are analyzed on agarose gels (0.8–1%) where under the optimal conditions a single, bright band of the expected size will be observed.

Thermally stable polymerases from different suppliers may vary in definition of units and optimum magnesium concentration. Reaction conditions may therefore vary depending on the particular enzyme used. Enzyme storage buffers may contain gelatin, bovine serum albumin or nonionic detergents to stabilize the polymerase.

In standard 100 μl reactions, 2.5 units of *Taq* polymerase are routinely used, with a recommended range of 0.5–5 units to test for the optimal amount. Stock dNTP solutions must be neutralized to pH 7.0 and their concentrations determined by spectrometry. These often come already prepared as part of a PCR kit. dNTPs are present in equimolar amounts of 20 to 200 pM each. Magnesium, critical to primer annealing and extension, should be optimized and present in the range of 0.5–2.5 mM.

KCl promotes primer annealing and should not exceed 50 mM. The reactions are buffered with Tris-HCl (10–50 mM, pH 8.3). Primer concentrations used in rDNA amplifications range from 0.1 to 1.0 µM with 0.1 ng to 1.0 µg DNA template added to a 100 µl reaction.

The conditions given below for rDNA amplification have been found to work well with most bacteria studied in our laboratory.

Post amplification

Often the goal of an rDNA amplification is subsequent sequence determination. We have found it convenient to utilize a commercially available kit designed for cloning PCR products (e.g., TA Cloning System; Invitrogen Corp., San Diego, CA). We have also had success with directed cloning using the restriction endonuclease sites synthesized into the primers and with blunt-ended cloning. The latter requires more PCR products and for this we have utilized the products obtained from four of the reactions described below. For pure culture studies, several clones should be pooled to prepare the sequencing template. This will minimize the possibilities of obtaining an artifact (i.e., a misincorporated nucleotide) of the PCR reaction. The cloned rDNAs are may then be readily sequenced (e.g., Sequenase; U.S. Biochemicals, Cleveland, OH). We have also obtained good results using the PCR products directly in thermocycle sequencing reactions (e.g., ABI thermocycle sequencing, Applied Biosystems Inc, Foster CA).

Procedures

Preparation of template DNA

The reader is referred to the first part of the manual for procedures used to isolate template DNA. High-molecular-weight (ca. 25 kb), phenol-chloroform extracted DNA will generally be of sufficient purity. If inhibitors are present in a DNA preparation, as might be expected in DNA obtained from an environmental sample, serially diluting (ten-fold) the preparation can often overcome the inhibition. We use approx. 100 ng of pure culture DNA in a 50 µl reaction. Cells obtained from pure cultures may also serve to provide template (see Note below).

rDNA amplification

Work at a clean area covered with fresh bench paper and observe the precautions given above. When preparing the reaction tubes also include the appropriate positive and negative controls (without the added template). DNA is added last to limit the potential for carry over.

Steps in the procedure

Thermal cycle parameters
Set the temperature controller to the following program: 94 °C for 1 min (denaturing), 56 °C for 1 min (annealing), and 72 °C for 2 min (extension). A cycle without timed ramps works well. Usually 25 to 30 cycles gives a good yield. The completed reaction is held at 4 °C (e.g., by linkage to a "soak" file).

Reaction mixture
The following components are added to an autoclaved 500 µl tube:

dH$_2$O:	64 µl.
10× PCR buffer:	10 µl.
MgCl$_2$ (25 mM):	8 µl.
dNTP mixture:	8 µl.
forward primer (10 µM):	2 µl.
reverse primer (10 µM):	2 µl.
template DNA (ca 100 µg/ml):	2 µl.

Mix gently, and quick spin in a microcentrifuge to bring the solution to the bottom of the tube. Add 50 µl of sterile mineral oil to the surface of the mixture and place in the temperature controller.

Enzyme addition
Start the temperature program. When the denaturing temperature has been reached add:

2.5 units of *Taq* polymerase in 4 µl ice-cold 1/10 TE.
Complete the thermal cycle program.

Agarose gel analysis
Analyze 5 µl of the completed reaction on a 0.8% agarose gel with size markers (e.g., 1 kb ladder; Bethesda Research Laboratories, Gaithersburg, MD).

Notes
- Ad 1. These annealing temperatures have worked well for us yet may be further optimized should greater specificity be required. Alternatively, the annealing temperature may be set to 37 °C when the exact sequence of the priming site is uncertain. The lower temperature diminishes the specificity of the primers and allows for mismatches in the template/primer hybrid.
- Ad 2. The volume of water may be adjusted to accommodate variations in concentrations of reaction components (e.g., in DNA, in the negative controls without added template, or when optimizing for magnesium).
- Ad 2. The 10× PCR buffer used does not contain magnesium. The final concentration of magnesium used in this reaction is 2 mM. Alternatively, the PCR buffer obtained with a kit (which often contains magnesium) may be used and the magnesium concentration optimized if needed.
- Ad 2. The dNTP mixture is prepared just before use by combining equal volumes of 10 mM solutions. Convenient pre-mixed nucleotides for PCR are also available (Boehringer, Mannheim, Indianapolis, IN).
- Ad 2. From colonies on a plate a small amount of cells (ca. 1 µl) are suspended in 1.0 ml 1/10 TE, vortexed, and alternatively frozen (—70 °C) and thawed (65 °C) three times. About 2 µl of this lysate is added to the reaction. This may also be adapted to cells harvested from small amounts of liquid culture.
- Ad 3. This "hot start" increases specificity of priming.
- Ad 3. We use *Taq* enzyme from Boehringer Mannheim.

Solutions
- TE
    - 10 mM Tris-HCl, pH 8.0.
    - 1.0 mM EDTA.
- 1/10 TE
    - a 1/10 dilution of TE in sterile distilled water.
- 10× PCR buffer
    - 100 mM Tris-HCl, pH 8.3.
    - 500 mM KCl.
- dNTP mixture
    - Prepared fresh by mixing equal volumes of 10 mM dATP, 10 mM dGTP, 10 mM dTTP, 10 mM dCTP.
- primers
    - primers are diluted in sterile distilled water.

References

1. Amann RI, Stromley J, Devereux R, Key R, Stahl DA (1992) Molecular and microscopic identification of sulfate-reducing bacteria in multispecies biofilms. Appl Environ Microbiol 58: 614–623.
2. Burggraf S, Stetter KO, Rouviere P, Woese CR (1991) *Methanopyrus kandleri*: an archaeal methanogen unrelated to all known methanogens System Appl Microbiol 14: 346–351.
3. De Rijk P, Neefs JM, Van de Peer Y, De Wachter R (1992) Compilation of small ribosomal subunit RNA sequences. Nucleic Acids Res 18: 4028.
4. Dobson SJ, Colwell RR, McMeekin TA, Franzmann PD (1993) Direct sequencing of the polymerase chain reaction-amplified 16S rRNA gene of *Flavobacterium gondwanense* sp. nov. and *Flavobacterium salegens* sp. nov.: two new species from a hypersaline antarctic lake. Int J Syst Bacteriol 43: 77–83.
5. Dorsch M, Stackebrandt E (1992) Some modifications in the procedure for direct sequencing of PCR amplified 16S rDNA. J Microbiol Methods 16: 271–279.
6. Eden PA, Schmidt TM, Blakemore RP, Pace NR (1991) Phylogenetic analysis of *Aquaspirillum magnetotacticum* using polymerase chain reaction-amplified 16S rRNA-specific DNA. Int J Syst Bacteriol 41: 324–325.
7. Edwards U, Rogall T, Blöcker H, Emde M, Böttger EC (1989) Isolation and direct complete nucleotide determination of entire genes: characterization of a gene coding for 16S ribosomal RNA. Nucleic Acids Res 17: 7843–7853.
8. Gargas A, Taylor JW (1992) Polymerase chain reaction (PCR) primers for amplifying and sequencing nuclear 18S rDNA from lichenized fungi. Mycologia 84: 589–592.
9. Giovannoni S (1991) The polymerase chain reaction. In: Stackebrandt E, Goodfellow M (eds) Nucleic Acids Techniques in Bacterial Systematics, pp. 177–203. John Wiley & Sons, Chichester.
10. Hiraishi A (1992) Direct automated sequencing of 16S rDNA amplified by polymerase chain reaction from bacterial cultures without DNA purification. Lett Appl Microbiol 15: 210–213.
11. Innis MA, Gelfand DH (1990) Optimization of PCRs. In: Innis MA, Gelfand DH, Sninsky JJ, White TJ (eds) PCR Protocols: A Guide to Methods and Applications, pp. 3–12. Academic Press, Inc. San Diego.
12. Kwok S (1990) Procedures to minimize PCR-product carry-over. In: Innis MA, Gelfand DH, Sninsky JJ, White TJ (eds) PCR Protocols: A Guide to Methods and Applications, pp. 142–145. Academic Press, Inc. San Diego.
13. Kwok S, Higuchi R (1989) Avoiding false positives with PCR. Nature 339: 237–238.
14. Lane DJ, (1991) 16S/23S rRNA sequencing. In: Stackebrandt E, Goodfellow M (eds) Nucleic Acids Techniques in Bacterial Systematics, pp. 115–147. John Wiley & Sons, Chichester.
15. Lane DJ, Pace B, Olsen GJ, Stahl DA, Sogin ML, Pace NR (1985) Rapid determination of 16S ribosomal RNA sequences for phylogenetic analyses. Proc Natl Acad Sci USA 82: 6955–6959.
16. Larsen N, Olsen GJ, Maidak BL, McCaughey MJ, Overbeek R, Macke TJ, Marsh TL, Woese, CR (1993) The ribosomal databas project. Nucleic Acids Res. 13:3021–3023.
17. Liesack W, Weyland H, Stackebrandt E (1991) Potential risks of gene amplification by PCR as determined by 16S rDNA analysis of a mixed-culture of strict barophilic bacteria. Microb Ecol 21: 191–198.
18. Medlin L, Elwood HJ, Stickel S, Sogin ML (1988) The characterization of enzymatically amplified eukaryotic 16S-like rRNA-coding regions. Gene 71: 491–499.
19. Myers TW, Gelfand DH (1991) Reverse transcription and DNA amplification by a *Thermus thermophilus* DNA polymerase. Biochem 30: 7661–7666.
20. Mylvaganam S, Dennis, PP (1992) Sequence heterogeneity between two genes encoding

16S rRNA from the halophilic archaebacterium *Haloarcula marismortui*. Genetics 130: 399–410.

21. Olsen GJ, Larsen N, Woese CR. (1991) The ribosomal RNA database project. Nucleic Acids Res 19supp: 2017–2021.

22. Pace NR, Stahl DA, Lane DJ, Olsen GJ (1986) The use of rRNA sequences to characterize natural microbial populations. Adv Microb Ecol 9: 1–55.

23. Saiki, RK (1989) The design and optimization of the PCR. In: Erlich HA (ed) PCR Technology: Principles and Applications for DNA Amplification, pp. 7–16. Stockton Press, New York.

24. Saiki R, Gelfand, DH, Stoffel S, Schraf SJ, Higuchi R, Horn GT, Mullis KB, Erlich HA (1988) Primer-direct enzymatic amplification of DNA with a thermostable DNA polymerase. Science 239: 487–491.

25. Saris PEJ, Paulin LG, Uhlén M (1990) Direct amplification of DNA from colonies of *Bacillus subtilis* and *Escherichia coli* by the polymerase chain reaction. J Microbiol Methods 11: 121–126.

26. Stahl DA, Amann R (1991) Development and application of nucleic acid probes. In: Stackebrandt E, Goodfellow M (eds) Nucleic Acids Techniques in Bacterial Systematics, pp. 205–248. John Wiley & Sons, Chichester.

27. Weisburg WG, Barns SM, Pelletier DA, Lane DJ (1991) 16S ribosomal DNA amplification for phylogenetic study. J Bacteriol 173: 697–703.

28. Wilson KH, Blitchington RB, Greene RC (1990) Amplification of bacterial 16S ribosomal DNA with polymerase chain reaction. Clinical Microbiol 28: 1942–1946.

29. Wisotzkey JD, Jurtshuk P, Fox GE (1990) PCR amplification of 16S rDNA from lyophilized cell cultures facilitates studies in molecular systematics. Curr Microbiol 21: 325–327.

30. Woese, CR (1987) Bacterial Evolution. Microbiol. Rev. 51: 221–271.

Molecular Microbial Ecology Manual **3.3.2**: 1-8, 1995.

Bacterial community fingerprinting of amplified 16S and 16-23S ribosomal DNA gene sequences and restriction endonuclease analysis (ARDRA)

ARTURO A. MASSOL-DEYA [1, 2], DAVID A. ODELSON [2, 3], ROBERT F. HICKEY [2, 4] and JAMES M. TIEDJE [2]

[1] *Department of Biology, University of Puerto Rico, Mayagüez, PR 00681, USA;*
[2] *NSF Center for Microbial Ecology, Michigan State University, E. Lansing, MI 48824, USA;* [3] *Department of Biology, Central Michigan University, Mt. Pleasant, MI 48859, USA;* [4] *Michigan Biotechnology Institute, Lansing, MI 48909, USA*

Introduction

The 16S and 23S rRNA genes have been utilized for phylogenetic analysis of both prokaryotic and eukaryotic organisms (see Section 3). In addition to direct comparison of the nucleic acid sequences [9], numerous groups have used the rapid method of polymerase chain reaction (PCR) amplification of this gene [6] as well as the complete rRNA locus [3, 4, 8] for a simple method for identification of bacterial genera and species. In these latter procedures, the amplified ribosomal gene (rDNA) is subjected to restriction endonuclease digestion; this has been termed ARDRA (Amplified Ribosomal DNA Restriction Analysis [8]). The resulting restriction fragment pattern is then used as a fingerprint for the identification of bacterial genomes. This method is based on the principle that the restriction sites on the RNA operon are conserved according to phylogenetic patterns.

Although ARDRA has been used for the characterization of bacterial isolates, in theory this method can also be used for analyzing mixed bacterial populations. If the rDNA fingerprints for individual bacteria in a community are sufficiently different, then one can examine the amplified products for a series of distinct patterns resulting from the different populations that make up the community. This method of Ribosomal DNA fingerprinting of communities could be used for a quick assessment of genotypic changes over time or between different locations reflecting different environmental conditions.

This method involves the use of a pair of universal priming sequences for the PCR amplification of either the 16S rRNA genetic loci or the intergenic regions of the 16S and 23S rRNA genes. The latter regions exhibit a large degree of sequence, length and frequency variation because the spacer region is not well conserved. In general, ARDRA using the 16S rRNA genes will result in a simpler pattern (3–5 bands per genome when 4 base site-specific restriction endonucleases are used) than ARDRA using the 16–23S region. Especially in the simpler case, one would expect it to be more difficult to resolve different but related organisms since they would share common restriction fragments. ARDRA using the 16–23S rDNA region will provide a more diverse template for restriction analysis resulting in more complex band patterns and potentially higher resolution of community members. This version of ARDRA may be useful in simpler communities that are composed of closely related populations. In highly diverse microbial communities, both primer approaches may either yield too complex a pattern or too little resolution of distinct bands. If this is the case, an alternative may be to use group specific primers for a given bacterial group or to carry out the DNA amplification under more discriminatory conditions (e.g. increasing stringency conditions during the annealing step of PCR). Further, resolution can also be obtained by hybridization with DNA probes specific for the target groups. Hence several modifications of this method can be adjusted to optimize for the community under study.

This PCR based technique provides a measure of structure and composition of microbial communities. As little as picogram quantities of DNA (1–10 cells) are satisfactory for this method. In turn, community dynamics can be monitored at the genotypic level without drastically disturbing the system by sampling. Additionally this method is free of the bias of culture dependent methods, although bias can still occur if DNA is not extractable or the primers do not attach. The method is not exhaustive since it normally detects only the more dominant members. This method can also be used on pure cultures to compare their patterns with patterns derived from a community. Furthermore, band patterns of pure cultures can be used for a quick presumptive identification if they match band patterns of taxonomically described strains. Whether used with isolates or communities, matching ARDRA band patterns cannot be used to unequivocally establish identity between strains since different but closely related strains (and phenotypes) can have the same band patterns. The value of the method lies with its speed and ability to evaluate differences in dominant phylogenetic groups present in a community.

Experimental approach

A pair of highly conserved flanking sequences are used for primer binding sites to amplify the 16S ribosomal genes from microbial systems (Fig. 1). The intragenic PCR product (~1500 bp) is then used as substrate for restriction endonuclease digestion followed by gel electrophoresis. In order to facilitate analysis and resolution (i.e. the ability to distinguish among different populations) separate digestions with three different restriction enzyme digestions is recommended. The sensitivity should be maximized by concentrating the amplified rDNA or digested rDNA product before

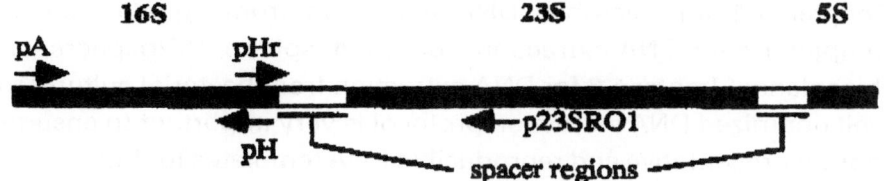

Figure 1. General schematic representation of the ribosomal RNA operon showing approximate localization of primer binding sites for PCR amplification. The spacer region can vary in length (e.g. number of tRNAs) and sequence among the different operons within a single organism [1].

electrophoresis. For the 16–23S rDNA intergenic region, amplification is carried out by using a pair of opposite highly conserved 16S and 23S ribosomal primers (Fig. 1) followed by restriction enzyme digestion and gel electrophoresis.

Primer sequences for the 16S rRNA gene amplification

Primer Designation	Position (*E. coli* 16S rRNA)	Sequence (5'-3')[7]
pA	19–38	AGA GTT TGA TCC TGG CTC AG
pH	1541–1581	AAG GAG GTG ATC CAG CCG CA

Primer sequences for the 16–23S rRNA intergenic amplification

Primer Designation	Position (*E. coli* 16-23S rRNA)	Sequence (5'-3')
pHr (reverse)	1518–1541	TGC GGC TGG ATC ACC TCC TT
p23SRO1	1069–1052	GGC TGC TTC TAA GCC AAC

Procedures

Template for PCR amplification

For mixed microbial cultures, it is important to maximize cell lysis before amplification since lysis efficiency is unequal for different bacterial types and physiological stages. Cell lysis by using repetitive (5 times) freeze (dry ice-ethanol bath) and thaw (80 °C water bath) cycles is often satisfactory for PCR. Methods for extraction of template DNA from microbial communities is suggested in other sections (see

Chapter 1.1 for microbial DNA extraction from aquatic sources; Chapter 1.4 for DNA extraction from phytosphere, rhizosphere, and rhizoplane; Chapter 1.6 for DNA extraction from bacterial cultures). A well optimized DNA extraction protocol is very important to ensuring more representative and reproducible DNA templates for PCR.

Steps in the PCR amplification
1. Prepare master solution containing:

ddH$_2$O	86.5 μl
10× PCR Buffer	10.0 μl
100× dNTP's	1.0 μl
100× primer χ	1.0 μl
100× primer γ	1.0 μl
Taq DNA polymerase (5 U/μl)	0.5 μl

2. Add 1 μl (approximately 100 ng) of bacterial or community DNA, or cell suspension (approximately 10 to 10^4 cells). Total reaction volume is 100 μl.
3. Amplification with the following temperature profile:

	Temperature	Time	
Denature	92 °C	2 min, 10 s	
Melt	92 °C	1 min, 10 s	
Anneal	48 °C	30 s	(30–35 cycles)
Extend	72 °C	2 min 10 s	
Final extend	72 °C	6 min 10 s	
Soak	4 °C	hold	

4. After amplification, 5 to 10 μl is electrophoresed on 0.7% agarose gel made in TAE buffer to examine the amplified products (Fig. 2).

Notes
1a. Maintain all solutions on ice.
1b. Reagents should be added in the order indicated.
1c. Always include a non-DNA control (e.g. water).
1d. Primers χ & γ are pA & pH for 16S rDNA amplification or pHr & p23SRO1 for 16–23S rDNA amplification.
1e. Taq DNA polymerase from different sources may behave differently because of different formulations, assay conditions, or unit definitions.
2. When cells are used as templates, they can be pre-treat (e.g. freeze & thaw) directly in the PCR vials containing 86.5 μl ddH$_2$O with the PCR buffer. The master solution of nucleotides, primers and polymerase are then added before amplification.

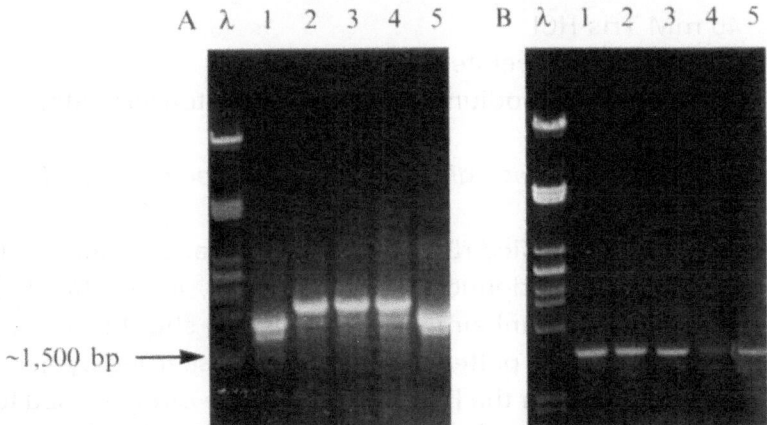

~1,500 bp ⟶

Figure 2. Products of PCR amplification from microbial communities samples. (A) The 16S rDNA amplification yielded a 1,500 bp fragment from an axenic culture (lane B) as well as from bioreactor samples (lanes C to F). (B) The 16–23S rDNA amplification of the intergenic region of the same samples yielded fragments of different sizes. Phage lambda (l) DNA digested with PstI was used as the size marker (lanes A) with the base pairs indicated on the left.

3a. Optimal cycle temperatures may vary with different thermal cycler models. The indicated set of temperatures and times were determined empirically for the Perkin Elmer thermal cycler model 9600.

3b. When cells are used as templates, melting steps should be performed at 94 °C for 3 min. In addition, annealing should be done at 52 °C.

3c. When DNA is used as template, 48 °C provides low stringency annealing conditions, but discriminatory enough to avoid nonspecific amplification.

Solutions

- 10× PCR buffer
    - 500 mM KCl
    - 100 mM Tris-HCl, pH 8.3
    - 15 mM $MgCl_2$
- 100×dNTP's solution
    - 25 mM of each nucleotide (ATP, TTP, CTP and GTP)
- Primers
    - 100 µM stock solution in TE buffer (10 mM Tris, 0.1 mM EDTA, pH 8)
- AmpliTaq DNA polymerase
    - 5 Units/µl (Perkin-Elmer Cetus, Norwalk, CT)
- 1× TAE electrophoresis buffer (pH 7.5)

- 40 mM Tris-HCl
- 5 mM sodium acetate
- 1 mM EDTA (disodium ethylenediaminetetraacetate)

Restriction endonuclease digestion and gel electrophoresis

Digestion of the amplified rDNA product with tandem tetrameric site-specific restriction endonucleases (e.g. *Hae*III, *Mse*1, *Hinf*1, *Sau*3A, *Hpa*II; Boehringer Mannheim, Indianapolis, IN) should yield a restriction enzyme digestion pattern which can be discerned by gel electrophoresis. An aliquot of the PCR product can be directly used for each digestion. Digest in the appropriate restriction buffer (generally provided by all commercial companies in 10× concentration when enzymes are purchased) at optimal temperature for 2–3 h. As noted above with community level amplification it is useful to concentrate the amplified product. This can be accomplished by ethanol precipitation (1/2 volume 3 M sodium acetate with 2 volumes of cold 100% ethanol at –20 °C for 30 min to overnight [5] before or after digestion. The amount of DNA to be loaded per well is dependent of the limiting bands which need to be resolved. The least amount of DNA that can be consistently detected with ethidium bromide staining is about 10 ng. DNA amounts greater than 100 ng will not be resolved as a sharp clean band. The rDNA restriction fragments are then electrophoresed on 4% (w/v) NuSieve agarose gels (FMC, Rockland, ME) at 7 volts/cm containing 0.5X TAE, and stained in 0.5 µg/ml ethidium bromide solution for 20 min (Fig. 3). Finally, DNA banding patterns database and numerical analysis can be performed with analytical hardware/software package such as AMBIS MicroPM$_{TM}$ based system (Microbiology Pattern Matching) [2].

Acknowledgements

This work was supported by the National Science Foundation grant BIR-9120006 and by the National Institute of Health grant GM14047-04 to AMD.

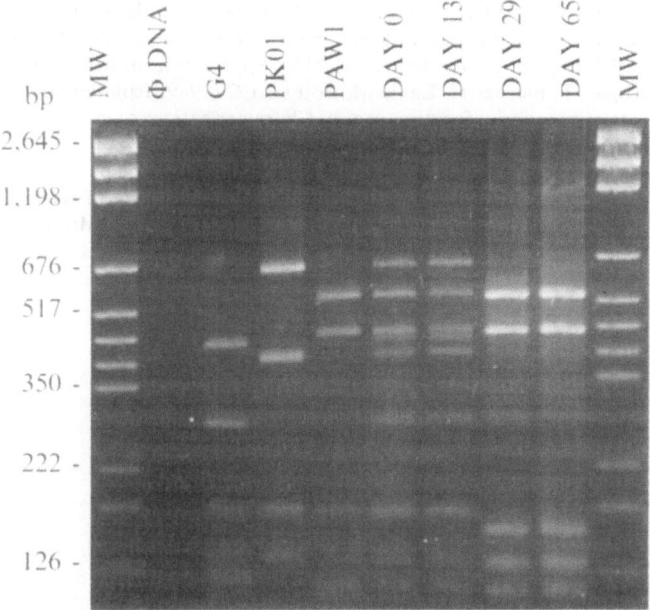

Figure 3. An example of a shift in community structure that was studied by the ARDRA method. Photograph of ethidium bromide stained agarose gel of ARDRA patterns from a biofilm community in a fluidized bed reactor sampled from 0 to 65 days of operation. The bioreactor was initially inoculated with strains: G4, *Pseudomonas cepacia*; PKO1, *P. pickettii*; and PaW1, *P. putida* The ARDRA patterns for these pure cultures are shown inn lanes 2-4. At day 19 groundwater containing its indigenous microorganisms was allowed to enter the reactor. The different ARDRA patterns in the biofilm after this date show that the invading strains displaced the inoculated strains. Lanes MW, molecular weight DNA markers with indicated base pairs on the left; lane Φ DNA, PCR amplification in water as negative control; all other lanes are *Hpa*II restriction endonuclease digestion of PCR amplified 16S rDNA.

References

1. Brosius J. Dull TJ, Sleeter. DD, Noller HF (1981) Gene organization and primary structure of a ribosomal RNA operon from *Escherichia coli.* J Mol Biol 148: 107–127.
2. Hook LA, Odelson DA. Bogardt AH, Hemmingsen BB, Labeda DP, MacDonell MT (1991) Numerical analysis of DNA restriction fragment length polymorphisms and whole-cell protein banding patterns: a means of bacterial identification at the species and subspecies level. Newsletter (USFCC) 21(3): 1–10.
3. Jayarao BM. Doré JJE, Baumbach GA. Matthews KR, Oliver SP (1991) Differentiation of *Streptococcus uberis* from *Streptococcus parauberis* by polymerase chain reaction and restriction fragments length polymorphism analysis of 16S ribosomal DNA. J Clin Microbiol 29: 2774–2778.
4. Jensen MA, Webster JA. Straus N. (1993) Rapid identification of bacteria on the basis of polymerase chain reaction-amplified ribosomal DNA spacer polymorphisms. Appl Environ Microbiol 59: 945–952.
5. Maniatis T, Fritsch EF, Sambrook J (1982) Molecular Cloning: A Laboratory Manual. New York: Cold Spring Harbor Laboratory Press.

6. Muyzer G, De Waal EC, Uitterlinden AG (1993) Profiling of complex microbial populations by denaturing gradient gel electrophoresis analysis of polymerase chain reaction-amplified genes coding for 16S rRNA. Appl. Environ. Microbiol. 59: 695–700.

7. Ulrike E, Rogall T, Blocker H, Emde M, Bottger EC (1989) Isolation and direct complete nucleotide determination of entire genes. Characterization of a gene coding for 16S ribosomal RNA Nucl Acid Res 17: 7843–7853.

8. Vaneechoutte M, Rossau R, De Vos P, Gillis M, Janssens D, Paepe N, De Rouck A, Fiers T, Claeys G, Kersters K (1992). Rapid identification of bacteria of the *Comamonadaceae* with amplified ribosomal DNA-restriction analysis (ARDRA). FEMS Microbiol. Lett. 93: 227–234.

9. Woese, C.R. 1987. Bacterial evolution. Microbial Rev 51: 221–271.

Molecular Microbial Ecology Manual **3.3.4**: 1–12, 1995
ⓒ 1995 *Kluwer Academic Publishers.*

Investigation of fungal phylogeny on the basis of small ribosomal subunit RNA sequences

YVES VAN DE PEER and RUPERT DE WACHTER

Department of Biochemistry, University of Antwerp (UIA), Universiteitsplein 1, B-2610 Antwerp, Belgium

Introduction

A study published in 1989 [15] on the evolutionary position of fungi within the eukaryotic realm as derived from the structure of small ribosomal subunit RNA comprised only 4 fungal SSU rRNA sequences (SSU rRNA is small ribosomal subunit RNA, i.e. the 16S rRNA of prokaryotes, the 18S rRNA of eukaryotes). At present, December 1993, more than 110 complete or nearly complete SSU rRNA sequences from fungi have been published. This increase can be ascribed to the fact that SSU rRNA has proven an excellent molecule for the reconstruction of phylogeny of bacterial as well as eukaryotic lineages, for reasons enumerated by Woese [41] and Sogin [30].

Fungi are decomposers in most ecosystems and make important contributions to the ecological balance of our world. Many of them form symbiotic relationships with other organisms, mostly plants, while they also constitute the majority of plant pathogens. Some also cause diseases in humans and animals. Finally, they have great importance in food and drug industry. Correct identification of fungi requires a dependable classification scheme, which should be based on the evolutionary relationships among species. In addition, since for many fungi, in particular yeasts, morphological characteristics are scarce, sequences constitute the main tool for studying their genealogy. In this chapter, we present a working scheme for the construction of evolutionary trees starting from nucleotide sequences. Furthermore, detailed evolutionary trees based on all presently known SSU rRNA sequences are shown for the "true" fungi, i.e. chytridiomycetes, zygomycetes, ascomycetes, and basidiomycetes.

Data collection, sequence alignment and tree construction

Data collection
Reconstruction of evolutionary relationships from SSU rRNA sequences or the corresponding genes is based on the systematic comparison of sequences available for the group under study. Therefore, the first concern is to collect all the sequences one wants to include in the analysis.

One possibility is to look for sequences of the gene under study in the EMBL [24] and GenBank [2] nucleotide sequence libraries. The major drawback is that sequences thus obtained are not aligned with one another and one needs to construct an alignment starting from scratch. However, software (recently reviewed by Chan *et al.* [8]) exists which takes as input a set of unaligned sequences and produces an alignment of them. These sequence alignment programs use a mathematical model to define what is considered an optimal alignment [8, 39]. One such program is CLUSTAL [17, 18] which runs on a wide variety of computer systems. However, it must be noted that the outcome of such multiple sequence alignment programs is dependent on the mathematical model used, on the degree of similarity between the sequences and on the weight given to insertions (gap penalty), the inclusion of which is necessary to align the sequences (see alignment of new sequences).

Alternatively, instead of collecting nucleotide sequences from databases such as EMBL or GenBank, sequences can also be retrieved from databases storing SSU rRNA sequences in aligned form. In our laboratory, a database on SSU rRNA structures is kept up to date and the aligned sequences are made available to the scientific community in computer readable form through "anonymous ftp" [23]. A copy of this database is also distributed by the EMBL nucleotide library at Heidelberg [24]. At the moment, this database contains 565 aligned eukaryotic sequences of which 114 are from fungi, oomycetes and slime moulds not included. A comparable database, also distributing sequences in aligned form is the RDP database in Urbana, Illinois, USA [22].

Alignment of new sequences
Even if a set of aligned sequences is available, alignment is necessary to include the newly determined sequences. As a result of insertions and deletions occurring during evolution, gaps have to be inserted in order to obtain an alignment where nucleotides derived from the same ancestral nucleotide occupy the same alignment position. In the process of alignment, the more conserved parts of the sequences serve as landmarks for aligning the more variable parts. Furthermore, in the case of SSU rRNA, knowledge of the secondary structure can provide additional information since boundaries of secondary structure elements tend to occupy homologous positions in different sequences, and should therefore be taken into account in the alignment process. The secondary structure of

SSU rRNA is well-established nowadays and detailed models are available [14, 23].

Sequences can be aligned and alignments can be edited using an ordinary word processor but special sequence editors which facilitate aligning have been developed, e.g. SeqEdit, available from the RDP database [22], MALIGNED [9], and ALMA [33]. The sequences in our database are aligned with the aid of a sequence editor called DCSE [10] that can handle several hundreds of sequences at a time and visualizes both primary and secondary structure features. New sequences can be aligned automatically but visual refinement of the alignment is possible as well.

Tree construction
Once an alignment has been established, evolutionary trees can be constructed. Many different methods exist for constructing trees from a sequence alignment. An important distinction is that between maximum parsimony, maximum likelihood and distance matrix methods. In this paper, we are confining ourselves to distance matrix methods since these allow the construction of evolutionary trees from large sequence alignments within reasonable time. For a detailed discussion of the other methods, we refer to the excellent review of Swofford and Olsen [32].

When a tree is constructed using a distance matrix method, the dissimilarity between each pair of sequences is calculated. Dissimilarity is defined as the number of alignment positions where the sequences contain a different nucleotide, divided by the total number of positions where both sequences contain a nucleotide. These dissimilarity values are then converted into evolutionary distances using a correction that compensates for the occurrence of multiple mutations. The correction most applied in evolutionary studies is probably the one of Jukes and Cantor [19], which starts from the simple assumption that all nucleotides in the alignment change at the same rate and that all nucleotides can be substituted by any other nucleotide with equal probability. Although it is generally agreed that both assumptions are violated in the evolution of real sequences, the Jukes and Cantor model performs well in most studies simulating the evolution of nucleic acid sequences. According to the Jukes and Cantor model, the evolutionary distance d between two sequences A and B is derived by the equation:

$$d_{AB} = -\frac{3}{4}\ln\left(1 - \frac{4}{3}f\right)$$

where f is the dissimilarity of sequences A and B.

Another correction is based on the two parameter evolution model of Kimura [20] which treats transitions and transversions separately. However, in studies based on rRNA data, differences obtained by applying both corrections are small. In the latter model the evolutionary distance is estimated as:

$$d_{AB} = -\frac{1}{2}\ln\left[\left(1 - 2p{-}q\right)\sqrt{1 - 2q}\right]$$

where p is the fraction of transitions and q is the fraction of transversions observed.

Once all evolutionary distances are computed, a tree can be inferred. Many different algorithms have been developed such as UPGMA [29], the method of Fitch and Margoliash [13], the neighbourliness method of Sattath and Tversky [27], the matrix optimization method of De Soete [11], and the neighbour-joining method of Saitou and Nei [25]. At the moment, neighbour-joining is one of the most popular tree construction methods and besides its efficiency in recovering the true tree topology in simulation studies [26, 31], its main advantage is its speed, which allows the processing of large sets of sequences. In general, neighbour-joining only needs a small fraction of the time required by maximum parsimony and maximum likelihood methods.

When the pattern of divergence among different taxa is studied, one would expect to observe the same tree topology, regardless of the tree construction method used. In practice, however, this is rarely the case. Even if the same algorithm is used consistently, the observed topology of divergence of the taxa often changes as the set of organisms chosen to represent that set of taxa is changed [36, 37]. Such artifacts are due to the stochastic nature of the mutation process, which results in events such as parallel mutations, local convergences and divergences, in short, statistical deviations from the mathematical model that forms the basis of tree construction. As a result, the tree topologies that are computed can only be considered as an approximation of the evolutionary history.

One method that can be used in order to estimate the reliability of an observed branching pattern, is bootstrap analysis. In the case of sequence data [12], this involves the resampling of alignment positions, with replacement, to create a series of samples (preferably 500 or more) of the same size as the original data. So, some nucleotides are dropped from the analysis while others are counted once, twice, and so on, as a result of random selection. The reliability of a cluster in the tree based on the real sequence data is given by the fraction of trees, based on the resampled data, containing a cluster of the same composition.

In our research group, a software package called TREECON [35] has been developed which implements all the items discussed above. Starting from a simple text file, containing nucleic acid sequences provided with gaps for mutual alignment, one can produce publishable trees in a simple and straightforward way. Several correction methods for converting dissimilarity into evolutionary distance are implemented, as well as several tree inferring algorithms. The whole package is menu driven to assure user-friendliness. It is also possible to convert file formats to formats used by other tree construction packages. TREECON is made available to other researchers and can be obtained by sending one high density floppy disk (1.44 Mbyte) to the authors. All evolutionary trees shown in this chapter were constructed with TREECON running on an IBM-compatible computer with a 80486 processor, using the DOS operating system.

Procedures

A practical scheme for constructing evolutionary trees based on fungal SSU rRNA sequence data, and using the TREECON software package [35] is given below:

- Fetch the fungal SSU rRNA sequences over electronic network. To do so, connect to host uiam3.uia.ac.be by ftp and log in as user "anonymous". The sequences can be retrieved from our database in a format similar to that required for the TREECON package. Since the latter format is very flexible, only minimal editing is required to construct an input file that can be used for tree construction. In this input file, sequences should be listed one after the other and should contain gaps for mutual alignment.
- If newly determined sequences have to be included in the analysis, these have to be added to the input file. Therefore, an option of the TREECON software package is to write the sequences of the input file in the form of an alignment, with the block size specified by the user. The alignment thus constructed can be handled by any ordinary word processor and the newly determined sequences can be aligned with the other sequences by adding gaps representing insertions and deletions. Finally, the alignment should be reconverted into a file in which the sequences are ordered sequentially. This, again, is no problem since the TREECON input file format is very flexible.
- Start the TREECON program, select the input file and select the sequences to be included in the analysis.
- Choose the desired method for converting dissimilarity into evolutionary distance and compute the distance matrix.
- Construct the evolutionary tree based on the computed distance matrix, e.g. by neighbour-joining.
- Root the evolutionary tree by selecting the outgroup.

- Draw the evolutionary tree on the screen, scale it to your own needs and save it as a text- or a PostScript file.
- Subject the tree to bootstrap analysis. Specify the number of resamplings needed.

Phylogeny of fungi

Of the four main divisions of "true" fungi, 114 SSU rRNA sequences are available at the moment, covering a wide range of different genera and families. Since recent literature discussing the evolutionary relationships between most of these genera is already quite extensive [1, 3–7, 16, 34, 40], the discussion in this chapter will be restricted to the main evolutionary lineages.

Phylogeny of chytridiomycetes and zygomycetes
Figure 1 shows an evolutionary tree based on all known SSU rRNA sequences of chytridiomycetes, zygomycetes, basidiomycetes, and ascomycetes. The red alga *Gracilaria lemaneiformis* was used as outgroup organism. The structure of the basidiomycete and ascomycete clusters is not shown but is detailed in Figs. 2 and 3. In the tree shown in Fig. 1, the chytridiomycete *Blastocladiella emersonii* forms the first line of divergence within the fungi, followed by the zygomycete *Mucor racemosus*. Next there is a divergence of, on the one hand, ascomycetes and basidiomycetes, and on the other hand, the remaining zygomycetes and chytridiomycetes. Based on this result, the chytridiomycetes seem polyphyletic, since *Blastocladiella* branches off separately from the other chytridiomycetes. The same is true for the zygomycetes, where *Mucor* representing the Mucorales forms a separate lineage from the endomycorrhizal zygomycetes which belong to the Endogonales. However, these results should be interpreted with caution, since for the Blastocladiales and Mucorales only one representative is known. The latter two species not considered, the zygomycetes and chytridiomycetes are clustered together and seem to form a monophyletic group, supported at a high level (> 95%) by bootstrap analysis. The branching order found within the zygomycetes is nearly identical with the result of Simon *et al.* [28], who studied diversification of endomycorrhizal fungi. Bruns *et al.* [7] found that the zygomycete

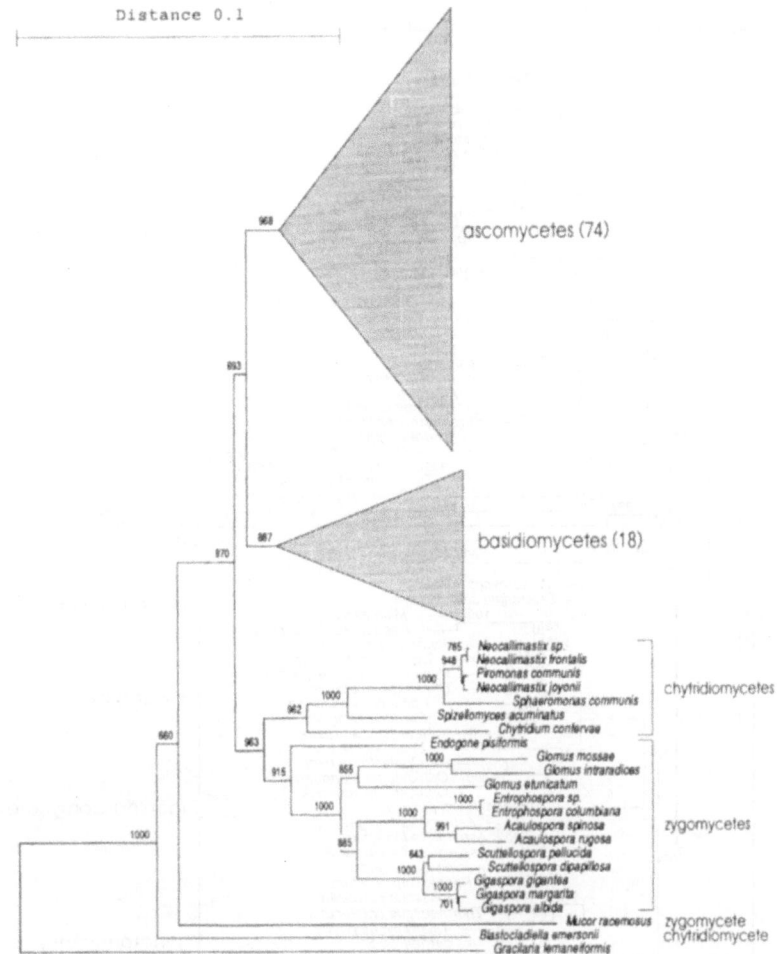

Figure 1. Neighbour-joining tree from all known SSU rRNA sequences of chytridiomycetes, zygomycetes, ascomycetes, and basidiomycetes. Distance values were calculated according to Jukes and Cantor [19]. In the process of distance calculation, the complete sequence alignment was used but insertions/deletions were not taken into account. The root was placed such as to equalize its distance to the outgroup species *Gracilaria lemaneiformis* and its average distance to all fungi. Ascomycetes and basidiomycetes are represented as isosceles triangles with a height approximately equal to the mean distance separating the species forming the cluster from the deepest branching point within that cluster, and a base proportional to the number of species. Chytridiomycetes are represented by 8 sequences and zygomycetes are represented by 14 sequences. The number of ascomycetes and basidiomycetes included is indicated between brackets. The distance between two organisms or groups of organisms, measured in substitutions per nucleotide, is obtained by summing the lengths of the connecting branches along the horizontal axis, using the scale on top. Bootstrap values higher than 50% (500/1000) are placed alongside the node considered.

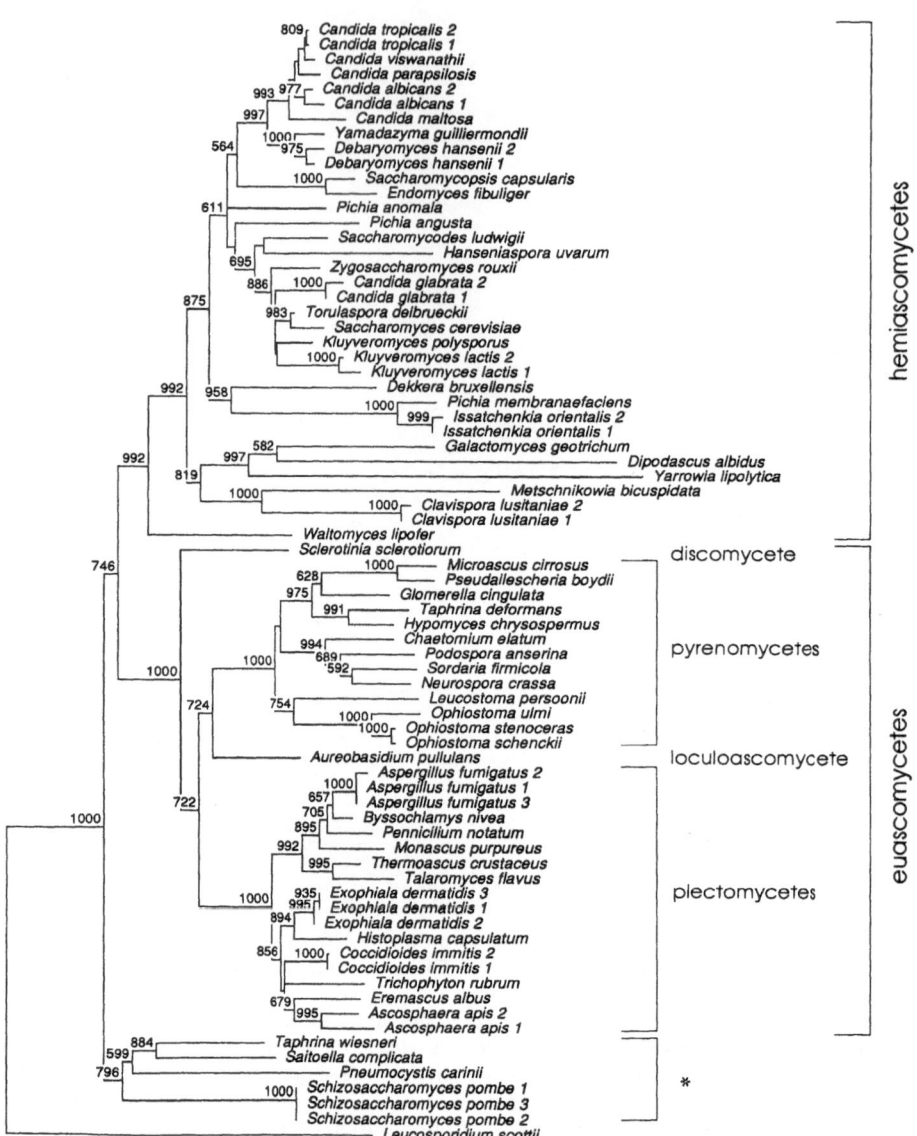

Distance 0.1

Figure 2. Evolutionary tree from all known SSU rRNA sequences of ascomycetes. The basidiomycete *Leucosporidium scottii* was used as outgroup organism. Construction and interpretation of the tree is as in Fig. 1. Numbers placed after the species names indicate that several strains of the same species were determined by different authors. The cluster indicated by an asterisk is formed by species whose exact taxonomic position is still undefined.

Endogone pisiformis clusters, rather unexpectedly, with the chytridiomycetes. In our analysis however, the latter species groups with the other zygomycetes with high confidence (> 90%).

Phylogeny of ascomycetes

Within the ascomycetes, shown in Fig. 2, two major lineages well supported by bootstrap analysis, can be discerned: the euascomycetes and the hemiascomycetes. A third branch, less supported by bootstrap analysis, groups the species *Taphrina wiesneri, Saitoella complicata, Pneumocystis carinii,* and *Schizosaccharomyces pombe.*

The euascomycetes are further divided into two major divisions, viz. the pyrenomycetes and the plectomycetes. *Sclerotinia sclerotiorum,* a representative of the discomycetes, and *Aureobasidium pullulans,* a member of the loculoascomycetes, form two separate branches within the euascomycetes. The hemiascomycete cluster consists entirely of yeasts and yeast-like fungi, and includes, among others, the heterogeneous genera *Candida* and *Pichia.* A detailed discussion about most of the species included in both the euascomycete and the hemiascomycete cluster is given in Wilmotte *et al.* [40].

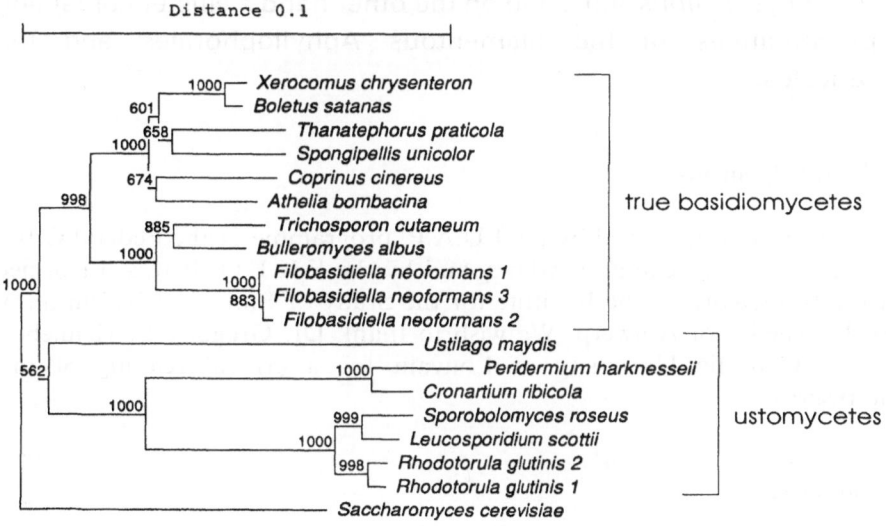

Figure 3. Evolutionary tree of the basidiomycetes. The ascomycete *Saccharomyces cerevisiae* was used to root the tree.

In recent classifications [21, 38], *Schizosaccharomyces* is considered a hemiascomycete. Based on ribosomal RNA data however, *Schizosaccharomyces* is nearly always found at the basis of the ascomycete cluster [7, 34, 40, this study]. Another fungus always found at the base of the ascomycete cluster is *Pneumocystis carinii*, the exact taxonomic position of which is still undefined.

Phylogeny of basidiomycetes

Figure 3 shows an evolutionary tree of the basidiomycetes. These are divided into two major lineages, viz. the ustomycetes (phragmo-basidiomycetes) and the true basidiomycetes (holobasidiomycetes). Both groups are supported at a high confidence level in bootstrap analysis, with the exception of *Ustilago maydis*, the position of which is less stable (cf. [40]).

Within the ustomycetes, three orders can be discerned: the Ustila-ginales represented by *Ustilago maydis*, the Sporidiales, comprising the yeasts *Leucosporidium scottii*, *Sporobolomyces roseus* and *Rhodotorula glutinis*, and the Uredinales, represented by *Cronartium ribicola* and *Peridermium harknesseii*. Within the true basidio-mycetes, two distinct lineages exist, both supported in 100% of the bootstrap samples: on the one hand a cluster containing members of the yeast order Filobasidiales comprising *Filobasidiella neoformans*, *Bulleromyces albus* and , and on the other hand a cluster containing representatives of the filamentous Aphyllophorales and the Agaricales.

Acknowledgements

This work was supported by the I.U.A.P. programme of the Federal Office for Scientific and Cultural Affairs, and by the F.K.F.O. It was performed in the framework of the Institute for the Study of Biological Evolution of the University of Antwerp. We wish to thank Dr. Grégoire L. Hennebert of the Catholic University of Louvain for a critical reading of this manuscript.

References

1. Barns SM, Lane DJ, Sogin ML, Bibeau C, Weisburg WG (1991) Evolutionary relationships among pathogenic *Candida* species and relatives. J Bacteriol 173: 2250–2255.

2. Benson D, Lipman DJ, Ostell J (1993) GenBank. Nucl Acids Res 21: 2963–2965.
3. Berbee ML, Taylor JW (1992) Two ascomycete classes based on fruiting-body characters and ribosomal DNA sequences. Mol Biol Evol 9: 278–284.
4. Bowman BH, Taylor JW, White TJ (1992) Molecular evolution of the fungi: human pathogens. Mol Biol Evol 9: 839–904.
5. Bowman BH, Taylor JW, Brownlee AG, Lee J, Lu S-D, White TJ (1992) Molecular evolution of the fungi: relationships of the basidiomycetes, ascomycetes, and chytridiomycetes. Mol Biol Evol 9: 285–296.
6. Bruns TD, White TJ, Taylor JW (1991) Fungal molecular systematics. Annu Rev Ecol Syst 22: 525–564.
7. Bruns TD, Vilgalys R, Barns SM, Gonzales D, Hibbett DS, Lane DJ, Simon L, Stickel S, Szaro TM, Weisburg WG, Sogin ML (1993) Evolutionary relationships within the fungi: analysis of nuclear small subunit rRNA sequences. Mol Phyl Evol 1: 231–241.
8. Chan SC, Wong AKC, Chiu DKY (1992) A survey of multiple sequence comparison methods. Bull Math Biol 54: 563–698.
9. Clark SP (1992) MALIGNED: a multiple sequence alignment editor. Comput Applic Biosci 8: 535–538.
10. De Rijk P, De Wachter R (1993) DCSE, an interactive tool for sequence alignment and secondary structure research. Comput Applic Biosci 9: 735–740.
11. De Soete G (1983) A least squares algorithm for fitting additive trees to proximity data. Psychometrika 48: 621–626.
12. Felsenstein J (1985) Confidence limits on phylogenies: an approach using the bootstrap. Evolution 39: 783–791.
13. Fitch WM, Margoliash E (1967) Construction of phylogenetic trees. Science 155: 279–284.
14. Gutell RR (1993) Collection of small subunit (16S- and 16S-like) ribosomal RNA structures. Nucl Acids Res 21: 3051–3054.
15. Hendriks L, Goris A, Neefs J-M, Van de Peer Y, Hennebert GL, De Wachter R (1989) The nucleotide sequence of the small ribosomal subunit RNA of the yeast *Candida albicans* and the evolutionary position of the fungi among the eukaryotes. Syst Appl Microbiol 12: 223–229.
16. Hendriks L, Goris A, Van de Peer Y, Neefs J-M, Vancanneyt M, Kersters K, Berny J-F, Hennebert GL, De Wachter R (1992) Phylogenetic relationships among ascomyetes and ascomycete-like yeasts as deduced from small ribosomal subunit RNA sequences. Syst Appl Microbiol 15: 98–104.
17. Higgins DG, Sharp PM (1988) CLUSTAL: a package for performing multiple sequence alignments on a microcomputer. Gene 73: 237–244.
18. Higgins DG, Bleasby AJ, Fuchs R (1992) CLUSTAL V: improved software for multiple sequence alignment. Comput Applic Biosci 8: 189–191.
19. Jukes TH, Cantor CR (1969) Evolution of protein molecules. In: Munro HN (ed) Mammalian Protein Metabolism, pp. 21–132. New York: Academic Press.
20. Kimura M (1980) A simple method for estimating evolutionary rate of base substitutions through comparative studies of nucleotide sequences. J Mol Evol 16: 111–120.
21. Kreger-van Rij NJW, Kurtzman CP (1984) Classification of the ascosporogenous yeasts. In: NJW Kreger-van Rij (ed) The Yeasts, a Taxonomic Study, pp. 25–32. Amsterdam: Elsevier.
22. Larsen N, Olsen GJ, Maidak BL, McCaughey MJ, Overbeek R, Macke TJ, Marsh TL, Woese CR (1993) The ribosomal database project. Nucl Acids Res 21: 3021–3023.
23. Neefs J-M, Van de Peer Y, De Rijk P, Chapelle S, De Wachter R (1993) Compilation of small ribosomal subunit RNA structures. Nucl Acid Res 21: 3025–3049.
24. Rice CM, Fuchs R, Higgins DG, Stoehr PJ, Cameron GN (1993) The EMBL data library. Nucl Acids Res 21: 2967–2971.
25. Saitou N, Nei M (1987) The neighbor-joining method: a new method for reconstructing phylogenetic trees. Mol Biol Evol 4: 406–425.

26. Saitou N, Imanishi T (1989) Relative efficiencies of the Fitch-Margoliash. maximum-parsimony, maximum-likelihood, minimum-evolution, and neighbour-joining methods of phylogenetic tree construction in obtaining the correct tree. Mol Biol Evol 6: 514–525.

27. Sattath S, Tversky A (1977) Additive similarity trees. Psychometrika 42: 319–345.

28. Simon L, Bousqet J, Lévesque RC, Lalonde M (1993) Origin and diversification of endomycorrhizal fungi and coincidence with vascular land plants. Nature 363: 67–69.

29. Sneath PHA, Sokal RR (1973) Numerical Taxonomy. San Francisco: Freeman WH.

30. Sogin ML (1989) Evolution of eukaryotic microorganisms and their small subunit ribosomal RNAs. Amer Zool 29: 487–499.

31. Sourdis J, Nei M (1988) Relative efficiencies of the maximum parsimony and distance-matrix methods in obtaining the correct phylogenetic tree. Mol Biol Evol 5: 298–311.

32. Swofford DL, Olsen GJ (1990) Phylogeny reconstruction. In: Hillis DM, Moritz C (eds) Molecular Systematics, pp. 411–501. Sunderland: Sinauer Associates.

33. Thirup S, Larsen NE (1990) ALMA, an editor for large sequence alignments. Proteins Struct Funct Genet 7: 291–291.

34. Van de Peer Y, Hendriks L, Goris A, Neefs J-M, Vancanneyt M, Kersters K, Berny J-F, Hennebert GL, De Wachter R (1992) Evolution of basidiomycetous yeasts as deduced from small ribosomal RNA sequences. Syst Appl Microbiol 15: 250–258.

35. Van de Peer Y, De Wachter R (1993) TREECON: a software package for the construction and drawing of evolutionary trees. Comput Applic Biosci 9: 177–182.

36. Van de Peer Y, Neefs J-M, De Rijk P, De Wachter R (1993) Evolution of eukaryotes as deduced from small ribosomal subunit RNA sequences. Biochem Syst Ecol 21: 43–55.

37. Van de Peer Y, Neefs J-M, De Rijk P, De Vos P, De Wachter R (1994) About the order of the major bacterial taxa. Syst Appl Microbiol 17: 32–38.

38. Von Arx JA, Van der Walt JP (1987) Ophiostomatales and Endomycetales. Stud Mycol 30: 167–176.

39. Waterman MS (1989) Sequence alignments. In: Waterman MS (ed) Mathematical Methods for DNA Sequences, pp. 53–92. Boca Raton: CRC Press.

40. Wilmotte A, Van de Peer Y, Goris A, Chapelle S, De Baere R, Nelissen B, Neefs J-M, Hennebert GL, De Wachter R (1993) Evolutionary relationships among higher fungi inferred from small ribosomal subunit RNA sequence analysis. Syst Appl Microbiol 16: 436–444

41. Woese CR (1987) Bacterial evolution. Microbiol Rev 51: 221–271.

Molecular Microbial Ecology Manual **3.3.5**: 1-22, 1995.

Sequence Databases

WOLFGANG LUDWIG

Department of Microbiology, Technical University of Munich, D-80290 Munich, Germany

Introduction

Comparative sequence analyses of coding and non-coding DNA and other macromolecules play a central part in biology in general and particularly in molecular ecology. Higher order structure and function of biomolecules are determined by their primary structures which are synthesised, processed and modified under the direct or indirect control of the corresponding genes and regulatory elements at the genomic level. Consequently, nucleic acid or protein sequence similarities often indicate structural and functional similarities and help to improve the understanding of complex biological structures and processes. Comparative sequencing in molecular ecology is used for phylogenetic analyses, identification, and monitoring of organisms after isolation and cultivation. Furthermore, their direct detection, quantification, and *in situ* monitoring within (complex) samples can be achieved by using modern sequence based probing techniques [17]. The identification of structural genes and control regions helps to study or predict physiological potentials and properties of microbial communities and to understand the roles of individual strains or species within these communities.

The ongoing development and worldwide application of rapid techniques of automated sample processing and sequence determination result in an enormous and permanently increasing daily output of sequence data. Researchers who want to analyse a new sequence want to know whether it is already known, whether similar sequences exist, or any additional information (e.g. functional assignment, higher order structure) related to similar sequences is available. The demand for structured sequence databases providing access for the scientific community is obvious. A primary use of such sequence databases is sequence data analysis by similarity searches. Several organizations in different countries are collecting, processing and distributing sequence data obtained from literature screening, direct submissions from research groups, and collaboration among the database centres.

The first systematic compilation of sequence data was undertaken by M. Dayhoff and colleagues at the National Biomedical Research Foundation in USA. The first edition of the ATLAS of Protein Sequence and Structure appeared in 1965 [5–7]. Later on, GenBank [2], EMBL [10, 12], and other sequence database centres collecting nucleic acid and protein sequences had been initiated. The PIR-International [13] and the SwissProt [3] databases had been founded with the aim to compile protein sequences. Special databases on ribosomal ribonucleic acid sequences had been established at the Universities of Illinois (Urbana, USA) [16] and Antwerpen (Belgium) [8, 20]. The history of sequence databases has recently been described by Smith [18] and by Burks *et al.* [4]. The major databases are regularly described in a special issue of Nucleic Acids Research.

Structure of the sequence databases

All major sequence database centres distribute the data in flat (ASCII) file formate some of them provide the data in special (compressed) computer readable formate. The sequences and additional information are stored as individual entries. In most cases, the information associated with the sequence is structured by hierarchical sets of records such as header record section, reference record section, general property record section, feature record section, and sequence record section. The header record sections usually contain descriptive information such as the type of molecule sequenced, the object described (gene), the biological source. Database tracking and accession information is also stored in this section. Whereas the individual databases use different sequence entry identifiers consisting of letter and/or numeral codes, the accession numbers of nucleic acid sequence entries are shared by the different databases. The typical structure of accession numbers is X00000, X represents an (uppercase) letter and the zeros a sequence of five numbers. The reference record section contains citations of the primary literature and information directly associated with these reports. Information on function, gene expression, and other characteristics that is not linked to sites or regions within the sequences is documented in the general property section. The feature record section compiles information that describes characteristics or functions of sites and regions within the particular sequence such as promotor regions, boxes, coding regions, introns-exons. However, the formates adopted by the major database centres differ with respect to completeness and arrangement of the accompanying information.

Accessibility of the sequence data

The major database centres distribute data and software for data extraction and analysis upon request on magnetic media, nowadays preferrably on CD-ROM at reasonable costs. Fileservers allow anybody with access to international computer networks to retrieve complete or partial databases or software via e-mail. Most of the database centres also provide services such as comparisons of user submitted sequences to the database. Requests and results are transmitted by e-mail. The e-mail addresses for requesting general information, data or services are given in the following with the description of the different databases. General information concerning the particular database and instructions for the use of the mail servers can usually be obtained by sending an e-mail message containing only the word help. This can be typed in the subject line or the message text field. Detailed information how to proceed is automatically returned to the requesters mail box. Those who have access to the network and FTP (file transfer protocol) networking software are able to download databases or software by direct anonymous FTP connection to computers at the respective database centres. As an example, the commands used in an FTP session on a SUN Sparc workstation (UNIX) for the connection to the host rdp.life.uiuc.edu of the Ribosomal Database Project are given in Fig. 1. After establishing the connection, the user is asked to login as anonymous. Usually, he then is asked to provide his e-mail address as a password. After anonymous login, the directories accessible to public can be screened and their contents transferred to the local directory of the user. In most cases, there are (ASCII) files in these directories containing information on the directory contents and instructions how to retrieve and to use them. These files are usually designated as README or similarily.

The European Bioinformatics Institute (EBI) Databases

The EMBL (European Molecular Biology Laboratory) Data Library [10, 12] was founded in 1980. Since September 1994 the sequence database is based at the European Bioinformatics Institute (EBI) located at the Hinxton Genome Campus, near Cambridge, UK [10]. Nucleotide sequences and additional information supplied by researchers or collected by literature scanning are stored. Bulk data from various European genome projects and large-scale sequencing initiatives now are fully automated incorporated. Sequence data from patents and patent applications are added to the data collection. This database is maintained in daily cooperation with GenBank and the DNA Databank of Japan (DDBJ). The SWISS-PROT protein sequence database [3] containing directly submitted protein sequences and those derived from nucleotide

```
mbitum9 ~> ftp rdp.life.uiuc.edu
```
```
Connected to rdp.life.uiuc.edu.
220 rdp FTP server (Version 2.0WU(10) Sat Jun 5 18:36:51 CDT 1993) ready.
Name (rdp.life.uiuc.edu:ludwig):
```
```
  anonymous
```
```
331 Guest login ok, send your complete e-mail address as password.
Password:
```
```
230-       The Ribosomal Database Project FTP Server
230-             (rdp.life.uiuc.edu)
230-
230- RELEASE 4.0        18 June 1994       (modified 18 June 1994)
230-
230-
230- All RDP data and programs are in the pub/RDP directory and its
230- subdirectories. Use "cd pub/RDP" to move to that directory.
230-
230- For general information about RDP data and services, get the
230- 00README files. For information about changes and updates, see the
230- 00RELEASE_NOTES.4.0 file.
230-
230- INSTRUCTIONS:
230-  You can navigate around the ftp site using standard ftp commands:
230-  ls                     == list directory contents
230-  cd dirname             == change to directory dirname
230-  get filename           == copy the file filename to your machine
230-  mget *                 == copy all files in this directory to your machine
230-  bin                    == change transfer type to binary (necessary for
230-                            "get"ting binary, executable, .tar and .Z files)
230-  bye                    == quit
230-
230- Data and additional services are available via the RDP email server.
230- Send an email message containing the words HELP COMPLETE to
230- server@rdp.life.uiuc.edu.
230-
230- Questions and comments about the server operation should be addressed to
230- Dr. Mike McCaughey (mrmike@geta.life.uiuc.edu).
230-
230- Questions and comments about sequences and alignments should be addressed to
230- Dr. Bonnie Maidak (rdp@phylo.life.uiuc.edu).
230-
230- The Ribosomal Database Project is funded by the National Science Foundation.
230-
230 Guest login ok, access restrictions apply.
```
```
200 PORT command successful.
150 Opening ASCII mode data connection for file list.
```
```
bin
etc
usr
pub
dev
rdp
incoming
226 Transfer complete.
40 bytes received in 0.0036 seconds (11 Kbytes/s)
```
```
ftp> cd pub
```
```
250 CWD command successful.
```
```
ftp> ls
```

MMEM-3.3.5/4

200 PORT command successful.

150 Opening ASCII mode data connection for file list.

RDP

226 Transfer complete.

5 bytes received in 0.00076 seconds (6.5 Kbytes/s)

ftp> **cd RDP**

250-Data and programs are provided by the Ribosomal Database Project

250-(RDP) at the University of Illinois at Urbana-Champaign with
250-contributions by individuals other than RDP staff members indicated
250-in each directory's 00Acknowledgements file.
250-
250-If you include data or results of data analysis from the RDP
250-in any publication, please cite:
250-   Niels Larsen, Gary J. Olsen, Bonnie L. Maidak, Michael J.
250-   McCaughey, Ross Overbeek, Thomas J. Macke, Terry L. Marsh
250-   and Carl R. Woese: The ribosomal database project. _Nucleic
250-   Acids Research_, Vol. 21 Supplement, p.3021-3023, 1993.
250-
250-Please see the file HELP_RDP_README for information on how you can help
250-the RDP improve and expand its services.
250-
250-Please read the file 00README
250- it was last modified on Sun Jun 19 18:31:45 1994 - 21 days ago
250-Please read the file 00README.ALL
250- it was last modified on Sun Jun 19 18:31:48 1994 - 21 days ago

250 CWD command successful.

200 PORT command successful.

150 Opening ASCII mode data connection for file list.

SSU_rRNA

LSU_rRNA
programs
.cache
00Acknowledgements
00README
00README.ALL
HELP_RDP_README
.cache.html
000README.ALL
00RELEASE_NOTES.4.0
226 Transfer complete.

148 bytes received in 0.0037 seconds (39 Kbytes/s)

ftp> **get 00README**

200 PORT command successful.

150 Opening ASCII mode data connection for 00README (4669 bytes).
226 Transfer complete.
local: 00README remote: 00README

4779 bytes received in 0.94 seconds (5 Kbytes/s)

ftp> **quit**

221 Goodbye.

mbitum9 ~>

Figure 1. Protocol of a file transfer from the Ribosomal Database Project centre by FTP-connection. Commands that were typed by the user at the local host are indicated by bold faced letters. Comments are given in boxes.

sequences is maintained in collaboration with the University of Genova. Examples of the formates in which nucleotide or protein sequences along with documentation are stored and distributed are shown and explained in Figs. 2 and 3. Furthermore, other important databases such as PROSITE (protein pattern database), TRNA (tRNA sequences), BERLIN (5S rRNA sequences) RRNA (small subunit rRNA sequences; see below), and PDB (Brookhaven protein 3D-structures) are collected and redistributed. Free biological software for commonly used computer systems is available at the EBI databases. Complete lists of available databases and software can be obtained from the EBI databases by network access or by surface mail (Table 1).

Table 1. EBI databases postal and e-mail addresses

Postal address:	European Bioinformatics Institute, Hinxton Genome Campus, Hinxton Hall, Hinxton, Cambridge CB10 IRQ, UK
Phone:	+44 (1223) 494400
Fax:	+44 (1223) 494468

Network addresses:

General inquiries	`datalib@ebi.ac.uk`
Data submission	`datasubs@ebi.ac.uk`
Data corrections, notification of data publication	`update@ebi.ac.uk`
Problems with network services	`nethelp@ebi.ac.uk`
Sequence retrieval and software down loading	`netserv@ebi.ac.uk`
Problems with software downloading, software submissions	`software@ebi.ac.uk`

New releases of databases and software on CD-ROM are published quarterly. Currently three volumes on disks contain the EMBL and SWISS-PROT sequence data bases and other biological data collections. The database information stored on the disks can be used by database searching and handling software included or available from third parties. Subscription forms are available at the addresses specified in Table 1.

The EBI electronic mail server enables anybody with access to international computer networks to download complete databases, individual entries from the sequence databases, free molecular biological software, and general information. The addresses and a selection of important mail messages are given in Table 2. The EBI anonymous FTP server supplements the EBI e-mail server giving access to all data available from the latter and more. The address for anonymous FTP connection as well as important directory paths are listed in Table 3.

Other access methods are gopher client connections to `gopher.ebi.ac.uk` and contacting the *World-Wide Web (WWW) server* using the Universal Resource Locator (URL) `http://www.ebi.ac.uk`

On-line database access is possible using the European Molecular

```
EMBL sequence identifier | Type of molecule | Database division | Sequence length
ID    TMEFTU     standard; DNA;        PRO;         1201 BP.
XX
Accession number
AC    M27479;
XX
Modification date
DT    02-FEB-1990 (Rel. 22, Created)
DT    02-FEB-1990 (Rel. 22, Last updated, Version 1)
XX
Sequence record summary
DE    T.maritima, elongation factor Tu (EF-Tu) gene, complete cds.
XX
Key word
KW    elongation factor.
XX
Biological source and taxonomy
OS    Thermotoga maritima
OC    Prokaryota; Bacteria; Thermotogales.
XX
RN    [1]
RP    1-1201
Bibliography
RA    Bachleitner M., Ludwig W., Stetter K.O., Schleifer K.H.;
RT    "Nucleotide sequence of the gene coding for the elongation factor
RT    Tu from the extremely thermophilic eubacterium Thermotoga
RT    maritima";
RL    FEMS Microbiol. Lett. 57:115-120(1989).
XX
SWISS-PROT accession number and sequence identifier
DR    SWISS-PROT; P13537; EFTU_THEMA.
XX
CC    Draft entry and printed copy of sequence for [1] kindly submitted
CC    by K.H.Schleifer, 13-SEP-1989.
XX
FH    Key            Location/Qualifiers
FH
FT    source         1..1201
FT                   /organism="Thermotoga maritima"
Nucleotides that code for protein
FT    CDS            1..>1201
FT                   /note="protein"
XX
Base composition
SQ    Sequence 1201 BP; 379 A; 277 C; 308 G; 237 T; 0 other;
Nucleotide sequence
      atggcgaagg aaaaatttgt gagaacaaaa ccgcatgtta acgttggaac gattggacat        60
      atcgaccacg gaaaatccac actgacagcc gctataacaa agtacctttc tctcaaggta       120
      cttgcccagt atattcctta cgaccagatc gacaaggccc ctgaagaaaa ggcaagagga       180
      atcaccatca acatcacaca cgttgagtat gagaccgaaa agagacacta cgctcatatt       240
      gactgtcccg gtcacgcgga ctacatcaag aacatgatca caggagcagc tcagatggac       300
      ggagccatcc ttgttgttgc cgcaaccgat ggtcccatgc cccagacaag agagcacgtg       360
      cttctcgcaa gacaggttga ggttccctac atgatcgtct tcataaacaa gacagacatg       420
      gttgacgatc ctgagctcat cgacctcgtc gagatggaag tgagagacct tctgagccag       480
      tacggttacc ctggagacga agtgccagtc ataagaggtt ctgctctgaa agccgtcgaa       540
      gctcctaacg atccgaatca cgaagcttac aaacccatcc aggagctcct cgacgctatg       600
      gataactaca ttcctgatcc tcagagagac gtcgataagc cgttcctcat gcccatcgaa       660
      gacgtgttct ccatcacagg aagaggaacg gttgttacag gaagaataga aagaggaaga       720
      atcagacccg gtgatgaagt tgagatcata ggtctcagct acgagatcaa gaagaccgtt       780
      gtgacgagtg tggaaatgtt cagaaaggaa ctcgatgaag gaatcgcagg agacaacgtt       840
      ggatgtctgc tcagaggaat cgacaaggat gaagttgaaa gaggacaggt tctcgcagct       900
      cccggaagca tcaaacctca caagaggttc aaggctcaga tctacgtttt gaagaaggaa       960
      gagggaggaa gacatacacc gttcacaaaa ggctacaagc ctcagttcta cataagaacc      1020
      gctgacgtta caggagaaat cgtaggactt cctgaaggtg tcgaaatggt catgcctgga      1080
      gaccacgtcg aaatggaaat agaactcatc taccctgtcg ctatcgaaaa gggacagaga      1140
      ttcgctgtaa gggaaggcgg aagaacagtt ggagctggtg tggttacaga agtcatcgag      1200
      t                                                                     1201
//
End of record marker
```

Figure 2. Flat file format of an EBI databases nucleic acid sequence entry. Short descriptions of the items are given within boxes.

```
        Sequence identifier          Type of molecule  Sequence length
ID    EFTU_THEMA       STANDARD;       PRT;    400 AA.
 Accession number
AC    P13537;
 Modification date
DT    01-JAN-1990  (REL. 13, CREATED)
DT    01-JAN-1990  (REL. 13, LAST SEQUENCE UPDATE)
DT    01-NOV-1990  (REL. 16, LAST ANNOTATION UPDATE)
 Description
DE    ELONGATION FACTOR TU (EF-TU) (FRAGMENT).
 Biological source and taxonomy
OS    THERMOTOGA MARITIMA.
OC    PROKARYOTA; NOT YET CLASSIFIED.
RN    [1]
RP    SEQUENCE FROM N.A.
 Bibliography
RA    BACHLEITNER M., LUDWIG W., STETTER K.O., SCHLEIFER K.H.;
RL    FEMS MICROBIOL. LETT. 57:115-120(1989).
 Comments
CC    -!- FUNCTION: THIS PROTEIN PROMOTES THE GTP-DEPENDENT BINDING OF
CC        AMINOACYL-TRNA TO THE A-SITE OF RIBOSOMES DURING PROTEIN
CC        BIOSYNTHESIS.
CC    -!- SUBCELLULAR LOCATION: CYTOPLASMIC.
 EMBL nucleic acid sequence accession number and identifier
DR    EMBL; M27479; TMEFTU.
DR    PROSITE; PS00301; EFACTOR_GTP.
KW    ELONGATION FACTOR; PROTEIN BIOSYNTHESIS; GTP-BINDING.
 Feature table data
FT    NP_BIND      19      26      GTP (BY SIMILARITY).
FT    NP_BIND      81      85      GTP (BY SIMILARITY).
FT    NP_BIND     136     139      GTP (BY SIMILARITY).
FT    NON_TER     400     400
 Amino acid sequence
SQ    SEQUENCE   400 AA;  44509 MW;  834970 CN;
      MAKEKFVRTK PHVNVGTIGH IDHGKSTLTA AITKYLSLKV LAQYIPYDQI DKAPEEKARG
      ITINITHVEY ETEKRHYAHI DCPGHADYIK NMITGAAQMD GAILVVAATD GPMPQTREHV
      LLARQVEVPY MIVFINKTDM VDDPELIDLV EMEVRDLLSQ YGYPGDEVPV IRGSALKAVE
      APNDPNHEAY KPIQELLDAM DNYIPDPQRD VDKPFLMPIE DVFSITGRGT VVTGRIERGR
      IRPGDEVEII GLSYEIKKTV VTSVEMFRKE LDEGIAGDNV GCLLRGIDKD EVERGQVLAA
      PGSIKPHKRF KAQIYVLKKE EGGRHTPFTK GYKPQFYIRT ADVTGEIVGL PEGVEMVMPG
      DHVEMEIELI YPVAIEKGQR FAVREGGRTV GAGVVTEVIE
 //
 End of record marker
```

Figure 3. Flat file format of an EBI databases SWISS-PROT protein sequence entry. Short descriptions of the items are given within boxes.

Table 2. Some important e-mail messages for the use of the EBI databases file server. E-mail address: netserv@ebi.ac.uk.[1], %%% should be replaced by dos, mac, unix, or vms.[2], %%%%%% should be replaced by the corresponding accession number

Command	Description
HELP	Detailed help files containing general information and instructions
HELP software	Listing of available software and instructions
GET %%%[1]_software	Down loading of software
GET NUC: %%%%%%[2]	Sequence retrieval
GET DOC: datasub.txt	Data submission form

Table 3. Some important subdirectories and their contents available by anonymous FTP connection to the host `ftp.ebi.ac.uk` or `193.62.196.6`

Directory path	Content
`/pub/help`	General information and instructions
`/pub/doc`	Information brochures and documents, submission form
`/pub/databases/embl`	Complete releases of the EMBL nucleotide sequence database
`/pub/databases/swissprot`	Complete releases of the SWISSPROT protein sequence database
`/pub/software`	Free molecular biological software

Biology Network (EMBnet). EMBnet national nodes are accessible via Internet in most European countries (Table 4). These nodes provide daily updated sequence databases, a collection of molecular biological databases, and biocomputing services, including user support and training. Usually for a nominal fee users are enabled to analyse database and own sequences applying comprehensive sequence analysis software such as the GCG package [9] via remote login (telnet) to one of the addresses listed in Table 4.

An alternative for on-line access is the use of the SRS package [11] provided by EMBL and the Norwegian EMBnet node. After opening a

Table 4. National EMBnet nodes, fax numbers and email addresses of the respective contact persons

EMBnet node	Contact person	fax number	email address
Austria	Grabner		`grabner@embdec.bcc.univie.ac.at`
Belgium	R. Herzog	+32–2–6509762	`rherzog@ulb.ac.be`
Denmark	H. Møller	+54–86–131160	`hum@biobase.aau.dk`
Finland	R. Harper	+385–0–4572302	`harper@csc.fi`
France	P. Dessen	+33–1–69333013	`dessen@coli.polytechnique.fr`
Germany	W. Chen	+49–6221–422333	`genius@embnet.dkfz heidelberg.de`
Greece	B. Savakis	+30–81–231308	`savakis@myia.imbb.forth.gr`
Israel	L. Esterman	+972–8–344113	`lsestern@weizmann.weizmann.ac.il`
Italy	M. Attimonelli	+39–80–484467	`attimonelli@mvx36.csata.it`
Netherlands	J. Noordik	+31–80–652977	`noordik@caos.caos.kun.nl`
Norway	R. Lopez	+47–22694130	`rodrigol@biomed.uio.no`
Spain	L. Pezzi	+341–585–4506	`carazo@cnbvx3.cnb.uam.es`
Sweden	P. Gad	+46–18–551759	`gad@perrier.embnet.se`
Switzerland	R. Dölz	+41–61–2672078	`embnet@comp.bioz.unibas.ch`
United Kingdom	A. Bleasby	+44–925–603100	`bleasby@dl.ac.uk`

terminal connection to odin.embl-heidelberg.de or biomed. uio.no a menu driven environment can be used for data analysis.

Three e-mail servers for remote database searches are operated by the EBI databases: BLITZ, Mail-FASTA, and Mail-Quick-Search which are specified in Table 5. The message mailed should contain the query sequence and service specific parameters. Detailed instructions are provided in response to a mail with the word HELP to the respective e-mail address (Table 5).

A special submission form has to be used for sequence data submission by e-mail (Table 1) or on Macintosh or IBM-PC compatible diskettes by post (Table 1). Submission forms are printed in the first issue each year of Nucleic Acids Research or can be obtained by e-mail or anonymous FTP (Tables 2 and 3). Authorin software (Macintosh and MS-DOS) for easy preparing completed submission forms are available from the file server (Table 2) or by anonymous FTP (Table 3).

Table 5. Short descriptions and addresses of the EBI databases e-mail servers for remote database searching

Email address	Command	Search type
blitz@ebi.ac.uk	BLITZ	Searches best local similarities of protein query sequences to sequences in the SWISS-PROT protein sequence database
fasta@ebi.ac.uk	Mail-FASTA	Searches for local similarities of protein and nucleotide sequence similarities against the corresponding databases
quick@ebi.ac.uk	Mail-Quick-Search	Searches for very similar sequences in the nucleotide sequence databases

The GenBank Database

GenBank™ [2] at NIH (National Institutes of Health) is one of the major sequence databases. It has been initiated in 1982 and is based at NCBI (National Centre for Biotechnology Information) since October 1992. The NCBI is part of the NLM (National Library of Medicine) located on NIH's Bethesda campus. The sequence data are collected from direct data submissions, NCBI journal scanning, and international collaboration with the EBI databases and DDBJ (DNA Data Bank of Japan). In addition to nucleotide sequences, data from protein sequence and structural databases, and from US and European patents are collected and distributed. MEDLINE abstracts of publications describing the sequences provide an important source of biological annotations for the sequence entries. An example of the flat file format in which GenBank distributes the sequence data along with further description and bibliography is shown and explained in Fig. 4. Other important databases such as Swiss-Prot (protein sequences), FlyBase (*Drosophila* sequences), PROSITE (protein sites and patterns), PIR (PIR-international protein sequences; see below) and PDB (Brookhaven protein 3D-structures) are maintained and redistributed by GenBank. Complete lists of the available databases and further information are available by network access and by surface mail (Table 6).

Full copies of the GenBank sequence data on CD-ROM are currently provided at a frequency of six releases per year. The contact addresses for CD-ROM orders are given in Table 6. The data are available in three versions:

NCBI-GenBank. This version provides a full release of the previous data supplemented by new data in the flat file formate shown in Fig. 4. Translations of coding regions are shown in feature tables. The sequence data are organized into divisions. No retrieval software is included.

Entrez. Sequences and references are provided on a set of currently two respective disks: *Entrez:* Sequences and *Entrez:* References. The first disk contains the nucleotide and protein sequence data of the various databases distributed by GenBank, whereas the second disk contains a bibliographic subset of MEDLINE references to papers describing the sequences. Retrieval software for easy database searching of the linked sequence and bibliographic data sets is included for the Macintosh and for PC-compatible systems running Microsoft Windows. Source codes or executables for an X11 version of the software for VMS and Unix platforms is available through anonymous FTP (Table 7).

Table 6. The GenBank postal and network addresses

Postall address:	National Center for Biotechnology Information
	Bldg. 38A, Rm. 8S–803, 8600 Rockville Pike, Bethesda MD 20894, USA
Phone:	+301 496 2475. CD-ROM orders: +202 783 3238
Fax:	+301 480 9241. CD-ROM orders: +202 512 2233
E-mail addresses:	
General inquiries	info@ncbi.nlm.nih.gov
Data submission	gb-sub@ncbi.nlm.nih.gov
Data corrections, updates	update@ncbi.nlm.nih.gov
Submission software	authorin@ncbi.nlm.nih.gov
E-mail servers	retrieve@ncbi.nlm.nih.gov
	blast@ncbi.nlm.nih.gov
Network *Entrez*	net-info@ncbi.nlm.nih.gov
FTP addesss:	ncbi.nlm.nih.gov
	130.14.25.1

Sequence length	Type of molecule	Genbank division	Modification date

LOCUS   TMOEFTU   1201 bp   ds-DNA   BCT   15-DEC-1989

Sequence record summary

DEFINITION   T.maritima, elongation factor Tu (EF-Tu) gene, complete cds.

Accession number

ACCESSION   M27479

KEYWORDS   elongation factor.

Biological source

SOURCE   T.maritima DNA.
  ORGANISM   Thermotoga maritima
             Prokaryotae; Gracilicutes; Scotobacteria; Anaerobic gram-negative
             straight, curved and helical rods; Bacteroidaceae.

Bibliography

REFERENCE   1 (bases 1 to 1201)
  AUTHORS   Bachleitner,M., Ludwig,W., Stetter,K.O. and Schleifer,K.H.
  TITLE   Nucleotide sequence of the gene coding for the elongation factor Tu
             from the extremely thermophilic eubacterium Thermotoga maritima
  JOURNAL   FEMS Microbiol. Lett. 57, 115-120 (1989)

Level of annotation and review

  STANDARD   full automatic
COMMENT
             Draft entry and printed copy of sequence for [1] kindly submitted
             by K.H.Schleifer, 13-SEP-1989.

NCBI sequence identifier

             NCBI gi: 154836

FEATURES   Location/Qualifiers

Nucleotides that code for protein

     CDS   1..>1201
                     /note="protein; NCBI gi: 154837."

Reading frame

                     /codon_start=1

Amino acid sequence of the encoded protein

                     /translation="MAKEKFVRTKPHVNVGTIGHIDHGKSTLTAAITKYLSLKVLAQY
                     IPYDQIDKAPEEKARGITINITHVEYETEKRHYAHIDCPGHADYIKNMITGAAQMDGA
                     ILVVAATDGPMPQTREHVLLARQVEPVYMIVFINKTDMVDDPELIDLVEMEVRDLLSQ
                     YGYPGDEVPVIRGSALKAVEAPNDPNHEAYKPIQELLDAMDNYIPDPQRDVDKPFLMP
                     IEDVFSITGRGTVVTGRIERGRIRPGDEVEIIGLSYEIKKTVVTSVEMFRKELDEGIA
                     GDNVGCLLRGIDKDEVERGQVLAAPGSIKPHKRFKAQIYVLKKEEGGRHTPFTKGYKP
                     QFYIRTADVTGEIVGLPEGVEMVMPGDHVEMEIELIYPVAIEKGQRFAVREGGRTVGA
                     GVVTEVIEX"
     source   1..1201
                     /organism="Thermotoga maritima"

Base composition

BASE COUNT   379 a   277 c   308 g   237 t

Nucleotide sequence

ORIGIN
        1 atggcgaagg aaaaatttgt gagaacaaaa ccgcatgtta acgttggaac gattggacat
       61 atcgaccacg gaaaatccac actgacagcc gctataacaa agtacctttc tctcaaggta
      121 cttgcccagt atattcctta cgaccagatc gacaaggccc ctgaagaaaa ggcaagagga
      181 atcaccatca acatcacaca cgttgagtat gagaccgaaa agagacacta cgctcatatt
      241 gactgtcccg gtcacgcgga ctacatcaag aacatgatca caggagcagc tcagatggac
      301 ggagccatcc ttgttgttgc cgcaaccgat ggtcccatgc cccagacaag agagcacgtg
      361 cttctcgcaa gacaggttga ggttccctac atgatcgtct tcataaacaa gacagacatg
      421 gttgacgatc ctgagctcat cgacctcgtc gagatggaag tgagagacct tctgagccag
      481 tacggttacc ctggacgaga agtgccagtc ataagaggtt ctgctctgaa agccgtcgaa
      541 gctcctaacg atccgaatca cgaagcttac aaacccatcc aggagctcct cgacgctatg
      601 gataactaca ttcctgatcc tcagagagac gtcgataagc cgttcctcat gcccatcgaa
      661 gacgtgttct ccatcacagg aagaggaacg gttgttacag gaagaataga aagaggaaga
      721 atcagacccg gtgatgaagt tgagatcata ggtctcagct acgagatcaa gaagaccgtt
      781 gtgacgagtg tggaaatgtt cagaaaggaa ctcgatgaag gaatcgcagg agacaacgtt
      841 ggatgtctgc tcagaggaat cgacaaggat gaagttgaaa gaggacaggt tctcgcagct
      901 cccggaagca tcaaacctca caagaggttc aaggctcaga tctacgtttt gaagaaggaa
      961 gagggaggaa gacatacacc gttcacaaaa ggctacaagc ctcagttcta cataagaacc
     1021 gctgacgtta caggagaaat cgtaggactt cctgaaggtg tcgaaatggt catgcctgga
     1081 gaccacgtcg aaatggaaat agaactcatc taccctgtcg ctatcgaaaa gggacagaga
     1141 ttcgctgtaa gggaaggcgg aagaaccagtt ggagctggtg tggttacaga agtcatcgag
     1201 t
//

End of record marker

Figure 4. Flat file formate of an GenBank database entry. Short descriptions of the items are given within boxes.

Table 7. Some important GenBank database subdirectories and their contents available by anonymous FTP connection to the host ncbi.nlm.nih.gov or 130.14.25.1.[1] % should be replaced by mac (Macintosh) or dos (MS DOS)

Directory path	Content
/repository	Author maintained databases
/genbank	Full database release in flat file formate
/genbank/daily	Cumulative update file
/ncbi-asn1	Full database release in ASN.1 formate
/toolbox	NCBI Sofware tools
/pub	Public domain software
/pub/authorin/%[1]	Downloading of the Authorin program

NCBI-sequences. The data distribution disks contain the integrated sequence data set in the ISO ASN.1 standard data description format. This version is primarily intended for software developers. No retrieval software is included.

The GenBank electronic mail servers provide full database sequence retrieval, IRX-based text searching, and (BLAST) [1, 14] sequence similarity searching to users with access to electronic mail. The corresponding addresses and a selection of important mail messages are specified in Table 8.

Files containing the full release or daily updates of the GenBank database are available to users on the Internet via anonymous FTP. The data can be downloaded optionally either in the flat file formate or the

Table 8. Some important e-mail messages for the use of the GenBank e-mail servers. [1], % should be replaced by the specified database; [2], % should be replaced by the search string (accession number, author name, key word, locus name, etc.); [3], % should be replaced by the BLAST program version (protein or nucleotide sequence search); [4], % should be replaced by the nucleic acid or protein sequence in FASTA formate

E-mail address	Command	Description
retriev@ncbi.nlm.nih.gov		
	HELP	Detailed help files containing general information and instructions
	DATALIB %[1]	The sequence retrieval server accepts text
	BEGIN	string(s) as queries, runs an IRX search, returns
	%[2]	matching record(s)
blast@ncbi.nlm.nih.gov		
	HELP	detailed help files containing general information and instructions
	PROGRAM %[3]	The BLAST server accepts a query sequence,
	DATALIB %[1]	runs similarity search, and returns the results
	BEGIN	
	%[4]	

ASN.1 formate. Software tools for handling the latter formate and for the development of ASN.1 applications are also provided on the FTP server. The addresses and some directory paths are shown in Table 7.

Network *Entrez* and BLAST, client/server versions of the *Entrez* molecular sequence retrieval and BLAST sequence similarity searching systems, respectively, are available for the general use over the Internet. On line access of the clients to the servers via Internet is needed (Table 7). End-user sites have to be registered at NCBI. Information on registering hosts can be obtained by requests to the e-mail address net-info@ncbi.nlm.nih.gov. Network *Entrez* client programs are available for Macintosh, MS Windows, SUN Sparc, SGI IRIX, and DEC VMS platforms on the corresponding FTP server. Electronic mail and Internet addresses for obtaining general information, downloading of the *Entrez* software, and establishing TCP/IP connections are listed in Tables 6 and 7.

Recently GenBank has made its services accessible through *World-Wide Web (WWW)* server (http://www.ncbi/nlm/nih.gov.) WWW provides client interfaces for Macintosh, PC, Unix, and other systems. General information and client software can be retrieved by anonymous FTP to ftp.ncsa.uiuc.edu in the /Web directory.

A special submission form has to be used for sequence data submission by e-mail or on Macintosh or IBM-PC compatible diskettes by post (Table 6). Authorin software (Macintosh and MS-DOS) for easy preparing completed submission forms are available from the file server (Table 6) or by anonymous FTP (Table 7).

The PIR-International databases

The PIR-International protein sequence database [13] was established as an internationally distributed data collection effort in 1987. The PIR international database centres, the National Biomedical Research Foundation (NBRF) Washington, the Martinsried Institute for Protein Sequences (MIPS), and the Japan International Protein Information Database (JIPID), produce a single protein sequence database. The protein sequence data are extracted from the literature or received by direct submission from the researchers. Protein coding regions of nucleic acid sequences are translated into the corresponding protein sequences. Additional information describing the sequences in accordance with current biological understanding is stored along with the sequences. The data will be organized by similarity and evolutionary relationships. Biochemical information is displayed in a consistent and comprehensive form. The further goal of the PIR sequence data bases is to merge information from different sources on the same sequence, to detect, record, and if possible resolve discrepancies, resulting in a more reliable and comprehensive presentation of the information and elimination of redundancy in the data collection. The data are stored and distributed in several databases such as PIR1 (proteins are assigned to groups that are similar over the majority of their lengths), PIR2 (verified data), PIR3 (preliminary entries), NRL_3D (3D sequence-structures), PATCHX (sequence data that have not yet been fully processed by PIR-International), and Alignment (aligned protein sequences that share at least 55% sequence similarity).

The PIR and other databases are distributed quarterly on CD-ROM. The ATLAS Multidatabase Information Retrieval System is also included. It provides simultaneous access to multiple databases. Strings such as accession numbers, authors, organisms names can be searched in these databases. The ATLAS program runs on various computer platforms such as VAX/VMS, Open VMS, DEC OSF/1, DEC ULTRIX, SunOS, Silicon Graphics, Macintosh, and MS/DOS. The CD-ROM releases also contain the FASTA program for sequence similarity searches. The PIR and NRL_3D databases in VAX/VMS and ASCII formates as well as retrieval software are also available on 9-track tape, on TK50 and TK70 streaming tape cartridges, and on DAT 4 mm cartridges. The addresses for getting general information and subscribing for CD-ROM or tape releases are given in Table 9.

Request servers for information and data retrieval as well as database searching using the ATLAS and/or FASTA programs via e-mail are

Table 9. Postal and network addresses of the PIR-International database centers

The PIR International centers

MIPS, Martinsried Institut für Proteinsequenzen, Max Planck Institut für Biochemie
Postal address: D-82152 Martinsried, FRG
Phone: +49 89 8578 2657
Fax: +49 89 8578 2655
Network addresses:
  General inquiries   mips@ehpmic.mips.biochem.mpg.de

PIR Technical Services Coordinator, National Biomedical Research Foundation
Postal address: 3900 Reservoir Road, NW, Washington, D.C. 20007, USA
Phone: +1 202 687 2121
Fax: +1 202 687 1662
Network addresses:
  General inquiries   pirmail@nbrf.georgetown.edu
  Request server   fileserv@nbrf.georgetown.edu

JIPID, Japan International Protein Information Database, Science University of Tokyo
Postal address: 2669 Yamazaki, Noda 278, Japan
Phone: +81 471 239778
Fax: +81 471 221544
Network addresses:
  General inquiries   tsugita@jpnsut31.bitnet

available at the US PIR-International database centre (Table 9). On-line access is provided by the US and European PIR International centres upon request. Further information can be obtained by contacting the addresses specified in Table 9.

The PIR and NRL_3D databases are available for anonymous FTP, WAIS and Gopher access from the University of Houston Gene-Server. The corresponding address and directory path information are given in Table 10.

Table 10. PIR-International database subdirectories and their contents available by anonymous FTP connection to the host ftp.bchs.uh.edu or 129.7.2.24

Directory path	Content
/pub/gene-server	Databases, software and general information
/pub/gene-server/pir	PIR-International protein sequence database

Databases of small and large ribosomal subunit RNA sequences

A database of small ribosomal subunit RNA (SSU rRNA) [20] was established at the University of Antwerp in 1984 and yearly releases are available since 1986. A database of large subunit rRNA structures (LSU rRNA) is provided since 1994 [8]. 16S and 23S like rRNA sequences are weekly collected from the GENBANK database and the EBI databases. The sequences are processed by using a set of appropriate programs with respect to primary structure similarity and higher order structure prediction. The data are stored in form of an alignment along with the postulated secondary structure pattern in encoded form. Additional information on the taxonomy of the corresponding organisms, the completeness of the individual sequences (only sequences which comprise at least 70% of homologous residues of the *Escherichia coli* 16S rRNA are included), accession numbers, and bibliography are stored and distributed with the sequence data. Software for sequence data editing (DCSE) and phylogenetic tree reconstruction (TREECON) [19] are available for MS-DOS and VAX-VMS platforms.

The database (one release of per year) as well as software are available on magnetic media from the database maintainers (Table 11). The sequence data base is also distributed on CD-ROM by the EMBL Sequence Data Library (RRNA, see above). Each rRNA sequence is stored in a separate file including symbols indicating gaps that had been introduced to align homologous residues, and in addition symbols indicating secondary structure elements. An example of the flat file format is given in Fig. 6. Lists of organisms and additional information are also provided in separate files (taxo.ssu, taxo.lsu).

The complete data base and software can be downloaded by FTP connection to the address specified in Table 11 or from the EBI databases (Table 3).

```
                    Accession number
ENTRY              A48314         #type complete
Sequence record summary
TITLE              elongation factor Tu - Thermotoga maritima
Modification date
DATE               31-Dec-1993 #sequence_revision 31-Dec-1993 #text_change
                   08-Jul-1994

Accession number
ACCESSIONS         A48314
Bibliography
REFERENCE          A48314
   #authors        Bachleitner, M.; Ludwig, W.; Stetter, K.O.; Schleifer, K.H.
   #journal        FEMS Microbiol. Lett. (1989) 57:115-120
   #title          Nucleotide sequence of the gene coding for the elongation
                   factor Tu from the extremely thermophilic eubacterium
                   Thermotoga maritima.
   #accession      A48314
                   ##status preliminary
                   ##molecule_type DNA
                   ##residues 1-400 ##label BAC
GenBank accession number of the corresponding DNA sequence entry
                   ##cross-references GB:M27479
CLASSIFICATION     #superfamily elongation factor Tu homology
FEATURES
   13-139                     #domain elongation factor Tu homology #label EF1
SUMMARY       #length 400   #molecular_weight 44509  #checksum 8629
Amino acid sequence
SEQUENCE
                  5         10        15        20        25        30
        1 M A K E K F V R T K P H V N V G T I G H I D H G K S T L T A
       31 A I T K Y L S L K V L A Q Y I P Y D Q I D K A P E E K A R G
       61 I T I N I T H V E Y E T E K R H Y A H I D C P G H A D Y I K
       91 N M I T G A A Q M D G A I L V V A A T D G P M P Q T R E H V
      121 L L A R Q V E V P Y M I V F I N K T D M V D D P E L I D L V
      151 E M E V R D L L S Q Y G Y P G D E V P V I R G S A L K A V E
      181 A P N D P N H E A Y K P I Q E L L D A M D N Y I P D P Q R D
      211 V D K P F L M P I E D V F S I T G R G T V V T G R I E R G R
      241 I R P G D E V E I I G L S Y E I K K T V V T S V E M F R K E
      271 L D E G I A G D N V G C L L R G I D K D E V E R G Q V L A A
      301 P G S I K P H K R F K A Q I Y V L K K E E G G R H T P F T K
      331 G Y K P Q F Y I R T A D V T G E I V G L P E G V E M V M P G
      361 D H V E M E I E L I Y P V A I E K G Q R F A V R E G G R T V
      391 G A G V V T E V I E
 ///
End of record marker
```

Figure 5. Flat file formate of the PIR-International databases. Short descriptions of the items are given within boxes.

Table 11. Postal and network addresses of the database on small and large ribosomal subunit RNA at the University of Antwerpen

The database on small ribosomal subunit RNA	
Postal address:	Department of Biochemie, Universiteit Antwerpen, Universiteitsplein 1, B-2610 Antwerp, Belgium
Network addresses:	
e-mail	dwachter@reks.uia.ac.be
	rrna@reks.uia.ac.be
FTP connection	uiam3.uia.ac.be
	143.169.8.11

```
Accession number
acc:M21774
Source
src:MSB 8
Strain
str:DSM 3109
Taxonomy
ta1:Bacteria
ta2:Thermotogales
ta3:
ta4:
ta5:
Bibliography: Authors
aut:Achenbach-Richter,L., Gupta,R., Stetter,K.O. and Woese,C.R.
Bibliography: Title
ttl:Were the original eubacteria thermophiles?
Bibliography: Journal
jou:System. Appl. Microbiol.
Bibliography: Date
dat:1987
Bibliography: Volume
vol:9
Bibliography: Pages
pgs:34-39
Type of RNA
mty:SSU
Organism
seq:Thermotoga maritima
Nucleotide sequence
[illegible]
//
End of record marker
```

Figure 6. Flat file formate of a 16S rRNA sequence entry from the database of small ribosomal subunit RNA structures (University of Antwerp). Short descriptions of the items are given within boxes. Secondary structure elements are indicated by brackets. Files containing the helix numbering are available e.g. Helixssu.EUK.

The Ribosomal Database Project

The Ribosomal Database Project (RDP) [16] was initiated and is maintained by Carl Woese and coworkers at the University of Illinois (Urbana, USA). Large and small subunit rRNA sequence data are drawn from major sequence databases (GENBANK [2] and EBI [10]), other rRNA sequence collections [8, 20], and from direct submissions provided by the researchers to RDP. The sequence data are stored and distributed in aligned form. The entries are arranged according to the phylogenetic relationships of the corresponding organisms. Additional information concerning the source and taxonomy of the respective organisms, the method of sequence determination, bibliography, and other relevant data are provided with the sequences. An example of the flat file formate of a single sequence entry is given in Fig. 7. A phylogenetic tree containing all prokaryotes for which small ribosomal subunit rRNA sequence data have been deposited at RDP is available in different formates. Secondary structure models in Postscript format are supplied for many sequences [15, 16]. New releases of the sequence and other data are obtainable at a rate of four per year. Besides the database, software is developed, collected and distributed by RDP. Editors for sequence data input, handling and analysing are offered for UNIX (AE2 and GDE) and VAX/VMS (SeqEdit) platforms. Treeing (fastDNAml) and tree drawing programs (TreeTool) are available as well. Postal and network addresses as well as phone and fax numbers to contact the RDP are listed in Table 12.

The RDP electronic mail server offers data and software access. Furthermore, some analytic functions are included in the server. After submission of full or partial rRNA sequence data (including oligonucleotide probe data) upon request these are aligned, checked or phylogenetically analysed and the results returned via e-mail. The address and some useful mail messages are listed in Table 13.

The RDP FTP-server provides access to the database of aligned sequences, the secondary structur models, the phylogenetic trees and the software

Table 12. The Ribosomal Database Project postal and network addresses

The Ribosomal Database Project,

Postal address:	Bonnie L. Maidak (curator)
	Dept. of Microbiology, University of Illinois, 131 Burrill Hall,.
	407 South Goodwin Avenue, Urbana, IL 61801, USA
Phone:	+217 333 5866
Fax:	+217 244 6697
Network addresses:	
E-mail server	rdp@phylo.life.uiuc.edu
FTP connection	rdp.life.uiuc.edu
	128.174.228.13

```
Sequence identifier│ Length icl. alignmentgaps        Type of molecule     Modification date
LOCUS      Tt.maritim    2688 bp     RNA                 RNA            15-JUN-1989
Source of sequence          Strain
DEFINITION  Thermotoga maritima str. MSB8.
REFERENCE   1
Bibliography
   AUTHORS   Achenbach-Richter,L., Gupta,R., Stetter,K.O. and Woese,C.R.
   TITLE     Were the original eubacteria thermophiles?
   JOURNAL   Syst. Appl. Microbiol. 9, 34-39 (1987)
COMMENT      Organism information
Source of strain
                Culture collection: DSM 3109
                Sequence information (bases 1 to 2688)
Accession number
                Corresponding GenBank entry: M21774 (Release 77.0)
                Source of sequence: CR Woese lab
Sequence composition
BASE COUNT     327 a     430 c     564 g     241 t    1126 others
Aligneed nucleotide sequence
ORIGIN
        1  ~~~~~~~~~~ ~~~~~~~~~~ ~~~~~~~~~~ ~~~UAUAUGG AGGGUUU-GA U-CCUGGCUC
       61  AGGGUGAAC- GC-UGGCGGC G-UG-C-CUA ACACAUGCAA GU-CGAGCGG G---------
      121  ---------- ---------- ---------- --------GG AA-ACUCCC- -UUCG--GGG
      181  AGGAGUAC-- ---------- ---------- ---------- ---------- ------CC--
      241  -AGCGGCGG- -ACGGGUGAG U---AAC-AC GUGGGUAA-- CCU-GC-CCU CCGGAGG-GG
      301  GAU-AA-CCA GG-G-GAAA- CCCUGG-CUA AUACCCCAU- ACG-CUCCA- -U-C--AAC-
      361  --GCAA---G UU--GG-UGG AG-GAAA-GG GGC------- -------GUU U---------
      421  ----GCCCC- G--CCG---- ------GAGG --AGGGGCCC GCGG-CCC-A UCA-G-GUA-
      481  G---UUGGUG -GG-GUAAC- GG-CCCACCA AGCCGACGAC GGGUAGCCGG CCU-G-AGA-
      541  GGGU-GGUCG GCCACAGGGG CA-CUGA-GA -CACG-GGCC CCA-CUCC-U ACG--GGAG-
      601  GC-AGC-A-G -UGGGGAAUC UUGGACAAU- GGGG-GAAA- CCCUGAUCCA GCGACGCCGC
      661  G-UGC-GGGA CGAA-GCCC- -UU-CG-GG- GUGUAAA--- ---------- ----------
      721  -CCGCUGUG- GC-GGGG-GA A-GA-AUAAG G-UAGGGAG- ----G-AAA- -----UGCCC
      781  UA-CCG-AU- GA-CGGU-A- CCC-C-GCU- AGAAAGC-CC C-GGCUAACU ACGU-GCCAG
      841  CAGCCGCGGU AAU-ACGUAG -GGGGCAAGC GUUACCCGGA UUUACUGGGC GUAAAGGGGG
      901  CG-UAGGCGG CCU-GGUGUG -UCGG-AUGU G-AAAUC-CC ACGG-CUCAA -CCGUGG-GG
      961  C-UGCA-UCC GA-AAC-UAC CA-GGC-UU- GGGGG----- CGGUAGAGGG AGAC-GGAAC
     1021  UGCCGGU-GU A-GGG-GUGA AA-UCC-GUA GAU-AUCGGC AGG-AAC-G- CCGG--U-G-
     1081  GGGAAGCCGG --UCUCCUG- -----GGCCG ACCCGACGC UGAGG--CCC GAAA-GC-CA
     1141  GGG-GAGCAA ACCGG-AUUA GAUAC-CCGG GUAG-UCCUG -GCUGUAAAC GAUGCCCA-C
     1201  UAGGUGUGGG GGG------- UAA------- ---------- ---------- -------UCC
     1261  CUCCGUGCUG AAGCUAACGC GU-UAAGUGG GCCGCCUG-G GGAGUACGCC C-GCAA-GGG
     1321  U-GAAACUC- AAA-GGAAUU G-ACGGGGG- CCC--GC-A- -CA-AGCGGU GGAGCG-UGU
     1381  GGU-UUAAUU GG-AUGCUAA G-CCAAGAA- CCUUA-CCAG GGC-UUGACA -UG-------
     1441  ---C-C-GGU G---GUACCU -CCCC-GAAA -GGG-GUAGG GA-CCCAGUC C--UU-----
     1501  ---------- ---------- ---------- ---------- ---------- ----------
     1561  ---------- ---------- ---------- ------CG-- GGACUGGG-- -AGCCG-GCA
     1621  ---------- ---------- ---------- ---------- ---------- CAGGUGGUGC
     1681  ACGGCCGUCG UCAGCUCGUG CC-GUGA-GG UGU-UGGG-U UAAGU-CCCG CAACGAGCGC
     1741  AACCCCUG-C C--CCUAG-- UUGC-CAG-C GG-------- ---------- ----------
     1801  -UUCGG---- ---------- ---------- ---------- ---------- ----------
     1861  ---------- ---------- ---------- ---------- ---------- ----------
     1921  ---------- ---------- ---------- ---------- ---------- ----------
     1981  ---------- ---------- ---------- ---------- ---------- ----------
     2041  ---------- ---------- ---------- ---------- ---------- ----------
     2101  ---------- ---------- ---------- ----CCGGG- CACU-C-U-A GG-GGGAC-U
     2161  GCCGGC-GA- CGA-GCCG-- GAGG-A-AGG -AGGGG-AUG ACGUC--AGG UACUCGU-G-
     2221  CC-CCUUAUG -CC-CU-GGG CGACACACGC GCUACAAUGG -GCGGUA--C -AAUGGGU-U
     2281  -GC-GA-CCC CG-CGAGGGG ---------- -GA-GCCAA- UCCCC---AA AGCCG-CCC-
     2341  UCAGUUCGGA -UCGCAGGC- UGCAACCC-G CCUGCG-UGA AGCCGGAAU- CGCUAGUA-A
     2401  UCGC-GGA-U C-AGCCAU-G CCGCGGUGAA UACGUUCCCG GGCCUUGUAC ACACCGCCCG
     2461  UCACG--CCA CCCGA-GUCG GGGGCUCCC- GAAGACACC- UAC-C-CC-A A-CCC-----
     2521  ---------- ---------- --------GA AA-------- ---------- ----------
     2581  --------G GGAGG-GGGG GUGUCGAGGG AGAACCUGG- CG-AGG-GGG GCGAAGUCGU
     2641  AACAAGGUAG CCGUACCGGA AGGUGCGGCU GGAUCACCUC CUUUCU~~
//
End of record marker
```

Figure 7. Flat file formate of the Ribosomal Database Project containing an individual 16S rRNA sequence entry. Short descriptions of the items are given within boxes.

Table 13. Some useful e-mail messages for the use of the **RDP** e-mail server. E-mail addrss: `server@rdp.life.uiuc.edu`

Command	Description
ALIGN_SEQUENCE	returns submitted sequence(s) in aligned form
CHECK_PROBE	searches for targets of a specified probe sequence
HELP	detailed help files containing general information and instructions
SIMILARITY_RANK	Returns a list of the most similar sequences to the submitted
SUBSCRIBE	Returns periodically notifications about new data and services

Table 14. Some important RDP subdirectories and their contents available by anonymous FTP connection to the host `rdp.life.uiuc.edu` or `128.174.228.13`

Directory path	Content
pub/RDP	General information and instructions
pub/RDP/programs	Software
pub/RDP/sequences/SSU	Alignemts, trees, secondary structure models of small ribosomal subunit rRNAs
pub/RDP/sequences/LSU	Alignments, secondary structure models of large ribosomal subunit rRNAs

collection. The addresses as well as helpful directory paths are shown in Table 14.

The RDP submission form can be obtained from RDP by anonymous FTP.

Acknowledgements

The author wishes to thank H.-W. Mewes and R. Fuchs for providing unpublished manuscripts.

References

1. Altschul SF, Gish W, Miller W, Meyers EW, Lipman DJ (1990) Basic local alignment search tool. J Mol Biol 215: 403–410.
2. Benson D, Boguski M, Lipman DJ, Ostell J (1994) GenBank. Nucl Acids Res 22: 3441–3444.
3. Bairoch A, Boeckmann B (1994) The SWISS-Prot protein sequence data bank: current status. Nucl Acids Res 22: 3578–3580.
4. Burks C, Cinkosky MJ, Fisher WM, Gilna P, Hayden JE-D, Keen GM, Kelly M, Kristofferson D, Lawrence J (1992) GenBank. Nucl Acids Res 20: 2065-2069.
5. Dayhoff MO, Eck RV, Chang MA, Sochard MR (1965) Atlas of Protein Sequence and Structure. Silver Spring, MD: National Biomedical Research Foundation.

6. Dayhoff MO (1972) Atlas of protein sequence and structure, Vol. 5, National Biomedical Research Foundation, Washington, D.C.
7. Dayhoff MO (1979) Atlas of protein sequence and structure, Vol. 5, Supplement 3, National Biomedical Research Foundation, Washington, D.C.
8. De Rijk P, Van de Peer Y, Chapelle S, De Wachter R (1994) Database on the structure of large ribosomal subunit RNA. Nucl Acids Res 22: 3495–3501.
9. Devereux J, Haeberli P, Smithies O (1984) A comprehensive set of sequence analysis programs for the VAX. Nucl Acids Res 12: 387–395.
10. Emmert DB, Stoehr PJ, Stoessner G, Cameron GN (1994) The European Bioinformatics Institute (EBI) databases. Nucl Acids Res 22: 3445–3449.
11. Etzold T, Argos P (1994) SRS – an indexing and retrieval tool for flat file data libraries. Comput Appl Biosci 9: 49–57.
12. Fuchs R, Cameron GN (1993) EMBL data library databases and tools. In: Bishop, Rawlings (eds) Nucleic Acid and Protein Sequence Analysis – A Practical Approach. Oxford: Oxford University Press, in press.
13. George DG, Barker WC, Mewes H-W, Pfeiffer F, Tsugita A (1994) The PIR-International Protein Sequence Database. Nucl Acids Res 22: 3569–3573.
14. Gish W, States DJ (1993) Identification of protein coding regions by database similarity searche. Nature Genet 3: 266-272.
15. Gutell RR, Gray MW, Schnarre MN (1993) A compilation of large subunit (23S-like and 28S-like) ribosomal RNA structure. Nucl Acids Res 21: 3055-3074.
16. Maidak BL, Larsen N, McCaughey MJ, Overbeek R, Olsen GJ, Fogel K, Blandy J, Woese CR (1994) The ribosomal database project. Nucl Acids Res 22: 3485–3487.
17. Schleifer KH, Ludwig W, Amann R (1993) Nucleic acid probes. In: Goodfellow M, McDonnell O (eds) Handbook of New Bacterial Systematics, pp. 463–510. London/New York: Academic Press.
18. Smith TF (1990) The history of the genetic sequence databases. Genomics 6: 702–707.
19. Van de Peer Y, De Wachter R (1992) TREECON: a software package for the construction and drawing of evolutionary trees. Comput Applic Biosci 9: 177–182.
20. Van de Peer Y, Van den Broeck, De Rijk P, De Wachter R (1994) Database on the structure of small ribosomal subunit RNA. Nucl Acids Res 22: 3488–3494.

Molecular Microbial Ecology Manual **3.3.6**: 1-15, 1995
© 1995 *Kluwer Academic Publishers.*

In situ identification of micro-organisms by whole cell hybridization with rRNA-targeted nucleic acid probes

RUDOLF I. AMANN

Department of Microbiology, Technical University of Munich, D-80290 Munich, Germany

Introduction

The sequencing of 16S and 23S ribosomal RNA (rRNA) molecules is currently the gold standard for the classification of new microbial isolates. Comparative analyses of these sequences are for the first time in the history of microbiology facilitating the reconstruction of universal phylogenetic trees [38]. Among many other important findings the work of Carl Woese and his colleagues demonstrated that only certain (by far not all) phenotypic/physiological groups of micro-organisms are monophyletic (e.g., methanogenes, cyanobacteria, spirochetes). About 10 years ago it has been proposed to use an rRNA approach for studies in microbial ecology [21]. The microbial diversity should be analyzed in a cultivation-independent way by direct rRNA sequence retrieval, whereas nucleic acid probes complementary to rRNA or rRNA genes should be the tools to monitor population dynamics in the environmental samples. By their own nature rRNA-targeted probes track genotypes which are not necessarily linked to one phenotype. Microbial ecologists who want to apply this approach to investigate correlations between community structures and functions should be aware of this fact and design or apply rRNA-targeted probes accordingly.

rRNA molecules are ideal targets for nucleic acid probes for several reasons [30]: (i) they are functionally conserved molecules present in all organisms; (ii) the primary structures of 16S and 23S rRNA molecules are composed of sequence regions of higher and lower evolutionary conservation; (iii) 16S rRNA sequences have already been determined for a large fraction (currently close to 50%) of the validly described bacterial species and (iv) their natural amplification with high copy numbers per cell (usually more than 10,000) greatly increases the sensitivity of rRNA-targeted probing. This last point is also the reason for the reliable detection of individual microbial cells by a combination of relatively insensitive fluorescently monolabeled, rRNA-targeted oligonucleotides and standard epifluorescence microscopy. The same combination fails to detect nucleic acids present at lower copy numbers per cell, e.g., mRNA or chromosomal genes.

By comparison of a newly retrieved rRNA sequence to the continuously growing rRNA data bases [17,20] oligonucleotide probes can be designed in a directed way. Due to the patchy evolutionary conservation of rRNA primary structures the specificity of rRNA-targeted oligonucleotide probes can be tailored to the needs of the investigator reaching from the subspecies [31] to the kingdom [1,14] level. Intermediate group-specific oligonucleotides complementary to sequence regions characteristic for phylogenetic entities like genera, families or subclasses have been successfully used for rapid identification of bacteria [4,9,18]. Such probes have the potential to close the gap between standard identification schemes and our current perception of natural classification of bacteria. Molecular taxonomists have placed, e.g., the majority of Gram-negative bacteria in the alpha-, beta-, gamma-, and delta-subclasses of the class *Proteobacteria* (formerly called purple bacteria [38]). A set of group-specific oligonucleotide probes for major phylogenetic sublineages of micro-organisms allows a crude characterization of the community structure in a certain ecosystem. Though not targeting defined physiological groups of micro-organisms such a probe set is very helpful, e.g., in the detection of cultivation dependent biases in the analysis of microbial communities [35].

Probe design

The design of rRNA-targeted, group- or species-specific probes should consider several aspects. These have been reviewed in detail elsewhere [30]. Here just a short list that summarizes theoretical points and practical experiences of the author:

a. Any probe will only be as good as the sequence data base and alignment used for its construction. Most important is of course the correctness of the target sequence.
b. A prerequisite for successful probe design is a thorough knowledge of the phylogeny of the organism(s) of interest; it will not be possible to find a target region on the rRNA molecules that distinguishes a poly-phyletic assemblage like the "pseudomonads" from "non-pseudo-monads".
c. As soon as target and non-target sequences have been specified a window of 18 (± 3) nucleotides is shifted through the sequence alignment to locate a region where all non-target sequences have mismatches and, in case of a group-specific probe, target sequences are identical.
d. Number, position and quality of the mismatch(es) are of prime import-ance for the practical value of a probe. One centrally located mismatch in an 18mer oligonucleotide can be sufficient for discrimination.

e. The location of the probe target site in a 16S or 23S rRNA secondary structure model has to be exactly determined. Both paired (helix) and unpaired (loop) regions can serve as targets. However, care should be taken that a probe is not complementary to both halves of a long helix (> 4 nucleotides). This would automatically result in self-complementarity of parts of the probe sequence which could severely influence its performance in the hybridization.

f. Inaccessibilities of certain target sites during whole cell hybridization have been reported [4]. These are likely caused by higher order structures, e.g., protein-rRNA interactions in the fixed ribosomes. A probe binding to extracted, purified rRNA does not necessarily yield good signal strength with whole fixed cells. If possible, probe target sites should be selected that have been used successfully in other whole cell hybridization experiments.

Based on a good rRNA sequence data base the probe selection should be performed in a computer-assisted way using appropriate software. The selected target region should be compared to available rRNA sequences in regular time intervals to implement new sequences. At the Department of Microbiology of the Technical University München Dr. Wolfgang Ludwig and coworkers are currently developing a program package that includes all necessary tools. The ARB program package for handling and analyzing rRNA sequence data has been designed for SUN OpenWindows. Aligned rRNA sequences together with higher order structure information, documentation and a phylogenetic tree are stored in a central database. The PROBE tool searches potential target sites for sequence-specific probes and evaluates them by comparison with the complete database. A tool for consensus analysis to find group-specific signatures is under development. The tree stored in the database will be shown on the screen and will be used for selecting groups of sequences for further analysis by the PROBE tool.

Additionally, the probes should be further tested by dot blot hybridization against nucleic acids of multiple reference organisms ("phylogrid analysis" [9,18]). Thereby additional target strains and, e.g., closely related nontarget species for which rRNA sequences are not yet available can be quite rapidly examined. One has to be aware that probe specificities and sensitivities are strongly dependent on the exact hybridization conditions (for a detailed review see [30]). Parameters like hybridization and washing temperatures, concentrations of monovalent cations and denaturing agents (e.g., formamide) have to be carefully optimized. Only if a probe works nicely with known reference organisms it can be applied with good confidence to complex environments. However, as long as the real microbial diversity in a certain sample is unknown, which is currently the case in most ecosystems [6,12,13,29,33,37], probes cannot be "verified". Consequently, rRNA-targeted probes should be regarded as tools being subject to refinement.

Applications and limitations

The combination of PCR-assisted sequence retrieval and fluorescent oligonucleotide probing has been used to assign rRNA sequences and thereby phylogenetic affiliations to individual cells of hitherto uncultured bacterial endosymbionts of protozoa [3,10,11,27,28] and of magnetotactic bacteria [7,25,26]. Other applications encompassed enumeration and analysis of spatial distribution of bacterial populations in activated sludge [34,35] and in biofilms [3,19,23,32]. Flow cytometry facilitates high resolution, rapid, automated identification of microorganisms in hybridized cell suspensions [1,36].

Limitations of whole cell hybridization with fluorescent rRNA-targeted probes can originate from low cellular ribosome contents of target organisms and from background fluorescence of the samples [15]. A direct correlation between the growth rates of bacterial cells, the average ribosome contents, and the probe-conferred fluorescence has been observed [8,36]. This can be used to estimate growth rates of individual cells in situ [22]. However, with decreasing growth rate the hybridization signal can quickly approach the detection limit of a regular epifluorescence microscope or flow cytometer [36]. This has prompted studies to develop more sensitive whole cell hybridization techniques.

Signal amplification by hybridization with multiple monolabeled oligonucleotides is possible (2), but often restricted to two– or threefold increase by the limited availability of target sites with identical specificity within the rRNA molecules. Whole cell identification of fixed bacterial cells was achieved with digoxigenin- and enzyme-labeled oligonucleotides [5,39]. If detection is (indirectly or directly) via enzymatic transformation of a suitable substrate probe-conferred markers can be detected very sensitively. Applications may be hindered by limited permeability of fixed cells (especially of Gram-positive bacteria) for the relatively large anti-digoxigenin-antibody or oligonucleotide-enzyme conjugate. Nevertheless, non-fluorescent enzyme-linked probe assays may be the only way to identify microorganisms in environments with strong autofluorescence like plant tissues.

Experimental approach

Most microorganisms are made permeable for oligonucleotide probes by fixation with aldehydes (formalin, paraformaldehyde, glutaraldehyde) and/or alcohols (methanol, ethanol). This was initially shown by hybridization of whole fixed cells with radioactively labeled, rRNA-targeted probes [14]. The probe-conferred radioisotopes have to be visualized by quite time-consuming autoradiography. Fluorescent-dye labeled oligonucleotide probes also hybridize specifically to the intracellular target sites on the ribosomes [2,8]. They are advantageous because of their superior spatial resolution and the instantaneous detectability in epifluorescence microscopes. Over the last years most whole cell hybridization studies have been performed with fluorescent oligonucleotides. The following manual describes in detail the labeling and purification of oligonucleotides with fluorescent dyes, the fixation of cells, their immobilization on glass slides and the use of whole cell hybridization for identification of individual cells. The protocols have been applied successfully to many micro-organisms. Representative examples are shown in Figs. 1 and 2. However, the reader should be aware that new specimens may require adjustments both in sample preparation and probing.

Procedures

Synthesis of fluorescent oligonucleotides

Note

Oligonucleotides are chemically synthesized (e.g., Applied Biosystems, Foster City, CA, USA). In the last coupling cycle an aminohexylphosphate-linker (i.e., Aminolink 2, Applied Biosystems) can be attached to the 5' end of the oligonucleotide. After cleavage from the column with ammonia (32%; Merck, Darmstadt, Germany) the resulting primary amino-group can be coupled with various activated substrates. Labeling reagents include biotin, digoxigenin, enzymes (alkaline phosphatase, horseradish-peroxidase) and fluorescent dyes (e.g., fluorescein, tetramethylrhodamine, Texas Red). Biotin, digoxigenin and fluorescent dyes are mostly available as isothiocyanates or N-hydroxysuccinimidyl esters. Recently oligonucleotides directly labeled at the 5' end with fluorescein became available from custom synthesis laboratories. They also yield good results in whole cell hybridization.

Fluorescent dye-oligonucleotide coupling

Solutions
- 1 M Sodium carbonate-bicarbonate-buffer, pH 9.0.
- Carboxyfluorescein-N-hydroxysuccinimidyl ester (Boehringer Mannheim, Mannheim, Germany), or
- Tetramethylrhodamine-isothiocyanate (Molecular Probes Inc., Eugene OR, USA).
  both dissolved to 5 mg/ml in dimethylformamide (Merck, Darmstadt, Germany).

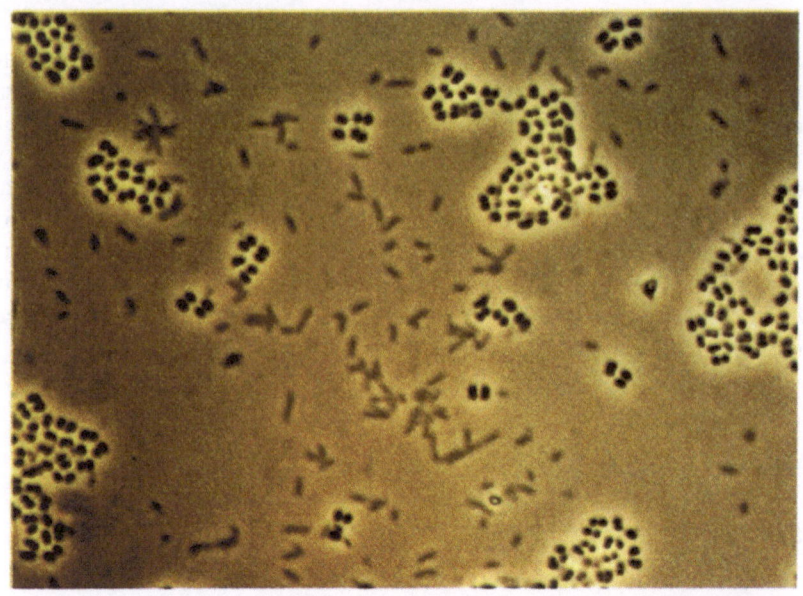

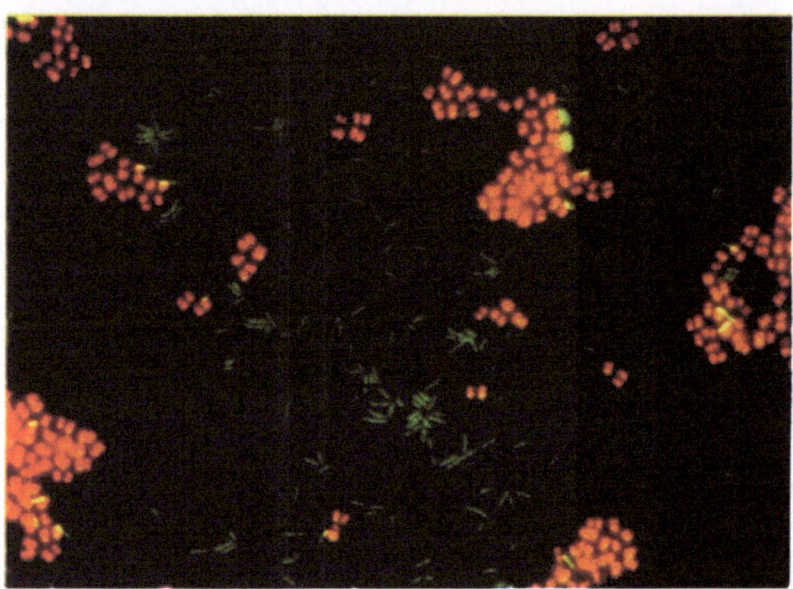

Figure 1. Whole cell hybridization of an artificial mixture of *Micrococcus sedentarius*, *Streptococcus thermophilus* and *Brevundimonas diminuta* (formerly *Pseudomonas diminuta*) with probes HGC and NHGC. A phase contrast (upper panel) and epifluorescence micrograph (double exposure with Zeiss filter sets 09 and 15; lower panel) are shown for one identical microscopic field. The tetramethylrhodamine labeled probe HGC (an oligonucleotide probe targeting the group of Gram-positive bacteria with a high DNA G+C content) hybridized specifically to the clusters of *Micrococcus sedentarius*, whereas fluorescein-labeled probe NHGC (complementary to a region of the 23S rRNA conserved in

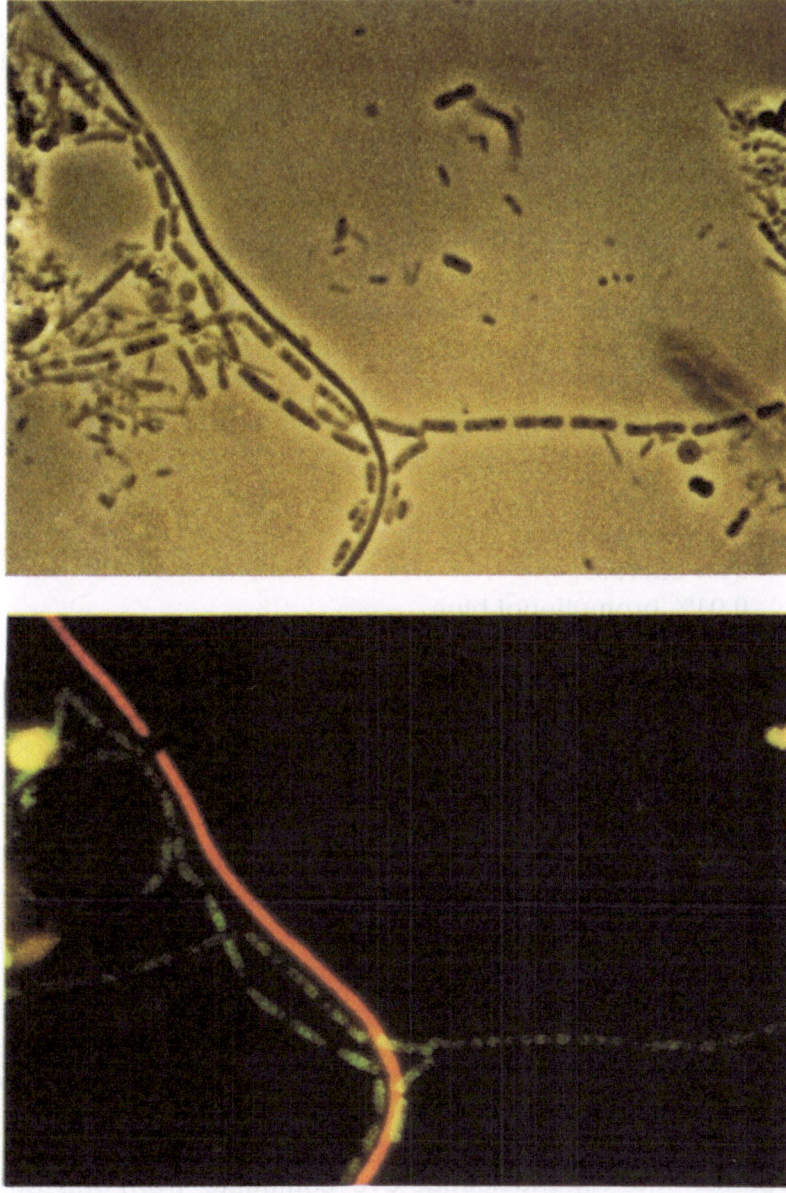

Figure 2. Simultaneous in situ identification of the filamentous bacteria *Sphaerotilus natans* (green) and Eikelboom type 021N (red) in activated sludge [34].

---

all *Bacteria*, except for the Gram-positives bacteria with a high DNA G+C content) bound to the rod-shaped *Brevundimonas diminuta* cells and the ovoid *Streptococcus thermophilus* cells (less frequent).

1. Add, in order, to a 1.5 ml Eppendorf tube:
   175 µl Aminolinked oligonucleotide (100 µg) dissolved in ddH_2O.
   50 µl Sodium carbonate buffer.
   25 µl Carboxyfluorescein-NHS-ester (Boehringer Mannheim, Germany) dissolved in dimethylformamide.
2. Mix and incubate in the dark at room temperature for 8 to 16 h.

Purification of fluorescent oligonucleotides

Solutions
− Sucrose loading buffer
   20% sucrose.
   0.01% bromphenol blue.
   1 mM EDTA.
− 1 × NNB
   0.14 M Tris base.
   0.045 M Boric acid.
   0.003 M EDTA.
− TE buffer
   10 mM Tris/HCl.
   1 mM EDTA.
− 15% acrylamide solution in 1 × NNB.

Note
See Sambrook et al. [24] for preparation of nondenaturing polyacrylamide gels. Any vertical electrophoresis setup (length approx. 20 cm, spacer 1–2 mm) with an appropriate comb (wells should hold > 100 µl) is suitable.

1. Separate the oligonucleotide-dye conjugate from the bulk of unreacted dye by passing the reaction mixture through a NAP-5 Sephadex G25 column (Pharmacia, Freiburg, Germany) equilibrated with ddH_2O. The oligonucleotide (labeled as well as unlabeled) will elute from the column after approximately 1 ml of ddH_2O. The eluate is fractionated and OD260 is monitored.

Unreacted dye does not elute in the first 2 milliliters. Pool appropriate fractions.

2. Reduce the effluent volume to 50 µl in a Speed-Vac concentrator (Bachofer, Reutlingen, Germany).
3. Add 50 µl of sucrose loading buffer and load on a 15% non-denaturing polyacrylamide gel (20 × 20 cm). Separate by electrophoresis in 1 × NNB at about 300 V (15 V/cm) for 2 to 3 h.
4. Remove the gel from the apparatus and place on a plastic-film wrapped thin layer plate impregnated with a UV indicator (Analtech Inc., Newark, DE, USA). Visualize nucleic acids by shadowing the plate by illumination with a short wave (254 nm) hand held lamp. The higher molecular weight fluorescent dye-oligonucleotide-conjugate (migrating slower than unreacted oligonucleotide) can be selectively visualized by long wave UV illumination (366 nm).
5. Excise and crush the region of the gel containing the fluorescent oligonucleotide.
6. Passively elute the conjugate in 800 µl TE buffer overnight. Recover colored TE buffer, add another 800 µl and continue elution for 3 h.
7. Pass the pooled eluate through a NENsorb20-column (DuPont, Wilmington DE, USA) for final purification and concentration [16] as recommended by the manufacturer.
8. Dry the conjugate (Speed Vac) and dissolve in about 100 µl ddH$_2$O.
9. Determine the yield and labeling efficiency spectrophotometrically.

Note
The ratio of absorbance at 260 nm (maximum for nucleic acids) and the peak absorbance of the fluorescent dye (fluorescein 490 nm, tetramethylrhodamine 550 nm) provide an estimate of probe yield and "quality". The 260 nm/490 nm or 260 nm/550 nm ratios should be about 3 for monolabeled 20mers. Final recovery of the fluorescent conjugate is approx. 25% of the starting material.

10. Dilute fluorescent probes to a concentration of 25 ng/µl with ddH$_2$O and store frozen in aliquots at −20 °C. For long-time storage and transport at room temperature probes are best lyophilized.

Cell fixation

Note

During hybridization the cells are exposed to elevated temperature, detergents and osmotic gradients. Thus, fixation is essential for maintaining the morphological integrity of the cells. A variety of fixatives have been evaluated. Glutaraldehyde results in considerable autofluorescence of the specimen. Autofluorescence is minimized by fixation in fresh (not older than 24 h) formaldehyde solutions. We routinely prepare a solution of 4% paraformaldehyde (Sigma, St. Louis, MO, USA) in phosphate buffered saline (PBS) as fixative. This fixation protocol was developed and optimized for Gram-negative bacteria. Probe permeability of paraformaldehyde fixed cells of certain Gram-positive bacteria is often limited (e.g., actinomycetes, streptococci). Here, probe penetration can be enhanced by lysozyme/EDTA treatment of fixed cells prior to hybridization [15], fixation by an ethanol series [23a], or by short time fixation in a alcohol/formaldehyde mixture [8]. Probe penetration into paraformaldehyde fixed cells of *Methanococcus* could be increased by increasing the SDS concentration in the hybridization buffer to 1% [39].

Solutions

- 1 × phosphate buffered saline (PBS)
    130 mM sodium chloride.
    10 mM sodium phosphate buffer.
    pH 7.2.
- 3 × PBS
    390 mM NaCl.
    30 mM sodium phosphate buffer.
    pH 7.2.
- 4% paraformaldehyde (Sigma, St. Louis, MO, USA) in PBS.
- 50%, 80% and 98% ethanol (Sigma).

Preparation of the paraformaldehyde fixative
1. Heat 65 ml of ddH$_2$O to 60 °C.
2. Add 4 g of paraformaldehyde.
3. Add one drop of 2 M NaOH solution and stir rapidly until the solution has nearly clarified (should take only 1–2 min).
4. Remove from heat source and add 33 ml of 3 × PBS.
5. Adjust pH to 7.2 with HCl.
6. Filtrate solution through a 0.2 µm filter.
7. Quickly cool down to 4 °C and store in the refrigerator or on ice.

Paraformaldehyde fixation of microbial cell suspensions
1. Add three volumes of paraformaldehyde fixative to one volume of sample and hold for 1 to 3 h at 4 °C.
2. Pellet fixed cells by centrifugation (5000 × g) and remove fixative.
3. Wash cells in 1 × PBS and then resuspended in 1 × PBS to a final concentration of 10^8–10^9 cells/ml.
4. Add one volume of ice-cold ethanol and mix.
5. Fixed cells may now be spotted on glass slides or may be stored in the freezer (–20 °C) for several months.

Note
Not all cells from environmental samples might be pelleted by centrifugation. This should be checked by membrane filtration of the supernatant. Probe-conferred fluorescence is determined by the rRNA content of fixed cells which is again influenced by the growth rate [8,36]. Initial experiments should therefore be performed with rapidly growing cells.

Immobilization of fixed cells on microscope slide

Note
We routinely use slides on which a hydrophobic coating separates six glass surface windows (Paul Marienfelder KG, Bad Mergentheim, Germany). The hydrophobic coating prevents mixing of probes applied to adjacent spots on the slide. Thereby positive and negative controls can be performed simultaneously on the same slide.

Pretreatment of Microscope Slides
1. Clean slide surface by soaking in a warm detergent solution for one hour, rinse thoroughly with dH_2O and air dry.
2. Coat cleaned slides with gelatin by dipping them in a warm (70 °C) solution of 0.1% gelatin (Baker Chemical Co., Philipsburg NJ, USA), 0.01% chromium potassium sulphate (Aldrich Chemical Corp., Milwaukee, WI, USA).
3. Allow to air dry in a vertical position.

Immobilization of fixed microbial cells on gelatin-coated slides
1. Spread approximately 3 µl of the fixed cell suspension on a gelatin-coated slide over an area of about 5 mm in diameter.
2. Allow the smears to air dry.
3. Dehydrate the cells by successive passages through 50, 80 and 98% ethanol washes (3 min each). Slides can be stored dry at room temperature indefinitely.

Whole cell hybridization

Note

Hybridization must be carried out in a properly sealed moisture chamber to prevent evaporative concentration of the hybridization solution which might result in nonspecific binding of the fluorescent probe to the cells. A 50 ml polypropylene screw top tube (Corning Glass Works, Corning, NY, USA) serves as a convenient and portable hybrization chamber.

Materials
- 50 ml polypropylene screw top tube.
- Whatman 3MM paper.
- Hybridization Buffer

   0.9 M sodium chloride.

   0.01% sodium dodecylsulphate.

   20 mM Tris/HCl.

   X% formamide.

   pH 7.2.

1. Soak a slip of Whatman 3MM paper in hybrization buffer and place in tube.
2. Allow the chamber to equilibrate for several minutes at the hybrization temperature.
3. For each spot to be hybridized mix 8 µl of hybridization buffer with 1 µl of fluorescent probe.
4. Spread 9 µl of hybridization buffer/probe mix on each spot of fixed cells.
5. Quickly transfer slide to the prewarmed moisture chamber and hybridize for 2 h.

Note

The temperature of hybridization depends upon the dissociation temperature (T_d) of the oligonucleotide and must be empirically optimized. We have successfully used probes ranging from 15 to 25 nucleotides in length and hybrization temperatures ranging from 37 to 55 °C. Higher temperatures damage the fixed cells. If a higher stringency is required formamide (up to 50%) can be added to the hybridization solution. This might also improve the hybridization results, likely by altering higher order structures in the rRNA molecules or ribosomes.

6. Remove slide from moisture chamber and immediately stop hybridization by rinsing the probe from the slides with hybridization buffer prewarmed to hybridization temperature.

7. Transfer the slide in polypropylene screw top tube filled with 50 ml hybridization buffer; incubate for 20 min at hybridization temperature.

Note
The stringency of this washing step can be increased by lowing the sodium chloride concentration.

8. Remove salts by shortly dipping slide in ddH_2O, shake away excess water and air dry.
9. Mount slides in a glycerol/PBS mountant with a pH > 8.5 (e.g., Citifluor, Citifluor Ltd., Canterbury, UK) and view with an epifluorescence microscope equipped with suitable filter sets, e.g., Zeiss filter sets nos. 09 for fluorescein, 15 for tetramethylrhodamine.

References

1. Amann RI, Binder BJ, Olson RJ, Chisholm SW, Devereux R, Stahl DA (1990) Combination of 16S rRNA-targeted oligonucleotide probes with flow cytometry for analyzing mixed microbial populations. Appl Environ Microbiol 56: 1919–1925.
2. Amann RI, Krumholz L, Stahl DA (1990) Fluorescent-oligonucleotide probing of whole cells for determinative phylogenetic, and environmental studies in microbiology. J Bacteriol 172: 762–770.
3. Amann RI, Stromley J, Devereux R, Key R, Stahl DA (1992) Molecular and microscopic identification of sulphate-reducing bacteria in multispecies biofilms. Appl Environ Microbiol 58: 614–623.
4. Amann R, Springer N, Ludwig W, Görtz H-D, Schleifer K-H (1991) Identification in situ and phylogeny of uncultured bacterial endosymbionts. Nature (London) 351: 161–164.
5. Amann R, Zarda B, Stahl DA, Schleifer K-H (1992) Identification of individual prokaryotic cells by using enzyme-labeled, rRNA-targeted oligonucleotide probes. Appl Environ Microbiol 58: 3007–3011.
6. DeLong EF (1992) Archaea in coastal marine environments. Proc Natl Acad Sci USA 89: 5685–5689.
7. DeLong EF, Frankel RB, Bazylinski DA (1993) Multiple evolutionary origins of magnetotaxis in bacteria. Science 259: 803–806.
8. DeLong EF, Wickham GS, Pace NR (1989) Phylogenetic stains: ribosomal RNA-based probes for the identification of single microbial cells. Science 243: 1360–1363.
9. Devereux R, Kane MD, Winfrey J, Stahl DA (1992) Genus and group-specific hybridization probes for determinative and environmental studies of sulphate-reducing bacteria. System Appl Microbiol 15: 601–610.
10. Embley TM, Finlay BJ, Brown S (1992) RNA sequence analysis shows that the symbionts in the ciliate *Metopus contortus* are polymorphs of a single methanogen species. FEMS Microbiol Lett 97: 57–62.
11. Embley TM, Finlay BJ, Thomas RH, Dyal PL (1992) The use of rRNA sequences and fluorescent probes to investigate the phylogenetic positions of the anaerobic ciliate *Metopus palaeformis* and its archaeobacterial endosymbiont. J Gen Microbiol 138: 1479–1487.

12. Fuhrman JA, McCallum K, Davis AA (1992) Novel major archaebacterial group from marine plankton. Nature (London) 356: 148–149.
13. Giovannoni SJ, Britschgi TB, Moyer CL, Field KG (1990) Genetic diversity in Sargasso Sea bacterioplankton. Nature (London) 345: 60–63.
14. Giovannoni SJ, DeLong EF, Olsen GJ, Pace NR (1988) Phylogenetic group-specific oligodeoxynucleotide probes for identification of single microbial cells. J Bacteriol 170: 720–726.
15. Hahn D, Amann RI, Ludwig W, Akkermans ADL, Schleifer KH (1992) Detection of micro-organisms in soil after in situ hybridization with rRNA-targeted, fluorescently labelled oligonucleotides. J Gen Microbiol 138: 879–887.
16. Johnson MT, Read BA, Manko AM, Pappas G, Johnson BA (1986) A convenient new method for desalting, deproteinizing, concentrating DNA or RNA. Biotechniques 4: 64–70.
17. Larsen N, Olsen GJ, Maidak BL, McCaughey MJ, Overbeek R, Macke TJ, Marsh TL, Woese CR (1993) The ribosomal database project. Nucleic Acids Res 21: 3021–3023.
18. Manz W, Amann R, Ludwig W, Wagner M, Schleifer K-H (1992) Phylogenetic oligodeoxynucleotide probes for the major subclasses of proteobacteria: problems and solutions. System Appl Microbiol 15: 593–600.
19. Manz W, Szewzyk U, Eriksson P, Amann R, Schleifer KH, Stenström T-A (1993) in situ identification of bacteria in drinking water and adjoining biofilms by hybridization with 16S and 23S rRNA-directed fluorescent oligonucleotide probes. Appl Environ Microbiol 59: 2293–2298.
20. Neef JM, Vandepeer Y, DeRijk P, Chapelle S, DeWachter R (1993) Compilation of small ribosomal subnit RNA structures. Nucleic Acids Res 21: 3025–3049.
21. Olsen GJ, Lane DJ, Giovannoni SJ, Pace NR, Stahl DA (1986) Microbial ecology and evolution: a ribosomal RNA approach. Ann Rev Microbiol 40: 337–365.
22. Poulsen LK, Ballard G, Stahl DA (1993) Use of rRNA fluorescence in situ hybridization for measuring the activity of single cells in young and established biofilms. Appl Environ Microbiol 59: 1354–1360.
23. Ramsing NB, Kühl M, Jörgensen BB (1993) Distribution of sulphate-reducing bacteria, O_2 and H_2S in photosynthetic biofilms determined by oligonucleotide probes and microelectrodes. Appl Environ Microbiol 59: 3820–3849.
23a. Roller C, Wagner M, Amann, R, Ludwig W, Schleifer K-H (1994) In situ probing of Gram-positive bacteria with high DNA G+C content using 23S rRNA-targeted oligonucleotides. Microbiol 140: 2849–2858.
24. Sambrook J, Fritsch EF, Maniatis T (1989) Molecular Cloning. Second edition. Cold Spring Harbor Laboratory Press, Cold Spring Harbor USA.
25. Spring S, Amann R, Ludwig W, Schleifer, N Petersen K-H (1992) Phylogenetic diversity and identification of nonculturable magnetotactic bacteria. System Appl Microbiol 15: 116–122.
26. Spring S, Amann R, Ludwig W, Schleifer KH, Gemerden H van, Petersen N (1993) Dominating role of an unusual magnetotactic bacterium in the microaerophilic zone of a freshwater sediment. Appl Environ Microbiol 59: 2397–2403.
27. Springer N, Ludwig W, Amann R, Schmidt HJ, Görtz H-D, Schleifer K-H (1993) Occurrence of fragmented 16S rRNA in an obligate bacterial endosymbiont of Paramecium caudatum. Proc Natl Acad Sci USA 90: 9892–9895.
28. Springer N, Ludwig W, Drozanski V, Amann R, Schleifer K-H (1992) The phylogenetic status of Sarcobium lyticum, an obligate intracellular bacterial parasite of small amoebae. FEMS Microbiol Lett 96: 199–202.
29. Stackebrandt E, Liesack W, Goebel BM (1993) Bacterial diversity in a soil sample from a subtropical Australian environment as determined by 16S rDNA analysis. FASEB J 7: 232–236.
30. Stahl DA, Amann RI (1991) Development and application of nucleic acid probes in bacterial systematics (p. 205–248). In: Stackebrandt E, Goodfellow M. (eds) Sequencing

and Hybridization Techniques in Bacterial Systematics. John Wiley and Sons, Chichester, England.

31. Stahl DA, Flesher B, Mansfield HR, Montgomery L (1988) The use of phylogenetically based hybridization probes for studies of ruminal microbial ecology. Appl Environ Microbiol 54: 1079–1084.

32. Szewzyk U, Manz W, Amann R, Schleifer K-H, Stenström T-A (1994) Growth and in situ detection of a pathogenic *Escherichia coli* in biofilms of a heterotrophic water bacterium by use of 16S- and 23S-rRNA-directed fluorescent oligonucleotide probes. FEMS Microbiol Ecol 13: 169–175.

33. Torsvik V, Goksoyr J, Daae FL (1990) High diversity of DNA of soil bacteria. Appl Environ Microbiol 56: 782–787.

34. Wagner M, Amann R, Kämpfer P, Aßmus B, Hartmann A, Hutzler P, Springer N, Schleifer KH (1994) Identification and *in situ* detection of gram-negative filamentous bacteria in activated sludge. System Appl Microbiol 17: 405–417.

35. Wagner M, Amann R, Lemmer H, Schleifer KH (1993) Probing activated sludge with oligonucleotides specific for proteobacteria: inadequacy of culture-dependent methods for describing microbial community structure. Appl Environ Microbiol 59: 1520–1525.

36. Wallner G, Amann R, Beisker W (1993) Optimizing fluorescent in situ hybridization of suspended cells with rRNA-targeted oligonucleotide probes for the flow cytometric identification of micro-organisms. Cytometry 14: 136–143.

37. Ward DM, Weller R, Bateson MM (1990) 16S rRNA sequences reveal numerous uncultured micro-organisms in a natural community. Nature (London) 345: 63–65.

38. Woese CR (1987) Bacterial evolution. Microbiol Rev 51: 221–271.

39. Zarda B, Amann R, Wallner G, Schleifer KH (1991) Identification of single bacterial cells using digoxigenin-labelled, rRNA-targeted oligonucleotides. J Gen Microbiol 137: 2823–2830.

Molecular Microbial Ecology Manual **4.1.3**: 1–19, 1995
© 1995 *Kluwer Academic Publishers.*

Immunofluorescence colony-staining (IFC)

JIM W.L. VAN VUURDE and JAN M. VAN DER WOLF
DLO Research Institute for Plant Protection (IPO-DLO), Binnenhaven 12, P.O. Box 9060, 6700 GW Wageningen, The Netherlands

Introduction

The detection of culturable target bacteria at low levels in soil is often unreliable by the variable interference of non-target bacteria on the detection level. For model studies this problem may be solved by using a mutant strain of the wild-type target bacterium with resistance against an antibiotic in combination with a selective medium containing that antibiotic [7]. Immunofluorescence colony-staining (IFC) was developed for the detection of culturable wild-type target bacteria in complex substrates with a high microbial background [17]. IFC is based on a combination of isolation and serology, and can be directly linked to polymerase chain reaction (PCR) for confirmation of positive colonies [16]. IFC can be applied for the detection of target bacteria in sample extracts [4,23], but also for in situ detection on roots [14]. Isolation and serology, which form the basis for IFC, are first briefly discussed together with PCR which was used for verification of IFC-positive colonies.

Dilution plating on agar media makes it possible to obtain pure cultures from suspected colonies, for which a reliable ecological characterization and taxonomical identification can be made. The detection level is ca. 10^1 colony forming units (c.f.u.) per ml sample at a low prevalence of interfering non-target bacteria. However, in substrates with a high level of background organisms, the detection level of the target bacteria may only be 10^4 to 10^6 c.f.u. per ml. Test samples can be agar-mixed (pour) plated to improve the recovery of target bacteria by reducing the interference of non-target bacteria. For most target bacteria with an average colony growth rate, one cm^2 agar layer can host ca. 10,000 background colonies and still allow high recoveries of the target [23]. Besides the high recovery, pour plating also provides good conditions for the serological detection of culturable target bacteria by fixing the colonies grown in the agar matrix.

Serological detection is based on the specific recognition of the target by anti-target antibodies. Immuno-isolation procedures, e.g., based on trapping on an antibody-coated solid phase [19,22] or by trapping through immunomagnetic beads [5,26] can be used to enhance the detection level by removing background bacteria prior to isolation, antibody or gene probe detection. Especially the use of antibodies conjugated with a fluorescent dye in immunofluorescence (IF) microscopy has shown its value for the detection of target bacteria in complex environs (see Chapter 4.2.). However, IF-staining does not discriminate between culturable and non-culturable cells. For selective immunostaining of target colonies in pour plates, fluorochrome-labelled antibodies proved more successful than those with enzyme or gold labelling [20].

Methods directed towards nucleic acid sequences such as polymerase chain reaction (PCR), have the advantage over serology that typical nucleic acid sequences on DNA or RNA of the target can be used to develop probes (see Section 2). Antibodies can only be used against immunogenic products of expressed genes. PCR has a similar detection level as IF for the detection of *Pseudomonas fluorescens* in soil extracts [11]. The standard PCR is not quantitative and does not discriminate between culturable and non-culturable targets. Application of PCR on large series of samples from complex substrates is limited because of the need to remove first PCR-inhibiting substances by laborious DNA extraction techniques. However, for relatively clean samples as cells from colonies on or in agar media, DNA extraction is not required. Therefore, PCR forms a valuable tool for the verification of a target colony in a poor plate. If needed, Southern blotting or RFLP-analysis (see Chapter 3.4.1) can be used to characterize the amplified product in order to improve the specificity.

Principle of IFC

IFC [18] combines pour plating, to detect the target at a low detection threshold by reduced interference of background organisms, with immunofluorescent staining to differentiate the target from the non-target colonies. Semi-(s)elective media checked for high recovery rates of the target are recommended for pour plating. Plates are incubated 1 to 3 days to obtain small colonies in the medium. The detection level is determined by the diameter of the petri dish and the selectivity of the medium. Wells (16 mm diameter) of 24-wells tissue-culture plates are used for routine application on large series of samples. The detection level of this format was determined for soft rot *Erwinia* spp. at ca. 10^2 c.f.u. per ml [23]. Culturable bacteria (target or cross-reacting) can be isolated directly from IFC-positive colonies for confirmation. When cross-reacting bacteria are isolated, the specificity of IFC can in general be optimized by absorption of the antiserum with the cross-reacting strain or by making the medium selective against these bacteria, e.g., by using a suitable antibiotic not affecting the target bacterium.

Check on false positives by direct confirmation of colonies

The specificity of IF and PCR will be dependent on the specificity of the probe (antibody or gene probe) and of the assay format. Complex natural substrates such as soils will contain a large variety of indigenous bacterial species, which may include cross-reacting bacteria. The majority of indigenous species will be present in low numbers (e.g., between 1 and 10^4 cells per gram). Consequently, the risk on a cross-reaction will increase with a decrease of the detection level. Reliable positive results with antibody or gene probes based assays at low detection levels can only be obtained when positive results can be confirmed.

To confirm positive IFC-results the following protocols were developed for cells taken directly from an IFC-positive colony (see Fig. 2):

(1) Re-isolation for pure-culture characterization. Target bacteria can be identified by reference procedures and can be tested for ecological traits. Isolation from IFC-positive colonies also provides an efficient tool for selective screening for and isolation of cross-reacting bacteria.

(2) Sampling of fluorescent cells for testing in PCR [16] showed better potential for routine confirmation of IFC-positive colonies than isolation, but lacks the possibility to obtain pure cultures for further research.

Application

IFC was used for the detection of a variety of target bacteria at a level of 10^2–10^3 target cells per ml of sample using 16 or 25 mm diameter wells of tissue-culture plates, e.g., *Erwinia chrysanthemi* and *E. c.* subsp. *atroseptica* in cattle slurry [23] and potato peel extract [4], and a *Pseudomonas fluorescens* strain in soil [8]. The detection level of IFC for the extremely slow growing *Clavibacter michiganensis* subsp. *sepedonicus* was 10^4 cells per ml cattle slurry manure using 9 cm petri dishes [10]. Colonies of a cross-reacting bacterium from the cattle slurry could be morphologically distinguished from those of *C. m.* subsp. *sepedonicus*. Isolation from IFC-positive colonies proved very efficient to select cross-reacting bacteria by screening millions of background colonies with antiserum against *E. c.* subsp. *atroseptica* [25] and *E. chrysanthemi* [15] and *Xanthomonas campestris* pv. *pelargonii* [25].

Application of the in situ enrichment of target bacteria was useful for studies of the colonization of roots with *E. chrysanthemi* [14] and with *Pseudomonas* sp. [9]. IFC enabled in planta studies of the infection process of potato stems with *E. c.* subsp. *atroseptica* by embedding hand cut transections for in situ enrichment prior to immunofluorescence staining [24].

IFC-protocols for population studies and in situ detection

The various protocols for IFC are organized as follows:

Immunofluorescence colony-staining (IFC) of sample extracts
Procedures for the detection and quantification of target bacteria in sample extracts after pour plating: After incubation for colony growth, immunofluorescent staining can be applied for dried or fresh agar pour platings. The dried agar procedure (see protocol 1.1.) often results in a more brilliant staining of the IFC-positive colonies and less background fluorescence. Dried agar preparations can be stored for months prior to staining. The dried agar technique is also recommended for the preparation of positive and negative reference preparations (see protocol 1.3.).

The use of non-dried agar pour plates (see protocol 1.2.) is recommended if isolation of culturable cells from IFC-positive colonies is intended, e.g., to check for cross-reacting strains.

The presence of culturable and non-culturable target bacteria can be studied with micro-colony IFC [21]. This technique allows to distinguish non-dividing cells from culturable cells, which form colonies, using a 40× objective with long working distance (see protocol 1.4.).

Double staining of two different target bacteria for simultaneous detection in the same preparation with a green and a red fluorescent conjugate is described in protocol 1.5.

Preparation of samples for in situ IFC
In situ detection of target bacteria on roots, in transections of plant parts or on substrate particles is described in protocol 2.

Verification of IFC-positive colonies
The isolation of culturable bacteria from IFC-positive colonies (see protocol 3.1.) for confirmation of pure cultures may be less successful for dried agar plates as compared to non-dried agar plates (e.g., for *E. chrysanthemi*, N.J.M. Roozen, pers. com.). When confirmation is intended by pure cultures, isolations from IFC-positive colonies from non-dried agar plates are recommended (see protocol 1.2.). For routine application we recommend the use of dried agar and confirmation for the presence of the target by PCR (see protocol 3.2.).

Fluorochrome conjugation of antibodies
Purification of immunoglobulins (mainly IgG-type) prior to conjugation is described in protocol 4.1. The conjugation of antibodies with the fluorochrome fluorescein iso-thiocyanate (FITC) is described in protocol 4.2., and that with Texas red in protocol 4.3.

Procedure

1. Immunofluorescence colony-staining (IFC) of sample extracts

1.1. Procedure for dried agar plates (based on [18])

Steps in the procedure
1. Add each test sample (max. 100 µl) to a 16 mm diameter well of a tissue-culture plate. Add 300 µl of agar medium at 45 °C to each well. Swirl the plate gently clockwise and counter-clockwise for a homogeneous distribution of the bacteria in the agar, or use an ELISA-plate shaker.
2. Incubate at optimal conditions for the target bacterium. Stop the incubation when pinhead sized colonies are formed (1 to 2 days for most bacteria).
3. Open the multi-well plate and dry the agar into a thin film by blowing warm air over the agar surface (hair dryer ca. 50 cm above plate ca. 3 h; incubator at 35 °C with circulating air ca. 6 h).
4. Add 400 µl of diluted FITC-conjugated antiserum to the well and incubate while gently shaking at room temperature for ca. 18 h (overnight). The conjugate is diluted with PBST.
5. Remove the non-bound conjugate from the wells. Wash by two 15 min rinses with 1 ml of PBST. Remove excess PBST from the petri dish after the last washing. Removal of liquid from the wells can be done with a pipette or a tip connected with a vacuum aspirator.
6. Place the well with the stained agar film under a microscope with incident blue light and 2.5–10× objective and 4–6.3× eye piece magnification. Inspect the plate for colonies with green fluorescence.
7. Confirm the identity of IFC-positive colonies by puncturing the colony under the microscope with a capillary tube followed by plating on a medium (see protocol 3.1.) or PCR (see protocol 3.2.).

Notes
- Ad 1. The format of the test can be adapted to a different detection level by using a well of, e.g., 6-, 12-, or 48-wells tissue-culture plate or petri dishes with 60 or 90 mm diameter.
- Ad 1. A final agar concentration of 0.8 to 1.0% after mixing with the sample is recommended.

- Ad 3. A ventilated incubator provides the most regular drying and so helps to avoid cracking of the medium. Dried pour plates can be stored several weeks in a container with silica gel at 4 °C before colony-staining is done.
- Ad 4. Depending on experimental conditions such as antiserum dilution, incubation with the fluorescent conjugate can often be reduced to 4 h. The working dilution of the conjugate should be determined in titration experiments for the microscopical set-up to be used. Working dilutions of 50 to 100 times were in general used in our experiments. The staining brilliance of colonies of pure cultures of the target is often less intense than that of target colonies against a background of saprophytes. Diluted conjugate from a stock containing glycerin may result in a weaker staining.
- Ad 4. Non-target organisms can be counterstained, e.g., 45 min with ethidium bromide, a red fluorescent dye [8].
- Ad 6. Since blue light is used for FITC, the objectives of a standard light microscope can replace the more expensive special UV objectives. The highest fluorescence intensity is obtained for objectives with a relatively high numeric aperture and low eyepiece magnification.
- Ad 6. Interpretation of test results should be based on comparison with those of the positive and the negative control, which can be prepared according to protocol 1.3. Small parts of the agar films of these controls can be stained simultaneously with each test series as a standard reference.
- Ad 6. Fading of stained colonies during observation with 10× or lower objectives is very limited compared with fading of IF-stained cells observed at much higher objective magnifications. Reading of IFC-stained agar films can best be done directly after staining. However, stained preparations stored dry for several months at 4 °C often still showed brilliant fluorescence when they were rewetted with PBST before observation.

Materials and solutions
- ELISA plate shaker.
- Ventilated incubator or warm air dryer.
- Microscope with incident blue light (490 nm).
- Polystyrene petri dishes.
- 24-wells tissue-culture plates.
- Agar medium.
- Fluorescein iso-thiocyanate (FITC)-conjugated antiserum.
- Phosphate buffered saline + 0.1% Tween 20 (PBST) (per liter)
    - 8.8 g NaCl.
    - 2.9 g $Na_2HPO_4.12H_2O$.
    - 0.36 g KH_2PO_4.
    - 0.2 g KCl.
    - adjust pH to 7.2 with NaOH.
    - add 1 ml Tween 20.

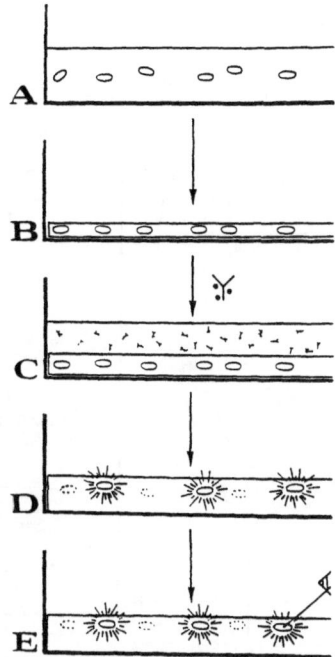

Figure 1. Diagram of immunofluorescence colony-staining (IFC) (Van Vuurde [18], scheme reproduced with permission of APS press). A. Growth of colonies in agar medium. B. Preparation of agar film by drying the medium with warm air. C. Incubation with FITC-conjugated antibodies. D. Observation with incident blue light at low magnification. E. Isolation of bacteria from an IFC-positive colony.

1.2. Procedure for non-dried agar plates

For some target bacteria (e.g., *Erwinia chrysanthemi*), isolation from IFC-positive preparations has a higher success rate when the medium is not dried before staining and cells are picked up with a fine capillary tube as soon as possible after the staining. The steps in the procedure differ from that of the dried agar (1.1.) on the following points:

Modified steps in procedure 1.1.
4. The incubation temperature during staining should be kept below 20 °C to reduce secondary growth of bacteria.
5. The stained preparations should be washed 2 times ca. 4 h with sterile PBST to reduce the relatively high background fluorescence of the medium in non-dried agar.

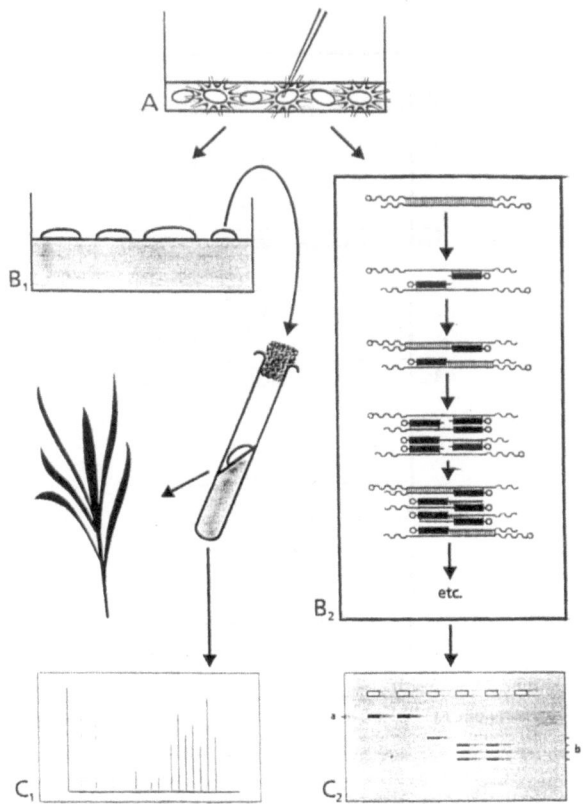

Figure 2. Scheme of confirmation of IFC-positive colonies by isolation of pure cultures or by PCR (scheme reproduced from IPO-DLO Annual Report 1992). (A) Fluorescent colonies in pour plates and (B1) verification by isolation from IFC-positive colonies to obtain a pure culture for, e.g., testing of ecological behavior or (C1) characterization by fatty acid profiling. (B2) Verification by PCR (and RFLP if needed) prior to (C2) characterization by gel electrophoresis.

7. Just before isolation washing with sterile PBST is recommended to remove part of the external microbial contamination on the agar.

1.3. Preparation of standard positive and negative reference preparations

Procedure to prepare standard tests preparations for titration, comparative experiments and reference controls. The films can be stored for several month dry at low temperature in the presence of silica gel.

Steps in the procedure

1. Prepare a bacterial suspension from a 24 or 48 h old pure culture on a tube or plate by suspending the bacterial smear from an öse (see Fig. 1) in 5 ml of 1/4-strength Ringer's with 0.1% Tween 20 (RT) in a 20 ml test tube. Suspend the bacterial smear by rubbing against the glass at the liquid surface. Vortex for 30 s.
2. Adjust the bacterial density to ca. 10^8 cells per ml (e.g., check with barium sulphate reference tubes [6; p. 98] or by optical density.
3. Prepare a dilution series by transferring 0.5 ml suspension to 4.5 ml RT and to 4.5 ml negative sample extract in a 20 ml test tube. Vortex and repeat the ten-fold dilution steps till a dilution of 10^5 cells per ml.
4. Add 100 µl of 10^7, 10^6 and 10^5 cells per ml in RT to a 20 ml sterile test tube and add 12 ml agar medium at 45 °C while stirring. Pour plate in a 9 cm plastic petri dish.
5. Incubate 24 to 48 h till small colonies are just visible in the lower concentrations (smaller than pinhead size).
6. Dry the plates with current warm air till they are completely dry (hair dryer ca. 50 cm above plate ca. 3 h; incubator at 35 °C with circulating air ca. 6 h).
7. Remove the dried agar films from the plate by pressing the edge with a small flat spatula to allow air below the agar film and the plastic and continue by gently lifting it with the spatula.
8. Cut areas of ca. 5 by 5 mm for use in a 24 wells plate.
9. Store the remaining agar film cool (4–10 °C) and dry (with dehydrated silica gel in an airtight box).

Notes
- Ad 3. Extracts of samples with natural background organisms, spiked with the target, often show a more brilliant fluorescence of the target colony. Spiked sample preparations are recommended as positive controls together with pure culture preparations.
- Ad 7. When the removal of the agar film gives problems, the agar layer can best be placed on a plastic sheet before drying. After drying the sheet can be peeled off.

Materials and solutions (see also 1.1)
- 9 cm petri dishes.
- 20 ml test tubes.

1.4. Procedure to determine culturable and non-culturable cells

Steps in the procedure
1. Mix the sample and the agar, and prepare a layer of ca. 1 mm thick on a microscope slide.
2. Incubate at optimal conditions for the target bacterium to form micro-colonies (5 to 50 cells per colony).
3. Dry and stain the agar layer as described in protocol 1.1. Wash carefully to prevent washing away the agar film.
4. Mount the stained agar film with mounting buffer under a cover glass and observe the slide at ca. 400× magnification.

Notes
- Ad 1. Alternatively, a well of a multi-well microscope slide (Cel-line), a chamber of a tissue culture chamber/slide (Lab-Tek, Miles Scientific), or a well of a tissue-culture plate or 96-well polystyrene flatbottom microplate can be used.
- Ad 2. For most plant pathogenic bacteria an 8 h or overnight incubation is suitable. This can be checked by, e.g., phase-contrast microscopy.
- Ad 4. Objectives with a cover glass correction will have a better resolution to observe cells and colonies in the agar.
- Ad 4. Preparation made in the well of a microtiter or tissue-culture plate can be observed for bacterial cells through the bottom of the inverted plate with a long distance objective (e.g., Olympus LWDCDPL 40×/N.A. 0.55 with 1.9 mm working distance).

Materials and solutions
- microscope slides.
- cover glass (e.g., 24 × 60 mm).
- mounting buffer.
    - PBS pH 7.6 (one part).
    - glycerin (9 parts).

1.5. Double staining

Steps in the procedure
The steps in the procedure differ from that of FITC-conjugate for the dried agar (1.1.) on the following points:
4. Add 200 µl of each conjugate to the well, two times less diluted with 0.01 M PBST than the working dilution, and incubate with gently shaking for ca. 18 h (overnight).

6. Observe the stained agar film with blue light for green fluorescent FITC-marked colonies, with green light for red fluorescent Texas-red marked colonies, or with yellow light for simultaneous observation of FITC- and Texas red-marked colonies.

Notes
- Ad 6. Check unstained sample controls for the presence of red autofluorescent colonies.

Materials and solutions (see also 1.1.)

- FITC-labelled antibodies.
- Texas red-labelled antibodies.
- Filterblocks for FITC (maximum excitation 495 nm, emission 525 nm) and for Texas red (maximum excitation 596 nm, emission 615 nm) or a combination filter block for simultaneous observation (e.g., from Omega or Zeiss).

2. Preparation of samples for in situ IFC

The method is developed for a fast and very sensitive in situ detection of colonization patterns of target bacteria on roots, in transections of plant material and on substrate particles.

Steps in the procedure
1. Prepare plant material free of adhering substrate particles.
2. Use sterile filter paper to blot washed material till all adhering water is removed and/or to press the plant material in a flat shape (e.g., seedlings).
3. Spread the object on a thin layer of a suitable agar medium in a petri dish or well of a tissue-culture plate and cover it with a thin layer of still liquefied agar medium (45 °C).
4. Incubate 24 to 48 h till small colonies of the target can be expected.
5. Dry the agar layer with a warm air stream.
6. Perform IFC staining as described in 1.1.

Notes
- Ad 1. Roots can be cleaned from most of the adhering soil by gently washing away the rhizosphere soil with a fine artist brush.
- Ad 6. To save conjugate, large preparations like root systems or leaves in agar can best be stained after cutting away the non-relevant parts of the agar film. Diluted conjugate can be used again after centrifugation for 10 min at 10,000 × g and filter sterilization. Store at 4 °C and check working titre before use.

Materials and solutions

See also 1.1

3. Verification of IFC-positive colonies

Isolation of fluorescent colonies for identification can be done by puncturing colonies with a capillary needle, followed by dilution plating of collected cells.

This isolation procedure is often complicated by the contamination of collected fluorescent bacteria with (high numbers of) non-target cells, which makes laborious and time consuming purification of antibody-reacting bacteria necessary. Furthermore, some bacterial species easily die in the agar films during the staining procedure or storage of the (dried) agar films, as was shown for *Erwinia chrysanthemi*.

Therefore also an alternative method was developed for rapid characterization of fluorescent colonies based on polymerase chain reaction (PCR) [16].

3.1. Procedure for pure cultures from cells of IFC-positive colonies

Steps in the procedure
1. Wash the preparation several times with PBST to remove bacteria present on the surface.
2. Place a fluorescent colony in the center of the microscope field of a low magnifying objective (2.5–4×).
3. Puncture the colony with a needle tip or a fine glass capillary tube.
4. Transfer the sample to a small tube to prepare a suspension in ca. 120 µl of PBST.

5. Pour plate 5 and 45 µl in 16 mm wells and surface plate 50 µl on a differential medium for the target bacterium (90 mm diameter petri dish, inoculate 2 more plates with the rod used to distribute the 50 µl on the first plate).
6. Incubate the media.
7. Inspect the differential medium for colonies with the phenotype of the target.
8. Check these colonies, e.g., using IF cell-staining and perform identification on pure cultures.
9. If no target colonies are found on the spread plates, perform IFC on the non-dried agar pour plates (see 1.2.).
10. Inspect the pour plates for IFC-positive colonies. If the prevalence of IFC-positive colonies is high:
    a. Inspect the spread plates again for colonies of the target with an atypical phenotype or for cross-reacting bacteria and repeat step 8 or
    b. Repeat step 1 to 8 for IFC-positive colonies in the non-dried pour plates.

Notes
- Ad 1. If isolation of culturable cells from IFC-positive colonies was intended for the experiment, use preferably non-dried (replicate) plates for isolation. The isolation procedure is now optimized by using non-dried agar medium and an incubation at 17 °C during IFC-staining and washing.
- Ad 2. A stereo microscope with fluorescent light illumination facilitates sampling of cell from fluorescent colonies (e.g., Wild M3 with Intralux 6000 glass fiber light tube and blue light fluorescence filter set).
- Ad 3. Select a relatively large IFC-positive colony located relatively free of background colonies.
- Ad 8. The percentage successful re-isolation will be determined, e.g., by the density of non-target colonies per cm^2, the colony size, the rate of survival of cells in the target colony during the preparation process and the contamination of the outside of the agar gel after the IFC staining. Secondary growth of target and non-target organisms will occur during the conjugate incubation.

Materials and solutions (see also 1.1)
- Extended Pasteur pipets to serve as a capillary needle

3.2. Characterization of fluorescent colonies with polymerase chain reaction (PCR)

With PCR IFC-positive colonies can be confirmed for culturable and non-culturable target cells without the need for DNA extraction. PCR is a rapid detection method which can be performed in one working day. Dying of cells during staining or storage will not affect PCR directly, as long as the DNA stays intact.

Steps in the procedure
1. Rewet dried agar films in PBST for at least 5 min in advance to the collection of fluorescent colonies from the agar films.
2. Punch fluorescent colonies from the agar using a Pasteur pipet with a bended end (ca. 90 °).
3. Transfer the punches to an Eppendorf tube with demiwater or a buffer suitable for the PCR. The volume of liquid should be twice the sample volume for the PCR, but at least 15 µl.
4. Boil the punches during 5-15 min.
5. Centrifuge during 3 min in an eppendorf centrifuge at maximal speed.
6. Use the supernatant for PCR. The amplification product can be analyzed by standard techniques (dot blot, gel electrophoresis, RFLP and Southern blotting).

Notes
- Ad 2. During the incubation with the FITC-conjugate, viable target cells from surface colonies may spread and cause secondary colonization of the surface of the agar films. This outside contamination, will be visible at higher objective magnification. These target cells may cause 'false positive' reactions in PCR when contaminating samples from negative colonies in the same preparation. Secondary colonization of the surface due to secondary growth can be reduced by addition of 0.05 mg/ml NaN_3 to the antibody conjugate dilution buffer, but then isolation by dilution-plating will not be possible.
- Ad 6. The detection level can be affected by non-target colonies or sample components in the agar punches. This may incidentally cause false negative reactions.

4. Conjugation of antibodies with fluorochrome

4.1. IgG purification from crude antiserum (based on [13])

Steps in the procedure

1. Add 20 ml of sodium acetate buffer to 10 ml of crude antiserum and dialyse against 500 ml of sodium acetate buffer for 4 h. Repeat two more times with fresh buffer.
2. Adjust the suspension to room temperature (20–25 °C) and add dropwise 0.82 ml of octanoic acid under vigorous stirring of the suspension. Continue stirring for 30 min.
3. Centrifuge for 10 min at 7,700 × g.
4. Dialyse the supernatant against 300 ml of potassium phosphate buffer for at least 2 h at 4 °C. Repeat one more time with fresh buffer.
5. Adjust the suspension to room temperature and add, while stirring, an equal volume of a saturated ammonium sulphate solution at room temperature. Continue stirring for 5 min.
6. Centrifuge for 10 min at 7,700 × g.
7. Resuspend the sediment in 5 ml of distilled water.
8. Dialyse against 125 ml of sodium chloride solution for at least 4 h at 4 °C. Repeat four more times with fresh buffer.
9. If the suspension is not clear, centrifuge for 10 min at 12,100 × g.
10. Adjust protein (IgG) concentration to 6 mg/ml (E280 = 1.34 for 1 mg/ml) by diluting with sodium chloride solution.
11. Use directly for conjugation or store in portions of 6 mg at −20 °C.

Notes
- Ad 1. Purified IgG for conjugation should not contain NaN_3.
- Ad 1. IgG's can be alternatively purified by 3 subsequent ammonium sulphate precipitations [1] or with a protein A/G affinity column for small quantities (e.g., MabTrap GII, Pharmacia).

Materials and solutions
- Centrifuge.
- Dialysis tube.
- Potassium phosphate buffer (per liter).
    - 5.8 g K_2HPO_4.
    - 2.27 g KH_2PO_4.

- Adjust pH to 7.2 with 1 N NaOH or 1 N HCl.
- Sodium acetate buffer (0.06 M).
  - 5.28 g $C_2H_3NaO_2$ to 1 l distilled water.
  - Adjust pH to 4.8 with 3.4 ml acetic acid per l distilled water.
- Sodium chloride solution (per liter).
  - 8.5 g NaCl.
  - Adjust pH to 7.5 with 1 N NaOH or 1 N HCl.

4.2. FITC-conjugation of IgG (based on [1])

Procedure for 10 ml IgG
1. Dissolve 5 mg FITC in 8 ml 0.1 M carbonate buffer.
2. Add the FITC solution dropwise to 10 ml IgG (6 mg/ml) while stirring.
3. Add 2 ml of 0.85% NaCl.
4. Adjust the pH to 9.5 with 1 M NaOH.
5. Incubate overnight at 4 °C in a dark place while stirring.
6. Separate conjugate from unbound FITC with Sephadex G25, e.g., a disposable PD-10 column.
7. Check fluorochrome to protein ratio (F/P).
8. Store at 4 °C or better in small portions at −20 °C (avoid regular freezing and defrosting).

Notes
- Ad 7. The molar F/P of the conjugate can be determined [3] based on the absorbance ratio of FITC (A_{495}) to protein (A_{280}). A ratio of 0.5–1.0 is recommended [2].

Materials and solutions
- Magnetic stirrer.
- Sephadex G25 column or a disposable PD 10 column (Pharmacia).
- IgG purified at a protein concentration of 6 mg/ml (E280 = 1.34 for 1 mg/ml).
- FITC isomer 1 (Sigma No. F7250).
- Carbonate buffer (0.1 M)
- 10.6 g Na_2CO_3 in 1 l distilled water.
- Adjust pH to 9.5 with 8.4 g of $NaHCO_3$ in 1 l distilled water.
- 1 M NaOH.

4.3. Texas red-conjugation of IgG (based on [12])

Procedure for 2 ml IgG

1. Dilute IgG with borate buffer till a concentration of 2 mg/ml.
2. Dialyse 24 h at 4 °C against borate buffer.
3. Dissolve 400 μg Texas red in 2 ml of IgG (2 mg/ml) while stirring.
4. Incubate 2 h at room temperature.
5. Separate conjugate from unbound Texas red with a Sephadex G25 column or a PD-10 column, and elute with 0.01 M PBS.
6. Check fluorochrome to protein ratio (F/P).
7. Store at 4 °C, or better in small portions at –20 °C (avoid regular freezing and defrosting).

Notes

– Ad 5. The F/P of the conjugate can be determined [3] based on the absorbance ratio of Texas red ($A596$) to protein (A_{280}). A ratio of 0.5–1.0 is recommended [2]. A too high F/P ratio may cause reduced fluorescence by self-quenching for Texas red.

Materials and solutions

– Dialysis tube.
– Magnetic stirrer.
– Sephadex G25 column or a disposable PD-10 column (Pharmacia).
– IgG purified at a protein conc. of 2 mg/ml (E280 = 1,34 for 1 mg/ml).
– Texas red (Sigma No. S3388).
– Borate buffer (0.025 M) (per liter).
– 4.76 g $Na_2B_4O_7 \cdot 10H_2O$ to 900 ml of distilled H_2O.
– Adjust pH to 9.0 with ca. 46 ml 0.1 M HCl.

References

1. Allan E, Kelman A (1977) Immunofluorescent stain procedures for detection and identification of *Erwinia carotovora* var. *atroseptica*. Phytopathology 67: 1305–1312.
2. Goding JW (1986) Monoclonal antibodies: Principles and practice. Academic Press, London.
3. Goldman M (1968) Fluorescent antibody methods. Academic Press, New York.
4. Jones DAC, Hyman LJ, Tumeseit M, Smith P, Pérombelon MCM (1994) Blackleg potential of potato seed: determination of tuber contamination by *Erwinia carotovora* subsp. *atroseptica* by immunofluorescence colony staining and stock and tuber sampling. Ann Appl Bact 124: 557–568.

5. Jones JB, Van Vuurde JWL (1990) Magnetic immunoisolation of *Xanthomonas campestris* pv. *pelargonii*. In: Klement Z (ed) Proc 7th Int Conf Plant Path Bact, pp. 883–888. Budapest, Hungary.

6. Klement Z, Rudolph K, Sands DC (1990) Methods in phytobacteriology, p. 568. Akademia Kiado, Budapest.

7. Kloepper JW, Beauchamp CJ (1992) A review of issues related to measuring colonization of plant roots by bacteria. Can J Microbiol 38: 1219–1232.

8. Leeman M, Raaijmakers JM, Bakker PAHM, Schippers B (1991) Immunofluorescence colony staining for monitoring pseudomonads introduced in soil. In: Beemster ABR, Bollen BJ, Gerlagh M, Ruissen MA, Schippers B, Tempel A (eds) Biotic interactions and soil-borne diseases, pp. 374–380. Elsevier, Amsterdam.

9. Raaijmakers JM (1994) Microbial interactions in the rhizosphere, pp. 43–55. Utrecht University, Utrecht (PhD thesis).

10. Roozen NJM, Van Vuurde JWL (1991) Development of a semi-selective medium and an immunofluorescence colony-staining procedure for the detection of *Clavibacter michiganensis* subsp. *sepedonicus* in cattle manure slurry. Neth J Pl Path 97: 321–324.

11. Smalla K, Cresswell N, Mendonca-Haglar LC, Wolters A, Van Elsas JD (1993) Rapid DNA extraction protocol from soil for polymerase chain reaction-mediated amplification. J Appl Bact 74: 78–85.

12. Titus JA, Haugland R, Sharrow SO, Segal DM (1982) Texas red, a hydrophylic, red-emitting fluorophore for use with fluorescein in dual parameter flow microfluorometric and fluorescence microscopic studies. J Immunol Meth 50: 193–204.

13. Tobias I, Maat DZ, Huttinga H (1982) Two Hungarian isolates of cucumber mosaic virus from sweet pepper (*Capsicum annuum*) and melon (*Cucumus melo*): identification and antiserum preparation. Neth J Pl Path 88: 177–183.

14. Underberg HA, Van Vuurde JWL (1990) In situ detection of *Erwinia chrysanthemi* on potato roots using immunofluorescence and immunogold staining. In: Klement Z (ed) Proc 7th Int Conf Plant Path Bact, pp. 937–942. Budapest, Hungary.

15. Van der Wolf JM, Van Beckhoven JRCM, De Boef E, Roozen NJM 1993. Serological characterization of fluorescent Pseudomonas strains cross-reacting with antibodies against Erwinia chrysanthemi. Neth J Pl Pathol 99: 51–60.

16. Van der Wolf JM, Van Beckhoven JRCM, De Vries PhM, Raaijmakers JM, Bakker PAHM, Bertheau Y, Van Vuurde JWL (1994) Polymerase chain reaction for verification of fluorescent colonies in immunofluorescence colony-staining. In: Van der Wolf JM Evaluation of serological methods for detection of *Erwinia chrysanthemi* in potato peel extracts, pp. 67–79. Utrecht University, Utrecht ((PhD thesis).

17. Van Vuurde JWL (1987) New approach in detecting phytopathogenic bacteria by combined immunoisolation and immunoidentification assays. EPPO Bull 17: 139–148.

18. Van Vuurde JWL (1990a) Immunofluorescence colony staining. In: Hampton R, Ball E, De Boer S (eds) Serological methods for detection and identification of viral and bacterial plant pathogens, pp. 299–305. APS Press, St. Paul.

19. Van Vuurde JWL (1990b) Immuno-isolation. In: Hampton R, Ball E, De Boer S (eds.) Serological methods for detection and identification of viral and bacterial plant pathogens, pp. 331–338. APS Press, St. Paul.

20. Van Vuurde JWL (1990c) Immunostaining of colonies for sensitive detection of viable bacteria in sample extracts and on plant parts. In: Klement Z (ed) Proc 7th Int Conf Plant Path Bact, pp. 907–912. Budapest, Hungary.

21. Van Vuurde JWL (1991) Immunofluorescence colony-staining (IFC) and immuno-fluorescence cell-staining as tools for the study of rhizosphere bacteria. WPRS Bulletin 14(8): 215–222.

22. Van Vuurde JWL and Van Henten C (1983) Immunosorbent immunofluorescence microscopy (ISIF) and immunosorbent dilution-plating (ISDP): New methods for the detection of plant pathogenic bacteria. Seed Science Technol 11: 547–559.

23. Van Vuurde JWL, Roozen NJM (1990) Comparison of immuno fluorescence colony-staining in media, selective isolation on pectate medium, ELISA and immunofluorescence cell staining for detection of *Erwinia carotovora* subsp. *atroseptica* and *E. chrysanthemi* in cattle manure slurry. Neth J Pl Path 96: 75–89.

24. Van Vuurde JWL, De Vries PhM, Lopez MM (1994a) Colonization of potato stems by *Erwinia* spp., including in planta detection in stem sections with immunofluorescence colony-staining (IFC) after in situ enrichment in agar medium. Proc 8th Int Conf Plant Pathogenic Bacteria, INRA, pp. 747–752. Versailles.

25. Van Vuurde JWL, De Vries PhM, Roozen NJM (1994b) Application of immunofluorescence colony-staining (IFC) for monitoring populations of *Erwinia* spp on potato tubers, in surface water and in cattle manure slurry. Proc. 8th Int Conf. Plant Pathogenic Bacteria, INRA, pp. 741–746. Versailles.

26. Vermunt AEM, Franken AAJM, Beumer RR (1992) Isolation of salmonellas by immunomagnetic separation. J Appl Bact 72: 112-118.

Molecular Microbial Ecology Manual **4.1.8**: 1–12, 1995
© 1995 *Kluwer Academic Publishers.*

Fluorescent staining of microbes for total direct counts

JAAP BLOEM

*DLO Research Institute for Agrobiology and Soil Fertility (AB-DLO), P.O. Box 129,
NL-9750 AC Haren, The Netherlands*

Introduction

Bacteria and fungi are the primary decomposers of dead organic matter
and play an important role in food webs and nutrient cycling. In studies
into the effects of management and pollution on the structure and
functioning of ecosystems, reliable estimates of microbial numbers,
biomass and activity are needed. Since 1970 epifluorescence microscopy
has become the major technique for direct enumeration of microbes in
water and soil. In principle, a known amount of water or homogenized soil
suspension is placed on a known area of a microscopic slide, the micro-
organisms are stained with a fluorescent dye, and numbers are tallied with
a microscope. Biovolumes and biomass can be estimated from lengths and
widths. The frequency of dividing cells (FDC), i.e., the percentage of cells
showing an invagination, can be used as an index of the in situ specific
growth rate of bacteria in water [17] and in soil [8]. Metabolically active
fungal hyphae can be estimated after staining with fluorescein diacetate
(FDA) which becomes fluorescent when it is enzymatically hydrolyzed [32].
The FDA method has been used also for bacteria. However, not all
bacteria are able to take up FDA. The same limitation applies to other
viability probes for bacteria [10].

   Introductions to practical fluorescence microscopy have been given by
Ploem and Tanke [26] and by Taylor and Salmon [34]. Extensive reviews
on microscopic techniques in microbial ecology were recently given by Fry
[15], Gray [16] and Frankland et al. [14]. Water samples are usually
preserved with about 2% formaldehyde and soil suspensions with about 4%
formaldehyde (final concentrations). If fixed samples are stored at 2–4 °C
for longer than a week significant losses of bacteria may occur [10,35].
Prepared slides can be stored longer in a refrigerator or a freezer. For total
bacterial counts in water samples acridine orange (AO) [18] and 4',6-
diamidino-2-phenylindole (DAPI) [27] are the most commonly used stains.
AO intercalates between the stacked bases of DNA and RNA but it also
binds to other cellular constituents, detritus and clay. DAPI is more

specific and binds to double stranded DNA, especially to the portions rich in adenine plus thymine. Therefore, DAPI produces less non-specific background staining than AO and has also been preferred for sediment samples [15,28]. DAPI has been reported to yield about 40% lower estimates of cell volumes and sometimes also about 30% lower numbers than AO [33]. Apparently DAPI stains mainly the interior of the cells, whereas AO also stains cell walls and cells with very little or no DNA. In liquid cultures with actively growing bacteria any stain can be used for counting. In field samples, however, there is a continuous range from viable active cells to dead organic particles resulting in great differences in fluorescence intensity of stained particles. It must be realized that microscopic counting is more or less subjective because different persons use different "intuitive thresholds" to discriminate between living cells and background particles. This may result in a factor two or three difference in estimates of bacterial numbers in the same samples. The use of automatic image analysis facilitates much more quantitative and objective estimates of bacterial numbers and volumes because now thresholds and criteria for detection of cells can be quantified and standardized [6,12,31].

Counting of microorganisms in soil is more difficult than in water because of the presence of soil particles and more non-specific background staining. A variety of preparation and staining methods have been used for soil bacteria [15]. Using confocal laser scanning microscopy and automatic image analysis Bloem et al. [10] compared differed methods. We found no significant differences in counts with AO, fluorescein isothiocyanate (FITC) [4] and 5-(4,6-dichlorotriazine-2-yl) aminofluorescein (DTAF). We do not prefer AO because of the high background staining, which is higher in clay and loam than in sand. AO is an cationic stain which adsorbs to mainly negatively charged clay, humic complexes, acidic polysaccharides, glycosaminoglycans, galactosaminoglycans, liposomes and negative phospholipids [25]. Also with DAPI, Hoechst 33258 (bisbenzimide), ethidium bromide and propidium iodide we found a very high background staining in soil, which may be due to their cationic structure [3]. The anionic stains FITC and DTAF gave much less background staining and images with good contrast. These stains bind covalently to neutral amino groups of proteins, especially on cell surfaces. DTAF has been reported to have a greater purity and stability than FITC [7,10]. A nucleic acid stain which yielded good results in soil is differential fluorescent stain (DFS), a mixture of europium(III) thenoyltrifluoroacetonate (europium chelate) and the disodium salt of 4,4'-bis(4-anilino-6-bis(2-hydroxyethyl)amino-S-triazine-2-ylamino)2,2'-stilbene disulphonic acid which is also called calcofluor white or fluorescent brighter (FB) [1,2]. Europium chelate stains DNA and RNA red, and FB stains cellulose and polysaccharides (cell walls) blue. Blue cells are assumed to be inactive or dead. Red cells are assumed to be active because in these cells the fluorescence is dominated by the europium chelate, indicating a higher nucleic acid content and a higher growth rate. In a field study Bloem et al. [11] observed that the fraction of presumably active bacteria varied between 15 and 80%. The highest values coincided with peaks in the bacterial growth rate as indicated by relatively high frequencies of dividing cells and peaks in bacterial numbers. A disadvantage of DFS is that the europium cannot be detected with a confocal laser scanning microscope for automatic image analysis [10]. Europium is a phosphorescent dye [13] which starts light emission microseconds after excitation. This is too slow for a confocal laser scanning microscope, but it is no problem for conventional epifluorescence microscopy.

Bacteria in water samples are counted after collection on membrane filters [18]. Soil bacteria can be counted on a filter or directly on a glass slide in a soil smear [4]. A smear is prepared by drying 10 µl of a homogenized soil suspension on a microscopic slide. We prefer soil smears because we found less background staining and less fading of fluorochromes in smears than on filters [10]. In contrast with filters, soil smears can be stored at 2 °C in the dark for a year without loss of fluorescence intensity. Moreover, soil films in smears are completely flat which is a great advantage for automatic image analysis.

Here procedures are described for total direct counts of bacteria in water samples with DAPI and AO, and in soil smears with DTAF and DFS (Fig. 1). Also procedures are given for estimates of total hyphal length of soil fungi with FB [37] and of hydrolytically active hyphae with FDA [32].

Procedures

Staining of bacteria in water with DAPI

Solutions
– All solutions must be filtered through 0.2 µm pore-size membrane filters before use. Skin contact with and inhalation of all stains (powder) and of formaldehyde must be avoided.
– Irgalan Black (Acid Black No. 107; Ciba-Geigy Corp.), 2 g l^{-1} in 2% (v/v) acetic acid, to pre-stain polycarbonate membrane filters. Stain for at least 3 min and rinse with particle-free water. Also Sudan Black B (Merck) can be used: 67 mg l^{-1} in 50% ethanol for 18 h, dissolve first in 95% ethanol [15]. Black polycarbonate filters are also commercially available.
– DAPI stock solution (1000 mg l^{-1}), can be kept for a year at –20 °C in the dark.

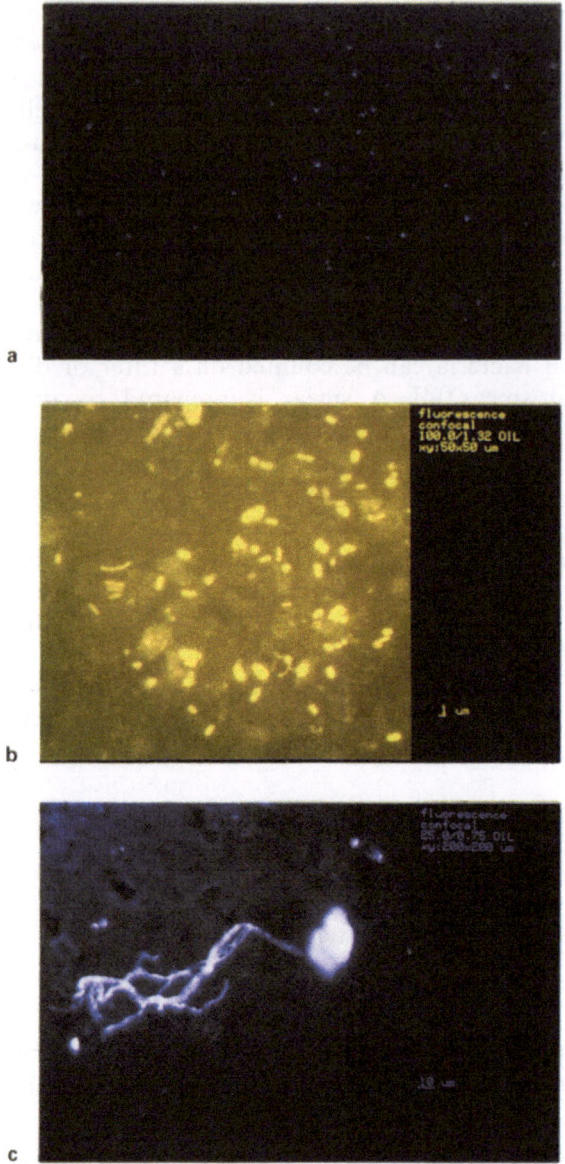

Figure 1. DAPI-stained aquatic bacteria on a filter (a), DTAF-stained soil bacteria in a smear (b), and FB-stained fungal hyphae from soil on a filter (c). Photomicrographs were made using a conventional epifluorescence microscope (a) and a confocal laser-scanning microscope (b, c).

- DAPI work solution (10 mg l^{-1}) obtained by 1:100 dilution of the stock, can be stored for a week at 4 °C in the dark.

Procedure
1. Fix the sample with 2% formaldehyde, e.g., by adding 1 ml of 37% formaldehyde to 19 ml sample.
2. Dilute the sample with particle-free water (e.g., 1:50) to obtain eventually about 25–50 cells per field of view.
3. Take 5 ml of the sample and add 0.5 ml DAPI work solution. This results in a final concentration of 1 mg l^{-1}. Concentrations from 0.01 to 20 mg l^{-1} have been used [15], optimization may be necessary. Stain for 15 min in the dark.
4. Filter through a 0.2 µm pore-size black polycarbonate filter (e.g., Nuclepore, 25 mm diam). A 0.45–1 µm pore-size cellulose filter can be used as a backing under the polycarbonate filter to spread the vacuum for a random distribution of bacteria on the filter. With a 25 mm diam filter, at least 5 ml must be filtered.
5. Place the filter on a microscopic slide, put a very small drop of non-fluorescent immersion oil on the filter and mount a coverslip. Avoid pressure on the coverslip. When the oil has spread the coverslip can be sealed with nail varnish. The slides can be stored for at least 24 weeks at 4 °C [27] or –20 °C in the dark.

Staining of bacteria in water with AO

Solutions
- Irgalan Black as with DAPI.
- AO stock solution (300 mg l^{-1}), can be kept at 4 °C in the dark for at least a week.

Procedure
The same as with DAPI except:
3. Take 5 ml of the sample and add 0.5 ml AO stock solution. This results in a final concentration of 30 mg l^{-1}. Concentrations from 5 to 200 mg l^{-1} have been used [15], optimization may be necessary. Stain for 5 min in the dark.
4. Filter as above and rinse by filtering 3 times 5 ml particle-free water.

MMEM-4.1.8/5

5. Mount the filter with immersion oil on a microscopic slide as described above. With AO we obtained good results only when Cargille (type B) immersion oil was used (R.P. Cargille Laboratories Inc., Cedar Grove N.J., 07009, USA). Other non-fluorescent immersion oils and glycerol seemed to dissolve some stain resulting in a bright green background which made counting impossible [10]. The slides are often stored for weeks at −20 °C in the dark.

Staining of bacteria in soil with DTAF

Solutions
– Buffer solution consisting of 0.05 M Na_2HPO_4 (7.8 g l^{-1}) and 0.85% NaCl (8.5 g l^{-1}), adjusted to pH 9.
– Stain solution consisting of 2 mg DTAF dissolved in 10 ml of the buffer [30].
– Filter all solutions through a 0.2 μm membrane before use. The stain solution should not be stored for longer than a day.

Procedure
1. Clean microscopic slides with 70% ethanol to remove hydrophobic material from the glass surface. Stick a piece of tape (4 × 2 cm), with a hole of 12 mm diameter in the center, on each slide. Use self-adhesive plastic tape (e.g., 4 cm × 25 m) with a removable non-adhesive backing, for instance plastic which is normally used to cover books (book shop) or "Transpaseal Matt Clear" available from drafting supply firms [29]. Holes can be punched in the tape with a 12 mm diam punch from an ironware shop. Alternatively, printed microscope slides (Bellco Glass Inc., Vineland NJ, USA, or Cel-Line Associates Inc., Newfield NJ, USA) can be used.
2. Homogenize 20 g of soil and 190 ml particle-free water in a blender (e.g., Waring, New Hartford, Conn., USA) for 1 min at maximum speed (20,000 r.p.m.).
3. Fix 9 ml of the soil suspension by adding 1 ml 37% formaldehyde (3.7% final concentration) and mix on a test tube mixer (Vortex). Thus, the suspension contains 90 mg soil ml^{-1}. Fine-textured soils (clay) may require a further 3- to 10-fold dilution to prevent masking of bacteria by soil particles [10].
4. Resuspend the fixed soil suspensions. After 2 min of settling to

remove coarse particles, smear 10 µl soil suspension (0.9 mg soil) evenly in the hole of the tape on a glass slide [4]. The water-repellent tape keeps the suspension in a defined area of 113 mm^2. The slides are placed on tissue in a plastic tray.

5. After air-drying, peel off the tape and flood the spots of dried soil film with drops of stain solution for 30 min at room temperature. To prevent drying, the tissue is moistened before, and the trays are covered during staining.

6. Rinse the slides 3 times for 20 min with buffer and finally for a few seconds with water, by putting them in slide holders and passing them through 4 baths. After air-drying mount a coverslip with a small drop of immersion oil. The edges of the coverslip can be sealed with nail varnish. The slides can be stored at 2 °C for at least a year [10].

Staining of bacteria in soil with DFS

Solutions
– DFS is prepared by dissolving 3.5 g l^{-1} europium chelate (Kodak cat no. 1305515, Eastman Fine Chemicals, Rochester NY, USA) and 50 mg l^{-1} fluorescent brightener (e.g., Sigma no. F6259, Sigma Chemical Co., St. Louis MD, USA) in 96% ethanol. After a few minutes, when the powder has dissolved completely, dilute to 50% ethanol with an equal volume particle-free water. In order to avoid high counts in blanks (preparations without soil added) due to precipitation of europium chelate, it is better to prepare the DFS solution one day before use and to filter the stain through a 0.2 µm pore-size membrane immediately before use. With all stains blanks should be checked regularly.

Procedure
As with DTAF except:
5. After air-drying, stain the slides for 1 h in a bath with DFS solution. Flooding with drops of DFS may lead to precipitation of europium chelate, resulting in fluorescent spots which may be confused with bacteria.

6. Rinse the slides 3 times for 5 min in baths with 50% ethanol. After air-drying mount a coverslip with a small drop of immersion oil. The slides can be stored at 2 °C for at least a year.

Counting

Count bacteria with an epifluorescence microscope at 1000–1250 × magnification in a dark room. DAPI and DFS are observed under UV illumination (BP 340–380 nm exciter filter, 400 nm beam splitter and LP 430 nm barrier filter), and AO and DTAF under blue illumination (BP 450–490 exciter filter, 510 nm beam splitter and LP 520 nm barrier filter). Usually it is sufficient to count bacteria in 10 fields of view (enclosed by an ocular grid). The fields should cover the slide randomly, e.g., by selecting fields along 2 central transects at right angles [15]. The specimen must not be observed while fields are being changed. Besides total numbers, cell lengths and widths can be estimated using an ocular micrometer and dividing cells can be counted as an index of the in situ specific growth rate. Lengths and widths can also be determined from photomicrographs projected on a screen [21]. It is convenient to use a PC with appropriate software beside the microscope for tallying counts and for instantaneous calculation of results and statistics [9].

Bacterial numbers per ml water or per g soil B are calculated with:

$$B = (N/X) (A/B) (1/S)$$

where N = the number of bacteria counted, X = the number of fields of view (grids) counted, A = the area of the slide covered by sample, should be checked microscopically [22], B = the area of the field of view, to be measured with an object stage micrometer and S = the amount of sample on the slide.

Biovolumes V can be calculated from length L and width W using the equation [19]

$$V = (\pi/4) W^2 (L - W/3)$$

The amount of carbon can be estimated from the biovolume using a specific carbon content of 3.1×10^{-13} g C μm^{-3} [15]. However, it must be noted that published biovolume-to-carbon conversion factors range from 0.5 to 16.1×10^{-13} g C μm^{-3} [20,24].

Staining of total fungal hyphae in soil

Solutions
- Irgalan Black: see the section on "staining of bacteria in water with DAPI".
- Fluorescent brightener (e.g., Sigma no. F 6259, Sigma Chemical Co., St Louis MD, USA): 2 g l^{-1} dissolved in particle-free water.

Procedure
1. Prepare a fixed soil suspension as described above for bacteria.
2. Resuspend the fixed soil suspension (90 mg soil ml^{-1}) and dilute with water in 3 steps: 1:5, 1:5 and 1:5, resulting in 5 ml suspension containing 3.6 mg soil. Some soils may require further dilution [37].
3. Add 5 ml of the fluorescent brightener solution (1 g l^{-1} final concentration) [23,37] and stain for 2 h at room temperature in the dark.
4. Filter 4 ml of the suspension (1.44 mg soil) through a black 0.8 µm pore-size polycarbonate filter (e.g., Nuclepore 25 mm diam).
5. Rinse the filter 3 times with 3 ml water.
6. Mount the filter on a slide with a small drop immersion oil. The slides can be stored for at least 2 months at 2 °C in the dark [10].

Staining of active fungal hyphae in soil

Solutions
- Fluorescein diacetate (FDA): 2 g l^{-1}, dissolved in acetone. This stock solution is stored at −20 °C. Before use the stock solution is diluted 1:18 with particle-free water to 111 µg ml^{-1}.
- Phosphate buffer pH 7.5 made of 800 ml of solution A and 200 ml of solution B, where A is 11.9 g l^{-1} $Na_2HPO_4 \cdot 2H_2O$ and B is 9.08 g l^{-1} KH_2PO_4.

Procedure
1. Prepare a fresh soil suspension by homogenizing 15 g soil with 190 ml filter-sterilized buffer for 1 min at maximum speed (20,000 r.p.m.) in a blender.
2. Dilute the suspension (75 mg soil ml^{-1}) with buffer in 3 steps: 1:10,

1:5 and 1:5, resulting in 5 ml suspension containing 1.5 mg soil.
3. Add 0.5 ml FDA solution (10 µg ml^{-1} final concentration), mix thoroughly and stain for 3 min [32].
4. Filter the suspension through a black 0.8 µm pore-size poly-carbonate filter.
5. Mount the filter on a slide with a small drop immersion oil and count immediately.

Counting

Count fungi under an epifluorescence microscope at about 400 × magnification. Fluorescent brightener is observed with UV, and FDA with blue illumination. In one or two transects over the filter (about 50–100 fields) hyphal lengths are estimated by counting the number of intersections of hyphae with (all) the lines of the counting grid [15].

The hyphal length H (µm grid^{-1}) is calculated as:

$$H = I \, \pi \, A/2L$$

where I = number of intersections per grid, A = grid area and L = total length of lines in the counting grid.

The total length of fungal hyphae F (m g^{-1} soil) is calculated as:

$$F = H \, 10^{-6} \, (A/B) \, (1/S)$$

where H = hyphal length (µm grid^{-1}), 10^{-6} = conversion of µm to m, A = area of the filter covered by sample, B = area of the grid and S = amount of soil on the filter.

Biovolumes V can be calculated from length L and width W using the equation [19]

$$V = (\pi/4) \, W^2 \, (L - W/3)$$

The amount of carbon can be estimated from the biovolume using a specific carbon content of 1.3×10^{-13} g C µm^{-3} [5,36].

References

1. Anderson JR, Slinger JM (1975a) Europium chelate and fluorescent brightener staining of soil propagules and their photomicrographic counting-I. Methods. Soil Biol Biochem 7: 205–209.
2. Anderson JR, Slinger JM (1975b) Europium chelate and fluorescent brightener staining of soil propagules and their photomicrographic counting-II. Efficiency. Soil Biol Biochem 7: 211–215.
3. Arndt-Jovin DJ, Jovin TM (1989) Fluorescence labelling and microscopy of DNA. In: Taylor DL, Wang YL (eds) Methods in Cell Biology, Vol 30. Fluorescence Microscopy of Living Cells in Culture. Part B. Quantitative fluorescence microscopy-imaging and spectroscopy, pp. 417–448. Academic Press, San Diego.
4. Babiuk LA, Paul EA (1970) The use of fluorescein isothiocyanate in the determination of the bacterial biomass of grassland soil. Can J Microbiol 16: 57–62.
5. Bakken LR, Olsen RA (1983) Buoyant densities and dry-matter contents of micro-organisms: conversion of a measured biovolume into biomass. Appl Environ Microbiol 45: 1188–1195.
6. Bjørnsen PK (1986) Automatic determination of bacterioplankton biomass by image analysis. Appl Environ Microbiol 51: 1199–1204.
7. Blakeslee D, Baines MG (1976) Immunofluorescence using dichlorotriazinylaminofluorescein (DTAF). I. Preparation and fractionation of labelled IgG. J Immunol Methods 13: 305–320.
8. Bloem J, de Ruiter PC, Koopman GJ, Lebbink G, Brussaard L (1992a) Microbial numbers and activity in dried and rewetted arable soil under integrated and conventional management. Soil Biol Biochem 24: 655–665.
9. Bloem J, van Mullem DK, Bolhuis PR (1992b) Microscopic counting and calculation of species abundances and statistics in real time with an MS-DOS personal computer, applied to bacteria in soil smears. J Microbiol Methods 16: 203–213.
10. Bloem J, Bolhuis PR, Veninga MR, Wieringa J (1995a) Microscopic methods for counting bacteria and fungi in soil. In: Alef K, Nannipieri P (eds) Methods in Applied Soil Microbiology and Biochemistry. Academic Press, London, in press.
11. Bloem J, Lebbink G, Zwart KB, Bouwman LA, Burgers SLGE, de Vos JA, de Ruiter PC (1994) Dynamics of microorganisms, microbivores and nitrogen mineralization in winter wheat fields under conventional and integrated management. Agric Ecosyst Environ 51: 129–143.
12. Bloem J, Veninga MR, Shepherd J (1995b) Fully automatic determination of soil bacterium numbers, cell volumes and frequencies of dividing cells by confocal laser scanning microscopy and image analysis. Appl Environ Microbiol 61: 926–936.
13. Davidson RS, Hilchenbach MM (1990) The use of fluorescent probes in immunochemistry. Photochem Photobiol 52: 431–438.
14. Frankland JC, Dighton J, Boddy L (1990) Methods for studying fungi in soil and forest litter. In: Grigorova R, Norris JR (eds) Methods in Microbiology. Volume 22. Techniques in Microbial Ecology, pp. 343–404. Academic Press, London.
15. Fry, J.C. 1990. Direct methods and biomass estimation. In: Grigorova R, Norris JR (eds) Methods in Microbiology. Volume 22. Techniques in Microbial Ecology. Academic Press, London, pp. 41–85.
16. Gray, T.R.G. 1990. Methods for studying the microbial ecology of soil. In: Grigorova R, Norris JR (eds) Methods in Microbiology. Volume 22. Techniques in Microbial Ecology, pp. 309–342. Academic Press, London.
17. Hagström Å, Larsson U, Hörstedt P, Normark S (1979) Frequency of dividing cells, a new approach to the determination of bacterial growth rates in aquatic environments. Appl Environ Microbiol 37: 805–812.
18. Hobbie JE, Daley RJ, Jasper S (1977) Use of nuclepore filters for counting bacteria by fluorescence microscopy. Appl Environ Microbiol 33: 1225–1228.

19. Krambeck C, Krambeck H-J, Overbeck J (1981) Microcomputer assisted biomass determination of plankton bacteria on scanning electron micrographs. Appl Environ Microbiol 42: 142–149.
20. Kroer N (1994) Relationships between biovolume and carbon and nitrogen content of bacterioplankton. FEMS Microbiol Ecol 13: 217–224.
21. Lee S, Fuhrman JA (1987) Relationships between biovolume and biomass of naturally derived marine bacterioplankton. Appl Environ Microbiol 53: 1298–1303.
22. Lebaron P, Trousellier M, Got P (1994) Accuracy of epifluorescence microscopy counts for direct estimates of bacterial numbers. J Microbiol Methods 19: 89–94.
23. Morgan P, Cooper CJ, Battersby NS, Lee SA, Lewis ST, Machin TM, Graham SC, Watkinson RJ (1991) Automated image analysis method to determine fungal biomass in soils and on solid matrices. Soil Biol Biochem 23: 609–616.
24. Norland S, Heldal M, Tumyr O (1987) On the relation between dry matter and volume of bacteria. Microb Ecol 13: 95–101.
25. Paul, JH (1982) Use of Hoechst dyes 33258 and 33342 for enumeration of attached and planktonic bacteria. Appl Environ Microbiol 43: 939–944.
26. Ploem JS, Tanke HJ (1987) Introduction to Fluorescence Microscopy. Royal Microscopical Society, Microscopy Handbooks 10. Oxford University Press, Oxford.
27. Porter KG, Feig YS (1980) The use of DAPI for identifying and counting aquatic microflora. Limnol Oceanogr 25: 943–948.
28. Schallenberg M, Kalff J, Rasmussen JB (1989) Solutions to problems in enumerating sediment bacteria by direct counts. Appl Environ Microbiol 55: 1214–1219.
29. Schmidt EL, Paul EA (1982) Microscopic methods for soil micro-organisms. In: Page AL, Miller RH, Keeney DR (eds) Methods of Soil Analysis. Part 2. Chemical and Microbiological Properties. 2nd edition, pp. 803–814. American Society of Agronomy, Madison WI, USA.
30. Sherr BF, Sherr EB, Fallon RD (1987) Use of monodispersed, fluorescently labeled bacteria to estimate in situ protozoan bacterivory. Appl Environ Microbiol 53: 958–965.
31. Sieracki ME, Johnson PW, Sieburth J McN (1985) Detection, enumeration, and sizing of planktonic bacteria by image-analyzed epifluorescence microscopy. Appl Environ Microbiol 49: 799–810.
32. Söderström BE (1977) Vital staining of fungi in pure cultures and in soil with fluorescein diacetate. Soil Biol Biochem 9: 59–63.
33. Suzuki M, Sherr EB, Sherr BF (1993) DAPI direct counting underestimates bacterial abundances and average cell size compared to AO direct counting. Limnol Oceanogr 38: 1566–1570.
34. Taylor DL, Salmor ED (1989) Basic fluorescence microscopy. In: Taylor DL, Wang YL (eds) Methods in Cell Biology. Volume 29. Fluorescence Microscopy of Living Cells in Culture. Part A. Fluorescent Analogs, Labelling Cells and Basic Microscopy, pp. 207–237. Academic Press, San Diego.
35. Turley CM, Hughes DJ (1994) The effect of storage temperature on the enumeration of epifluorescence-detectable bacterial cells in preserved sea-water samples. J Mar Biol Ass UK 74: 259–262.
36. Van Veen JA, Paul EA (1979) Conversion of biovolume measurements of soil organisms grown under various moisture tensions, to biomass and their nutrient content. Appl Environ Microbiol 37: 686–692.
37. West, AW (1988) Specimen preparation, stain type, and extraction and observation procedures as factors in the estimation of soil mycelial lengths and volumes by light microscopy. Biol Fertil Soils 7: 88–94.

Molecular Microbial Ecology Manual **4.1.9** 1–15, 1995

Slide immunoenzymatic assay (SIA)

EVERLY CONWAY DE MACARIO and ALBERTO J.L. MACARIO
Wadsworth Center for Laboratories and Research, New York State Department of Health, and Department of Biomedical Sciences, School of Public Health, The University at Albany, Albany, NY 12201–0509, U.S.A.

1. Introduction

The use of antibodies as molecular probes for microbes and their antigens has a long tradition [19,23,24]. Poly- and monoclonal antibody probes have been used for microbial identification, characterization, classification, quantification, and manipulation for many years. The development of the enzyme-linked immunosorbent assay (ELISA) [8] has made the use of antibody probes in microbiology and microbial ecology popular and rewarding. The quantitative slide immunoenzymatic assay (SIA) [6] is a miniature ELISA with specific advantages and applications. SIA is simple, sensitive, and versatile [2,3,18], provides reproducible results, and can be automated, (Fig.1) [2,4]. SIA technology can be used for a variety of purposes [1,2,9,18,22–24]. These include identification and quantification of microbes in axenic cultures, in mixed cultures, and in complex microbial communities [4,6,7,14–16,24]. For these purposes, SIA can be used alone or in combination with immunofluorescence and other techniques [14]. SIA allows also detection and quantification of soluble and micro-particulate antigens secreted or released by microbes (microbial footprints as it were) into the culture medium, or into the environment (water and soil, for example) [2–4,6,7,23]. More recently, SIA technology has been adapted for molecular biology, to detect bacteria by nucleic-acid hybridization [5].

Three main elements are required for the immunologic identification of microbes and their antigens: the antibody probe (mono- or polyclonal, calibrated, of pre-defined specificity spectrum); reference antigen (standardized and made permanent, i.e., large quantities in a form that would last stored unaltered, for example formalin-fixed bacteria); and the assay (calibrated, all steps standardized, tested by comparing results with other methods). In this chapter, we will describe SIA as it is performed with polyclonal antibody probes.

The three elements mentioned above are standardized and remain stable over many months or years. They are a constant. The variable factor is the sample, which may be one of several types: cells, subcellular fractions, and soluble antigens; or cell suspensions, and solid aggregates or consortia. From the latter, cell suspensions may be prepared for testing, soluble antigens may be extracted, or blocks for histologic sectioning may be generated for analysis [10,16,17,20,21]. This sample variety notwithstanding, a satisfactory level of uniformity can be achieved by preparing the samples following a set of standard rules and techniques which will yield comparable samples. This is particularly important when complex microbial communities are examined for microbial identification and quantification of specific subpopulations.

In this chapter, cell suspensions will be the only type of sample considered.

Procedures

2. Preparation of antibody probes

2.1. Immunogen preparation (axenic culture): formalinized bacteria

1. Five ml cultures should be grown to an absorbance of 0.4 or greater at 660 nm, 1 cm cuvette.
2. Centrifuge at 15,000 × g for 20 min at 4 °C.
3. Remove the supernatant with a Pasteur pipet and discard.
4. Resuspend the pellet with a newly made fixative solution of 4 ml formalin 37% (in H_2O) plus 96 ml of 0.85% NaCl to an absorbance of 0.7 or greater at 660 nm, 1 cm cuvette. (As a rough guide, one ml of formalin/saline solution for a cell pellet from 5 ml of culture). Use good-quality fresh formalin 37%. Salts concentration and/or composition may be modified as needed to preserve special organisms (e.g., halophilic micro-organisms). Omit formalin for non-formalinized samples (antibiotics may be added in this case).
5. Mix very thoroughly and gently using a Pasteur pipet, avoiding formation of air bubbles (Do NOT use a vortex!).
6. Place the formalinized cells in a leak-proof glass tube, conveniently labeled and sealed for shipment or storage at 4 °C.
7. Wash cells to get rid of the fixative before use.

2.2. Immunization of rabbits

2.2.1. Preparation of cell suspension for injection (immunogen)
1. Prepare cell suspension (immunogen) for injection by centrifuging 1.0 ml of formalinized bacterial cells for 2 min at

15,000 × g in a microcentrifuge. Discard supernatant and retain pellet.
2. Wash the cell pellet 2 times with 1.0 ml sterile phosphate-buffered saline (PBS) by re-suspending the cell pellet carefully and thoroughly using a Pasteur pipet. Centrifuge again as in step 2 and repeat once more.
3. Re-suspend the pellet in 0.5 ml sterile PBS.

2.2.2. Adjuvant; preparation of mixture antigen-adjuvant

Prepare antigen-adjuvant mixture by suspending 0.5 ml of the washed pellet in 0.5 ml of the Hunter's TiterMAX (CytRx Corp). The mixture of cells with adjuvant is mixed carefully and thoroughly by repeatedly drawing it and ejecting it with a 5 ml sterile syringe in the same tube.

2.2.3. First inoculation

The total amount of cells-adjuvant mixture is distributed in equal volumes in three sites of the rabbit's neck, subcutaneously.

2.2.4.. Second inoculation (boost)

It is done the same way as the first one, but 28 days after the first and without adjuvant.

2.3. Antiserum preparation

2.3.1. Blood collection

Blood is removed by cannulation of the ear's central artery. Blood is collected in 10 ml Monoject sterile blood collection tubes (Krackeler Scientific) following the recommendations of the Institutional Animal Care and Use Committee of not more than 7.5% of the total blood volume when bleedings are weekly, 10% when bleedings are bi-weekly and 15% when the bleedings are monthly.

2.3.2. Separation of serum

1. To separate the blood clot from the serum, centrifuge at 1,000 × g (GLC-2B, Sorvall) for 20 min in Monoject- sterile blood-collection tubes. The serum rises to the top of the tube.
2. Remove the serum using a sterile 10 ml pipet and transfer it to a 50 ml sterile conical tube.

3. Transfer the serum to a sterile 20 ml syringe with a 18g 1 ½ in. needle and pass through a 0.22 μm filter (Gelman Sciences).
4. Aliquot into Dram vials and store at −70 °C.

2.3.3. Antibody titration (for antigen see 3.1.1)

1. To titrate the antibody, dilute the antiserum in 1000 μl complete medium (CM+). Start with a 1:200 dilution (i.e., use 5 μl antiserum removed with a sterile pipet).
2. Remove 500 μl, using a sterile 1 ml pipet, of the 1:200 dilution and add to 500 μl of fresh CM+, to make a 1:400 dilution.
3. Keep performing these 1:2 dilutions until the desired series of dilutions is completed.
4. Assay each dilution with the reference methanogen by indirect immunofluorescence (IIF) [12] and by SIA (see 3.2).

2.4. Polyclonal antibody probe preparation

2.4.1. Calibration

The S-probe [11,12] is prepared from the first bleeding serum of a single rabbit obtained 14 days after the second inoculation (boost) of antigen. Different sera from different rabbits or from different bleedings of the same rabbit are not mixed. The antiserum is titrated as described above (see I.C.c.). The S-probe is the highest dilution of the titration curve's plateau. The working (W) probe for SIA is derived from the S-probe [15] (see 3.2.1).

2.4.2. Determination of specificity spectrum of the S-probe

Use a panel of antigens relevant, or expected to be relevant to the kind of samples to be examined [11–13]. For example, if the ecosystem to be studied is expected to contain methanogens and some other bacteria associated with them, the probe ought to be tested with reference micro-organisms representing the methanogens and the other bacteria. Thus, the spectrum of reactivities of the probe is pre-determined with regard to microorganisms that one might find in the sample.

3. Quantitative slide immunoenzymatic assay (SIA)

3.1. Antigen

3.1.1. Reference
Axenic culture well characterized physiologically, morphologically, phylogenetically, etc. [11,13]. Prepared as the Immunogen (see 2.1).

3.1.2. Sample
(a) New isolate (axenic culture) or enrichment culture (only two or three different species present), same as Immunogen (see 2.1), or (b) complex mixture with several species (e.g., sample from an anaerobic bioreactor, or from soil, landfill, etc.).

3.1.2.1. Preparation of a cell suspension (E.g., sample from an anaerobic bioreactor)
1. 10 ml of sample should be collected in a 15 ml sterile, graduated centrifuge tube.
2. Centrifuge at 15,000 × g for 20 min at 4 °C.
3. Remove the supernatant with a Pasteur pipet and discard.
4. Resuspend the pellet with a newly made fixative solution of 4 ml formalin 37% (in H_2O) plus 96 ml of 0.85% NaCl to an absorbance of 0.7 or greater at 660 nm, 1 cm cuvette. (As a rough guide, one ml of formalin/saline solution for a cell pellet from 5 ml of culture). Use good-quality, fresh formalin 37%. Salts concentration and/or composition may be modified as needed to preserve special organisms (e.g., halophilic organisms). Omit formalin for non-formalinized samples (antibiotics may be added in this case).
5. Mix very thoroughly and gently using a Pasteur pipet, avoiding formation of air bubbles (Do NOT use a vortex!).
6. Place the formalinized cells in a leak-proof glass tube, conveniently labeled and sealed for shipment or storage at 4 °C.
7. Wash cells to get rid of the fixative before use (see 3.1.2.2 below, steps 5–8).

3.1.2.2. Preparation of a cell suspension from a solid sample (E.g., upflow anaerobic sludge blanket uasb bioreactor granules). Solid samples may be prepared for immunologic analysis in various ways. For example, a cell suspension may be prepared [10], or the sample

may be processed for sectioning to obtain thin (or ultrathin) histologic sections [17]. Here, only the preparation of cell suspensions by
mechanical means will be described.

1. Add 1 ml of wet granules to 5 ml of PBS at pH 7.2, and disrupt
   with a tissue grinder (Krackeler Scientific, Inc.) to form a homogeneous suspension.
2. Transfer the entire suspension to a sterile 10 ml test tube.
3. Wash the remaining material off the walls of the tissue grinder
   with an additional 5 ml of PBS and add to the initial suspension.
4. Formalinize half of the cell suspension (see 3.1.2.1, steps 1–7);
   store the rest in PBS at 4 °C.
5. Centrifuge 200 µl of formalinized cell suspension at 15,000 × g in
   a microfuge centrifuge (Eppendorf model 5415C).
6. Remove the supernatant and store at −70 °C for future use, if
   necessary, e.g., to measure soluble antigens (see Note 1 below).
   The pellet is washed by resuspending it carefully and thoroughly
   with a Pasteur pipet in 1 ml of PBS and centrifuging it again.
7. Repeat wash two more times.
8. Adjust the final volume in PBS so that 5 µl of sample covers 80%
   of an SIA slide circle.

Note:

Soluble and microparticulate antigens

Although assays for this type of antigens are not described here, the following is
pertinent to help the reader envision the limitations and advantages of IIF and SIA.
While IIF has the great advantage of allowing morphologic identification of microbes
and their colonies in biofilms and granular consortia [10,12,14–16,20], it cannot be
used to study soluble and microparticulate antigens, non-visible with the light
microscope. These antigens are usually abundant in the fluid phase of ecosystems
that naturally include a fluid phase. These antigens are also abundant in extracts from
cells, consortia, and solid ecosystems. It is therefore important to measure antigens
undetectable by light microscopy and immunologic techniques that depend on it. SIA
is amenable to measuring particulate antigens (regardless of whether or not they are
visible with the light microscope), as well as soluble antigens [2,3,6,9,18,23]. SIA can
be automated for quantitative measurements of antigen more readily than IIF [2,4].
However, to obtain direct counts of microbes, IIF is more straightforward than SIA.
The latter requires calibration each time a new type of microbe has to be identified
[14,15].

3.2. Procedure

3.2.1. Coating SIA-slide circles

Coat three circles of SIA slides (Cel-Line Associates) per probe and three for negative and positive controls with antigen. Each circle is 3 mm in diameter.

1. Use a cell suspension, in PBS that will give about 80% coverage when using 5 µl of sample.
2. Dry the 5 µl of sample on the SIA slide circle using a Tissue Tek II slide dryer (Lab-Tek Products).
3. Pass the back of the slide containing the dried suspension through the flame of the Bunsen burner 2–3 times very briefly to fix the cells to the slide, allow to cool.
4. Wash with glass-distilled water (GDW) to remove any residual salt crystals, and dry again in the slide dryer.
5. Check the circle coverage by the cells, with a light microscope.

Note

The cell concentration (i.e., number of cells per circle) is critical. It must be adjusted to obtain spectrophotometric (absorbance) readings of the final reaction within the optimal range established for the system under analysis. Calibration of the assay is necessary [14,15]. For this purpose, optimal concentrations of antigen (i.e., the reference microbe corresponding to the antibody probe) and antibodies have to be determined. The optimal antigen concentration is that which gives a spectrophotometric reading within the safe, reliable range of the instrument, when tested with the optimal antibody dilution (usually 2–3 twofold higher than the S-probe). This optimal dilution for SIA may be called the working (W) probe. Optimal concentrations of antibody and antigen are arrived at by testing several combinations of them, which allows one also to establish a correlation between cell numbers and reading values. From these values a standard curve may be derived, useful to determine the number of cells in a sample by interpolating the reading within the curve. Even in the absence of such a calibration for accurate determination of cell numbers, one may use SIA for direct quantitative comparisons between samples. For example, one can determine the time-course quantitative profile of a given microbial species in a single ecosystem by testing sequential samples and plotting the readings versus time. In practice, this approach has proven useful to monitor methanogenic subpopulations in bioreactors and to assess the effect of perturbations to the biomass on the size of the subpopulations [16].

3.2.2. Incubation with the W-probe

1. Add 10 µl of the W-probe for each sample being tested and positive controls.
2. Add 10 µl of normal rabbit serum at the same dilution as that of the W-probe to the corresponding negative controls.
3. Place the slide in a humid chamber, at 23 °C, for 30 min. Evaporation must be reduced to a minimum.
4. After the incubation, wash the circles evenly using a gentle flow of GDW from a plastic tube. Do not allow to dry.

3.2.3. Incubation with the second antibody

1. Dilute the second antibody (GAR-IgG-PO), to the pre-tested optimal level (usually 1:1000) in CM+ (see section 4, etc.).
2. Add 10 µl of the diluted second antibody to the circles which previously had been exposed to the W-probe or the normal rabbit serum at the same dilution as the W- probe and incubate for 30 min in a humid chamber at 23 °C.
3. Wash with GDW and dry.

3.2.4. Incubation with the substrate for the enzyme

The substrate is o-phenylenediamine (OPD) for peroxidase, or para-nitrophenyl phosphate (PNPP) for alkaline phosphatase (see section 4, etc).

1. To prepare the substrate for peroxidase (o-phenylenediamine, OPD) for development of yellow color, add 10 µl of the substrate solution (OPD 1 g/ml, in 0.1 M citric acid buffer, pH 4.5, and 1.0 µl of 30% hydrogen peroxide) to all circles being tested on the SIA slide. To prepare the substrate for alkaline phosphatase (p-nitrophenyl phosphate, PNPP) for development of brown color, add 10 µl of the substrate solution (PNPP 1 g/ml, in "Sigma 104" buffer containing 0.05 M Na_2CO_3, pH 9.8, and 0.001 M $MgCl_2$. For 1 l add 5.3 g Na_2CO_3 and 0.2 g $MgCl_2 \cdot 6H_2O$. Adjust pH to 9.8 with HCl and bring to 1 l with H_2O.) to all circles being tested on the SIA slide.
2. Place the slide in a humid chamber, so the drops do not evaporate.
3. The reaction is then timed and appearance of yellow (or brown) color is visually determined.

3.2.5. Spectrophotometric (absorbance) readings

These are done with a vertical-beam spectrophotometer (provided by Dynatech Laboratories, for example) with SIA slide holder and carrier (Microsample Holder and Carrier Therefor, U.S. Patent No. 4,682,890; Microcircle System, U.S. Patent No. 4,682,891), see Fig. 1 [2,4]. Time of reading (absorbance at 450 nm for peroxidase, and at 410 nm for alkaline phosphatase) is usually every 15 min after addition of substrate. The SIA-slide was devised to fit within the format of the micro-titer plates and readers: each SIA-slide has 24 flat reaction areas or circles distributed to match the position of the wells of the micro-titer plates, and four slides (96 circles) match the entire plate (96 wells) as shown in Figs. 1a–c. The system is modular.

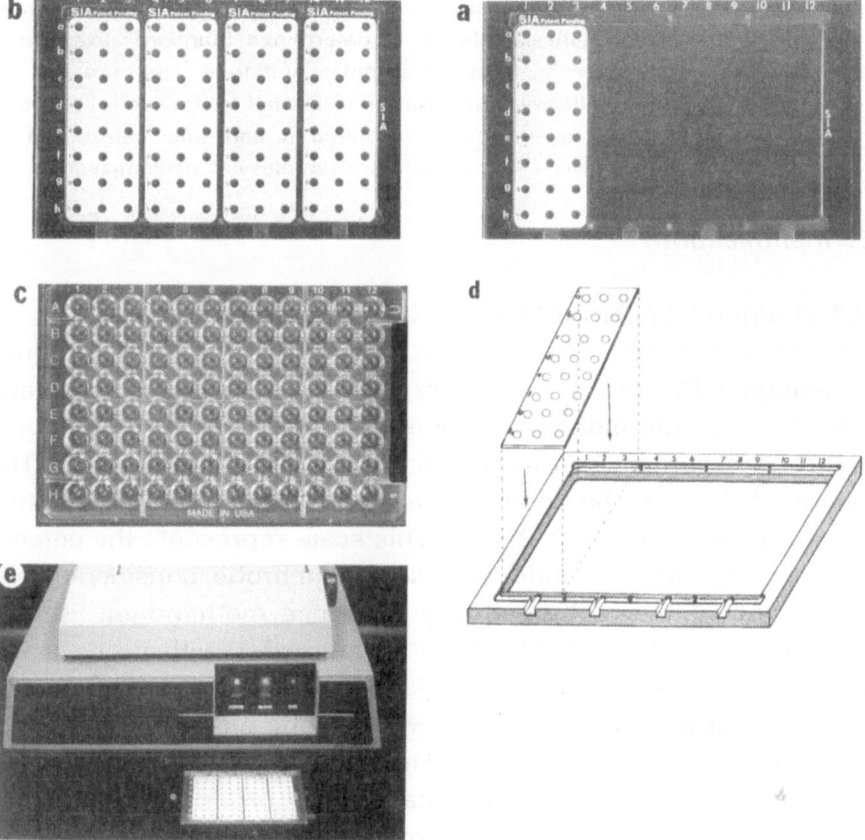

Figure 1. One (a) or four (b) SIA-slides, whose flat reaction areas (circles) are distributed to match the wells of a micro-titer plate (c) have been placed on to the SIA-slide holder as shown in (d) for reading with a vertical beam spectrophotometer (e).

The desired number of SIA-slides from 1 up to 4, depending on the number of samples to be tested, is fitted onto the slide holder as demonstrated in Fig. 1d. Then the holder with the slide, or slides, is placed on the reader's platform or carrier as if it were a micro-titer plate, for spectrophotometric reading (Fig. 1e). Readings can be done manually, one circle at a time, or automatically as for micro-titer plates.

Note
- Negative control. Circles in which serum from a non-immune rabbit at the dilution of the W-probe is substituted for the probe. Usually one takes three circles for each sample and three circles for the reference microbe. Absorbance values read in these circles are averaged and subtracted from the readings in the other circles with sample or positive control.
- Positive control. Calibrated to provide a certain reading for a defined cell number when cell concentration in the sample is to be determined. Standard curves can be derived to interpolate results from the samples and thus determine the cell concentration in them. Otherwise, they can be calibrated to provide a 100% level against which sample readings can be compared and thus categorized as containing more or less cells. Likewise, different samples can be compared.

3.3. Identification

3.3.1. Antigenic fingerprinting

This method is based on the use of a panel of W-probes for reference methanogens [11,13,15]. To identify immunologically a new isolate, or a reactive morphotype in an enrichment culture or complex mixture, the sample has to be tested with all the W-probes. The reaction of the microbe under analysis with the probes is quantified as zero (no reaction), 1, 2, 3, or 4. This scale represents the percent reaction of the microbe under analysis with a probe, considering that the reaction of the probe with the reference methanogen used to generate the probe is 100%. Thus a value of 1 with probe No. 1 means that the new isolate being identified reacts 25% as compared with the homologous reaction between probe No. 1 and reference methanogen No. 1. A value of 2 will indicate 50% reaction, and so on. These values are tabulated horizontally in the order of the probes in the panel, which results in a series of data (the antigenic fingerprint) that can be compared position by position with those of each reference methanogen for which an antigenic fingerprint exists.

3.3.2. Comparative analysis of antigenic fingerprints.

Once the antigenic fingerprint of the microbe under analysis has been determined, it has to be compared with the fingerprints of reference methanogens [11,13]. The latter may be arranged in a reference chart so that it is easy to visually compare the fingerprints by a simple manual method. For example, if the fingerprint of the microbe under analysis is written on a piece of transparent plastic, in exactly the same way as the reference fingerprints are printed on the chart, the plastic can be placed on top of the chart and slid slowly from top to bottom. This will allow a comparison of the fingerprint in the plastic with each one of those in the chart. Thus, it will be determined whether the fingerprint of the microbe under analysis has an identical counterpart in the chart, or a similar one, or it has no resemblance to any of them. In the former case, one may conclude that the microbe under analysis is antigenically identical (i.e., antigenically very closely related) to the reference methanogen showing the same fingerprint as its own. Similarly, degrees of antigenic relatedness are assessed to identify the microbe under analysis. If a reference fingerprint is found that is identical or similar to the new one, one can infer that the microbe under analysis belongs to the same group of the reference methanogen whose fingerprint it reproduces completely or partially. The degree of relatedness of newly identified microbes with the members of the genus to which the most similar reference organism belongs can also be estimated [13,21]. For this purpose, the fingerprinting data are rearranged in a different sequence that matches a matrix formed by the phylogenetic organization of the microbes related to the new one. In this matrix, the fingerprints of all the members of a genus are adjacent to one another, and genera of the same families are also adjacent to one another. When the new antigenic fingerprint is inserted into this fingerprint matrix near its most similar counterpart, one can easily see to which group the newly identified microbe belongs and which member(s) of the group it resembles.

3.3.3. Auxiliary tests

Other tests may be performed to help microbial identification. Among these tests, the most important is indirect immuno-fluorescence (IIF) as part of a set of complementary methods, i.e., SIA-constellation [14].

4. Reagents, instruments, supplies, solutions

4.1. Reagents

- Formalin (formaldehyde), 37% (Catalog # 5014; Mallinckrodt, Inc., Paris, KY).
- Labelled antibody goat anti-rabbit (GAR-IgG-PO; PO, horseradish peroxidase; Catalog # 55689; Organon Teknika-Cappel Corp., Durham, NC).
- Labelled antibody goat anti-rabbit (GAR-IgG-AP; AP, calf intestinal alkaline phosphatase; Catalog # 59298; Organon Teknika-Cappel Corp., Durham, NC).
- o-phenylenediamine (OPD) (Catalog # P-1526; Sigma Chemical Co., St. Louis, MO)
- p-nitrophenyl phosphate (Sigma 104) (Catalog # 104-0; Sigma Chemical Co., St. Louis, MO).
- Hunter's TiterMAX (Catalog # R-1; Vaxcel™, Inc., Norcross, GA)
- Citric acid, monohydrate (Catalog # A104–500; Fisher Scientific Co., Pittsburgh, PA).

4.2. Instruments

- 5 and 10 µl Eppendorf pipets (Eppendorf, Germany).
- Rainin automatic pipet (Rainin, Woburn, MA).
- SIA slides, 18 (for IIF) or 24 (for SIA) circles/slide (Catalog # 10-609 and 10-754, respectively; Cel-Line Associates, Newfield, NJ).
- Microsample Holder and Carrier for the Microcircle System (Wadsworth Center for Laboratories and Research, Albany, NY).
- Tissue Tek II slide dryer (Lab-Tek Products, Naperville, IL).
- Bunsen burner (Fisher Scientific Co., Pittsburgh, PA).
- Spectrophotometer (Gilford Instrument Laboratories, Inc., Oberlin, OH).
- Minireader II (Dynatech Laboratories, Inc., Alexandria, VA).
- Microfuge centrifuge (Eppendorf, Germany).
- Tissue Grinder (Krackeler Scientific, Inc., Albany, NY).
- Light microscope with bright-field and phase-contrast optics (Carl Zeiss, Germany).
- Humid chamber (Krackeler Scientific, Inc., Albany, NY).
- Sorvall GLC-2B centrifuge (Dupont Instruments, Newtown, CT).

4.3. Supplies

- Eppendorf pipet tips (Eppendorf, Germany).
- Pasteur pipettes (Krackeler Scientific, Inc., Albany, NY).
- 1 cm quartz and glass cuvettes (Gilford Instrument Laboratories, Inc., Oberlin, OH).
- Test tubes, 12 × 75 mm (VWR Scientific, San Francisco, CA).
- 1.5 ml microfuge tubes (Bio-Rad Laboratories, Richmond, CA).
- 15 ml, 50 ml conical tubes (Corning Glass Works, Corning, NY).
- 1 ml, 3 ml, 5 ml, 10 ml, 20 ml syringes (Becton-Dickinson Co., Rutherford, NJ).
- 16g 1 ½ in., 18g 1 ½ in., 25g & in., 26g ⅝ in. needles (Becton-Dickinson Co., Rutherford, NJ).
- 0.22 µm filters, low protein binding (Gelman Sciences, Ann Arbor, MI).
- Dram vials (Krackeler Scientific, Inc., Albany, NY).
- 1 ml, 5 ml, 10 ml plastic pipettes (Becton-Dickinson Co., Rutherford, NJ).
- Monoject sterile blood collection tubes (Catalog # 24-1-301710; Krackeler Scientific, Inc., Albany, NY).
- Leak-proof glass tubes with screw cap (Krackeler Scientific, Inc., Albany, NY).

4.4. Solutions

- Phosphate-Buffered Saline (PBS), pH 7.2: 6.8 g NaCl, 1.48 g Na_2HPO_4, 0.43 g KH_2PO_4 and 1000 ml GDW.
- Complete Medium (CM+): 10.29 g Dulbecco's Modified Eagle Powder (Gibco BRL, Grand Island, NY), 3.6 g Dextrose, 3.81 g $NaHCO_3$, 11.4 ml penicillin/streptomycin, 11.4 ml non-essential amino acids, 0.67 g l-glutamine, 114 ml gamma globulin-free horse serum, 57 ml calf serum and 1000 ml GDW.
- Fetal Calf Serum (FCS) (Sigma Chemical Co., St. Louis, MO).
- 0.1 M citric acid buffer, pH 4.5: 21 g citric acid, monohydrate and 1000 ml GDW, adjust pH with NaOH.
- 30% hydrogen peroxide.
- Glass-Distilled Water (GDW).
- Antibody probes (prepared as described in this protocol).
- Normal rabbit serum (Sigma Chemical Co., St. Louis, MO).
- Immersion oil (Carl Zeiss, Germany).

Note

Antibodies, supplies, and solutions must be sterile. All manipulations should follow the rules for maintenance of sterility.

Acknowledgement

We thank James S. Swab for his help. This work was supported in part by grant No. 706-RIER-BEA-85 from NYSERDA.

References

1. Cerrone MC, Kuhn, RE (1991) Description of a urease-based microELISA for the analysis of limiting dilution microcultures. J Immunol Meth 138: 65–75.
2. Conway de Macario E, Jovell RJ, Macario AJL (1985) The slide immunoenzymatic assay: a simple laboratory tool with multiple applications. BioTechniques 3: 138–145.
3. Conway de Macario E, Jovell RJ, Macario AJL (1987) Slide immunoenzymatic assay (SIA): improving sensitivity to measure antibodies when samples are very small and dilute and antigen is scarce. J Immunoassay 8: 283–295.
4. Conway de Macario E, Jovell RJ, Macario, AJL (1987) Multiple solid phase for storage of dry ready-for-use reagents and efficient performance of immunoenzymatic and other assays. J Immunol Meth 99: 107–112.
5. Conway de Macario E, Jovell RJ, Macario, AJL (1990) Adaptation of the slide immunoenzymatic assay for quantification of DNA hybridization: SIA-DNA. BioTechniques 8: 210–217.
6. Conway de Macario E, Macario AJL, Jovell RJ (1983) Quantitative slide micro-immunoenzymatic assay (micro-SIA) for antibodies to particulate and non-particulate antigens. J Immunol Meth 59: 39–47.
7. Conway de Macario E, Macario AJL, Jovell RJ (1986) Slide immunoenzymatic assay (SIA) in hybridoma technology. Meth Enzymol 121: 509–525.
8. Engvall E, Perlmann P (1971) Enzyme-linked immunosorbent assay (ELISA): quantitative assay of immunoglobulin G. Immunochemistry 8: 871–874.
9. Gaveriaux C, Loor F (1988) IgE measurement in serum and culture supernatant. In: Pal SB (ed) Reviews on Immunoassay Technology, Vol. 2, pp. 175–186. Chapman and Hall, Inc., New York, U.S.A.
10. Koornneef E, Macario AJL, Grotenhuis JTC, Conway de Macario E (1990) Methanogens revealed immunologically in granules from five upflow anaerobic sludge blanket (UASB) bioreactors grown on different substrates. FEMS Microbiol Ecol 73: 225–230.
11. Macario AJL, Conway de Macario E (1983) Antigenic fingerprinting of methanogenic bacteria with polyclonal antibody probes. Syst Appl Microbiol 4: 451–458.
12. Macario AJL, Conway de Macario E (1985) A preview of the uses of monoclonal antibodies against methanogens in fermentation biotechnology: Significance for public health. In: Macario AJL, Conway de Macario E (eds) Monoclonal Antibodies against Bacteria, Vol. I, pp. 269–286. Academic Press, Inc., Orlando, FL.
13. Macario AJL, Conway de Macario E (1985) Monoclonal antibodies of predefined molecular specificity for identification and classification of methanogens and for probing their ecologic niches. In: Macario AJL, Conway de Macario E (eds) Monoclonal Antibodies against Bacteria, Vol. II, pp. 213–247. Academic Press, Inc., Orlando, FL.
14. Macario AJL, Conway de Macario E (1985) Antibodies for methanogenic biotechnology. Trends Biotechnol 3: 204–208.

15. Macario AJL, Conway de Macario E (1988) Quantitative immunologic analysis of the methanogenic flora of digestors reveals a considerable diversity. Appl Environ Microbiol 54: 79–86.
16. Macario AJL, Conway de Macario E, Ney U, Schoberth SM, Sahm H (1989) Shifts in methanogenic subpopulations measured with antibody probes in a fixed-bed loop anaerobic bioreactor treating sulfite evaporator condensate. Appl Environ Microbiol 55: 1996–2001.
17. Macario AJL, Visser FA, van Lier JB, Conway de Macario E (1991) Topography of methanogenic subpopulations in a microbial consortium adapting to thermophilic conditions. J Gen Microbiol 137: 2179–2189.
18. Monroe D (1985) The solid-phase enzyme-linked immunospot assay: Current and potential applications. BioTechniques 3: 222–229.
19. Palleroni NJ (1993) Some reflections on microbial ecology. In: Guerrero R, Pedros-Alio C (eds) Trends in Microbial Ecology, pp. 501–504. Spanish Society for Microbiology Press, Barcelona, Spain.
20. Schmidt JE, Macario AJL, Ahring BK, Conway de Macario E (1992) Methanogenic subpopulations in a thermophilic acetate–degrading granular consortium: Effect of Mg^{2+}. Appl Environ Microbiol 58: 862–868.
21. Sieburth JMcN, Johnson PW, Macario AJL, Conway de Macario E (1993) C_1 bacteria in the water column of Chesapeake Bay. II. The dominant O_2- and H_2S-tolerant methylotrophic methanogens, coenriched with their oxidative and sulphate reducing bacterial consorts are all new immunotypes and probably include new taxa. Mar Ecol Prog Ser 95: 81–89.
22. Smith KO, Ludwig MJ, Stigall BW, Boswell N (1991) Enzyme-linked immunosorbent assay (ELISA) for HIV antibody by a glass slide technique. J Immunol Meth 136: 239–246.
23. Tijssen P (1986) Enzyme immunoassays. In: Kohler RB (ed) Antigen Detection to Diagnose Bacterial Infections, Vol. I, pp. 99–136. CRC Press, Inc., Boca Raton, Florida, U.S.A.
24. Witzel K-P (1990) Approaches to bacterial population dynamics. In: Overbeck J, Chrost R (eds) Aquatic Microbial Ecology, pp. 96–128. Springer Verlag, Berlin, Germany.

Molecular Microbial Ecology Manual **5.1.1**: 1–22. 1995
© 1995 *Kluwer Academic Publishers.*

Natural transformation in aquatic environments

JOHN H. PAUL and HAYDN G. WILLIAMS

Department of Marine Science, University of South Florida, St Petersburg Campus, 140 Seventh Avenue South, St. Petersburg, Fl 33701, USA

Introduction

Transformation is a process in which competent cells take up DNA and incorporate it into their genome. It is one of the mechanisms by which genes may be spread through natural populations [27]. A wide variety of bacteria express competence (the physiological ability to take up DNA) during normal growth. Frischer et al. [9] found up to 16% of isolates taken from Tampa Bay were naturally competent. When competent cells come into contact with a source of DNA, this is bound to the cell and taken up. As well as DNA in free solution, competent bacteria can take up DNA associated with particulate matter [15,16,29], cellular debris [14], heat inactivated cells [23,30] and intact, live donor cells [1,23,24,28].

Transformation is generally most efficient with homologous chromosomal DNA. Once inside the cell it can be integrated into the host's chromosome by normal recombination processes [5,7]. However, in many cases almost any fragment of DNA may be taken up. Plasmid DNA without homology to the host may be recircularized by mismatch repair if multiple copies are taken up [6,10]. This would allow transformation between organisms of different genera. For example, Paul et al. [23] demonstrated transfer of a small non-conjugative plasmid from *E. coli* to a marine *Vibrio* by natural transformation.

Natural transformation may play an important role in the flow of genes through natural populations. One of the most commonly used approaches to study this type of gene transfer in nature has been to examine the behavior of model systems. Such systems often involve the transfer of an individual gene or group of linked genes to a predetermined recipient in a simple microcosm. For example Graham and Istock [8] examined transformation of *Bacillus subtilis* in soil microcosms, Stewart et al. [29] studied transfer of rifampicin resistance between strains of *Pseudomonas stutzeri* in sediment microcosms and Paul et al. [22] demonstrated the uptake of plasmid DNA by a marine *Vibrio* strain in marine water and sediment microcosms. Experiments usually compare the response of the system to various environmental conditions with idealized laboratory conditions. A variation on this approach is to perform open experiments in the field (*in situ*). Bale et al. [2,3] developed a method to follow transfer of a conjugative plasmid in river epilithon. The technique was then adapted to study natural transformation [30].

Chromosomal transformation using *Acinetobacter calcoaceticus* in river epilithon

Experimental approach

This article describes a method to demonstrate the transfer of the His+ gene from *Acinetobacter calcoaceticus* BD413 [14] to a histidine auxotroph HGW1521(pQM17) [30] by natural transformation in both laboratory and open field experiments. The low transfer frequencies and/or target populations that often occur in these *in situ* experiments can result in very low numbers of transformants (<10 cells/ml) that are difficult to detect. Therefore it is advisable to perform a series of assays starting with simple laboratory experiments and progressing in complexity to *in situ* experiments with recipients growing as part of the natural population. Competent bacteria and a source of transforming DNA are inserted into an aquatic environment with minimal disturbance. After a period to allow gene transfer to take place the bacteria are extracted and enumerated on various selective media so that the tranformation frequency (number of transformants per recipient) can be estimated. Whilst the method has been developed using *Acinetobacter* as a model organism, it may be possible to modify it for use with other naturally competent bacteria, genes or environments.

Procedures

Bacterial strains
A. calcoaceticus strain BD413 is a prototrophic (His+) soil isolate expressing very high levels of competence [14]. *A. calcoaceticus* strain HGW1521(pQM17) is a histidine auxotroph (His-) derived from BD413 [30]. It is spontaneously resistant to 100 µg/ml rifampicin and 100 µg/ml spectinomycin and carries the plasmid pQM17 which encodes resistance to 27 µg/ml mercury [24].

Source of transforming DNA
Transforming DNA may be presented in a variety of forms such as cultures containing live bacteria, suspensions of heat inactivated

cells, crude bacterial lysates, purified chromosomal DNA, purified plasmid DNA or plasmid multimers. The type of transforming DNA used can affect the characteristics and frequency of transformation under various conditions. For example, purified DNA preparations generally give higher transformation frequencies whereas live donor cells may allow transformation to occur in the presence of nucleases [1].

Whole cell preparations

1. The donor organism (BD413) is grown overnight in 50 ml of Luria broth (LB).
2. Harvest the culture by centrifugation (10,000 × g, 10 min) and wash twice in B22 salts solution. Resuspend the pellet in 50 ml of B22 salts solution.
3. The washed cell suspension can be used directly as a source of transforming DNA.

Heat inactivated cell suspensions

Pasteurize 1 ml of washed cell suspension in a sterile microfuge tube by heating to 72 °C for 2 h. Heat inactivated cells may be stored at 5 °C until needed.

Crude lysates

Spin down 1 ml of washed cell suspension in a microfuge tube and resuspend the pellet in 1 ml of Juni lysis buffer. Pasteurize the suspension by heating to 72 °C for 2 h and store at 5 °C until needed.

Notes
- Ad 1. Defined minimal media and/or antibiotic additions may be used instead of LB to maintain selective pressure for certain phenotypes.
- Ad 3. Whole cell preparations should used immediately and cannot be stored.

Solutions

- B22 salts solution (pH 7.2) [2]
    - 3.89 g/l KH_2PO_4.
    - 12.5 g/l K_2HPO_4.
    - 0.19 g/l $MgSO_47H_2O$.
    - 1.09 g/l $(NH_4)_2SO_4$.

- Succinate-B22 minimal medium (pH 7.2) [2]
    - B22 salts solution.
    - 10 g/l sodium succinate.
    - If solid medium is required add 15 g/l agar.
- Juni lysis buffer (pH 7.0) [14]
    - 0.05% (w/v) SDS.
    - 0.15 M NaCl.
    - 0.015 M sodium citrate.
- LB (pH 7.2)
    - 10 g/l Tryptone.
    - 5 g/l yeast extract.
    - 10 g/l NaCl.

Competent cells

In the case of *A. calcoaceticus*, a 16 h culture grown in LB should contain a large number of highly competent cells suitable for transformation assays. However, if the method is to be used with other organisms it is advisable to first determine the competence phase of that strain.

Steps in the procedure

1. Prepare an overnight culture of the recipient (e.g., HGW1521 (pQM17)) in 50 ml of LB.
2. Inoculate 500 ml of fresh LB in a 2 l flask to an optical density at 540 nm (A540) of 0.1 with the overnight culture and incubate with shaking at 20 °C.
3. Follow the optical density (A540). Remove enough culture to yield approximately 10^7 cells.
4. Filter the sample on to a 24 mm diam., 0.22 μm pore size nitro-cellulose filter. Filter the source of DNA (e.g., 1 ml of a crude lysate of BD413) on to a 24 mm diam., 0.22 μm pore size nitrocellulose filter. Place the two filters together so that the DNA and cells are in contact.
5. Place the filters on an agar plate (e.g., PCA) and incubate at 20 °C for 15 min.
6. After the incubation, retrieve the filters and resuspend cells in 4 ml of B22 salts solution containing DNaseI (50 μg/ml) by vortexing for about 30 s. Prepare a decimal dilution series ranging from 1 to

10^{-7} of the suspension in B22 salts solution containing DNaseI (50 µg/ml) and enumerate bacteria by plate counts on both media selective for recipients (PCA) and transformants (SE22). Incubate plates at 20 °C.

7. Repeat steps 3–6 at 1 h intervals until the culture is in late stationary phase. Adjust the volume of culture filtered each time according to the optical density to give approximately the same number of recipients.

8. For each mating calculate the transformation frequency as the number of transformants (c.f.u./ml on SE22)/the number of recipients (c.f.u./ml on PCA). Determine the optical density and growth phase at which the highest transformation frequencies were achieved. Use cells grown to this point of their growth phase as a source of competent cells in future experiments.

9. Preparing competent cells: Take 50 ml of culture expressing maximum competence as determined above. Harvest the culture by centrifugation (10,000 × g, 10 min) and wash twice in B22 salts solution. Resuspend the pellet in 50 ml of B22 salts solution.

Notes
- Ad 2. If more sample is needed for the experiment, replicate 500 ml cultures can be used.
- Ad 3. One ml of culture at an optical density of 1 should yield approximately 10^7 cells.
- Ad 5. If no transformants are formed within 15 min at any growth phase, try longer incubation periods (e.g., 1 h).
- Ad 6. DNaseI should prevent any further transformation.
- Ad 9. Competent cells should not be frozen and stored but freshly prepared for each experiment.

Solutions
- DNaseI stock solution
    - 50 mg/ml DNaseI, filter sterilize.
- PCA (standard plate count agar) (pH 7)
    - 2.5 g/l yeast extract.
    - 5 g/l pancreatic digest of casein.
    - 1 g/l glucose.
    - 10 g/l agar.
- Rifampicin (Rif) stock solution
    - 40 mg/ml Rifampicin.
    - 0.2 M NaOH.

- PCA + Rif (pH 7)
  - PCA.
  - 100 µg/ml rifampicin.
  - pH 7.
- SE22 (pH 7.2)
  - B22 salts solution.
  - 10 g/l sodium succinate.
  - 0.3 g/l EDTA.
  - 15 g/l agar.
  - 100 µg/ml rifampicin.
  - 27 µg/ml mercury.
  - 0.05% spent culture of BD413 grown in succinate B22 medium, then filter sterilized to remove cells.

Laboratory filter mating assay

1. Filter 10^7–10^8 competent cells (approximately 1 ml of an overnight culture) onto a 24 mm diam., 0.22 µm pore size nitrocellulose filter. Filter the source of DNA (equivalent to 1 ml of an overnight culture of donor) on to a 24 mm diam., 0.22 µm pore size nitrocellulose filter. Place the two filters together so that the DNA and cells are in contact.
2. Place the filters on an agar plate (e.g., PCA) and incubate at 20 °C for 24 h.
3. After the incubation, retrieve the filters and resuspend cells in 4 ml of B22 salts solution containing DNaseI (50 µg/ml) by vortexing for about 30 s. Prepare a decimal dilution series ranging from 1 to 10^{-7} of the suspension in B22 salts solution containing DNaseI (50 µg/ml) and enumerate bacteria by plate counts on selective media. Select recipient cells, HGW1521(pQM17) on PCA + Rif. If whole cell preparations of BD413 were used as donor, select donors on succinate-B22 minimal medium. Select prototrophic transformants of HGW1521(pQM17) on SE22.
4. Incubate the plates at 20 °C until individual colonies can be counted (typically about 2 days).

Notes

- Ad 3. If the culture forms a sticky mat of cells over the filter that is difficult to completely resuspend add 2 to 3 sterile glass beads (1–3 mm diam.) to the suspension prior to vortexing as this helps break up the biofilm.

- Ad 3. The DNaseI prevents further gene transfer from occurring whilst cells are plated out.
- Ad 3. Cell-to-cell transformation can occur on some selective media when live donors are used as a source of transforming DNA, resulting in overestimation of the transfer frequency [30]. The EDTA in SE22 prevents this.
- Ad 3. If low numbers of transformants are anticipated, filter 3 ml of the resuspended cell solution through a 0.25 µm, 45 mm diam. nitrocellulose filter and place the filter cell side up on a SE22 plate.

Transformation in beaker microcosms

1. Collect 500 ml of river water in a sterile bottle and several stones from the river bed. Choose stones (approximately 10 × 15 × 2 cm) with at least one smooth flat surface.
2. Scrub the stones with a stiff bristled brush to remove native epilithon and rinse in distilled water. Wrap scrubbed stone in foil and sterilize (121 °C, 20 min).
3. Prepare donor and recipient filters as for laboratory filter matings (step 1).
4. Place the filters on the surface of the sterile scrubbed stone. Cover them with a larger sterile paper filter (Whatman No.1) and secure with elastic bands. Put the stone in a sterile 1 l beaker containing 400 ml of freshly collected river water and incubate at 20 °C for 24 h.
5. After the incubation retrieve the filters and treat as for laboratory filter matings (steps 3–4).

In situ transformation assay

1. Transport cultures of competent cells and source of DNA to the river bank.
2. Prepare donor and recipient filters as for laboratory filter matings (step 1).
3. Place the filters on the surface of a sterile scrubbed stone. Cover them with a larger paper filter and secure with elastic bands. Put the stone in a large mesh size net bag which is then placed midstream on the river bed secured to a metal stake.
4. Attach a minimum/maximum thermometer to the stake to monitor the temperature of the river. Allow matings to proceed for 24 h.
5. After the incubation use a sterile plastic bag to retrieve the stone.

Remove the filters and treat them as for laboratory filter matings (steps 3–4).

Notes
- Ad 1. Pre-prepared competent cells can be transported to the river on ice. However, do not transport or store competent cells for long periods on ice as this will result in a loss of competence. An alternative method is to grow and prepare competent cells at the river site if laboratory facilities are available locally.
- Ad 3. To prevent contamination through handling, wear rubber gloves and wash them first in 70% ethanol and then in the river. Sterilize net bags and elastic bands by immersing in 70% ethanol and then wash thoroughly in the river.
- Ad 3. Additional stones collected locally and placed in the bag can help prevent the bags being moved by the current.
- Ad 5. Handle only the outside surfaces of the bag, use it as a sterile glove with which to pick up the stones. It is best to plate out cells as soon as they are removed from the river. If this is not possible transfer the suspension on ice to the laboratory.

Transformation of bacteria incorporated into the epilithon

Recipient and donor cultures are incorporated into growing epilithon (*in situ*) on separate stones, then placed together to allow gene transfer to occur. Transformants are directly enumerated from the biofilm.

Steps in the procedure

1. Collect smooth stones approximately 10 × 15 × 2 cm with at least one flat surface from the river bed and place aside on a sterile surface.
2. Prepare donor and recipient filters as for laboratory filter matings (step 1) but do not place filters together.
3. Place each filter, face down on a separate stone. Cover with a larger paper filter, secure with elastic bands and put each stone in a large mesh size net bag.
4. Secure bags to metal stakes in separate parts of the river, ensuring donors and recipients can not come into contact with each other. Incubate the stones in the river for 24 h to allow development of a biofilm.
5. Retrieve the stones. Mark the area on the stone where each filter is with a diamond marker and remove the filters. Gently wash the stones by immersing in river water. Place the two stones together so that the area exposed to the source of DNA and the area

exposed to the recipient are in contact. Hold the stones together with elastic bands.

6. Return the stone to the river and incubate for a further 24 h.
7. Collect the stones in a sterile plastic bag. Separate the stones, place a rubber ring around the area of the stone harboring the recipient to form a well and add 3 ml of B22 salts solution containing DNaseI (50 µg/ml). Scrub the area within the well using a short, stiff bristled stencil brush for at least 3 min. Try to resuspend as much of the epilithon as possible. Transfer the suspension to a sterile bottle and vortex for 20 s.
8. Treat suspension as for laboratory matings. Calculate the transfer frequency as for laboratory filter matings.

Notes
- Ad 1. Both donor and recipient stones should have flat surfaces that can be placed together with as much contact as possible. If no suitable stones are available locally slate disks [25] can be used.
- Ad 7. Sterilize the brush and rubber ring by washing first in 70% ethanol, then in sterile B22 salts solution. If the suspension leaks away from the well, use a pipette to transfer liquid back into the well.

Confirming putative transformants

Colonies growing on the plates selective for transformants (SE22) must be shown to be transformants and not spontaneous mutants or indigenous organisms.

1. Test for known unselected characteristics. Pick 10–50 well isolated putative transformant colonies with a sterile toothpick and streak onto PCA + spectinomycin (100 µg/ml). Discard any isolates that are not resistant to spectinomycin.
2. Juni [14] suggested transformation assays could be used as taxonomic tests for certain organisms. True transformants should be able to re-transform other recipients. Prepare a crude lysate of the putative transformants as described above. Spread 100 µl of lysate over the surface of a succinate-B22-minimal medium plate and allow to dry. Spread 100 µl of a washed suspension of recipient (HGW1521(pQM17)) over the plate. To confirm that the lysate solution was sterile, spread 100 µl on to a PCA plate and to confirm the recipient was auxotrophic spread 100 µl of washed cell suspension on to a succinate-B22 minimal medium plate.

Incubate the plates at 20 °C for 24 h. If the putative transformant was an *Acinetobacter* it should have transformed the recipient to prototrophy, forming a confluent lawn on the plate. The cells only and lysate only plates should be clear of growth.

3. If molecular probes are available to either the recipient or the gene transferred, it should be confirmed that the probe will hybridize to putative transformants.

Control experiments

1. To confirm transformants were produced by gene transfer and not spontaneous mutation repeat transformation assays substituting distilled water for the source of transforming DNA. For transformation to be considered to have occurred the transformation frequency must be significantly higher than the frequuency obtained when no source of transforming DNA is added.

2. To confirm transformation only occurred during the incubation phase (e.g., in the river) repeat the assay omitting the incubation period, i.e., as soon as the source of DNA and recipients are placed in contact, immediately remove, resuspend and plate out them out.

Calculating transfer frequencies

The transformation frequency is expressed as transformants per recipient. Each experiment should be repeated in triplicate, using a separate overnight culture of both donor and recipient for each replicate. For each mating, calculate the c.f.u./ml of donors, recipients and transformants in the final suspension from the plate counts;

$$\text{Transformation frequency} = \frac{\text{number of transformants (c.f.u./ml on SE22)}}{\text{number of recipients (c.f.u./ml on PCA + Rif)}}$$

The overall transformation frequency is then calculated as the mean transformation frequency obtained from the replicate experiments. Mean transformation values can be compared by student t-tests or analysis of variance [11,26].

Plasmid transformation using marine high frequency of transformation recipients

Experimental approach

Natural plasmid transformation is a less well-studied subelement of the field of natural transformation. Chromosomal transformation involves transfer of genes between closely related species, and usually requires homologous recombination [27]. Plasmid transfer also requires some type of homology for recircularization of the plasmid, which is believed to enter the cell as a linear and possibly single-stranded molecule [4]. To ensure self-homology, we use plasmid multimers made in vitro as transforming DNA. The procedure described below involves the Inc Q/P4 plasmid pQSR50 [19], which is a Tn5 containing derivative of R1162 [18]. The plasmid encodes kanamycin and streptomycin resistance. The recipient used is the High Frequency of Transformation (HfT) *Vibrio* strain WJT-1C [9], but other HfT *Vibrio* strains such as MF1-C, MF4-C, and JT-1 can also be used. The strains are grown into stationary phase, and the DNA added either in seawater or marine sediment. The transformants are recovered by plating on selective media, and verified by molecular probing with a probe made to the neomycin phosphotransferase gene (*nptII*) of Tn5 [9].

Procedures

Strains
Vibrio WJT-1C [9] grows on ASWJP+PY medium [20] at 28 °C, and can be maintained on plates stored at 4 °C for several months or stored in 50% glycerol/ASWJP+PY at −80 °C. *E. coli* RM1259 (pQSR50) was the source of the plasmid, and was grown on LB supplemented with 50 µg/ml kanamycin and 25 µg/ml streptomycin.

Transforming DNA
Large scale plasmid preparations were prepared from 4 × 500 ml cultures of *E. coli* RM1259 as previously described [9]. Plasmid multimers were produced by the protocol described below, which involves digestion at a unique restriction site followed by ligation

under conditions which favor concatemerization. In our experience, the protocol cannot be successfully scaled up. The plasmid is digested in individual 10 µg quantities and then ligated. Typically we perform 10 such individual reactions at once.

Steps in the procedure
1. Digest pQSR50 by adding the following to a sterile 0.5 ml micro-fuge tube:
   A) 10 µg plasmid.
   B) 4.5 µl 1% molecular biology grade bovine serum albumin.
   C) 4.5 µl 10 × EcoR1 reaction buffer.
   D) 5 µl EcoR1 (~15 U/µl) (as much as 8 µl may be used).
   E) Sterile distilled water to bring total volume to 45 µl.
2. Vortex well. Digest at 37 °C for 3.5 h. Remove 3 µl to assess digestion by 1% agarose gel electrophoresis, leave remainder at 37 °C until gel has finished running.
3. If digestion is complete, set aside another 3 µl for the second gel.
4. Denature the EcoR1 by heating at 70 °C for 20 min. Cool to 15 °C.
5. Set up the ligation as follows:
   A) 39 µl EcoR1 digested plasmid.
   B) 20 µl sterile distilled water.
   C) 15 µl T4 DNA Ligase Buffer, 5×.
   D) 16 µl ligase.
6. Incubate overnight at 15–16 °C.
7. Assess degree of multimerization by running on a 0.4% agarose gel, being sure to run high molecular weight standards, and the digested but not ligated pQSR50.
8. Add 5 µl sterile 0.5 M EDTA. Pasteurize multimers before use by heating at 70 °C for 2 h.

Preparation of recipient cells
1. A culture of *Vibrio* WJT-1C (25 ml) is grown overnight at 28–30 °C on a gyrotatory shaking platform (150–200 r.p.m.).
2. Cells are harvested by centrifugation at 10,000 × g for 10 min at 20 °C.
3. The cells are resuspended in 20 ml of media lacking peptone and yeast extract (ASWJP) and used immediately.

Water column transformation assay

1. Collect seawater from the environment to be examined.
2. Add 0.5 ml (oceanic microcosm) or 1.0 ml (estuarine microcosm) of the competent cell suspension to 24.5 ml or 24.0 ml, respectively, of the seawater to be investigated in a sterile disposable 60 ml centrifuge tube.
3. If desired, add nutrients. These can be added as a solution of sterile filtered peptone and yeast extract, at a range of concentrations from 0.1 to 5 mg peptone/ml, and 0.02 to 1 mg yeast extract/ml.
4. Add 5 μg of transforming DNA (pQSR50 multimers as prepared above) or pasteurized calf thymus DNA (controls).
5. The mixture is incubated for the desired length of time on a gyrotatory shaker set at 3–5 r.p.m. and at 25–30 °C.
6. The incubations are harvested by centrifugation when obvious growth has occurred (i.e., estuarine microcosms in the presence of nutrients) or by filtration onto sterile 47 mm 0.2 μm Nuclepore filters.
7. Cells on the filters are resuspended by placing the filters into 5 ml ASWJP in a 15 ml conical centrifuge tube and vortexing vigorously for 1 to 2 min. Cell pellets from centrifugation are resuspended in 5.0 ml ASWJP with vortexing.
8. Aliquots of this suspension are diluted and plated on ASWJP+PY for enumeration of total CFU, and on ASWJP+PY plus 500 μg/ml kanamycin, 1 mg/ml streptomycin, and 5×10^{-6} M amphotericin B to enumerate transformants.
9. Plates are incubated 24–48 h to detect growth of transformants. The unique colony morphology of the HfT strains usually enables enumeration in the presence of the indigenous marine flora.
10. Presumptive transformants are verified by colony hybridization. Sterile MSI Magnagraph Nylon 66 filters (85 mm diameter) are used to lift colonies. The filters are then placed colony side up and grown on ASWJP+PY plus kanamycin and streptomycin for 48 h at 28 °C.
11. Colonies on filters are lysed and the DNA immobilized on the filters [23] and probed with the *nptII* gene probe [9].

- Ad 1. Untreated seawater, filter sterilized seawater, or autoclaved, sterile filtered seawater may be used, depending upon the needs of the study. Autoclaved seawater should always be filtered to remove precipitates which will bind DNA.
- Ad 3. The use of nutrients results in higher transfer frequencies in most instances.
- Ad 5. The typical incubation time for a transformation assay is overnight (16 h). Shorter times may be used (one to several hours). However, if very low frequencies are expected (as for experiments with the natural population present) addition of nutrients may enable detection of transfer.

Solutions

ASWJP (Recipe for one liter)

1. To 900 ml distilled or deionized water add 22.05 g NaCl, 9.84 g $MgSO_4$.
2. Add the following stock solutions;

Stock solution	Component(s)	Concentration (g/l)	Volume (ml)
#1	KCl	55.0	10.0
	$NaHCO_3$	16.0	
#2	KBr	8.0	10.0
	$SrCl_2$	3.4	
#3	Sodium silicate	4.0	1.0
#4	NH_4NO_3	1.6	1.0
#5	NaF	2.4	1.0
#6	$Na_2 \cdot HPO_4$	8.0	1.0
#7	$CaCl_2H_2O$	238.0	10.0
#44	Na_2EDTA	3.0	10.0
	$FeCl_3 \cdot 6H_2O$	0.384	
	$MnCl_2 \cdot H_2O$	0.432	
	$CoCl_2 \cdot 6H_2O$	0.002	
	$ZnCl_2$	0.0315	
	$CuCl_2$	0.025	
	H_3BO_3	0.342	

All of the stock solutions are made in deionized water and stored in polypropylene bottles at 4 °C. Stock solution #44 is a trace metal solution made by dissolving all components separately. It should be a clear, light yellow solution. Upon refrigeration, precipitates form with time which do not affect the performance of the medium.

3. Bring volume to 1 liter with distilled water.
4. For ASWJP+PY, add 5 g peptone, 1 g yeast extract.
5. For agar, add 15 g agar.
6. Sterilize by autoclaving.

Sediment transformation assay
Sediment transformation assays have been previously performed in flow-through columns [12,15,16] using 5 ml (or equivalent) syringes. Sands used in such columns have usually been autoclaved, washed, and "precharged" with transforming DNA. We have performed side-by-side comparisons of such columns with the much simpler,"plug" method described below and found comparable results.

Steps in the procedure
1. Add 3 cm^3 of sediment to a sterile 15 ml conical centrifuge tube.
2. Prepare HfT recipients as above ("preparation of recipient cells") except that 30 ml culture is required per sediment plug rather than 25 ml and that this volume of cells is resuspended in 100 µl ASWJP after the final wash. For example, if six sediment plugs are to be used, 180 ml of culture is required, resuspended in 600 µl ASWJP.
3. To each plug, 100 µl of the above recipient cell suspension and 15 µg DNA (plasmid multimers or control DNA) is added.
4. If nutrients are to be added, a concentrated stock is made (50 mg/ml peptone and 10 mg/ml yeast extract) and 100 µl is added.
5. The sediment mixture is stirred with a sterile pipette.
6. Plugs are incubated overnight (or the desired length of time in a time course study) at 28 °C (or the desired temperature if temperature is a factor being investigated).
7. Sediment is resuspended in 5 ml ASWJP with vigorous vortexing for 2 min.
8. Aliquots of the mixture are serially diluted and plated on ASWJP+PY for total CFU and on ASWJP+PY containing 500 µg/ml kanamycin, 1 mg/ml streptomycin, and 5×10^{-6} M amphotericin B for putative transformants.
9. Transformants are enumerated as above and verified by colony hybridization.

– The sediment to be used can be autoclaved or non-sterile, depending on the purpose of the study. We have only observed transfer in autoclaved sediments. If non-sterile sediments are to be used, there will be a high level of resistance to kanamycin and streptomycin amongst the indigenous flora. Hopefully, colony hybridization would identify transformants.

Plasmid transfer to the indigenous flora by natural transformation

The transfer of plasmid DNA to the indigenous marine flora presents several challenges not encountered when using cultivated recipients. Natural plasmid transformation is an infrequently occurring process even in defined cultures where the physiology of competence development is understood. The proportion of the population which is naturally competent is unknown for natural samples, as well as the proportion in which an IncQ/P4 plasmid can replicate. Because natural habitats rarely have more than 10^6 cells/ml, it is necessary to concentrate the microbial populations from water column samples. A second approach we have successfully used is to investigate environments where the microbial population has been concentrated by marine invertebrates, such as the tissues of filter feeders such as sponges, the guts of deposit-feeders such as holothurians (sea cucumbers), and the mucus of filter feeders such as scleratinian corals. Both approaches have yielded transfer to the indigenous flora in certain cases.

Concentration of microbial populations from water column samples

Microbial populations from 20 to 100 liters of seawater are concentrated using a Membrex Benchmark Rotary Biofiltration device fitted with a 400 cm^2 100 kd filter set up in the recirculation configuration [13]. The concentrated cell suspension (termed the retentate) is typically 35 to 50 ml, with an efficiency of recovery of ~80%.

Concentration of microbial populations from sponge tissues

1. Sponge tissue samples (10 g wet wt.) are cut into 2 cm^3 pieces with a sterile scalpel.
2. The sponge tissue pieces are homogenized in 50 ml sterile ASWJP in a model 909-1 Beadbeater (230 ml capacity; Biospecs Products, Bartlesville, OK) in an ice bath jacket for 5 min.

3. Sponge tissue pieces are removed by centrifugation at 800 × g for 1 min.

Concentration of microbial populations from coral mucus
1. Coral mucus is collected by SCUBA divers or snorkelers using 60 ml syringes with no needles. The mucus/seawater mixture is collected by drawing slowly on the plunger while moving the syringe orifice across the coral surface.
2. One liter of mucus is further concentrated to 35–50 ml by vortex flow filtration using a Benchmark Rotary Biofiltration device [13].

Concentration of microbial populations from holothurian guts
1. Holothurians (sea cucumbers) can be collected from shallow subtropical bays such as Florida Bay (check permit requirements with federal, state, and local authorities before collecting!).
2. Using a sterile scalpel, dissect cucumber longitudinally and remove gut intact. The gut contents (primarily sediments) are exuded into a sterile 60 ml centrifuge tube.
3. Extract 20 cm^3 of gut content material with an equal volume of ASWJP by vortexing vigorously.

Transformation assays using indigenous marine bacteria as recipients
1. Prepare plasmid multimers (pQSR50) as above.
2. Filter 5 ml of bacterial suspension (Membrex retentate, sponge extract, or gut content extract) through a sterile 47 mm 0.2 μm Nuclepore filter.
3. Place filter, cell spot side up, on an ASWJP+PY agar plate.
4. Add 4 μg plasmid multimers in 100 μl of 4.2 mM $MgCl_2$ by carefully spreading the DNA over the cell spot.
5. It is critical to have a control treatment consisting of five ml bacterial suspension filtered similarly and overlaid with calf thymus DNA.
6. Incubate the desired length of time (usually 16 to 20 h) at 25–30°C.
7. Resuspend cells by placing filter in 10 ml ASWJP+PY in a 125 ml sterile Erlenmeyer flask and shaking on a gyrotatory shaker at 150 r.p.m. for 1 h at room temperature.
8. Serially dilute and plate on non-selective media to enumerate

total CFU and on ASWJP+PY containing 500 µg/ml kanamycin, 1 mg/ml streptomycin, and 5×10^{-6} M amphotericin B to enumerate potential transformants.

9. Perform colony hybridization using the *nptII* probe. Pick hybridizing colonies, grow on selective media, and extract plasmid via a miniprep to verify plasmid acquisition [17].

Notes

There is usually a background level of resistance to kanamycin and streptomycin in most marine microbial communities. Therefore, it is imperative to verify plasmid acquisition, both by colony hybridization of controls and treatment plates and also by selecting individual hybridizing colonies for further study. Miniprep, restriction analysis, and Southern hybridization of these clones may still yield equivocal results. Restriction profiles of pQSR50 are often changed, either from marine bacterial methylation systems or by plasmid rearrangement [10].

Intergeneric plasmid transformation using E. coli donor cells and HfT recipients

Most workers in molecular biology are familiar with artificial transformation of plasmid DNA using *E. coli* cells rendered competent by chemical or physical methods. Such cells are not naturally competent. We have demonstrated transfer of non-conjugative plasmids between *E. coli* donor cells and HfT *Vibrio* recipients by a contact dependant, DNase sensitive process [23]. We feel that such a process may be the major mechanism of transformation in marine and aquatic environments, because of the lability of dissolved ("free") DNA [21].

Strains and plasmids

Table 1 shows the plasmids and *E. coli* strains used as plasmid donors. All *E. coli* strains are grown in LB medium [17] supplemented with the various antibiotics (usually kanamycin and streptomycin at 50 and 25 µg/ml, respectively). *Vibrio* JT-1 is a double antibiotic resistant chromosomal mutant of *Vibrio* WJT-1C and serves as plasmid recipient. It is grown on ASWJP+PY media at 28–30 °C in the presence of 500 µg/ml nalidixic acid and 150 µg/ml rifampicin.

Table 1. Plasmids and bacterial strains used in intergeneric, contact-dependent natural plasmid transformation

Strain or plasmid	Relevant characteristics	Source
Plasmids		
r1162	su^{r}str^{r}mob$^+$	[18]
pQSR50	r1162::Tn5 km^{r}str^r	[19]
pLV1013	km^{r}str^r *xylE* c1857	[31]
E. coli donor strains		
RM1259(pQSR50)	MV10 K12 C600 km^r str^r	[19]
RM1208(r1162)	MV10 K12 C600 str^r	[19]
ED8564(pLV1013)	lac$^-$ met$^-$ thi$^-$ hsdr^-_m −r^-_k km^r str^r xylE	[31]
Recipient strain		
Vibrio JT-1	nal^{r}rif^r	[23]

Preparation of donor and recipient cells

1. *E. coli* donor and recipient cells are grown under conditions described above overnight in gyrotatory shaking incubators (~150–200 r.p.m.).
2. Cells are harvested at 10,000 × g for 10 min at 20 °C and washed twice in growth media lacking antibiotics (ASWJP+PY for JT-1 and LB for *E. coli* strains).
3. Cells are resuspended in growth media lacking antibiotics and used immediately in transformation assays.

Protocol for intergeneric transformation

Broth matings

1. *E. coli* donor cells (1 ml) are mixed with 1 ml JT-1 recipient in a sterile 15 ml conical centrifuge tube.
2. A DNase control is set up identically, with the addition of 200 units of DNaseI.
3. The tubes are incubated at 30 °C statically for the desired length of time (up to 16 h).
4. Aliquots of each DNase control and treatment tube are serially diluted and plated on the appropriate media. For example, for crosses involving *E. coli* RM1259 as donor and *Vibrio* JT-1 as recipient, aliquots are plated on LB plus kanamycin and streptomycin to enumerate donors, ASWJP+PY plus nalidixic acid and rifampicin to enumerate recipients, and ASWJP+PY plus kanamycin, strepto-

mycin, nalidixic acid, and rifampicin (KSNR) to enumerate transformants.
5. Plasmid acquisition is verified by colony hybridization of at least
one KSNR plate as described above.

Filter matings

1. One ml of donor cells is mixed with 1 ml of recipient cells and immediately filtered onto a sterile 47 mm, 0.2 µm Nuclepore filter.
2. A DNase control is set up similarly, except that after filtration, 200
units of DNaseI is dribbled over the cell spot.
3. The filter is incubated cell side up on an ASWJP+PY plate for the
desired length of time (usually 16 h) at 28–30 °C.
4. The filter is removed from the plate and added to 5–10 ml
ASWJP+PY and vigorously vortexed for 2 min to resuspend cells.
Aliquots are diluted and plated as for broth matings, above.

Notes

− Ad 1 It is recommended to pre-wet the filter by filtering 5 ml of medium prior to
filtration of the cell mixture.
− Ad 2. The DNase added to the control plate completely inhibits transformation. If
'transformants' are detected in the DNase control they would result from
spontaneous mutation or some other form of gene transfer such as conjugation.

Matings in seawater

1. One ml of donor and recipient cell suspensions prepared as described above are added to 4 to 25 ml of seawater in a sterile 60
ml disposable centrifuge tube.
2. If nutrients are to be added, concentrated peptone and yeast extract are added to a final concentration of 0.1 and 0.02 mg/ml,
respectively.
3. Incubations are at room temperature or the desired temperature
for the desired length of time (usually 16 h).
4. Aliquots of the incubations are serially diluted and plated on the
media described above.

Notes

Again, seawater to be used can be nonsterile, sterile filtered, or autoclaved, sterile
filtered. The volumes can be scaled up and the experiments performed *in situ* using
Fenwall Gas Permeable Tissue Culture Bags (Fenwall Scientific). It may be difficult to
enumerate the donor population on LB containing kanamycin and streptomycin

because of the high level of indigenous resistant organisms found in some environments. However, in our experience, there are usually no indigenous organisms that can grow on the KSNR plates.

Matings in sediment

1. Prepare donor cells and recipient cells as above but resuspend in 1/10 volume of grown media.
2. Add 3 cm^3 sediment to a sterile disposable 15 ml conical centrifuge tube.
3. Add an additional 1.5 ml ASWJP to cover the sediment.
4. Incubate for the desired time (usually 16 h) at the desired temperature.
5. Add 3.5 ml ASWJP and vortex vigorously for 2 min.
6. Dilute and plate as for broth matings.

References

1. Albritton WL, Setlow JK, Slaney, L (1982) Transfer of *Haemophilus influenzae* chromosomal genes by cell-to-cell contact. J Bacteriol 152: 1066–1070.
2. Bale MJ, Fry JC, Day MJ (1987) Plasmid transfer between strains of *Pseudomonas aeruginosa* on membrane filters attached to river stones. J Gen Microbiol 133: 3099–3107.
3. Bale MJ, Day MJ, Fry JC (1988) Novel method for studying plasmid transfer in undisturbed river epilithon. Appl Environ Microbiol 54: 2756–2758.
4. De Vos WM, Venema G (1981) Fate of plasmid DNA in transformation of *Bacillus subtilis* protoplasts. Mol Gen Genet 182: 39–43.
5. De Vos WM, Venema G (1982) Transformation of *Bacillus subtilis* competent cells: identification of a protein involved in recombination. Mol Gen Genet 187: 439–445.
6. Doran JL, Bingle WH, Roy KL, Hiratsuka K, Page WJ (1987) Plasmid transformation of *Azotobacter vinelandii* OP. J Gen Microbiol 113: 2059–2072.
7. Dubnau D (1991) Genetic competence in *Bacillus subtilis*. Microbiol Rev 55: 395–424.
8. Graham JB, Istock CA (1978) Genetic exchange in *Bacillus subtilis* in soil. Mol Gen Genet 166: 287–290.
9. Frischer ME, Thurmond JM, Paul JH (1990) Natural plasmid transformation in a high-frequency-of-transformation marine *Vibrio* strain. Appl Environ Microbiol 56: 3439–3444.
10. Frischer ME, Stewart GJ, Paul JH (1994) Plasmid transfer to indigenous marine bacterial populations by natural transformation. FEMS Microb Ecol 15: 127–136.
11. Fry JC (1989) Analysis of variance and regression in aquatic bacteriology. Binary 1: 83–88.
12. Jeffrey WH, Paul JH, Stewart GJ (1990) Natural transformation of a marine *Vibrio* species by plasmid DNA. Microbial Ecol 19: 259–269
13. Jiang SC, Thurmond JM, Pichard SL, Paul JH (1992) Concentration of microbial populations from aquatic environments by vortex flow filtration. Mar Ecol Progr Ser 80: 101–107.
14. Juni E (1972) Interspecies transformation of *Acinetobacter*: Genetic evidence for a ubiquitous genus. J Bacteriol 112: 917–931.

15. Lorenz MG, Wackernagel W (1990) Natural genetic transformation of *Pseudomonas stutzeri* by sand-adsorbed DNA. Arch Microbiol 154: 380–385.
16. Lorenz MG, Aardema BW, Wackernagel W (1988) Highly efficient genetic transformation of *Bacillus subtilis* attached to sand grains. J Gen Microbiol 134: 107–112.
17. Maniatis T, Fritsch EF, Sambrook J (1982) Molecular cloning: a laboratory manual. Cold Spring Harbor Laboratory, Cold Spring Harbor, New York.
18. Meyer R, Hinds M, Brosch M (1982) Properties of R1162, a broad-host-range, high-copy-number plasmid. J Bacteriol 150: 552–562.
19. Meyer R, Laux R, Boch G, Hinds M, Bayly R, Shapiro JA (1982) Broad-host-range IncP-4 plasmid R1162: effects of deletions and insertions on plasmid maintenance and host range. J Bacteriol 152: 140–150.
20. Paul JH (1982) Use of Hoechst dyes 33258 and 33342 for enumeration of attached and planktonic bacteria. Appl Environ Microbiol 43: 939–944.
21. Paul JH, Jeffrey WH, David AW, DeFlaun MF, Cazares LH (1989) Turnover of extracellular DNA in eutrophic and oligotrophic freshwater environments of southwest Florida. Appl Environ Microbiol 55: 1823–1828.
22. Paul JH, Frischer ME, Thurmond JM (1991) Gene transfer in marine water column and sediment microcosms by natural plasmid transformation. Appl Environ Microbiol 57: 1509–1515.
23. Paul JH, Frischer ME, Thurmond JM (1992) Intergenic natural plasmid transformation between *E. coli* and a marine *Vibrio* species. Mol Ecol 1: 37–46.
24. Rochelle PA, Day MJ, Fry JC (1988) Occurrence, transfer and mobilization in epilithic strains of *Acinetobacter* of mercury-resistance plasmids capable of transformation. J Gen Microbiol 134: 2933–2941.
25. Rochelle PA, Fry JC, Day MJ (1989) Plasmid transfer between *Pseudomonas spp.* within epilithic films in a rotating disc microcosm. FEMS Microbiol Ecol 62: 127–136.
26. Sokal RR, Rohlf FJ (1981) Biometry. Freeman, San Francisco, USA.
27. Stewart GJ, Carlson CA (1986) The biology of natural transformation. Ann Rev Microbiol 40: 211–235.
28. Stewart GJ, Carlson CA, Ingraham JL (1983) Evidence for an active role of donor cells in natural transformation of *Pseudomonas stutzeri*. J Bacteriol 156: 30–35.
29. Stewart GJ, Sinigalliano CD, Garko KA (1991) Binding of exogenous DNA to marine sediments and the effect of DNA/sediment binding on natural transformation of *Pseudomonas stutzeri* strain ZoBell in sediment columns. FEMS Microbiol Ecol 85: 1–8.
30. Williams HG, Day MJ, Fry JC (1992) Natural transformation on agar and in river epilithon. In: Gauthier MJ (ed) Gene Transfers and Environment, pp. 69–76. Springer Verlag, Berlin Heidelberg.
31. Winstanley C, Morgan JAW, Pickup RW, Jones JG, Saunders JR. (1989) Differential regulation of lambda p_L and p_R promoters by a cI repressor in a broad-host-range thermoregulated plasmid marker system. Appl Environ Microbiol 55: 771–777.

Molecular Microbial Ecology Manual **5.2.3**: 1-10, 1995.
© 1995 *Kluwer Academic Publishers.*

Detection of gene transfer in the environment: Conjugation in soil

ERIC SMIT and JAN DIRK VAN ELSAS

Institute for Plant Protection (IPO-DLO), P.O. Box 9060, 6700 GW Wageningen, The Netherlands

Introduction

Gene transfer in soil by conjugation can be studied in several different ways: a) by introducing donor (with a plasmid) and recipient bacteria into soil and check for plasmid transfer by selecting for the recipient and the plasmid, b) by introducing only the donor bacterium into the soil and investigate transfer to indigenous bacteria using a specific donor counter-selection method, c) by doing tri-parental mobilization and exogenous plasmid isolation assays. This chapter will mainly focus on *in situ* transfer experiments in soil involving donor and recipient or donor and indigenous bacteria. The exogenous plasmid isolation and tri-parental mobilization methods involve the use of a recipient or a strain with a mobilizable plasmid to detect plasmids in the soil bacterial community and will be described elsewhere in this manual.

Self-transmissible plasmids can mediate their own transfer since they carry both an origin of transfer (*oriT*) and transfer genes (*tra*), which are necessary for the formation of mating pairs, nicking of the plasmid and transfer of the DNA [1]. The plasmids are grouped in socalled incompatibility groups and plasmids of some groups (IncP, IncN, IncW and IncX) can mobilize non-conjugative plasmids, carrying *oriT* to a recipient [17]. Recently, transfer in the opposite direction was detected, e.g. certain plasmids were able to mobilize a plasmid into the (donor) strain which contained the self-transmissible plasmid. This process was called retro-transfer [8]. The occurrence of retro-transfer might have implications for transfer of recombinant DNA from introduced bacteria to indigenous ones.

There is ample literature on gene transfer between bacteria via conjugation in soil [2, 3, 6, 22]. Successful conjugal transfer depends on close contact between donor and recipient cells under favourable conditions of temperature, pH, and nutrient availability. Experiments in which donor bacteria carrying transferable selectable markers, either or not accompanied by recipients, were added to soil, have provided more direct evidence for the occurrence of conjugal gene transfer [10, 12, 14, 16, 18, 19, 21]. Conjugal transfer in soil has been shown to be affected by many soil parameters such as temperature, pH, organic matter content, clay content, water or nutrient availability. Briefly, the presence of nutrients, montmorillinitic clay minerals, plant roots and the absence of a competing microflora have been shown to promote conjugal plasmid transfer. Transconjugants could be detected after the addition of nutrients or after soil sterilization [10, 18, 20]. Similarly, the presence of montmorillinite clay [5] or bentonite clay [20] resulted in higher numbers of transconjugants. Plant roots also resulted in enhanced numbers of transconjugants both in donor to recipient and donor to indigenous bacteria transfer experiments [11].

Plasmid transfer between donor and recipient bacteria

The procedure to investigate plasmid transfer between donor and recipient is relatively straightforward, however there are a few points which have to be checked. For instance plasmid transfer could occur on the trans-conjugant selective plates instead of in the soil [12]. Apparently this socalled plate mating takes place depending on which donor or recipient bacteria are used. Before starting an experiment one should test if plate mating can occur (see procedures), in addition it should be controlled in the experiment. Using a *Pseudomonas fluorescens* strain, plate mating could be inhibited by using nalidixic acid as one of the antibiotics for donor counterselection (see also "plate mating assay") [12]. Mutation controls of both donor and recipient should also be done to check for the occurrence of spontaneously resistant bacteria growing on the transconjugant selective plates.

Plasmid transfer from donor to indigenous bacteria

One of the major problems in studying plasmid transfer to indigenous bacteria is donor counterselection. Before the transconjugants can be plated out the donor should be eliminated, because since the donor has the same plasmid and resistances it will overgrow the plates. There are several methods to eliminate the donor strain: a) the use of markers that are not expressed in the donor [18], b) the use of an auxotrophic donor; c) the use of a donor with an inducible host-killing gene [9]; d) the use of a donor which does not survive in the soil [4]; e) the use of a bacteriophage to lyse the donor [14]. In the procedures a protocol will be given for phage mediated donor counterselection (see Fig. 1) since the other methods have not yet proven to be applicable. For instance, in plasmid transfer studies in the epilithon the application of an auxotrophic donor proved to be inadequate [7].

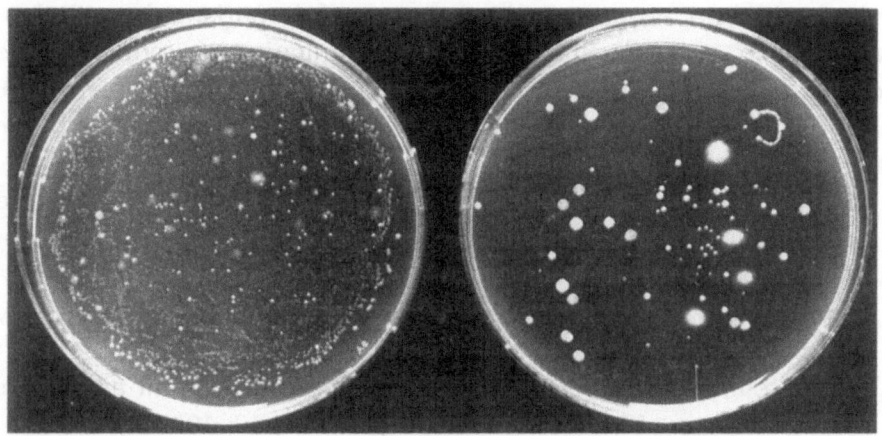

Figure 1. A 100-fold diluted soil suspension containing 10^7 *Pseudomonas fluorescens* (RP4p) donor cells was plated onto LB agar containing ampicillin and tetracycline (50 mg/l). Left plate: no phage treatment prior to plating; right plate: sample incubated with phage lysate (10^9 pfu/ml, 20 min) prior to plating. The left plate is completely covered with small donor colonies, the right plate shows well developed colonies of potential indigenous trans-conjugants[13].

In experiments to assess plasmid transfer in soil, many colonies are routinely found on transconjugant selective plates. These colonies can be either bacteria from soil naturally resistant to the marker genes used, or donor cells which escaped the counterselection or real transconjugants. To distinguish the first from the latter two, colony filter hybridization using a specific DNA sequence, present on (or cloned into) the plasmid, as a probe can be applied [13] (see also Chapter Jacobsen Chapter 1.4.1). The second step is to differentiate between donor and transconjugants. An extra marker in the donor, like for instance resistance to rifampicin, can facilitate this differentation.

Procedures

Selection of markers and plasmids

In order to study plasmid transfer from donor to recipient, the appropriate strains and plasmids should be chosen. Antibiotic resistance markers are particularly suitable for donor counterselection and to select for the plasmid. Moreover, the use of a combination of resistance is important to suppress the soil bacteria from growing on the plates. Some antibiotics such as rifampicin, nalidixic acid and kanamycin and tetracyline are more useful markers in soil than others such as ampicillin or chloramphenicol with regard to the occurrence of high levels of natural resistance (See Chapter 6.1.6: Antibiotic resistance genes as markers for soil studies). An antibiotic resistant mutant to be used as recipient can be made as follows:

1. Culture the potential recipient strain to maximal cell density in the appropriate medium.
2. Wash the strain 2× in 0.85% NaCl solution, and take up in such a volume of 0.85% NaCl that the cell density will be approximately 10^9 cells per ml.
3. Plate several dilutions onto agar plates with different antibiotics using a range of concentrations (e.g. 10, 25, 50, 100 mg/l).
4. Pick colonies from the highest concentration and streak them onto fresh plates and make sure that they are not contaminants by, for instance, using the BIOLOG (Hayward, CA, USA) or API (Montalieu, Vercieu, France) identification systems.
5. When resistance against higher concentrations (or other antibiotics) is required, the colonies can be cultured again and plated on plates with higher antibiotic concentrations.

It is advisable to select for a double mutant resistant against two antibiotics as recipient, since the use of two antibiotics reduces the number of spontaneous resistant mutants of the donor in the plasmid transfer experiments.

Filter mating assay

A filter mating assay should be done to test if, and with which frequency, the plasmid will be transferred to the recipient. Transfer of the plasmid could be impaired due to restriction activity in the recipient, or detection of transfer could fail due to lack of expression of the marker genes on the plasmid in the recipient.

1. Check plasmid transfer by adding 0.1 ml of washed cultures of donor and recipient on a 0.22 μm bacterial filter (e.g. MF membrane filter, Millipore, Bedford, MA, USA) on LB (or other appropriate) agar. Add the strains also separately on filters as controls on mutation.
2. Incubate overnight at a temperature adequate for growth of the strain with the lowest optimal temperature (e.g. 27 °C–37 °C).

3. Put the filters into tubes containing 5 ml of 0.85% NaCl solution and vortex to dislodge the bacteria.
4. Plate appropriate dilutions on media with antibiotics to enumerate donor, recipient and transconjugant cells.

Plate mating assay

Depending on the strains used, the plasmid can be transferred after the cells are plated on the transconjugant selective medium. This might result in an overestimation of plasmid transfer in environmental samples and therefore the occurrence of plate mating should always be checked in a control experiment [12].
1. Washed suspensions of cultures of donor and recipients strains are placed on separate filters (for details see above procedure) and incubated.
2. Filters are sampled separately and the dilutions are put together on the transconjugant selective plates and plated out immediately.
3. Colonies appearing are a result of plate mating.

   In most cases plate mating can be prevented by using the antibiotic nalidixic acid in combination with a resistant recipient. Nalidixic acid prevents plasmid transfer by inhibiting gyrase activity [12].

Plasmid transfer between introduced donor and recipient bacteria in soil

The following protocol is routinely used in our laboratory to investigate plasmid transfer between donor and recipient *Pseudomonas fluorescens* strains in soil [13]. The exact medium and antibiotics used depends on the strains, plasmids and markers that are going to be used.
1. Collect soil samples and dry them to 10% below the water content at field capacity (pF 2).
2. Add washed cell suspensions of the donor culture to the soil in such a volume that half of the lost water volume is replaced, and that the final cell numbers will be between 10^7 and 10^8 per gram soil.
3. Mix thoroughly and wait approximately 1 hour before adding the

recipient strain in the same volume and numbers as the donor.

4. Plant seedlings or make other additions to the soil samples and incubate the soil samples in a humid chamber to prevent the soil from drying.

5. Take (10 g) soil samples, put them into Erlenmeier flasks containing 95 ml of 0.1% sodiumpyrophophate solution and 10 g of gravel (dia. 2–4 mm), and shake the Erlenmeyer flasks for 10 min at 190 rpm in a gyrotory shaker.

6. Make 10-fold dilutions of the soil slurry and plate onto the appropriate medium with the appropriate antibiotics to enumerate donor, recipient and transconjugant cell numbers.

7. Incubate the plates for 2–5 days (check regularly) at 27 °C–37 °C depending on the strains. Carefully compare numbers of cfu on transconjugant selective plates with those on the controls for mutation and plate mating.

Take care to plate low (or undiluted) samples for the transconjugant enumeration, since transconjugant cell numbers will generally be much lower then donor and recipient counts.

Plasmid transfer to indigenous bacteria in soil using a bacteriophage for donor counterselection

The following protocol has been successfully used in studies on the transfer of different plasmids from *Pseudomons fluorescens* to indigenous bacteria in soil and rhizosphere [12, 14]. Isolation of a specific phage and preparation of a crude phage lysate is described in ref 14.

1. The soil samples are inoculated with the donor strain as described above in steps 1–4.

2. Soil samples (10 g) are added to Erlenmeyer flasks containing 95 ml of sterilized 0.1% sodiumpyrophosphate solution and 10 g of gravel, and the Erlenmeyer flasks are shaken for 10 min at 190 rpm in a gyrotory shaker.

3. Ten-fold dilutions are made for the enumeration of donor and the total number of culturable bacteria.

4. Sub-samples (0.5 ml) from the Erlenmeyer flasks are pipetted into Eppendorf tubes and mixed, by inverting the tubes, with 0.5 ml crude phage lysate (10^9 pfu/ml).

5. Tubes are incubated for 20 min at room temperature.

6. Samples of 0.1–0.2 ml are plated onto transconjugant selective plates.
7. Check the transconjugants for the presence of the plasmid by filter hybridization using a specific probe.
8. Check the transconjugants which reacted positively with the probe for not being the donor strain using the antibiotic resistance mutation of the donor or other characteristics.
9. Determine the species name of the transconjugants (using e.g. BIOLOG, Hayward, CA, USA) and confirm results by doing plasmid isolations or filter matings to a recipient.

Application of the method

Employing the procedure described above, transfer of the self-transmissible IncP plasmid RP4p and mobilization of the IncQ plasmid pSKTG between donor and recipient and donor and indigenous bacteria in soil microcosms planted with wheat plants was investigated [15]. As donor strains, *Pseudomonas fluorescens* R2f with RP4p, pSKTG or both was used, as recipient a *Pseudomonas fluorescens* R2f strain resistant to rifampicin and nalidixic acid was used. Plasmid transfer from donor to recipient was studied in filtermatings (Table 1) and in sterile soil (Table 2), and transfer to indigenous bacteria was investigated in soil microcosms planted with wheat plants (Table 3). In filtermatings the plasmid pSKTG was mobilized in the presence of RP4p, with a frequency similar to that of RP4p (approx. 10^{-2}), irrespective of the fact if both plasmids resided in the same (B) or in separate donor strains (A). In sterile soil, overall transfer frequencies were 100-fold lower, probably because the soil reduced cell-to-cell contact. Moreover, the mobilization frequency of pSKTG by RP4p was 100-fold lower when the plasmids were initially present in separate strains (A) as compared to when both plasmids were present in one strain (B). Comparable results were obtained in experiments aimed at assessing plasmid transfer to indigenous bacteria in soil and wheat rhizosphere (Table 3). The plasmid pSKTG was mobilized to a higher number of indigenous bacteria when RP4p was initially present in the same strain (B) as compared to the separate introduction (A).

Table 1. Transfer of plasmids RP4p and pSKTG between donor and recipient strains of *P. fluorescens* on filters on agar plates (4h) [from 15]*

| Tr | Log cfu/ml | | | | | | freq/d | | freq/r | |
| | Donor | | | Rec | Tran | | | | | |
	RP4p	pSKTG	both		RP4p	pSKTG	RP4p	pSKTG	RP4p	pSKTG
A	7.6	6.7	–	7.8	6.0^a	5.1^b	2.8×10^{-2}	2.3×10^{-2}	1.2×10^{-2}	1.9×10^{-3}
B	–	–	7.9	8.2	ND	6.2^c	ND	2.0×10^{-2}	ND	1.0×10^{-3}
C	–	8.8	–	8.5	–	[0]	–	[X]	–	[X]

* Tr = treatment, A: *P. fluorescens* R2f(RP4p), *P. fluorescens* R2f (pSKTG); B: *P. fluorescens* R2f (RP4p, pSKTG); C: *P. fluorescens* R2f (pSKTG); Rec: recipient (*P. fluorescens* R2fN) cfu; Tran: transconjugant cfu; freq/d: frequency per donor; freq/r: frequency per recipient [0]: below 10 cfu/ml; ND: not done; –: not relevant; [X]: below 10^{-8} ^a is significantly different from ^b and ^b is significantly different from ^c (P,0.05)

Table 2. Transfer of RP4p and mobilization of pSKTG between *P fluorescens* donor and recipient strains in sterile soil after 2 days* [from 15]

| Tr | Log cfu/g soil | | | | | | Freq | |
| | Donor | | | Rec | Tran | | | |
	RP4p	pSKTG	both		RP4p	pSKTG	RP4p	pSKTG
A	8.1	7.7	–	8.8	4.9^a	2.2^b	6.4×10^{-4}	2.7×10^{-6}
B	–	–	7.8	8.7	4.1^c	4.2^c	2.0×10^{-4}	2.0×10^{-4}
C	–	7.3	–	8.9	–	[0]	4.2×10^{-4}	–
D	7.9	–	–	8.8	4.5^a	–	–	–
E	–	–	–	8.8	[0]	[0]	–	–

* Explanation: Rec: *P. fluorescens* R2fN cfu; Tran: transconjugant cfu; both: plasmid RP4p and pSKTG present; Log numbers of cells per g of dry soil; Tr: treatment, A: *P. fluorescens* R2f(RP4p), *P. fluorescens* R2f(pSKTG); B: *P. fluorescens* R2f(RP4p, pSKTG); C: *P. fluorescens* R2f(pSKTG); D: *P. fluorescens* R2f(RP4p); E: no donor, *P. fluorescens* R2fN was included in all treatments, from 15 log cfu counts on day 0 (3 h after inoculation) were A: RP4p 7.8, pSKTG 7.7, rec. 8.5; B: both 8.2, rec. 8.4; C: pSKTG 7.6, rec. 8.5; D: RP4p 7.6, rec. 8.6; E: rec. 8.5 ^a is significantly different from ^b and ^c (P < 0.05); same letters are not significantly different (P < 0.05).

Table 3. Log numbers of indigenous transconjugants containing RP4p or pSKTG and donor cells 7 days after introduction of *P. fluorescens* R$_2$f with these plasmids in non-sterile Ede loamy sand planted to wheat in soil microcosms* [from 15]

Treatment		Log cfu/g soil				
		Donor			Indig. transconjugant	
		RP4p	pSKTG	both	RP4p	pSKTG
A	R	6.3	6.5	–	2.7 [a]	2.2 [a]
	B	6.4	6.4	–	[0]	2.1 [a]
B	R	–	–	5.0	3.0 [c]	4.2 [d]
	B	–	–	4.0	2.2 [a]	3.0 [c]
C	R	–	6.5	–	–	[0]
	B	–	6.4	–	–	[0]
D	R	6.3	–	–	4.3 [e]	–
	B	6.0	–	–	1.7 [f]	–
E	R	–	–	–	–	[0]
	B	–	–	–	–	[0]

* R: rhizosphere, B: bulk soil; Treatment A: *P. fluorescens* R2fR(RP4p), *P. fluorescens* R2fR(pSKTG); B: *P. fluorescens* R2fR(RP4p, pSKTG); C: *P. fluorescens* R2fR(pSKTG); D: *P. fluorescens* R2fR(RP4p); E: no strain added; log cfu/g soil on day 0: A: Donor R2f(RP4p) 7.7, donor R2f(pSKTG) 7.9; B: Donor R2f(RP4p; pSKTG) 6.7; C: Donor R2f(pSKTG) 7.5; D: Donor R2f(RP4p) 7.1; E: no donor added; [0] = below detection limit (approximately 10^2 cfu/g soil).
Values labeled with different letters were significantly different from each other ($P < 0.05$)

Acknowledgements

This work was sponsered by a grant from the EC-BIOTECH programme (BIO2CT-920921). We thank A.C. Wolters for critically reading the manuscript.

References

1. Clewell DB (1993) Bacterial Conjugation. New York: Plenum Publishing Corporation.
2. Fry JC, Day MJ (eds) (1990) Bacterial genetics in natural environments. Chapman and Hall, London.
3. Gauthier MJ (ed) (1992) Gene transfers and environment. Springer-Verlag, Berlin.
4. Henschke RB and Schmidt FRJ (1990) Plasmid mobilization from genetically engineered bacteria to members of the indigenous soil microflora. Curr Microbiol 20: 105–110.
5. Krasovsky VN, Stotzky G (1987) Conjugation and genetic recombination in *Escherichia coli* in sterile and non-sterile soil. Soil Biol Biochem 19: 631–638.
6. Levy SB, Miller R (eds) (1989) Gene Transfer in the Environment. McGraw-Hill, New York
7. Marchesi JR, Day MJ, Fry J (1993) Transfer of a recombinant broad host range plasmid to indigenous bacteria in microcosm experiments to simulate transfer to the epilithon. Abstract Book Bageco-4.
8. Mergeay M, Lejeune P, Sadouk A, Gerits J, Fabry L (1987) Shuttle transfer (or

retrotransfer) of chromosomal markers mediated by plasmid pULB113. Mol Gen Genet 209: 61–70.

9. Molin S, Klemm P, Poulsen LK, Gerdes K, Andersson P (1987) Conditional suicide system for containment of bacteria and plasmids. Biotechnology 5: 1315–1318.

10. Richaume A, Angle JS, Sadowski MJ (1989) Influence of soil variables on in situ plasmid transfer from *Escherichia coli* to *Rhizobium fredii*. Appl Environ Microbiol 55: 1730–1734.

11. Richaume A, Smit E, Faurie G, Van Elsas JD (1992) Influence of soil type on the transfer of RP4p from *Pseudomonas fluorescens* to indigenous bacteria. FEMS Microbiol Ecol 101: 281–292.

12. Smit E, Van Elsas JD (1990) Determination of plasmid transfer frequency in soil: consequences of bacterial mating on selective agar media. Curr Microbiol 21: 151–157.

13. Smit E, Van Elsas (1992) Methods for studying conjugative gene transfer in soil. In Wellington EMH, Van Elsas JD, eds. Genetic interactions among microorganisms in the natural environment pp. 113–126. Pergamon Press, London.

14. Smit E, Van Elsas JD, Van Veen JA, De Vos WM (1991) Detection of plasmid transfer from *Pseudomonas fluorescens* to indigenous bacteria in soil using bacteriophage ΦR2f for donor counterselection. Appl Environ Microbiol 57: 3482–3488.

15. Smit E, Venne D, Van Elsas JD (1993) Effects of co-transfer and retrotransfer on the mobilization of a genetically engineered IncO plasmid between bacteria on filters and in soil. Appl Environ Microbiol 59: 2257–2263.

16. Stotzky G (1989) Gene transfer among bacteria in soil. In: Levy SB, Miller RV (eds), Gene Transfer in the Natural Environment, pp. 165–222, McGraw-Hill, New York.

17. Thomas CM (1989) Promiscuous plasmids of gram-negative bacteria. London: Academic Press.

18. Top E, Mergeay M, Springael D, Verstraete W (1990) Gene escape model: transfer of heavy metal resistance genes from *Escherichia coli* to *Alcaligenes eutrophus* on agar plates and in soil samples. Appl Environ Microbiol 56: 2471–2479.

19. Van Elsas JD, Govaert JM, Van Veen JA (1987) Transfer of plasmid pFT30 between bacilli in soil as influenced by bacterial population dynamics and soil conditions. Soil Biol Biochem 19: 639–647.

20. Van Elsas JD, Trevors JT, Starodub ME (1988) Bacterial conjugation between pseudomonads in the rhizosphere of wheat. FEMS Microbiol Ecol 53: 299–306.

21. Wellington EMH, Cresswell N, Saunders VA (1990) Growth and survival of Streptomycete inoculants and extent of plasmid transfer in sterile and non-sterile soil. Appl Environ Microbiol 56: 1413–1419.

22. Wellington EMH and Van Elsas JD (eds) (1992) Gene Transfer Between Microorganisms in the Natural Environment. London: Pergamon Press.

Molecular Microbial Ecology Manual **5.3.2**: 1–12, 1995
© 1995 *Kluwer Academic Publishers.*

Phage ecology and genetic exchange in soil

PAUL R. HERRON

*School of Biological Sciences, University College of Swansea, Singleton Park, Swansea,
SA2 8PP, Wales, UK*

Introduction

The ecology of bacteriophages in soil is poorly understood with respect to
the distribution, activity and life cycle of their hosts [29]. Perhaps the main
reason for this is simply one of enumeration. Whilst it is relatively easy to
detect a given phage if one knows its propagative host, it is not possible
to observe those phages for which hosts are not known. It is estimated that
at present only a small fraction of bacteria in soil can be cultivated and
therefore the absolute numbers of bacteriophages within a soil sample
remain hidden. Recent studies in aquatic systems have shown using direct
counting procedures that the traditional phage assay may well
underestimate phage numbers by as much as two orders of magnitude [15].
Studies on phage ecology are further complicated by the fact that their
activity is dependent on the resources accessible to them, which, in the case
of obligate parasites, are the availability of active hosts. In soil, microbial
activity and abundance is discontinuous in space and time. Phage activity
is dependent on this and must therefore mirror this by rapid replication in
restricted growth periods and sites in order to maintain sufficient numbers
of infective phages in the soil environment for the survival of the
population. The concentration of active hosts in microsites may well
increase the likelihood of phage-host encounters, at least in the short term.
Unfortunately no information is available on the spatial distribution of
bacteriophages in soil and it is likely that counts derived from a given mass
of soil underestimate the concentration and significance in such sites.
Phages may be involved in either lysogenic or virulent associations with
their hosts in soil, however current detection procedures provide no
indication of total numbers in the environment as these methods rely on
the virulent reaction of the virus with its host. The importance of this point
was illustrated by Reanney [18], who estimated that free phages of *Bacillus
stearothermophilus* in soil, when detected by phage extraction and assay
techniques, comprised only 20% of the population, the remainder being
present as lysogens. Thus, even when working with a known host, the

numbers of free phages infecting that host in a soil sample can only be taken as an indication of the total number of bacteriophages present. If it is true that 80% of temperate phages are present in soil as prophages, it perhaps serves as an indicator of how active bacteriophages actually occur in soil. Due to the hostile conditions that can prevail in the soil environment, the ability of a bacteriophage to lysogenise its host obviously confers a greater capacity to survive environmental stress such as: proteolytic enzymes, abiotic stress etc. If one considers the evolution of bacteriophages through the reshuffling of gene modules [16,21,22,23], then the lysogenic state must be one of the most important states for the interaction of different phage genomes and their subsequent recombination into new genomes in the environment.

The relevance of phage mediated genetic exchange is unclear in the terrestrial environment, due to the lack of information on its occurrence [5,7–9,13,14,30]. There are two mechanisms by which bacteriophages can mediate bacterial gene transfer, firstly through phage conversion and secondly through transduction.

Phage conversion

DNA transfer by phage conversion can be defined as a change in the phenotype of a host, brought about by a gene within the genome of a temperate phage. This has been the most studied aspect of phage mediated gene transfer in terrestrial systems [7–9,13,14,30].

Transduction

Phage mediated genetic exchange by transduction may take one of three forms. Firstly, generalised transduction, when a defective phage, containing only donor bacterial DNA, infects a recipient host cell and its DNA subsequently integrates into the recipient chromosome or is maintained as a autonomously replicating plasmid. Secondly, specialised transduction, when a phenotypic change in the host is brought about by lysogenisation by a temperate phage; the host gene encoding that phenotypic change being attached to the phage's genome through incorrect excision of the prophage from the donor's chromosome. Thirdly, abortive transduction, where host chromosomal sequences, picked up by the phage, are injected into the recipient and expressed transiently before being diluted out by cell division if the transferred DNA is unable to replicate or recombine in the new host. Despite the current interest in bacterial gene transfer in the environment, transduction has not been extensively studied in soil when compared to transformation and conjugation, with only generalised transduction actually being demonstrated [5].

Experimental approach

Many experiments studying the terrestrial ecology of bacteriophages can be designed. At the simplest level this can be by taking a natural soil and counting specific groups of indigenous microorganisms and bacteriophages under different soil conditions. The progression to more complex gene transfer experiments needs more careful consideration of the approach that should be taken, in terms of the system used to monitor this exchange. Soil systems can be studied in several ways. An important point is that in a completely natural system, phage mediated genetic exchange will probably be a rare event due to its linkage to host activity, and as such x, steps may well have to be taken in order to increase the transfer frequency of these events to detectable level. This can be done by using either autoclaved or gamma irradiated soil to remove the indigenous community, and/or by amending the soil with nutrients to enhance host activity. Experiments involving gene exchange will probably require the use of cloned selective or elective marker genes which means that such experiments will, in all likelihood, be carried out in laboratory-based microcosms due to the restrictions that apply in most countries on the release of genetically engineered microorganisms into the environment. When carrying out any experiments in microbial ecology it is necessary to be aware of the statistical analysis required to obtain reliable data. To this end it is recommended that all microcosms should be studied in triplicate, samples from all microcosms should be taken in triplicate and that all samples should be assayed for phages or hosts in triplicate. If one monitors population dynamics over time the data then lend itself to examination by analysis of variance. This allows means to be separated through such statistical analyses as Tukey's least significant difference [3,17]. It is also recommended that control microcosms are included that contain (where appropriate) uninoculated soil, soil with either phage alone, donor alone or recipient alone to ensure that transductants/convertants are only detected in those microcosms containing the necessary inoculants.

Genetic exchange via phage conversion can be examined in several ways. Firstly, through addition of the phage lysate to the soil system and screening for marker expression in the host population, be it indigenous or inoculated. For example, P22 marked with the gene for bioluminescence (lux), was used to detect low numbers of Salmonella typhimurium by means of phage conversion in soil [26]. If one then goes on to examine the release of such a phage from an inoculated donor and the subsequent infection and lysogenisation of susceptible recipients, further controls must also be included to demonstrate that the phage movement takes place via the outlined mechanism and not through nonphage mediated chromosomal transfer or transposition. Whilst multiplicity of infection (ratio of phages to hosts) is important in optimisation of specialised transduction/phage conversion, it is critical to the detection of generalised transductants. If the ratio is high, generalised transductants may well not be detected due to subsequent lysis by nontransducing phage particles. The multiplicity of infection is easy to control when carrying out experiments involving a transducing phage and a recipient inoculum [5]. However in the case of a temperate donor and recipient it becomes much more difficult to control accurately. Perhaps the most elegant mechanism to avoid this problem was developed by Saye et al. [19,20] in an aquatic system. The recipient was inoculated carrying the transducing phage as a prophage, upon induction and release the phage was able to infect and pick up host DNA from a separate donor strain. This then allowed transducing particles to inject their DNA into unlysed recipients, which, still carrying the prophage, were resistant to lysis by nontransducing phage particles [19,20]. In terms of the numbers of inoculants added to a soil, a balance must be struck between ecologically relevant numbers and the titres necessary for phage infection to take place. It must be remembered that due to the heterogeneity and paucity of mixing of soil systems, a threshold level of both phages and hosts exists below which infection cannot take place; this is estimated to be somewhere between 10^2 and 10^3 colony forming units/g or plaque forming units/g [4,13].

Phage and host selection

Before embarking on a study of terrestrial phage ecology a good deal of care should be put into the selection of both phages and hosts. First, there are many phages in soil, and plaque morphology is a poor characteristic to base the identification of a bacteriophage on. If an indicator host is used that is not susceptible to any phages within a given sample with the exception of the introduced phages, it is possible to track that phage specifically [5]. However, this is unlikely to be the case with soil micro-organisms and as such it is probable that inoculated phages will not be able to be discriminated from the indigenous phage community unless molecular methods based on unique markers are used to identify plaques. This, however, can still create problems of false positives when one considers that many phages possess DNA modules that are shared between a variety of different phages [22]. There is no information on the evolution or to what extent different soil bacteriophages share genetic information so it is impossible to speculate on the likelihood of problems of this nature being encountered. It is possible, that through a thorough screening programme, a phage, specific for a given host that has a completely unique genome, could be found. However, such problems can be simply alleviated through the use of marker genes inserted into the phage genome [7–9,13,14], which allows the use of diagnostic probes and the potential phenotypic identification of prophages. This approach may also work for specialised transducing phages if the transduced genes to either side of the donor's phage integration site (attB) are known and diagnostic probes can be developed. This, of course, will only allow the identification of transducing phages and not the phage population as a whole. Generalised transducing phages are more problematic, as by their very nature, they do not form plaques, and as such cannot be detected in this manner.

Host selection, for use as donor or recipient, is in some ways easier, but in others more difficult. It is relatively easy to use a wide range of elective or selective markers in a host strain, through the use of cloning techniques (see Section 6 of this manual). It should, however, be noted that inoculated bacteria, be they genetically engineered or not, do not often survive and establish themselves successfully in the soil environment, especially if they are taken from laboratory culture [24]. It is therefore, recommended that indigenous hosts should be used if at all possible, and if they are marked through molecular techniques that this process should maintain the fitness of the host strain. In gene transfer experiments, well defined donors and recipients should be used that allow separation both at the phenotypic and the genetic level. This means that any phenotypic differences can be confirmed by molecular techniques. Also of critical importance is the marker with which the exchange will be monitored. Again, this should allow transductants/convertants to be detected on isolation plates and their subsequent confirmation at the molecular level (see Section 6 of this manual).

Procedures

Extraction of bacteriophages from soil

The isolation and quantification of bacteriophages in an environmental sample is of critical importance in the estimation of transduction frequency. Although, for the reasons outlined in the introduction, such gross measurements of phages, and indeed host numbers, cannot give a complete picture of phage-host population dynamics in situ. However, at the present time there are no suitable alternatives that allow the spatial distribution of phage or host activity in soil to be investigated. Advances in the field of in situ hybridisation based on diagnostic 16S rRNA probes may well prove useful in this area in the future, but the technique has so far not been conclusively demonstrated in a completely natural soil system [6].

Two methods are presented here, firstly for the isolation of chloroform resistant phages and secondly for the isolation of chloroform sensitive phages. They are methods C and D of Lanning and Williams [11]. We have used method C successfully for the isolation and enumeration of derivatives of the temperate actino-phage ΦC31 for many years and routinely experienced recoveries of ca. 85% of inoculated phages. Lanning & Williams [11] reported that for method C recoveries varied between 40 and 100%, depending on soil type, whilst method D invariably gave recoveries of about 100%; although this method is, of course, only useful for chloroform tolerant phages.

Isolation of chloroform resistant and sensitive bacteriophages (method C of Lanning & Williams [11])

1. Weigh out 20 g soil and transfer to a screw capped container.
2. Add 50 ml nutrient broth (pH 8.0) containing 0.1% (w/v) egg albumin. A 5% (w/v) stock solution of egg albumin should be made in advance, filter sterilised (0.22 μm) and 0.5 ml added to the preautoclaved nutrient broth solution (50 ml), just prior to use.
3. Shake soil/broth mix at 200 oscillations per minute (o.p.m.) for 30 min at 4 °C.
4. Stand at 4 °C for 16 h.
5. Transfer supernate to an Oakridge tube and centrifuge at 1200 × g for 30 min.
6. Filter supernate through a 0.45 μm Millipore filter.
7. Dilute filtered supernate in nutrient broth and assay.

Isolation of chloroform resistant bacteriophages (method D of Lanning & Williams [11])

Steps 1–5 as above.
6. Transfer 5 ml of supernate to a sterile glass vial.
7. Add 0.1 ml chloroform.
8. Shake on reciprocal flask shaker for 5 min at 200 o.p.m.
9. Stand at 4 °C for 2 min.
10. Remove aqueous layer, dilute in nutrient broth and assay.

Assaying phage populations

The most common approach to the assaying of phage populations is the use of the double agar layer technique [1] and whilst other methods are available that are based on electron microscopy [15], these have only been applied to the study of phage populations in the aquatic environment. For such methods to be used in the soil environment, further protocols for the purification of phages from soil particles must be developed. The assaying system described here is based on the protocol of Adams [1] using the modifications of Hopwood et al. [10] for the assaying of streptomycete phages. Other microorganisms will require subtle differences in approach, although it is felt that this procedure should work for phages of a broad range of terrestrial microorganisms.

1. Prepare 1.0 M $MgSO_4$ and 0.8 M $Ca(NO_3)_2$, if divalent cations are required for phage adsorption to the host.
2. Prepare 1.4% (w/v) nutrient agar containing 0.5% (w/v) glucose, autoclave and cool to ca. 50 °C. Add relevant divalent cations to the necessary concentration, if required. Pour ca. 7 ml into 5 cm petri dishes. For new phages 4 mM $Ca(NO_3)_2$ is recommended.
3. Prepare soft nutrient agar (0.7% [w/v] nutrient broth, 0.5% [w/v] glucose, 0.3–0.7% [w/v] agar and autoclave [plaque size is inversely proportional to agar concentration, we find that this range is adequate for the phages we have tested]). Add divalent cations to the necessary concentration, if required.
4. Cool soft nutrient agar to ca. 45 °C and add host cells to a concentration of 10^7–10^8 c.f.u./ml (the soft nutrient agar should look slightly turbid). Mix thoroughly.
5. Spot 100 µl of phage lysate onto nutrient agar plates.
6. Add 0.8 ml of soft nutrient agar to agar plates, swirl briefly to mix phages and host.
7. Leave to cool briefly and transfer plates carefully to an incubator and incubate for 12–18 h at the relevant optimum growth temperature depending on the host used.

Detection of host populations

Basic approaches to the extraction and viable counting of host populations in soil are discussed here. Other chapters in this manual describe methods for the detection of microorganisms using alternative techniques, such as nucleic acid extraction. Therefore these will not be covered here, although their use in combination with viability based methods is certainly recommended. Two approaches will be outlined here. The first is a simple procedure that enables large numbers of samples to be analyzed. The second, although more time consuming, gives enhanced levels of detection, coupled with the ability to separate different fractions of the soil microflora.

It is outside the scope of this article to describe different isolation media for all possible soil microorganisms. However, a few considerations can be outlined here. The first is that any heterotrophic microbial growth media will undoubtedly need the inclusion of antifungal agents. To this end, the use of 50 µg/ml cycloheximide (Stock solution: 25 mg/ml dissolved in distilled water; sterilised by autoclaving) and 50 µg/ml nystatin (Stock solution: 50 mg/ml dissolved in a small amount of 0.1 N NaOH and made up to the correct volume with sterile distilled water) are recommended. As stated above, in the event of the use of an inoculated host strain, phenotypic markers are needed to discriminate inoculants from the indigenous community. Many such markers, be they elective or selective, are now available and are discussed in Section 6 of this manual.

"Simple" extraction and enumeration of bacteria in soil [28]

1. Weigh 1 g of soil into a Universal tube.
2. Add 9 ml ¼ strength Ringers solution (g/l: NaCl, 2.25; KCl, 0.105; $CaCl_2$, 0.12; $NaHCO_3$, 0.05).
3. Shake (flask shaker) or vortex 10 min at maximum speed.
4. Serially dilute sample in ¼ strength Ringers solution and plate out on appropriate media.

Concentration method

The method listed here resulted from a development of earlier mechanical methods [2,12] and was designed to concentrate soil streptomycetes through dispersion of soil aggregates with an iminodiacetic acid ion exchange resin, desorption of bacteria from soil particles and their subsequent concentration by centrifugation [7–9,13,14]. As bacteria are concentrated from 100 g of soil, low numbers of can be detected, which is particularly useful when examining rare phenotypes; for inoculated streptomycetes it was found that as few as 0.1 c.f.u./g could be detected and 10 c.f.u./g accurately enumerated [7]. This method, with slight modifications, also works well for the isolation and concentration of unicellular bacteria [25,27].

1. Weigh 100 g soil into a Beckman 500 ml centrifuge pot, add 20 g Chelex-100 (Biorad) and 100 ml 0.1% (w/v) Sodium deoxycholate/ 2.5% (w/v) polyethylene glycol.
2. Shake 2 h on a Griffin wrist action shaker (4 °C).
3. Centrifuge for 30 s at 1,000 × g.
4. Filter supernate through a 5 cm diameter metal filter holder (Millipore) and store filtrate at 4 °C. No filter should be included in the holder, the filter holder alone is sufficient to remove the resin.
5. To pellet, add a further 100 ml of 0.1% (w/v) sodium deoxycholate/ 2.5% (w/v) polyethylene glycol, shake (1 h), centrifuge and filter as before. The pellet from the second centrifugation contains the mycelial fraction of streptomycete inoculants (Fraction A).
6. Pool filtrates and centrifuge at 3,000 × g. (15 min).
7. Remove supernate (Fraction B) and add 9 ml strength Ringers solution to the pellet and leave to resuspend overnight (Fraction C).

Fractions A and C almost exclusively contain streptomycete mycelia and spores, respectively. In Fraction B, a substantial proportion of inoculated unicellular bacteria is found, although their recovery is enhanced if Polyethylene Glycol (2.5%) is used as an eluent, rather than Polyethylene Glycol (2.5%)/sodium deoxycholate (0.1%) [25,27].

Confirmation of phenotypes

Soil contains vast numbers of different bacteria, which on isolation plates, may look very similar. It is therefore necessary that putative transductants or convertants are confirmed as such. Putative convertants or specialised transductants, that display the desired phenotype(s), should be picked off the initial isolation media, and after testing for purity, picked onto a lawn of an appropriate indicator strain, as if a phage assay was to be carried out. Upon incubation and subsequent growth, lysogens may spontaneously release phages that will appear as plaques on the indicator strain [7,30]. This approach may not always be successful for specialised transductants, if genes essential for lysis have been removed from the phage's genome during the initial excision from the donor's chromosome. The purified putative lysogens can then be probed with the marker gene or phage DNA by colony hybridisation [7,31]. Putative generalised transductants are more problematic as there is no possibility of screening for phage release. It is for this reason that suitable markers should be chosen in the first instance. One such possibility is the use of cloned genes, that are present in the donor but not the recipient. Putative transductants, after purification, can then be verified by probing with that marker gene.

Acknowledgements

This work was sponsored in part by a grant from the EC-BIOTECH programme (BIOT-CT91-0285).

References

1. Adams MH (1959) Bacteriophages. Interscience Publishers Inc. New York.
2. Fægri A, Torsvik VL, Goskøyr J (1987) Bacterial and fungal activities in soil: separation of bacteria and fungi by a rapid fractionated centrifugation technique. Soil Biol Biochem 9: 105–112.
3. Fry, JC (1989) Analysis of variance and regression and aquatic bacteriology. Binary 1: 83–88.
4. Germida JJ (1986) Population dynamics of *Azospirillum brasilense* and its bacteriophage in soil. Plant Soil 90: 117–128.
5. Germida JJ, Khachatourians GG (1988) Transduction of *Escherichia coli in soil*. Can J Microbiol 34: 190–193.
6. Hahn D, Amann RI, Ludwig W, Akkermans ADL, Schleifer KH (1992) Detection of micro-organisms in soil after in situ hybridistaion with rRNA- targeted, fluorescently labelled oligonucleotides. J Gen Microbiol 138: 879–887.
7. Herron PR, Wellington EMH (1990) New method for the extraction of streptomycete spores from soil and application to the study of lysogeny in sterile amended and nonsterile soil. Appl Environ Microbiol 56: 1406–1412.
8. Herron PR, Wellington EMH (1992) Extraction of *Streptomyces* spores from soil and detection of rare gene transfer events. In: Wellington EMH, van Elsas JD (ed) Genetic

Interactions Among Micro-organisms in Natural Environments, pp. 104–112. Pergamon Press, UK.

9. Herron PR, Wellington EMH (1994) Population dynamics of phage-host interactions and phage conversion of streptomycetes in soil. FEMS Microbiol Ecol 14: 25–32.

10. Hopwood DA, Bibb MJ, Chater KF, Kieser T, Bruton CJ, Kieser HM, Lydiate DJ, Smith CP, Ward JM, Schrempf H (1985) Genetic manipulation of streptomyces: a laboratory manual. The John Innes Foundation, Norwick, UK.

11. Lanning S, Williams, ST (1982) Methods for the direct isolation and enumeration of actinophages in soil. J Gen Microbiol 138: 2063–2071.

12. MacDonald RM (1986) Sampling soil microfloras: dispersion of soil by ion exchange and extraction of specific micro-organisms from suspension. Soil Biol Biochem 18: 399–406.

13. Marsh P, Wellington, EMH (1992) Interactions between actinophage and their streptomycete hosts in soil and the fate of phage borne genes. In: Gauthier, MJ (ed) Gene Transfers and Environment, pp. 135–142. Springer Verlag, Berlin.

14. Marsh P, Toth IK, Meijer M, Schilhabel MB, Wellington, EMH (1993) Survival of the temperate actinophage ΦC31 in soil and the effects of competition and selection on lysogens. FEMS Microbiol Ecol 13: 13–22.

15. Miller, RV and Sayler, GS (1992) Bacteriophage-host interactions in aquatic systems. In: Wellington EMH, van Elsas JD (ed) Genetic Interactions among Micro-organisms in the Natural Environment, pp. 176–193. Pergamon Press, UK.

16. Ogunseitan OA, Sayler GS, Miller RV (1992) Application of DNA probes to analysis of bacteriophage distribution patterns in the environment. Appl Environ Microbiol, 58: 2046–2052.

17. Petersen RG (1985) Separation of means. In: Design and analysis of experiments: Statistics Textbooks and Monographs, Vol 66, pp. 72–111. Marcel Dekker, New York, USA.

18. Reanney DC (1968) An assay for *Bacillus stearothermophilus* using thermophillic virus. New Zealand J Agricul Res 11: 763–770.

19. Saye DJ, Ogunseitan O, Sayler GS, Miller RV (1987) Potential for transduction of plasmids in a natural freshwater environment: effect of plasmid donor concentration and a natural microbial community on transduction in *Pseudomonas aeruginosa*, 53: 987–995.

20. Saye DJ, Ogunseitan OA, Sayler GS, Miller RV (1990) Transduction of linked chromosomal genes between *Pseudomonas aeruginosa* strains during incubation in situ in a freshwater habitat. Appl Environ Microbiol 56: 14–145.

21. Schmieger H (1990). Phage genetics and ecology. In: Fry JC, Day MJ (eds) Bacterial Genetics in Natural Environments, pp. 41–44. Chapman and Hall, UK.

22. Schneider J, Korn-Wendisch F, Kutzner HJ (1990) ΦSC623, a temperate actinophage of *Streptomyces coelicolor* Müller, and its relatives ΦSC347 and ΦSC681. J Gen Microbiol 136: 767–772.

23. Schneider J, Kutzner HJ (1989) Distribution of modules among the central regions of the genomes of several actinophages of Faenia and Saccharopolyspora. J Gen Microbiol 135: 1678.

24. Thompson IP, Young CS, Cook CA, Lethbridge G, Burns RG (1992) Survival of two ecologically distinct bacteria (Flavobacterium and Arthrobacter) in unplanted and rhizosphere soil. Soil Biol Bioch 24: 1–14.

25. Turpin PE, Dhir VK, Maycroft KA, Rowlands C, Wellington EMH (1992) The effect of *Streptomyces* species on the survival of Salmonella in soil. FEMS Microbiol Ecol 101: 271–280.

26. Turpin PE, Maycroft KA, Bedford J, Rowlands C, Wellington EMH (1993) A rapid luminescent based MPN method for the enumeration of Salmonella typhimurium in environmental samples. Lett Appl Microbiol 16: 24–27.

27. Turpin PE, Maycroft KA, Rowlands C, Wellington EMH (1993) An ion exchange based extraction method for the detection of Salmonellas in soil. J Appl Bacteriol 74: 181–190

28. Wellington EMH, Cresswell N, Saunders VA (1990) Growth and survival of streptomycete inoculants and extent of plasmid transfer in sterile and non sterile soil.

Appl Environ Microbiol 56: 1413–1419.

29. William ST, Manchester L, Mortimer AM (1986) The ecology of soil bacteriophage. In: Goyal SM, Gerba CP Bitton, G (eds) Phage Ecology, pp. 157–180. John Wiley, USA.

30. Zeph LR, Onaga MA, Stotzky G (1988) Transduction of *Escherichia coli* by bacteriophage P1 in soil. Appl Environ Microbiol 54: 1731–1737.

31. Zeph LR, Stotzky G (1989) Use of a biotinylated DNA probe to detect bacteria transduced by bacteriophage P1 in soil. Appl Environ Microbiol 55: 661–665.

Molecular Microbial Ecology Manual **6.1.7**: 1-17, 1995.
© 1995 *Kluwer Academic Publishers.*

Heavy metal resistances in microbial ecosystems

MAX MERGEAY

Flemish Institute for Technological Research, Mol, Belgium

Introduction

The importance of heavy metal or oxyanion resistance markers for ecological studies was recently recognized. Such markers include the plasmid borne resistances to mercury [1–6], zinc [7–9], copper [10–14], nickel [15–17] and arsenical compounds [18–19]. These markers and others (resistance to cadmium, chromate, cobalt, thallium, lead, antimony, silver, tellurite [20,21]) are representative of the bacterial adaptation to environments that contain high levels of heavy metals. The most typical of these environments are sediments or soils from industrial or mining areas which may contain up to 10% heavy metals and still provide viable counts of heterotrophic bacteria [7].

Other environments of interest are not as "extreme" regarding the adaptation to heavy metals but may contain local concentrations of heavy metals that are higher than normal: this is the case in the phyllosphere of cultivated plants which are treated with copper-containing fungicides [11] or in the manure of pigs which were fed with copper-based additives [14]. Another example is provided by clinical isolates of bacteria resistant to organomercurials or silver compounds that are topically used on external wounds [22].

Incidentally, it may be recalled that resistance to mercury, which is often carried by transposons, has also been regularly found in many antibiotic resistance or catabolic plasmids [23] outside any apparent selection pressure for mercury resistance. Another example for the absence of selection under certain environmental conditions is the organo-metal tributyltin (TBT), a highly toxic marine antifouling agent. When heterotrophic populations from TBT-polluted sediments and nonpolluted sediments from an estuary were compared, the latter demonstrated a greater level of resistance. The opposite was true for populations from two freshwater sediments [24]. The example from estuaries serves to illustrate that the isolation of highly resistant bacterial strains may not always reflect the constitution of the total population at a site impacted by a "chemical soup" of pollutants.

This chapter will focus on methods related to high metal resistances in aerobic heterotrophic bacteria. There are few data about adaptation to heavy metals in anaerobes. Microeucaryotes (yeasts,fungi,algae) may display some adaptations to heavy metals via metallothioneins but at concentrations (1–10 μM) two to four orders of magnitude lower than those found for resistant heterotrophic bacteria (1–30 mM). A special mention should be made of obligate acidophilic chemolithoautotrophs such as *Thiobacillus ferrooxidans* which may cope with concentrations of heavy metals at the molar level: however, although a specific adaptation to such high concentrations of heavy metals may be implied and should definitively deserve more scrutiny, the metabolism and the growth conditions do not allow an easy comparison with heterotrophic micro-organisms.

Bacterial heavy metal resistances as useful markers in soil ecology

Bacterial resistances to heavy metals and oxyanions may provide useful markers for ecological studies. Special interest for these markers arose recently in the perspective of studies about the environmental release (deliberate or accidental) of recombinant DNA via GEMs (genetically engineered microorganisms). *ars* (resistance to arsenite and arsenate) cassettes [19] are thought to be an useful substitute to antibiotic resistances as markers of the vectors that are used in recombinant DNA technology. Indeed the natural background in soil of bacteria that are resistant or tolerant to arsenic compounds is much higher than the background of bacteria which are resistant to antibiotics (de Lorenzo, pers. comm.). Therefore the danger to introduce an undesired marker through the environmental release of GEMs may be minimized. *tel* (resistance to tellurite) cassettes were also recently constructed in the same perspective (de Lorenzo, pers. comm.).

A *czc*CBAD gene cassette that is not expressed in *Escherichia coli* or in *Pseudomonas putida* but well in *Alcaligenes eutrophus* strains was used in experiments designed to follow gene escape or gene recruitment from GEMs in soil [8]. The same cassette proved also to be useful for exogenous isolation of plasmids in triparental matings that involved an *E. coli* donor, an plasmid free *Alcaligenes eutrophus* recipient and soil extracts. This allowed the selection of new broad-host range plasmids from soil bacteria. These plasmids were able to mobilize the *czc* cassette from *E. coli* to *A. eutrophus* and were thus selected on basis of their transfer and mobilization phenotypes [9].

Mer genes were the first to have been extensively studied in ecological studies [1–6]. Recently, they were also isolated from environmental bacteria of river epilithons in *P. putida* recipients in a process of exogenous isolation of environmental plasmids [25].

Basic principles for the design of a medium to assay the growth of bacteria in the presence of heavy metals

The choice of a growth medium is of major importance when defining bacterial resistance to a specific metal. Metals may interact with many components of the growth media, especially with phosphates, broth components and with organic acids used as carbon sources. It may be considered (Remacle, pers. comm.) that in most current media, up to 90% of the applied concentration of Zn, Cd, Co, Cu, etc. may be complexed and without effect on bacterial resistance. Broth containing media ("rich media") are especially misleading in this respect: the concentrations of heavy metals which exert a selective effect on bacterial growth are quite high and should not be used in the definition of bacterial resistances or sensitivities. This may be a source of problems with enteric bacteria which are most of the time grown in broth media or in media containing high phosphate levels.

Yet broth media (especially diluted broth) remain useful for the selection of resistant bacteria displaying different heterotrophic metabolisms. In order to minimize the problems linked with phosphates and complexation with heavy metals, a mineral minimal medium with little phosphate [26,27] was used for growth of bacteria in presence of heavy metals [27]. In this medium (TRIS minimal medium, see below), Tris replaces phosphate for buffering and ensures for the correct buffering capacity especially in the pH range of 7 to 9 which seems to be critical for many bacteria unable to use glucose and most of the sugars as carbon sources, e.g., the metal resistant bacteria belonging to the genus *Alcaligenes*. Phosphates are given at limiting concentrations, sometimes in the form of glycerophosphate.

With this minimal medium, the carbon sources have to be carefully chosen in order to not counterselect interesting bacteria: it was found that gluconate, lactate and succinate have allowed the selection and the growth of both glucose-utilizers and *Alcaligenes*-like bacteria. Therefore by allowing the selection of a rather broad spectrum of resistant bacteria, these carbon sources offer a good alternative to broth media. Similarly, glucose allows to select resistant genera which are distinct from *Alcaligenes*, while the use of long chain dicarboxylic acids (azelate, sebacate, suberate), which efficiently counterselect enterics and some fluorescent pseudomonads, helps to enrich for *Alcaligenes*-like isolates. Besides, many metal resistant *Alcaligenes* isolates share with the type strain CH34 [27,6] the capacity to grow in chemolithotrophic conditions. Therefore, it is obvious that varying carbon sources and growth conditions in Tris minimal medium offers possibilities to detect genetic transfer of metal resistances.

Although Tris minimal medium allowed the definition of minimum inhibitory concentrations (MIC) of heavy metals which were much lower than those observed with conventional media, it also displayed drawbacks with some metals during scaling up processes in fermenters [28]. It was indeed shown that the prolonged growth of resistant *Alcaligenes* in the presence of heavy metals promoted the bioprecipitation of zinc and cadmium in the form of carbonates and hydroxides. This bioprecipitation was the consequence of the substantial pH increase accompanying the plasmid-mediated mechanism of resistance against the metals. This mechanism is mainly based on the cation-proton antiporter afflux and is governed by *czc* (resistance to cadmium, cobalt and zinc) and *cnr* (resistance to cobalt and nickel) genes on the large plasmids of *A. eutrophus* CH34. Yet, the bioprecipitation was not or poorly observed with cobalt and nickel. The presence of Tris, which may form some complexes with heavy metals, was found to be a hindrance for the bioprecipitation or the biosequestration of these metals. Therefore, a medium deprived of Tris was used and found to substantially improve the bioprecipitation of Co, Ni and also Cu as well as the capacities of bioreactors to remove heavy metals from polluted effluents. However, absence of buffering decreased the growth rates.

Procedures

Definition of sensitivity and resistance thresholds

In order to evaluate the levels of resistance, two parameters may be used for practical purposes: Minimum Inhibitory Concentration (MIC) and Maximum Tolerable Concentration (MTC). It should be recalled once again that these parameters are not absolute but are defined relative to a specific medium. MIC in broth media are 2 to 5 times higher than in Tris minimal medium.

A simple protocol to evaluate MIC is as follows:
- Prepare plates containing various decreasing concentrations of metals, using a factor of 2 between two successive dilutions: for example 20, 10, 5, 2.5, 1.25, 0.625, 0.31, 0.16, 0.08 mM metal and control plates without metal.
- Plate in duplicate appropriate dilutions of a fresh overnight culture grown in minimal medium, centrifuged and rinsed in the same volume of a solution of $MgSO_4$ (10 mM).

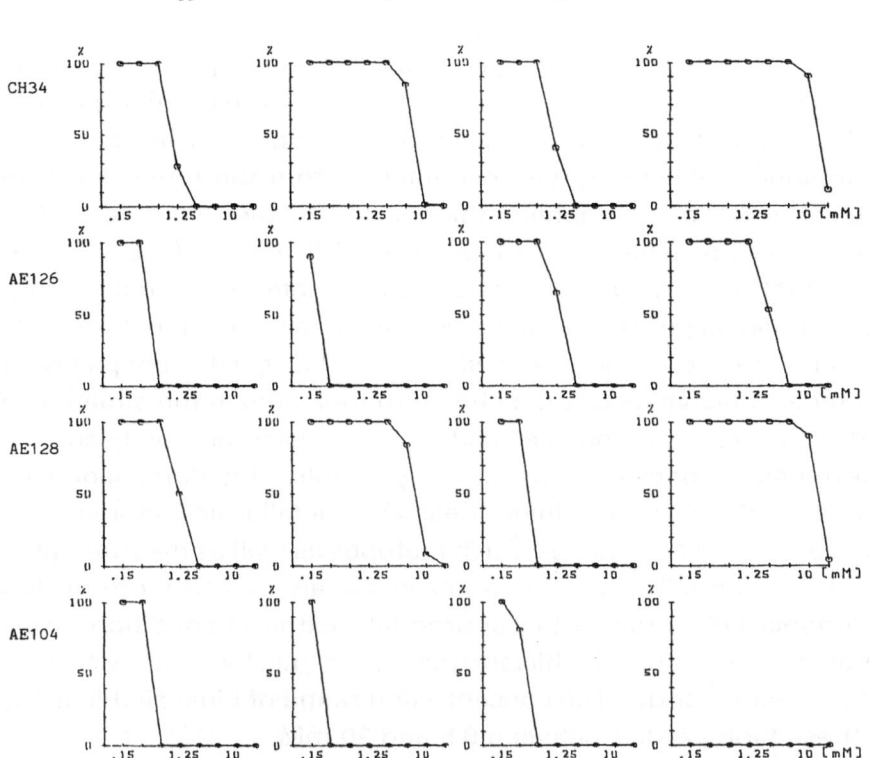

Figure 1. A. eutrophus CH34 derivatives. Viable colonies (in % of control) in function of heavy metal concentration. CH34: wild type (pMOL30, pMOL28, AE128: pMOL30, AE126: pMOL28, AE104: plasmid-free.

- Incubate 2 to 3 days and count.
- Plot the percentage of survival (control viable counts = 100%) as a function of the log of metal concentration. Double log representation is also possible. The MIC is given by the intersection of the survival curve with the horizontal axis (Fig. 1).

The MTC is the highest concentration of metal which do not affect the viable count and is used in selective media designed to promote the growth of a resistant bacterium by providing efficient counter-selection of undesired bacteria.

MIC should be measured in solid and in liquid media and be similar in both conditions. This allowed to check the quality of the gelose (agar). Indeed geloses of some brands may complex some heavy metals and bias the estimations of MIC.

Biotopes and bacterial genera of interest

Resistance levels in micro-organisms which correspond to a specific mechanism (from homoeostasis in the case of essential metals to resistance in the case of excessive concentrations or in presence of xenobiotic metals) may be encountered from the molar level (the special case of acidophilic thiobacilli) to the nanomolar level depending on genera and metabolism. All these mechanisms are important as they may ensure an appropriate biodiversity along a very broad chemical gradient. However, as far as heterotrophic bacteria are concerned, any study of the specific adaptation of bacteria to heavy metals may invite to look at extreme situations in order to avoid to work f.e. with passive tolerance of bacteria to intermediary concentrations of heavy metals. Therefore, sediments, wastes and soils from mining areas and metallurgic factories are of special interest as biotopes. Such biotopes as well as the rhizosphere of plants specially adapted to heavy metals (as to nickel in New Caledonia [15]) provide the appropriate extreme conditions which allow the selection of efficient resistance mechanisms, which are often plasmid borne. The concentration range of bioavailable metals in these biotopes lies between 0.5 and 20 mM.

Bacterial genera which have been encountered in the various searches for highly resistant bacteria include *Alcaligenes* (*eutrophus*, *xylosooxidans*) from the β-*Proteobacteria* [6,17,27], *Sphingomonas* (ex-*Pseudomonas paucimobilis*) [13, Hooyberghs, Diels and Mergeay, unpublished] from the α-*Proteobacteria*, *Pseudomonas* and *Klebsiella* from the α-*Proteobacteria* [16], *Rhodococcus* [29] and *Arthrobacter ilicis* from the Gram-positive bacteria [Margesin, Hooyberghs, Diels and Mergeay, unpublished]. Strains that were ranged from their color under the name of *Flavobacterium* without further taxonomic description were frequently encountered. Their storage or subcultivation were also rather fastidious.

Media

Rich media
Luria-Bertani Broth or other media for viable counts [30] may be used in selective conditions with very high concentrations of metal

Table 1. Metal concentrations

Metal	Salt	Medium	Concentration (mM)
Ag	$AgNO_3$	LB	0,5
		Tris-Min	0,8 – 1,5 – 3
Al	$Al_2(SO_4)_3$	Tris-Min	1 – 2
As	Na_2AsO_2	Tris-Min	2 – 3,2
	Na_2HAsO_4	Tris-Min	20 – 30
Bi	$Bi(NH_4)_3$	Tris-Min	1
Cd	$CdCl_2.5H_2O$	Tris-Min	0,8
			1 – 3
Cr	K_2CrO_4	Tris-Min	0,25 – 1 – 3
	$K_2Cr_2O_7$	Tris-Min	0,15 – 1,8
Co	$CoCl_2.6H_2O$	Tris-Min	0,5 – 0,8 – 1 – 2 – 5
Ce	$Ce(SO_4)_2$	Tris-Min	1
Cu	$Cu(NO_3)_2.3H_2O$	Tris-Min	0,8 – 1 – 1,2 – 1.8
		MGY	0,6 – 2
Hg	$HgCl_2$	LB	0,1 – 0,2
	Merbromine	LB	0,05 – 0,5
Ni	$NiCl_2.6H_2O$	Tris-Min	0,6 – 0,8 – 1 – 2 – 10
		LB	1 – 3 – 5
Pb	$Pb(CH_3COO)_2.3H_2O$	Tris-Min	0,3 – 0,5 – 3
Sb	$SbCl_3$	Tris-Min	0,8
		LB	0,3
Se	Na_2SeO_3	Tris-Min	5,0
Sn	$SnCl_2.2H_2O$	Tris-Min	5,0
Te	$HTeO_3$	Tris-Min	0,5 – 1 – 2 – 5
Tl	$TINO_3$	LB	0,07 – 0,15
		Tris-Min	0,06 – 0,16 – 2 – 5
Zn	$ZnSO_4.7H_2O$	LB	1 – 3 – 10 – 20
		Tris-ˆMin	0,6 – 0,8 – 1 – 2

as Zn^{++} 10 mM or Cu^{++} 18 mM [14]. Working concentrations in minimal media are much lower. Table 1 reports for every metal or oxyanion the used chemical species and relevant concentrations. However, as many soil bacteria of interest are oligotrophic it is recommended to use diluted broth for the viable counts.

Minimal media
– TRIS minimal medium [26,27] for 1 l:
  – Tris HCl, 50 mM, 6,06 g.
  – NaCl, 80 mM, 4,68 g.
  – KCl, 20 mM, 1,49 g.
  – NH_4Cl, 20 mM, 1,07 g.
  – Na_2SO_4, 3 mM, 0,43 g.
  – $MgCl_2.6H_2O$, 1 mM, 0,20 g.

- CaCl$_2$.2H$_2$O, 0,2 mM, 30 mg.
- Na$_2$HPO$_4$ 40 mg or 24 ml of a 1% solution.
    - Fe(III)NH$_4$ citrate; 10 ml of a solution of 48 mg/100 ml.
    - SL7 oligoelement solution 1 ml, pH 7.
    - Agar, 20 g.
    - carbon source (gluconate azelate, lactate: 0.2%).
    - Tris with sodium B glycerophosphate 294 mg instead of Na$_2$HPO$_4$.2H$_2$O.
- SL7 Oligoelements solution, for 1 l
    - HCl 25%, 10 mM, 1,3 ml.
    - ZnSO$_4$.7H$_2$O, 0,5 mM, 144 mg.
    - MnCl$_2$.4H$_2$O, 0,5 mM, 100 mg.
    - H$_3$BO$_3$ (boric acid), 1 mM, 62 mg.
    - CoCl$_2$.6H$_2$O, 0,8 mM, 190 mg.
    - CuCl$_2$.2H$_2$O, 0,1 mM, 17 mg.
    - NiCl$_2$.6H$_2$O, 0,1 mM, 24 mg.
    - Na$_2$MoO$_4$.2H$_2$O, 0,15 mM, 36 mg.

This low phosphate medium is especially recommended for *Alcaligenes* and related bacteria when cultivated in presence of heavy metals. To test the growth with carbon sources or on nitrate in anaerobiosis or in chemolithotrophic conditions the medium of Schatz et Bovell [31] will be preferred.

- MGY Mannitol Glutamate Yeast Extract [32], for 1 l
    - Na-glutamate (puriss), 1 g.
    - Yeast extract, 0.1 g.
    - NaCl, 0.2 g.
    - K$_2$HPO$_4$, 0.5 g.
    - MgSO$_4$.7H$_2$O, 0.2 g.
    - Mannitol, 10 g (or any appropriate carbon source).
- Agar, 15 g or 20 g/l.
    - Biotine, 1 µg.
    - Thiamine, 10 µg.

This medium is especially convenient in presence of copper for fluorescent pseudomonads, rhizobia and bacteria that use carbohydrates as carbon sources. MIC of Cu^{++} for *P. syringae* on MGY medium is similar to the MIC of Cu^{++} for metal resistant *Alcaligenes* on Tris minimal agar.

Sampling and processing of soil samples

Sampling occurred in aseptic conditions as much as possible: alcohol and flame sterilization of spade and sieve, use of sterilized recipients and plastic bags. Rough sieving was required to discard vegetal debris, little stones, conglomerates, brick waste, etc. Soil samples were stored in hermetic boxes at room temperature. If necessary, they were rehydrated with sterile distilled water.

2 g of sieved soil are weighed in a sterile erlenmeyer or in a cylindro-conical flask (125 ml) and were suspended in 18 ml 10 mM $MgSO_4$. The recipient was vigorously shaken for 2 h at room temperature in order to release a maximum of bacteria from soil particles. After decantation (1 h) supernatants were appropriately diluted and plated in duplicate or in triplicate.

Plates were incubated for three to four days. Colonies from non selective (viable counts) or from selective media were resuspended with toothpicks on 20 μL droplets of 10 mM $MgSO_4$ put on a grid and replicated on selective media provided with appropriate carbon sources, antibiotics or heavy metals, using a replicator made with inox nails fixed in epoxy resin poured inside a glass Petri plate.

Notes

Tips and tricks for some metals

Cadmium

Resistances higher than 1.5 mM (MTC) have not yet been reported or observed in heterotrophic bacteria. Soil isolates of *P. aeruginosa* are rather resistant to Cd^{++}. Cd-containing plates that were streaked with resistant bacteria should be conserved for storage at room temperature; at 4 °C, the resistance mechanism is not any more active and the toxicity will soon prevail. This laboratory observation has been generally confirmed for most of metals but is especially relevant for Cd^{++} and other non-essential highly toxic metals (Hg, Tl).

Cobalt

In cobalt resistant bacteria, MIC's up to 20 mM have been reported. Plasmid borne resistance mediated by *czc* (pMOL30) and *cnr* (pMOL28) increased the tolerance to cobalt 100× and 25×, respectively. On rich media, growth with high concentrations of cobalt may elicit the production of diffusible red-brown pigments, the nature of which is not known. In mining areas of Zaïre and Zambia, cobalt is present at high concentrations in sludges, mine wastes and together with copper acts as a selective agent in microbial ecosystems.

Copper

In Tris minimal medium, MICs of copper salts are rarely higher than 2 mM. A plate effect is observed with streaks of *Alcaligenes* isolates which could lead to some confusion in the determination of phenotypes: bacteria which are sensitive to Cu (plasmid-free derivatives) are completely inhibited in the streaks with high density of bacteria ("first line") but may produce some c.f.u. in less dense streaks, while resistant bacteria exposed to the same concentration of Cu would grow in any area. It is hypothesized that the normal mechanism for copper homoeostasis may be extremely efficient in the first stages of vegetative growth but not in stationary phase or when competition with other bacteria would be too acute. In these cases, a plasmid-borne resistance to Cu [10–13] would be of help. Another plate effect which was regularly encountered with *"Flavobacterium*-like" isolates is that colonies became light brown. No explanation has been found to date.

Mercury

Solutions of organomercurials should be freshly prepared and sterilized by filtration. Plates should be used quickly and should not be stored in the cold room. Glass Petri plates allow to avoid the formation of free radicals that may occur through the interaction of organomercurial compounds with the plastic plates (Summers, pers. comm.).

Nickel

Very high MTCs for this metal reaching 40 mM were observed in resistant bacteria. There are indications that in contrast to cadmium, cobalt and zinc, the nickel resistance marker of *cnr* from pMOL28 of *A. eutrophus* CH34) would be reasonably expressed in various heterologous bacterial hosts and would be a marker of interest in ecological studies. On rich medium plates containing 3 mM of Ni, the growth of streaked *cnr+* nickel-resistant derivatives of *A. eutrophus* suppressed in its neighbouring the growth inhibition of the nickel-sensitive strains. This effect has been also observed with cobalt and may be due to a kind of diffusible product.

Thallium

This metal is peculiarly toxic to heterotrophic bacteria: yet some resistance determinants (MIC not higher than 0.3 mM) were detected on the chromosome and the plasmids of *A. eutrophus* CH34. As thallium is not a frequent pollutant, it looked unlikely that the corresponding phenotype emerged from a selection pressure exerted by the presence of thallium. However we isolated a highly resistant *Alcaligenes*, strain

AB2, (MTC 2 mM) from sediments of waste water of the underground zinc and copper mine of Kipushi (Shaba, Zaïre) where ores rich in germanium, thallium and gallium had been regularly processed. The thallium resistance determinant of this strain, which is strongly related to *A. eutrophus* CH34, was found to be chromosomal.

Tin

Strains that are able to grow in presence of 75 mg/l of $SnCl_4$ may be considered as resistant [33]. For organotin, the diagnostic concentration is 15 mg/l of $(CH_3)_2$-$SnCl_2$ [33]. Plasmid mediated resistance is up to now still poorly documented (Wuertz, pers. comm.)

Zinc

Resistance or tolerance to zinc (MTC around 1 mM) seems to be more frequent than expected and to be encountered in a variety of environments which are not necessarily highly polluted. The zinc resistance phenotype corresponding to the *czc* operon (pMOL30 of *A. eutrophus* CH34) is not or very poorly expressed outside his original host. This was recently taken advantage in studies of gene escape in soil environments [7,8].

Basic genetic characterization of heavy metal resistance in environmental isolates

Basic taxonomy and reference strains

Ecological impact of metal resistance genes may be of different nature if these genes are chromosomal or plasmid borne. A preliminary genetic study of any environmental isolate that displayed a distinctive resistance phenotype may help in this way. First a rough taxonomic determination would be useful as it may point towards a reference strain or genus to which the environmental strain can be compared especially for MIC's of heavy metals. Reference strains to be used in this respect should be plasmid free and display a low restriction proficiency. This characteristics may be of importance for plasmid transfer assay. In that case, they should carry also a mutation for resistance to rifampicin (RpoB phenotype). This marker is indeed rather easy to select and useful for the purpose of counterselection. Good reference strains that have these characteristics are *Pseudomonas putida* KT2441R, *Escherichia coli* K12 C600 *hsdR rpoB* (or even HB101), *Alcaligenes eutrophus* CH34 AE805. (*Pseudomonas aeruginosa* PAO1 *rpoB* may be used maily for MIC's comparison: if this strain has to be used as a recipient in gene transfers, the mating should occur at 42 °C to circumvent restriction. However, this is not always applicable). Work to build other reference strains for genera as *Klebsiella*,

Sphingomonas, *Rhodococcus*, *Bacillus cereus*, *B. subtilis*, *Agrobacterium*, *Acinetobacter*, *Arthrobacter*, etc. would surely be rewarding. Once a special metal resistance phenotype was determined relatively to a reference strain (if suitable), the presence of plasmids has to be checked mainly by electrophoresis on agarose gels.

The correlation between the presence of plasmids and the metal resistance phenotype could be made in two possible ways:

– the curing of the plasmid with the loss of the resistance phenotype by chemical or physical (temperature) agents, or
– the conjugation mediated transfer or mobilisation of the corresponding genetic determinant into an appropriate reference strain. The combination of both evidences would give the most satisfactory conclusion.

Plasmid curing

In curing experiments with environmental isolates, it is crucial to rely on distinctive markers that were revealed by the taxonomic analysis as carbon sources, morphological or physiological characteristics, otherwise the cured derivative may be confused with any trivial contaminant. Table 2 reports working concentrations of some chemical curing agents that were efficient in Gram-negative soil isolates.

Table 2. Effective concentration range for some curing agents

Agent	Concentration (µg/ml)	
Mitomycine C	0.1–5.0	
Ethidiumbromide	500–800	
Acriflavine	10–50	
Novobiocine	500–1000	
Nalidixic acid	10–500	
Sodium dodecylsulphate	500–5000	

A useful procedure to find the critical concentration of the curing agent (e.g., with ethidium bromide or sodium dodecylsulphate) is to make successive dilutions of the curing agent by a factor two (as described above for the determination of metal MIC), to inoculate with the bacteria of interest and to let grow overnight. The tube corresponding to the first sublethal concentration will be examined for the presence of cured derivatives.

In some metal resistant strains of *Alcaligenes eutrophus* found in different geographical areas, attempts to cure the resident plasmids by rising the growth temperature above the optimal growth temperature interfered with the uncommon phenomenon of "Temperature Induced Mortality and Mutagenesis" (TIMM, also called Thermospontagenesis), which is characteristic of these isolates [34,35]. TIMM could indeed induce some rearrangements in megaplasmids of these strains. Nevertheless, rising growth temperature remains an useful means to cure plasmids [36].

In many Gram-positive bacteria, protoplasting and regeneration leads to curing of resident plasmids. The procedure that was described to cure plasmid borne resistance to cadmium in *Rhodococcus fascians* [29] is as follows: bacteria were treated with lysozyme (10 mg/ml) in TE buffer (50 mM Tris HCl, 20 mM EDTA (pH 7)) containing 20% polyethylene glycol. After a few minutes, cells were pelleted by centrifugation, washed in the same buffer (without lysozyme) and plated. This procedure produced plasmid curing in up to 75% of survivors [29].

Plasmid extraction

Physical evidence of the presence or absence of a plasmid is the obliged corollary of any successful curing or conjugation experiment.

Metal resistances are regularly found to be associated with large plasmids, which are often difficult to extract especially from poorly known bacteria. Most of the available extraction methods were developed for γ-*Proteobacteria* (enterics and fluorescent pseudomonads) [37] and might not be convenient for other bacteria. Therefore, we report here a procedure that was derived from the method of Kado et Liu [38]. This modified procedure (Taghavi, pers. comm.) was specifically devised to allow quantitative extraction of large plasmids in *Alcaligenes* and other β-*Proteobacteria* in the perspective of their analysis by restriction mapping and DNA hybridization according to Southern.

Protocol

- Inoculate from a single colony grown on (metal containing) selective medium 15 ml of Nutrient Broth (Difco) and let grow overnight.
- Harvest the cells in Nutrient Broth (Difco).
- Wash cells with 5 ml E buffer and centrifuge again.
- Add 2 ml lysis solution.
- Leave 20 min at room temperature under gentle agitation.
- Incubate 75 min at 68 °C.
- Add 0.5 ml 4 M NaCl.
- Add 6 ml phenol-chloroform (1:1) water saturated.
- Mix well for 30 min under gentle agitation.
- Centrifuge 30 min, 12,000 r.p.m. at 4 °C.

- Leave 2 h at 4 °C and take upper phase.
- Extract with 2 ml ether (evaporate resting ether by incubation at 37 °C).
- Centrifuge: 3 h, 40,000 r.p.m. at 4 °C (1.8 ml).
- Wash DNA pellet 3× with 70% ethanol.
- Resuspend DNA in 0.1 ml TE buffer.
  E buffer, 40 mM Tris-acetate, 2 mM EDTA, pH 7.9.
  Lysis solution, 40 mM Tris, 3% SDS, pH 12.6 with 5 N NaOH.

Plasmid-mediated conjugation

Conjugation experiments may be carried out in two ways, self-transfer and assisted transfer, both of them using reference strains described above as final recipients and the resistance to rifampicin as a donor counterselecting agent. Biparental matings that involve the environmental isolate and an appropriate recipient are used to evaluate the plasmid borne character of the considered heavy metal resistance phenotype. Assisted transfer through triparental matings or through gene capture (retrotransfer) is required when the plasmid is poorly or not self-transferable. The conjugating partner would contain a mobilizing plasmid as pRK2013 or pULB113. An *E. coli* resident, pRK2013 contains the replication origin of ColE1 and the Tra functions of IncP plasmids, mobilizes very efficiently a variety of plasmids and does not replicate outside its original host [39]. pULB113 (RP4::Mu3A) [40] is peculiarly useful when the mobilization may only occur through the Mu3A prophage-mediated formation of cointegrates.

Selection of metal resistant transconjugants should use the metal resistance marker that displayed a clear cut phenotype and at a concentration that should not exceed too much the MIC of the recipient. Twice the MIC is generally convenient for that purpose.

This is of special importance in heterologous matings because the expression level of many heavy metal resistance operons (*czc*, *cnr*, *pco*, *cop*, etc.) was often found to be decreased in hosts which are not closely related to the strain from which these operons originated. Therefore care should be taken to not exert too high a selection pressure.

Conjugation experiments may provide information of ecological value about the host range of metal resistance plasmids and the expression range of resistance markers. Exogenous isolation of mercury resistance plasmids is a striking illustration of this genetical and ecological approach [25]. An other promising system for expression in a variety of hosts is provided by the *ncc* and *xre* determinants for nickel resistance [41]. These two operons [*ncc*YXHCBAN (8 kb): resistance to Ni (up to 30 mM), Co and Cd; *nre* (1.8 kb): resistance to Ni (up to 3 mM)] are closely linked on a 14.5 kb fragment of the megaplasmid pTOM8 of *A. xylosoxidans* 31A. This fragment has been cloned in various broad host range vectors. *Ncc* is expressed in *A. eutrophus* and related strains while *nre* is expressed in *E. coli* and γ *Proteobacteria*. The *ncc-nre* cassette confered nickel resistance to a broad range of recipients [41, Q. Dong (pers. comm.)] which would be of interest for bioremediation purposes.

Acknowledgements

Work from our laboratory was supported by EEC contracts BAP-367B and BRIDGE and the Flemish Action Program for Biotechnology (VLAB-ETC-003).

References

1. Bale MJ, Fry JC, Day MJ (1988) Transfer and occurrence of large mercury resistance plasmids in river epilithon. Appl Environ Microbiol 54: 972–978.
2. Barkay T (1987) Adaptation of aquatic microbial communities to Hg^{2+} stress. Appl Environ Microbiol 53: 2725–2732.
3. Barkay T, Fouts DL, Olson BH (1985) Preparation of a DNA gene probe for detection of mercury resistance genes in gram-negative bacterial communities. Appl Environ Microbiol 49: 686–692.
4. Barkay T, Liebert C, Gillman M (1989) Environmental significance of the potential for *mer* (Tn*21*)-mediated reduction of Hg^{2+} to Hg^0 in natural waters. Appl Environ Microbiol 55: 1196–1202.
5. Barkay T, Liebert C, Gillman M (1989) Hybridization of DNA probes with whole-community genome for detection of genes that encode microbial responses to pollutants: *mer* genes and HG^{2+} resistance. Appl Environ Microbiol 55: 1574–1577.
6. Barkay T, Turner RR, Vandenbroek A, Liebert C (1991) The relationships of HG (II) volatilization from a freshwater pond to the abundance of *mer* gene pool of the indigenous microbial community. Microb Ecol 21: 151–161.
7. Bridges K, Kidson A, Lowbury EJL, Wilkins MD (1979) Gentamicin- and silverresistant *Pseudomonas* in a burns unit. Br Med J 1: 446–49.
8. Cooksey DA, Azad HR, Cha JS, Lim CK (1990) Copper resistance gene homologs in pathogenic and saprophytic bacterial species from tomato. Appl Environ Microbiol 56: 431–435.
9. Desomer J, Dhaese P, Van Montagu M (1988) Conjugative transfer of cadmium resistance plasmids in *Rhodococcus fascians* strains. J Bacteriol 170: 2401–2405.
10. Diels L, Van Roy S, Taghavi S, Doyen W, Leysen R, Mergeay M (1993) The use of *Alcaligenes eutrophus* immobilized in a tubular membrane reactor for heavy metal recuperation. Biohydrometallurgical Technologies, Edited by AE Torma, ML Apel, CL Brierly, The Minerals, Metals & Materials Society, pp. 133–144.
11. Diels, L, Mergeay M (1990) DNA probe-mediated detection of resistant bacteria from soils highly polluted by heavy metals. Appl Environ Microbiol 56: 1485–1491.
12. Dressler C, Kues U, Nies DH, Friedrich B (1991) Determinants encoding resistance to several heavy metals in newly isolated copper-resistant bacteria. Appl Environ Microbiol 57: 3079–3085.
13. Echols H, Garen A, Garen S, Torriani A (1961) Genetic control of repression of alkaline phosphatase in *E. coli* J Mol Biol 3: 425–438.

14. Gerhardt P, Murray RGE, Wood WA, Krieg NR (1993) Gene Mutation. in: Methods for general and molecular bacteriology. American Society of Microbiology, Washington DC.
15. Gerhardt P, Murray RGE, Wood WA, Krieg NR (1993) Plasmids. In: Methods for general and molecular bacteriology. American Society of Microbiology, Washington DC.
16. Hallas LE, Cooney JJ (1981) Effects of stannic chloride and organotin compounds on estuarine microorganisms. Dev Ind Microbiol 22: 529–535.
17. Hedges RW, Baumberg S (1973) Resistance to arsenic compounds conferred by a plasmid transmissible between strains of *Escherichia coli*. J Bacteriol 115: 459–460.
18. Herrero M, de Lorenzo V, Timmis KH (1990) Transposon vectors containing non-antibiotic resistance selection markers for cloning and stable chromosomal insertion of foreign genes in gram-negative bacteria. J Bacteriol 172: 6557–6567.
19. Kado CI, Liu ST (1981) Rapid procedure for detection and isolation of large and small plasmids. J Bacteriol 145: 1365–1373.
20. Keane PJ, Kerr A, New PB (1970) Crown gall of stone fruit. II. Identification and nomenclature of *Agrobacterium* isolates. Aust J biol Sci 23: 585–595.
21. Lebrun M, Loulergue J, Chaslus-Dancla E, Audurier A (1992) Plasmids in *Listeria monocytogenes* in relation to cadmium resistance. Appl Environ Microbiol 58: 3183–3186.
22. Mergeay M (1991) Towards an understanding of the genetics of bacterial metal resistance. Trends in Biotechnology 9: 17–24.
23. Mergeay M, Nies D, Schlegel HG, Gerits J, Charles P, Van Gijsgem F (1985) *Alcaligenes eutrophus* CH34 is a facultative chemolithotroph with plasmid-bound resistance to heavy metals. J Bacteriol 162: 328–334.
24. Powell B, Mergeay M, Christofi N (1989) Transfer of broad host range plasmids to sulphate-reducing bacteria. FEMS Lett. 59: 269–274.
25. Rouch D, Camakaris J, Lee BTO, Luke RKJ (1985) An inducible plasmid mediated copper resistance in *Escherichia coli*. J Gen Microbiol 131: 939–943.
26. Sadouk A, Mergeay M (1993) Chromosome mapping in *Alcaligenes eutrophus* CH34. Mol Gen Genet 240: 181–187.
27. Schatz A, Bovell C (1952) Growth and hydrogenase activity of a new bacterium, *Hydrogenomonas facilis*. J Bacteriol 63: 87–98.
28. Schlegel HG, Cosson JP, Baker JM (1991) Nickel-hyperaccumulating plants provide a niche for nickel-resistant bacteria. Bot Acta 104: 18–25.
29. Schmidt T, Schlegel HG (1989) Nickel and cobalt resistance of various bacteria isolated from soil and highly polluted domestic and industrial wastes. FEMS Microbiol Ecol 62: 315–328.
30. Schmidt T, Stoppel RD, Schlegel HG (1991) High-level nickel resistance in *Alcaligenes xylosoxydans* 31A and *Alcaligenes eutrophus* KTO2. Appl Environ Microbiol 57: 3301–3309.
31. Selifonova O, Burlage R, Barkay T (1993) Bioluminescent sensors for detection of bioavailable Hg (II) in the environment. Appl Environ Microbiol 59: 3083–3090.
32. Shapiro JA (1977) Appendix B: Bacterial plasmids. In: DNA insertion, elements, plasmids and episomes, Bukhari AI, Shapiro JA, Adhya SL (eds), pp 601–703.
33. Silver S, Misra TK (1988) Plasmid-mediated heavy metal resistances. Ann Rev Microbiol 42: 717–743.
34. Top E, De Smet I, Verstraete W, Dijkmans R, Mergeay M (1994) Exogenous isolation of mobilizing plasmids from polluted soils and sludges. Appl Environ Microbiol 60: 831–839.
35. Top E, Mergeay M, Springael D, Verstraete W (1990) Gene escape model: transfer of heavy metal resistance genes from *Escherichia coli* to *Alcaligenes eutrophus* on agar plates and in soil samples. Appl Environ Microbiol 56: 2471–2479.
36. van der Lelie D, Sadouk A, Ferhat A, Taghavi S, Toussaint A, Mergeay M (1992) Stress and survival in *Alcaligenes eutrophus* CH34: effects of temperature and genetic rearrangements. In: Gene transfers and environment. Gauthier MJ (ed.), Springer-Verlag.

37. Van Gijsegem F, Toussaint A (1982) Chromosome transfer and R-prime formation by an RP4:: Mini-Mu derivative in *E. coli, S. typhimurium, K. pneumoniae, Pr. mirabilis.* Plasmid 7: 30–44.
38. Williams JR, Morgan AG, Rouch DA, Brown NL, Lee BTO (1993) Copper-resistant enteric bacteria from United Kingdom and Australian piggeries. Appl Environ Microbiol 59: 2531–2537.
39. Wuertz S, Miller CE, Pfister RM, Cooney JJ (1991) Tributyltin-resistant bacteria from estuarine and freshwater sediments. Appl Environ Microbiol 57: 2783–2789.
40. Yang CH, Menge JA, Cooksey DA (1993) Role of copper resistance in competitive survival of *Pseudomonas fluorescens* in soil. Appl Environ Microbiol 59: 580–584.
41. Schmidt T, Schlegel HG (1994) Combined Nickel-Cobalt-Cadmium Resistance encoded by the *ncc* locus of *Alcaligenes xylosoxidans* 31A. J. Bacteriol. **176**, 7045–7054.

Molecular Microbial Ecology Manual **6.1.8**: 1–14. 1995

Biodegradation genes as marker genes in microbial ecosystems

BRUCE M. APPLEGATE [1,3], UDAYAKUMAR MATRUBUTHAM [3], JOHN SANSEVERINO [4] and GARY S. SAYLER [1,2,3]

[1] *Department of Microbiology and* [2] *The Program of Ecology,* [3] *Center for Environmental Biotechnology, University of Tennessee,* [4] *Technology Development Division, IT Corporation, Knoxville, TN 37932, USA*

Introduction

For the detection of specific biodegradative microorganisms in the environment, DNA hybridization has been extensively used and throughly reviewed [1, 4, 10]. More often, DNA hybridization can be applied in colony hybridization where the cells from any soil sample are washed off, cultured on nutrient containing agar plates, transferred to nylon membranes, lysed, their DNA fixed to the nylon membranes and subsequently hybridized with appropriate catabolic gene probe(s) [11]. DNA hybridization can also be used with directly extracted DNA from soil or any other source. Colony hybridization first requires culturing of cells while direct DNA extraction is done by *in situ* lysis of cells in the soil. Therefore, direct extraction of DNA overcomes the primary disadvantage in colony hybridization (culturability) as only a small fraction of the actual bacterial population in the soil is culturable. The DNA obtained through direct extraction from soil is usually blotted on nylon membrane and hybridized with appropriate catabolic gene probe(s) as in colony hybridization [7].

DNA hybridization techniques are often employed to determine the genetic potential of catabolic organisms in soil and also to estimate the frequency of specific catabolic gene(s) in any particular niche. As reported recently, the frequency of the naphthalene dioxygenase gene, *nahA*, was found to have a strong correlation with the presence of contaminant concentration in various Manufactured Gas Plant (MGP) soils [9].

DNA extracted directly from soil has primarily been used in DNA hybridization for determining gene frequencies. However, the direct DNA extraction method has only been applied for semi quantitative enumeration of gene frequencies as evident from the *nahA* gene study of MGP soils [9], and more appropriately in a qualitative capacity as there are numerous unknowns which limit the quantitative capacity of this technique. However, if a certain set of assumptions are made, direct DNA extraction can be used quantitatively. Two main variables are to be considered when using direct DNA extraction: lysis efficiency and extraction efficiency. Lysis efficiency can be further divided into two sub variables: overall and nondifferential efficiency. The latter of these two is the most significant one.

All direct DNA extraction techniques assume that all cell types in a soil sample are lysed with the same efficiency. This assumption is very crucial because if partial or preferential lysis occurred, the extracted DNA would not be representative of the population in the soil sample. Once the assumption of non-differential lysis is made the overall efficiency of lysis is not important and in most cases can be assumed to be 100%. The next variable to be considered is the extraction efficiency.

After cell lysis, the recovery of DNA can be estimated by the extraction efficiency of any particular procedure. The efficiency can be monitored by the addition of an internal DNA standard added to the sample after the cells have been lysed. This procedure involves the addition of a known amount of an exogenous DNA and following up its recovery by measuring its concentration in the final extracted DNA product. The recovery percentage of the internal DNA standard can thus be easily determined. This percentage is used as a correction factor to adjust the concentration of specific catabolic target genes or sequences to the concentration of total DNA extracted directly from soil. The internal standard also allows quantifying specific catabolic organisms among the total population and deriving the relative numbers of each biodegradative genotype in the soil sample.

An important aspect of using direct DNA extraction/DNA hybridization over colony hybridization is that the former enables normalization of the hybridization data. DNA extracted from different soil samples can be compared if they are normalized to the relative number of organisms originally present in the soil from which the total DNA was presumably extracted. Normalization of extracted DNA can be done by probing for 16S rRNA gene sequences in the total DNA. In the actual protocol, a universal oligonucleotide probe of 15 bases complimentary to the 16S rRNA gene of eubacteria (ACGGGCGGTGTGTRC), is allowed to hybridize with the extracted DNA blotted on a nylon membrane [13]. Then a ratio is arrived between the frequency of catabolic genes and the frequency of the 15 mer hybridizing sequences in the total DNA. In addition, if it is assumed that there are at least 5 copies of 16S rRNA genes per eubacterial cell then the 15 mer hybridization can be used as a way to estimate the overall number of eubacteria in the sample.

With the increasing number of available catabolic gene probes, it may be possible to ascertain the genetic structure of a biodegradative community. The probes can be used to dissect the community into various catabolic genotypes and compare them to other communities after normalization. The methods mentioned in the above section are described in detail with procedural steps, mathematical formulas and their purpose, in the remaining sections of this chapter.

Experimental approach

The primary method employed in detecting the presence of catabolic genes is DNA:DNA hybridization. A particular DNA sequence of interest is used as a gene probe in this method of hybridization. The gene probe is labeled either radioactively or nonradioactively, denatured and allowed to reassociate with putative complementary strands in any DNA sample. The sample may represent DNA in bacterial colonies (as in colony hybridization) or DNA by direct extraction from soil (as in slot or dot blots). Depending on the method of probe labelling, the reassociated DNA is then subjected to autoradiography or enzymatic assays. Positive DNA samples are directly scored in the case of colony hybridization, but are quantified with a standard curve for directly extracted DNA.

Probe generation

The advent of the polymerase chain reaction (PCR) has revolutionized molecular biology and probe development. The ability of PCR to geometrically amplify DNA has made PCR a valuable tool for generating gene probes. It allows production of large quantities of probe directly from chromosomal or extra chromosomal DNA purified from the bacterium carrying the desired gene without having to subclone the gene into a cloning vector. The advantage of using PCR generated templates in probe generation over DNA clones is the elimination of contaminating vector sequences in gene probes. With DNA clones, even after gel purification of DNA inserts the gene probes can still contain vector sequences, which are undesirable due to the potential of false positives.

The major drawback in using PCR for probe generation is that the DNA sequence of the probe of interest and/or its surrounding region has to be known to design the PCR primers. However using unsequenced gene probes to detect catabolic genes in the environment is risky in that the flanking sequences of the probe may code for conserved housekeeping genes like ribosomal RNA genes.

The basic premise for probe generation with PCR is the following. PCR primers are generally designed to produce the desired probe with a size of one kilobase or less. It is essential to have small probes containing the least conserved region of catabolic genes instead of the entire gene, as the latter may cause false positives after hybridization. One such typical DNA probe is for the gene encoding the multicomponent enzyme naphthalene dioxygenase. This gene has conserved regions coding for reductase or ferredoxin activity and also non-conserved region coding for the substrate binding subunit of the enzyme [12]. Therefore, if PCR primers were designed for the substrate binding region of the gene, the resulting probe will be very unique and specific as opposed to using the entire gene as a probe in hybridization.

Once the primers are chosen, the DNA template for probe generation is amplified with the DNA from the wild type strain carrying the specific catabolic gene as opposed to using the DNA from a clone. The reason for not using the clone as template in PCR is as previously discussed to avoid amplification of vector sequences. This is important as the DNA used as standards (described under standard sections) on a slot blot are usually from clones and not from the wild type strain. After estabilishing the conditions for the amplification of any specific fragment, the DNA is reamplified for at least three successive times to remove any background and non-specific sequences by dilution.

A single stranded radioactive probe then can be produced by asymetric linear amplification of the template DNA using only one of the primers and a radiolabeled nucleotide in this case dCTP(P^{32}). This reaction can be compared to primer extension with the exception that it is repeated for as many cycles the thermocycler is programed. The specific activity of the single stranded probe generated in this manner using a substituted radiolabelled nucleotide is high as the polymerase directly incorporates the radiolabel. The specific activity can be calculated theoretically with the following equation.

$$\left(\frac{\#dCTP}{probe} \right) \left(\frac{dpm}{\mu g\ dCTP} \right) = \left(\frac{dpm}{\mu g\ probe} \right)$$

Using the above value, the reaction efficiency can be calculated by TCA precipitation. The PCR method of probe generation is also effective when using non-isotopic labels such as digoxygenin (Genius, Boehringer Mannheim).

Procedures

Probe preparation

Polymerase Chain Reaction (Perkin Elmer Norwalk, CT)
- 10 µl 10× reaction buffer
- 2 µl each (10 mM) dATP, dGTP, dTTP, α-dCTP^{32}
- 2.5 µl primer of forward or reverse primer (5 ng/µl)
- 50 ng template DNA
- 0.5 µl *Taq* DNA polymerase (5U/ml)
- Adjust volume to 100 µl with dH$_2$O
- 38 cycles

Colony hybridization

Colony hybridization is very useful in examining the culturable catabolic organisms and allows the retrieval of these organisms for further examination. The organisms from any sample material are plated on non selective media usually quarter strength YEPG and subsequently probed for specific genes after fixing their DNA on to nylon membrane [11]. The number of positive organisms are then used in the enumeration of culturable organisms for any particular genotype based on the dilution factor. The basic colony hybridization protocol for an environmental sample is outlined below [11]:

1. One gram of soil is suspended in 8 ml of phosphate buffered saline (PBS) and 1 ml of sodium pyrophosphate (0.1%) and vortexed vigorously for 1 min.
2. After vortexing, a serial dilution of the suspension is spread on quarter strength YEPG agar plates to determine the number of culturable cells per gram of soil. Plates with an appropriate number of colonies (~ 100) are then subjected to colony hybridization.
3. The colonies are blotted on an 82 mm diameter nylon membrane by placing it on the plate for a minimum of one minute (note: if recovery of colonies is desired the nylon membrane and the petri dish need to be marked to locate the position of positive colonies).
4. A piece of Handi wrap is spread on a flat surface and 2 ml of lysis buffer is spotted on it (Handi wrap allows enough surface tension in the drop as to prevent it from spreading). The nylon membrane is placed with the colony side up on the drop for 2 minutes.
5. After lysis, place the membrane with colony side up on 2 ml of neutralization solution spotted on a Handi wrap as in the previous step.
6. After neutralization, the membrane is placed colony side up on Whatman 3 mm paper until dry.
7. After drying, the DNA on the membrane is fixed by baking the membrane in an 80 °C oven for one hour or by UV crosslinking. The latter can be performed by placing the membrane colony side down on a Handi wrap spread on a transilluminator and irradiating for about 1–4 min.

8. The membrane can be stored in a plastic bag at room temperature until hybridization.

Solutions

- Phosphate Buffered Saline (PBS)
  - Na_2HPO_4 1.2 g
  - NaH_2PO_4 0.8 g
  - NaCl   8.5 g
  - dH_2O 1 L
  - pH   7.6

- 0.25× YEPG Agar Medium
  - Yeast Extract   0.05 g
  - Dextrose   0.25 g
  - Polypeptone   0.50 g
  - NH_4NO_3 0.05 g
  - dH_2O 1 L
  - adjust pH to 7.0 with NaOH
  - Agar   17 g

- Lysis Buffer
  - NaOH   0.5 M
  - NaCl   1.5 M

- Neutralization Solution
  - Tris/HCl   0.5 M (pH 7.4)
  - NaCl   1.5 M

Slot blot preparation

Slot or dot blots are extensively used in analyzing DNA for complimentary sequences of specific target genes. The method basically consists of applying appropriate amounts of the DNA to a nylon membrane after alkali denaturation and fixing it to the membrane by baking or UV crosslinking. The blot also contains a set of DNA standards that allows the generation of a standard curve to quantify the target DNA. Finally the DNA is hybridized on the blot with an appropriate gene probe and is followed by detection and quantification.

Protocol

1. Samples and standard DNAs are prepared in TE buffer to a final volume of 100 µl in 1.5 ml Eppendorf tubes (with locking caps to prevent opening while boiling).
2. 400 µl of 0.5 N NaOH is added to the DNA samples resulting in a final concentration of 0.4 N NaOH in 500 µl.
3. The samples are then placed in a boiling water bath for 10 minutes, then immediately cooled on ice. They are stored on ice until ready for application to the nylon membrane. (Note: it might be necessary to spin the samples upon cooling to collect any condensed material on the lid or side of the tube.)
4. The samples are then loaded onto the slot/dot blot apparatus according to the instructions for the specific blotting apparatus that is being used.
5. After completion, the membrane is washed in 2× SSC.
6. It is then placed on a Whatman 3 mm filter paper and allowed to air dry.
7. The DNA is then fixed to the membrane either by baking at 80 °C or by UV crosslinking and stored at room temperature until ready for hybridization.

Solutions
- TE buffer
  - Tris/HCl   10 mM (pH 8.0)
  - EDTA   1 mM (pH 8.0)

- 2× SSC*
  - NaCl   17.53 g
  - Sodium citrate   8.82 g

Note
* Dissolve in 800 ml dH$_2$O and adjust the pH to 7.0 with few drops of 10 N NaOH. Bring up the final volume to 1 L. Aliquot to small volumes and sterilize by autoclaving.

Preparation of standards for colony and slot blot hybridization

Colony hybridization standards

Whenever performing hybridizations, it is important to include positive and negative controls both for the target and the probe sequences. The actual strain from which the target gene was isolated should preferentially be used as a positive control instead of an *E. coli* clone carrying the target gene or sequence on a vector. There are two reasons for this: first, it gives a control for lysis efficiency of the original strain, and second, the copy number of the target gene is maintained consistently. The original strain might carry one or few copies of the target gene based on whether the gene is located on the chromosome or on a plasmid whereas a clone usually carries the gene on a high copy number vector (200–300 copies) and hence would provide a stronger signal upon hybridization than the original strain. The copy number also influences the exposure time of the hybridized blots to X-ray films.

For negative controls any other strain that is similar to the original strain and carries no target gene or sequences should be used in colony hybridization.

1. Positive and negative control strains are cultured on petri plates with the appropriate media and blotted on the nylon membrane as described in colony hybridization section.
2. The blots are then cut into 6 equal sections (like the pieces of a pie) and stored.

Slot blot standards

Slot blot standards are prepared from DNA samples consisting of target catabolic genes or sequences. The target DNA sample is usually made from a clone of the target gene or the original strain containing the gene. However, it is convenient to use the DNA from the clone for DNA standards, as the amount of the standard DNA applied to a blot can be calculated from the target mass in the solution.

To create a clone of the target gene or sequence, PCR generated probe templates (described in the early sections of this chapter) can be directly cloned into a TA cloning vector (Invitrogen San Diego, Ca).

The TA cloning vector is designed with 5 prime thymidine overhangs which assists the cloning of PCR amplified fragments. *Taq* DNA polymerase preferentially adds adenine nucleotides to the 3 prime ends of amplified fragments [6]. The constructed target clone is transformed and amplified in an *E. coli* host strain. Then the recombinant plasmid is extracted and purified by ethidium bromide-cesium chloride ultracentrifugation [5]. The DNA concentration is determined by UV absorption at 260 nm. The following equation is used in calculating the amount of target/vector DNA that should be added to the blot.

$$\left(\frac{\text{kb target}}{\text{kb vector + target}}\right)\left(\frac{\text{total DNA μg}}{\text{μl}}\right) = \text{target DNA μg / μl}$$

Once the amount of target DNA per ml of the DNA solution has been determined, the solution is diluted to obtain 100 μl stock solutions of each of the following concentrations: 10 ng, 3 ng, 1 ng, 0.3 ng, 0.1 ng, and 0.03 ng. These concentrations are used in preparing a standard curve. The stock solutions are loaded on the slot blot in the following order: 10 ng, 1 ng, 3 ng, 0.1 ng, 0.3 ng, and 0.03 ng. This prevents formation of a contiguous smear on the blot and results in more distinct bands.

Methods such as densitometry, direct counting of radioactivity by scintillation can be employed to obtain values from the blot and to generate a standard curve.

Hybridization

The hybridization is performed after blotting and fixing the DNA to the nylon membrane. The temperature for the hybridization depends on the melting temperature (Tm) of the target gene being probed. It is usually maintained at 25 °C below the Tm of the target gene [2]. Thus it varies and is the responsibility of the researcher in deriving the appropriate hybridization temperature. The nylon membrane has positive charges and if they are not blocked the majority of the probe DNA will bind to the membrane instead to the target gene. There are many methods including, addition of salmon sperm DNA or powdered milk to block these positive charges. However, a hybridization solution developed by Church and Gilbert [3] provides an effective charge blocking like the other solutions. The hybridization solution consists of 7% sodium dodecyl sulfate (SDS) and has a long shelf life. However due to the high percentage of SDS pre-heating may be required before each use.

Protocol

1. The membrane is placed in a heat sealable pouch (Kapak/Scotchpak) and filled with the approriate amount of the hybridization solution. For a pint size (0.47 L) and a quart size (0.95 L) pouch 10 and 14 milliliters of the solution is used respectively. In the case of colony hybridization, a piece of the positive and negative control containing membrane is placed in each pouch.
2. The air bubbles are removed from the pouch over the edge of the lab bench and the pouch is sealed with a bag sealer.
3. The blots are prehybridized at the appropriate temperature by placing the pouch in a shaking water bath for at least 1 h.
4. The labeled probe is then added to the pouch by cutting a corner off and adding the probe to a final count of 10^6 dpm per ml. It is important not to introduce any air bubbles during the process.
5. The blots are hybridized at the appropriate temperature for a minimum of 8 hrs or for overnight by placing the pouch in a shaking water bath.
6. The membrane is removed from the hybridization solution and washed with high stringency wash buffer by placing the membrane in a shallow container containing the wash buffer and incubating the container at the same temperature used for hybridization. Four such washes are performed at 15 min intervals.
7. The blot is then placed on Whatman 3 mm paper and allowed to air dry. If the blot is to be stripped and reprobed, it should not be dried.
8. The dried membrane is subjected to autofluorography using two intensifying screens at −70 °C.

Solutions

- Hybridization solution
    - 0.5 M NaH_2PO_4
    - 1 mM disodium EDTA
    - 7% SDS
    - 1 L dH_2O
    - pH 7.2

- High Stringency Wash Buffer
    - 10 mM NaCl
    - 20 mM Tris
    - 1 mM disodium EDTA
    - 0.5% SDS
    - pH 7–8

Data analysis

The analysis of the hybridization data is crucial when using biodegradative genes as marker genes for the analysis of a microbial community. There are numerous calculations to be made in processing the information from the DNA directly extracted from the soil. While the calculations pertaining to colony hybridization are simple and straight forward, the following text will go through both the derivation and calculations used to determine numbers of organisms and the normalization of the data.

Colony hybridization

The calculation of the number of culturable organisms with a particular genotype is straight forward. The number of positive colonies is multiplied by the dilution factor as in the following equation:

$$(\text{\# of positive colonies})(\text{dilution factor}) = \frac{\text{\# organisms}}{\text{g sample}}$$

The results obtained from this can be reported by themselves or as the percentage of the total of culturable organisms. It is important to re-emphasize the word culturable as these results may significantly under estimate the population of the particular genotype.

Direct DNA extraction analysis

DNA extraction analysis is divided into three main areas: calculation of extraction efficiency, cell number determination, and normalization. At this point the assumptions that allow the use of these calculations need to be reiterated. The main assumption is that the cells are all lysed at the same efficiency. The classic technique using bead beating and high temperature SDS lysis, as described by Johnston *et al.* in this volume, is vigorous enough to warrant this assumption. The next assumption is that of overall lysis efficiency. This assumption is not as crucial as the previous one as the normalization of the data using the 16S oligo allows the comparison of samples. However if one wants to determine total biomass one can assume 100% lysis and use the oligo to give the total number of eubacteria present. The following sections will elucidate the calculations involved in this analysis.

Extraction efficiency

Extraction efficiency can be measured by the addition of an internal standard DNA to the sample, after the cell lysis step of most direct extraction protocols. A suitable DNA must be chosen for this purpose. The standard DNA must not contain sequences that are commonly present in the soil. A 500 bp fragment of lambda DNA has been chosen for this purpose as lambda phage are not readily found in the environment (exact DNA sequence of the lambda fragment is given in the chapter by Johnston *et al.* of this volume). Another very important aspect of using the 500 bp lambda DNA as an internal standard is one can probe with another piece of the phage to ascertain whether or not there is any lambda phage present in the soil [8]. If present, the amount of phage DNA must be subtracted to accurately determine the extraction efficiency. The 500 bp fragment was cloned into the TA cloning vector and subsequently amplified in *E. coli* and purified by ethidium-bromide ultracentrifugation. Before addition of the lambda fragment clone it is linearized by cleavage with an appropriate restriction endonuclease. This is done in case the extracted DNA is purified by ethidium-bromide ultracentrifugation. Uncut plasmid would super coil resulting in the separation of the extracted DNA and the standard. A 100 ng aliquot of lambda DNA is added to the sample (this is the amount of target DNA).

The equation for determining the extraction efficiency is shown below:

$$\frac{\text{ng } \lambda \text{ recoverd}}{100 \text{ ng } \lambda} = \% \text{ extraction efficiency}$$

Determination of cell numbers

Cell numbers can be quantitatively ascertained by determining the amount of target DNA hybridized with the probe. Once this is determined the following equations can be used to give the number of cells present with the particular gene. The first step is to determine the mass of one molecule of target DNA. This is determined using the following equation:

$$\text{ng of one target} = (\text{\# bp target}) \left(\frac{660g}{\text{mol bp}}\right)\left(\frac{1 \text{ mol bp}}{6.023 \times 10^{23} \text{ molecules}}\right)\left(\frac{10^9 \text{ ng}}{g}\right)$$

After the determination of the mass of one target molecule in nanograms as shown above this value can be plugged into the following equation along with the extraction efficiency to determine the number of cells containing that particular catabolic genotype.

$$\text{\# of cells} = \left(\frac{\text{ng hybridized}}{\text{ng of one target}}\right)\left(\frac{1}{\text{copy of \# target}}\right)\left(\frac{1}{\text{extraction efficiency}}\right)\left(\frac{1}{\text{sample fraction loaded}}\right)$$

Normalization can be done in two ways: the first approach is to relate the frequency of the catabolic gene to that of the value obtained using the 16S oligo probe. This approach allows the comparison of different samples to one another. However the alternative is to use the value obtained for the 16S hybridization to obtain the total number of eubacteria in the sample. This is done by assuming 5 copies of the 16S gene per cell . The standard in this case is a 1400 bp fragment of the 16S rRNA DNA from *Escherichia coli* which contains the complimentary sequence of the eubacterial oligo. The same basic equation is used as in calculating cell numbers except that the copy number is 5 as opposed to one in the case of most catabolic genes. Once this overall eubacteria cell number is obtained the data can be reported as the percentage of the community that is a particular catabolic genotype.

References

1. Atlas RM, Sayler GS, Burlage RS, Bej AK (1992) Molecular approaches for environmental monitoring of microorganisms. Biotechniques 12: 706–717.
2. Britten RJ, Graham DE, Neufield BR (1974) Analysis of repeating DNA sequences by reassociation. In: Grossman L, Moldave K (eds) Methods in Enzymology, Vol 29E, p. 363. New York: Academic Press.
3. Church GM, Gilbert W (1984) Genomic sequencing. Proc Natl Acad Sci 81: 1991–1995.
4. Holben WE, Jansson JK, Chelm BK, Tiedje JM (1988) DNA probe method for the detection of specific microorganisms in the soil bacterial community. Appl Environ Microbiol 54: 703–711.
5. Maniatis T, Fritsch EF, Sambrook J (1989) Molecular Cloning: A Laboratory Manual. New York: Cold Spring Harbor Press.

6. Mead DA, Pey NK, Herrnstadt C, Marcil RA, Smith LM (1991) A universal method for the direct cloning of PCR amplified nucleic acid. Biotechnology 9: 657–663.
7. Ogram A, Sayler GS, Barkay T (1987) The extraction and purification of microbial DNA from sediments. J Microbiol Meth 7: 57–66.
8. Sanger F, Coulson AR, Hong GF, Hill DF, Petersen GB (1982) Nucleotide sequence of bacteriophage λ DNA. Mol Biol 162: 729–773.
9. Sanseverino J, Werner C, Fleming J, Applegate B, King JMH, Sayler GS (1993) Molecular diagnostics of polycyclic aromatic hydrocarbon biodegradation in manufactured gas plant soils. Biodegradation 4: 303–321.
10. Sayler GS, Layton AC (1990) Environmental application of nucleic acid hybridization. Ann Rev Microbiol 44: 625–648.
11. Sayler GS, Shields MS, Tedford ET, Breen A, Hooper SW, Sirotkin KM, Davis JW (1985) Application of DNA-DNA colony hybridization to the detection of catabolic genotypes in environmental samples. Appl Environ Microbiol 49: 1295–1303.
12. Simon JM, Osslund TD, Saunders R, Ensley BD, Suggs S, Harcourt A, Suen W-C, Cruden DL, Gibson DT, Zylstra G (1993) Sequences of genes encoding naphthalene dioxygenase in *Pseudomonas Putida* G7 & *Pseudomonas sp* NCIB 9816–4. Gene 127: 31–37.
13. Stahl DA, Flesher B, Mansfield HR, Montgomery L (1988) Use of phylogenetically based hybridization probes for studies of ruminal ecology. Appl Environ Microbiol 54: 1079–1084.

Molecular Microbial Ecology Manual **6.2.1**: 1-16, 1995
© 1995 *Kluwer Academic Publishers.*

Design of microcosms to provide data reflecting field trials of GEMS

MARY A. HOOD and RAMON J. SEIDLER

Dept Cellular and Molecular Biology, the University of West Florida, Pensacola,
Fl 32514 US Environmental Protection Agency, Corvallis, OR 97333, U.S.A.

Introduction

The issue of releasing genetically engineered micro-organisms into the environment is not one of either we do or we do not. It has already been done [19]. There have been more than 1,000 applications for testing genetically engineered organisms in the environment, most of which have or will occur in the USA [28].

The primary purpose of these limited release experiments is to determine the efficacy of newly patented life forms. Thus, to date, while scientists have released hundreds of GEMs and 1 plants, almost no information has been gathered related to risk assessment. Risk assessment is essential if we are to prepare for inevitable undocumented releases of GEMs, such as accidental (or purposeful) spills down sink drains, etc.

Twenty years have passed since the first recombinant micro-organism was constructed, and for more than a decade GEMs have been used on a limited basis for bioremediation, agricultural/biocontrol technologies, and industrial reclamation and fermentation applications. But the long-term risks have yet to be calculated, and numerous scientists have expressed their concern. Seven years ago, Dr. R.R. Colwell summarized such apprehension by cautioning that "careful microcosm and mesocosm testing, as well as ample containment trials, should be done before a large scale field application can be done" [10]. Two questions relevant both to risk assessment and to any large scale field application are: (1) Can the risks of GEMs in the natural environment be assessed without actually releasing the agents into the environment? And (2) can the impact of a GEM on the environment be anticipated – is our knowledge of microbial ecology sufficient to predict with confidence the fate and longevity of an agent? It is with such questions in mind that the strategy of microcosm (and mesocosm) application has developed.

Since the 1970's, microcosms – or model ecosystems – have been used to ascertain the ecotoxicity of anthropogenic substances. Gillett and Witt [17] have defined microcosm as "a controlled, reproducible laboratory system that attempts to simulate the situation in a portion of the real world." Similarly, Wimpenny [46] recently defined microcosm as "a system that attempts to actually simulate, as far as possible, the conditions prevailing in the environment or part of the environment under study." Designs vary widely, but the intent of the microcosm is to faithfully simulate one or more aspects of the natural environment. These microcosm simulations are attempts to model both biotic and abiotic interactions, as well as fundamental ecological processes. Crucial to any microcosm experiment is monitoring and control during the experimental period [4]. Thus, these simulations are constructed with fewer variables than experiments conducted in the natural environment. Such simplification, however, is both an advantage and a disadvantage – the natural environment is anything but simple.

Cripe and Pritchard [11] wrote, microcosms are "laboratory systems that have been used by researchers to obtain environmentally relevant ecological information." Microcosms may be intact pieces of the ecosystem that are excised from the natural environment and brought into the laboratory with minimum disturbance with the hope that they reflect ecological processes occurring in that particular environment [16]. These kinds of microcosms are referred to as site specific. In contrast, other microcosms called "synthetic" or "constructed" microcosms include: (1) mixed flask cultures, (2) pond microcosms, (3) generic aquatic ecosystems, and (4) standard aquatic microcosms (as reviewed by Taub [43]). Synthetic microcosms may have the advantage of reproducibility and analytic ease (since they are composed of chemically defined media and the organisms are known), but they contain fewer organisms and the organisms may be strains that are laboratory adapted. Both types of microcosms, site specific and synthetic, lack large species abundance, and, because of size, fail to simulate large scale processes [43]. ·

Mesocosms are generally larger containerized systems placed in situ that attempt to reflect larger ecological processes and that measure more than just microbial processes [31]. They are supposed to fill the gap between laboratory microcosm studies and field release [9]. While designed to be placed in the field, the mesocosm functions in the same way a laboratory microcosm does. Thus, mesocosms may be considered a type of in situ microcosm. While it is a larger system and may reflect a higher level of complexity, the mesocosm has the same limitations as all enclosed systems.

Microcosms have been used most commonly to document survival, fate, die-off kinetics, and gene transfer among GEMs and other organisms that also may be added to the experimental model system. Very few experiments have been constructed to validate results with parallel studies conducted in confined situations using out-of-door test plots [8, 12]. Therefore, it has been easy for critics of microcosms to find fault with their value in biotechnology risk assessment of GEMs. Certainly microcosms can never fully mimic the natural environment because they cannot reproduce nature exactly. Scale limitations and the inability to include certain biological, chemical, and physical heterogeneities typical of the field are problems that the microcosm suffer from. Since they cannot be readily maintained, microcosms also suffer from exclusion of higher trophic levels; inclusion of mammals in microcosms can result in physical destruction and upsetting of the balance in physical and biological integrity within the microcosm. However, microcosms can be constructed to predict certain aspects of the field that are useful for fate and transport studies. Even if they cannot fully mimic the field, microcosms can be useful in risk assessments.

A significant advantage of a microcosm is the ease of experimental replication and design of appropriate controls such as temperature, moisture, light and nutrients. In addition, it is possible to obtain several samples from several ecosystems/locations and subject them to comparative, controlled experimental treatments at the same time within a microcosm. The capability for controlling environmental conditions allows one to experimentally determine impacts of specific environmental conditions on fate, survival and transport, and ecological effects of various recombinant organisms. In 1989, a workshop was sponsored to assemble scientists who discussed microcosm systems and regulatory endpoints relevant to environmental risk assessment of microbial products [12]. Fourteen microcosm systems were identified with various degrees of ecosystem complexity and performance characteristics.

Experimental approach

Figure 1 presents a general flow diagram of essential steps for designing release experiments using microcosm technology. Although microcosms have limitations and can never totally replicate the environment in terms of complexity, with the appropriate design, they can simulate certain aspects of the field and can, thus, be used to gain valuable information on risk assessment of GEMs. Within the discussion that follows, we will consider two aspects of microcosm applicability to GEM risk assessment: design and field validation.

Terrestrial microcosms

There are numerous designs of terrestrial microcosms [15,17], but most have several components in common: a boundary containing soil, instruments that document and control environmental conditions, and a sampling method – designed so that a finite amount of substance can be removed for analysis without disrupting any of the biological processes occurring within the unit.

We have found that control of basic environmental conditions such as

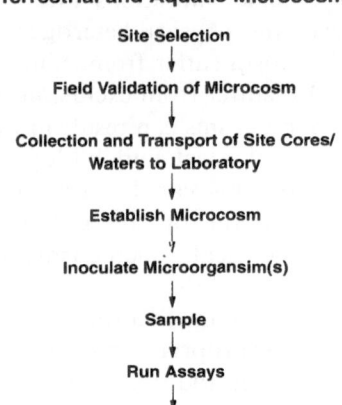

Terrestrial and Aquatic Microcosms

Site Selection
↓
Field Validation of Microcosm
↓
Collection and Transport of Site Cores/
Waters to Laboratory
↓
Establish Microcosm
↓
Inoculate Microorgansim(s)
↓
Sample
↓
Run Assays
↓
Analysis of Data

Figure 1. General flow diagram of essential steps for designing release experiments using microcosm technology.

soil moisture, temperature, and light exposure are crucial to experimental replication and repeatability of experimental findings [3,4]. In addition to documenting environmental conditions, one needs to document the kinds and numbers of organisms present and the physical/chemical properties of the soil. This would include gross soil composition such as percent sand, loam, clay, etc.; soil pH, TOC, total N, cation exchange capacity.

The distinction between natural versus artificial soil also needs clarification. Many investigators destroy the structure of soil by sieving prior to placing it into microcosms. Then the sieved soil is used to study leaching qualities or microbial transport. This creates a highly artificial system which has very little similarity to the natural system. Soil sieving disrupts soil texture and structure and will affect compaction and root distribution. The most desirable situation is to remove soil from the proposed release site and to maintain its integrity; this is especially important whenever transport and fate are primary experimental goals of a GEM release. If, however, the GEM is to be released into a plowed agriculture ecosystem, sieving the soil probably best represents the simulated condition. The use of an artificial soil may also have merits if the experiment is designed to validate a protocol in different regions of the world. In this manner, use of local soils will not add experimental variability to the protocol.

Aquatic microcosms

There are hundreds of aquatic model ecosystems described in the recent literature that vary in complexity from flasks containing water [30] to survival chambers [5] to large-scale sophisticated systems with flow-through, temperature, light, pH, DO and other controls [43]. Furthermore, there have been microcosms designed for numerous specific aquatic environments such as freshwater lake systems [23,41], sea water [30], river waters [45], and activated sludge [29], to list a few. However, the microcosm needed for risk assessment of GEMs must truly simulate the environment and the processes that are most relevant to the GEM under consideration [32]. In a review of site-specific aquatic microcosms as test systems for fate and effects of micro-organisms, Cripe and Pritchard [11] examined the major parameters in aquatic microcosm design. They discussed (1) microcosm size and scaling, (2) construction material, (3) light and temperature control, (4) water replacement, (5) mixing, and (6) sampling.

Comparison of the microcosm with the field is the most vital step in making the case for the reliability of data generated in microcosms. This process is called calibration or validation, and there have been several validations of microcosms [7,8,25,45].

Procedures

Terrestrial microcosms

Design

The soil-core microcosm [44] is a standard, site specific unit designed to study an intact component of the soil ecosystem. It is perhaps the only terrestrial microcosm system with validated performance conducted under laboratory, controlled environmental chambers, and field conditions [7,8]. Bolton and his colleagues compared four systems using the soil-core microcosm design: (1) intact cores at ambient (22 °C) laboratory conditions, (2) intact cores in a growth chamber with temperature fluctuations that simulate average field conditions, (3) field lysimeters, and (4) field plots. They used these systems to evaluate the fate and possible ecological effects of a rifampicin resistant pseudomonad [7,8,16].

Methods used to establish the soil-core microcosm have been described in detail [7,44] (Figs. 2 and 3). Briefly, the cores are typically 17.5 cm diameter and 60 cm in length and are removed from the soil intact by using steel coring devices containing polyethylene pipe (Fig. 2). Once transported to the laboratory, the surface soil may be removed if necessary and sieved. The surface soil taken from the top 10–15 cm of core soil can be readily amended with any number of supplements,

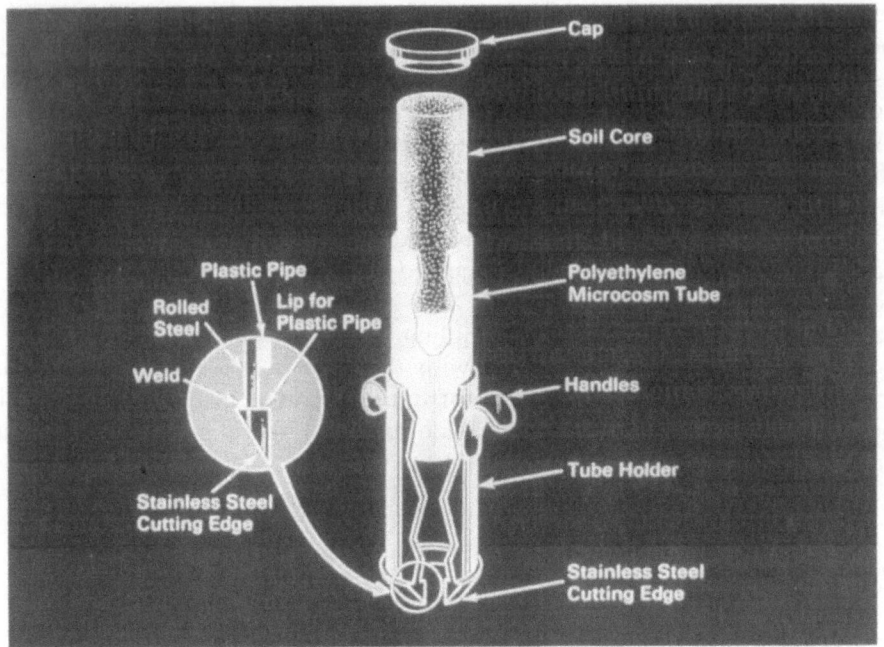

Figure 2. Diagrammatic representation of the apparatus used to remove an intact soil core. The polyethylene tube is inserted into the rolled steel pipe and pushed into the soil under high pressure. The soil can then be removed as an intact core contained within the polyethylene tube [44].

including fertilizer, plant matter, and the GEM itself. Seeds may also be planted after the top soil is added back to the column (Fig. 3).

Field validation

Soil-core microcosms were initially evaluated and validated for assessing the impact of utility wastes on primary productivity, uptake of metals, and disruption of cycling processes [44]. More recently, the same microcosm design was used to field calibrate ecological parameters resulting from the introduction of antibiotic resistant bacteria into an agricultural ecosystem [7,8]. Ecological parameters investigated included microbial viable counts of the added GEM, soil heterotrophs, fluorescent and total *Pseudomonas* sp., and diversity of heterotrophic bacteria.

Specifically, the four systems were inoculated with a rifampicin resistant rhizobacterium, *Pseudomonas* sp. RC1. Numerous comparisons and evaluations were conducted over time, soil depth, and stage of growth of wheat planted into the soil-cores. Briefly, the investigators verified that "similar " to "indistinguishable " micro-biological parameters were generally observed in all four systems, especially when microbial populations were compared at the same three-leaf stage of the wheat plant. Colonization of the rhizoplane by the introduced *Pseudomonas* RC1 in microcosms incubated at ambient temperature in the laboratory or in the growth chamber were similar to those in the field microcosms. However, conditions for indigenous microbial growth appeared more favorable in growth chamber microcosms as reflected by higher heterotrophic counts and greater species diversity than recorded in the field microcosms. In addition, ecosystem functional parameters such as soil dehydro-genase activity, plant biomass production, and ^{15}N-fertilizer uptake by wheat were different in the four systems, i.e., these parameters

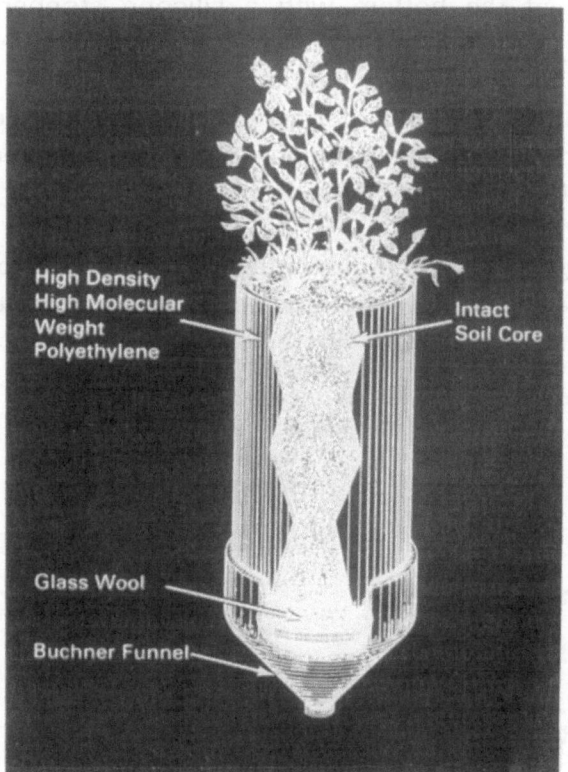

Figure 3. The intact soil core contained within the polyethylene tube can be fitted to a Buchner funnel for support. The surface can be planted with suitable specimens or allowed to grow the native flora present at the collection site. The GEM can be added to the core in any of several ways [7,8,16].

did not calibrate between laboratory and field. It is significant, however, that the introduced root-growth inhibiting rhizobacterium (*Pseudomonas* RC1) suppressed indigenous fluorescent pseudomonads on the rhizoplane, and the suppression was noted in both laboratory as well as field microcosms.

Aquatic microcosms

Design
Ecocore I and II: Ecocore I (Fig. 4) and II (Fig. 5) microcosms were first designed and used by microbial ecologists at the EPA Gulf Breeze Laboratory [11,37]. Ecocore I consisted of a 3.5 cm diameter × 40 cm length glass tube inserted into sediment to a depth of 8 cm and sealed with a silicone stopper; then the tube was removed and sealed again at the bottom with a silicone stopper for careful transport to the laboratory. Site water was added to give a volume of 175 ml, and an 18 gauge hypodermic needle was inserted to provide a gentle stream of air for aeration and mixing. Temperatures were kept constant by placing the cylinders in an incubator. A large number of cylinders could be used collectively providing the advantage of large replication.

Ecocore II (Fig. 5) consisted of a larger version of Ecocore I where a sediment core (up to 27 l) could be collected from the environment

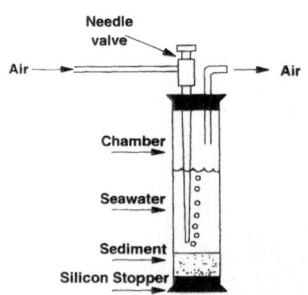

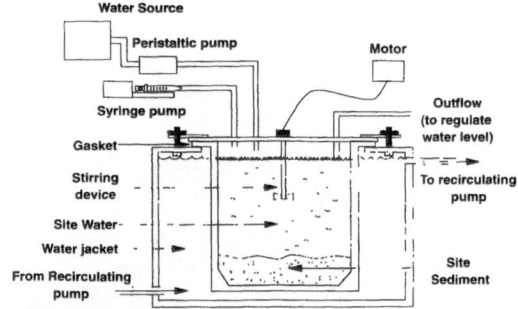

Figure 4. Diagrammatic representation of the aquatic-sediment microcosm called Ecocore I designed to provide ample replication of estuarine systems [11,37].

Figure 5. Diagrammatic representation of aquatic-sediment microcosm (Ecocore II) field validated for estuarine ecosystem [25].

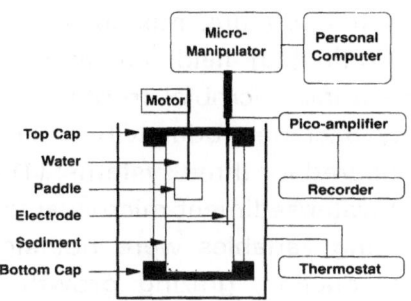

Figure 6. Diagrammatic representation of aquatic-sediment microcosm that has been field validated for freshwater river (Rhine River) ecosystem [45].

and returned to the laboratory. Core water was removed without disturbing the water sediment interface, and new site water was gently overlaid. Sealed with silicone stoppers and fitted with various monitoring probes, the vessels were maintained in a constant temperature water bath. Mixing was accomplished by a stirring rod operated by a small revolution motor. The system could be sealed, and in this case air was provided above the water-air interface. A flow-through system could be operated with a constant-level siphon used to maintain water at a preset level.

The following is a description of the microcosm used (Fig. 6) by the Netherlands Institute for Oceanic Sciences [45]: A plexiglas tube (vol 1.2 l) provided both corer and incubation chamber. Sediment cores were inserted into lake bottom sediments and removed to collect half sediment and half water. Various sealing devices, washers and O-rings were used to provide a closed system. Cores were returned to the lab within the same day. A stirring paddle mounted 3 cm above the water sediment interface provided mixing. The cylinders were placed in a water bath and maintained at a constant temperature. Using peristaltic pumps, filtered site water was added to the cylinders at a flow rate of 0.1 ml/h. Water was removed either by peristaltic pumps or allowed to pass through overflow holes.

Field validation
The two aquatic microcosms described above have been field validated [25,45]. These microcosms fit into the general category called site specific microcosms. They were comprised of replica cylinders containing site water and/or water and sediment and control mecha-

nisms for adjusting temperature, mixing, light and water replacement. Kroer and Coffin [25] field validated a modified Ecocore microcosm by measuring microbial trophic level interactions, i.e., primary productivity, bacteria productivity and grazing. These measurements were compared for three systems: (1) field sites, (2) water microcosm, and (3) water/sediment microcosms. Within these three systems, the following variables were not significantly different: bacterial carbon production, grazing growth efficiency (ratio of production to grazing), and bacteria abundance and activity (^{3}H-thymidine uptake). Although phytoplankton primary production varied significantly between the field site and the microcosms (probably due to the container effects), the differences were measurable and predictable. The study [25] concluded that because differences between field sites and microcosm could be quantified, such microcosms could be useful tools for risk assessment of GEMs and would be valid for up to 19 days.

Wagner-Dobler et al. [45] measured nutrient levels, photosynthetic potential, community structure of heterotrophic bacteria, thymidine incorporation rates and respiratory activity of the sediment community in microcosms with water and sediments and in field sites. These ecosystem parameters were indistinguishable for the first 2 weeks, but by 3 and 4 weeks small differences were detected although photosynthetic potential, heterotrophic bacterial community structure (depth profiles), and oxygen microgradiants were not different among the microcosm and field sites. They concluded that the microcosms would be valid models of aquatic sediments for up to 4 weeks.

Endpoints relevant to risk assessment

The following endpoints have been used or suggested for use in comparing terrestrial or aquatic microcosms with the field. After each parameter, a current reference is given that describes a method commonly employed.

Primary parameters ("A" refers to those applicable to aquatic microcosms; "T" refers to those applicable to terrestrial microcosms): biomass (A,T)

microbial biomass (A,T) total bacteria - AODC [20].

total viable bacteria (A,T) [24].

plankton (A) [18].
biomass plants (T) [7].
metabolic activity (A)
   heterotrophic activity-tritiated thymidine uptake [34] or
   ^{14}C-labeled organic-carbon compounds [32].
grazing rates (A) [47].
primary productivity (A) - chlorophyll a or ^{14}C-carbonate uptake [33].
biogeochemical parameters (A,T)
   carbon-total and organic [42].
   nitrogen-ammonium, nitrate, nitrite levels [42].
   dissolved oxygen- D.O. meters and probes.
   phosphate [33].
enzyme activity (A,T)
   dehydrogenase [36].
   chitinase [21].
   phosphatase [36].
bacterial community structure (A,T)[45].
species diversity index (A,T) [1].
mycorrhizal colonization (T) [40].
numbers and species diversity (A,T) [1]
   nitrogen fixers and rates - acetylene reduction method [35].
   nitrifiers [22].
   denitrifiers [27].
   fluorescent pseudomonads [38].
   cellulolytic bacteria [39].
   chitinolytic bacteria [21].
   lipid analysis [14].

Sampling procedures and frequencies

Terrestrial

Sampling from terrestrial habitats presents the challenge of optimizing and standardizing methods for each experimental situation. Recovery methods vary in efficiency, and satisfactory procedures are often specific to the particular plant/microbe situation. Several methods are available, for example, for recovering microorganisms from leaf surfaces [13]. For enumeration from soil, we suggest that more than one suspending buffer be used to optimize GEM recovery – the composition of soil components that influence organism recovery vary tremendously. Thus, GEM recovery in soils can be influenced by the suspending medium, by the GEM quantities per soil volume, and by how long the mixture is shaken, as well as by other factors. Methods to monitor GEM survival and fate in the insect/plant/soil ecosystems have also been described [3]. Recovery from insects presents the challenge of preventing the destruction of the sample, as well as the need to discriminate between surface contamination and gut colonization.

Sampling frequencies are highly dependent upon the nature of the test system and the purpose of the experiment. Changes that occur in viable counts, gene transfer to form transconjugants, and organism transport all occur rapidly, within days rather than weeks. On the other hand, if the purpose is to determine GEM survival time, one may want to sample over a period of months [7,8]. In either situation, when using microcosms the issue of destructive sampling, edge effects, and depletion of sample material are always concerns.

Aquatic

Sampling frequencies are, of course, dependent on the logistics in sample analysis (what sample load can be handled) and the design of the microcosm, but usually sampling reflects a daily pattern. If a sample is taken from a microcosm that remains intact after sampling, then the size of the microcosm will play a role in number of samples. However, if the microcosm is sampled using the entire system, the number of samples will depend on replica cores. Replica samples are important to ensure reliability, and in the two field validated microcosm studies described previously, 3 and 4 replica samples were taken [25,45]. The area of sampling in aquatic microcosms would include: (1) water-air interface, (2) water column, (3) sediment-water interface, and (4) sediment.

Statistical analyses

Statistical evaluations may include any number of comparative analyses including analysis of variance, least significant difference, t-tests and difference in means. Such procedures can be found in a variety of statistical texts. Also, an excellent review of statistics used in microbial ecology is given in Atlas and Bartha [2].

How GEMS are released

Terrestrial microcosms

The release of GEMs into terrestrial microcosms should be conducted so as to best simulate the manner of inoculation that would be expected to occur in an out-of-doors release. Thus, GEMs can be coated onto seed surfaces, aerosolized onto plants with suitable devices, inoculated into soil or onto seeds in soil with pipettes, or by similar variations of these procedures [15]. The goal should be to inoculate the GEM at high density so as to give it suitable opportunity to colonize and accomplish the intended goal. Such concentrations normally would be 10^8 to 10^9 cells/ml. There may be instances where repeated inoculations would increase the effectiveness of the GEM, and these should also be considered.

Aquatic microcosms

The manner in which the GEM is released into aquatic microcosms would depend upon the type of microcosm, but most studies have added the GEM directly to the water column by pipetting a liquid suspension of the organism. It is assumed that this would represent the most likely manner in which the microorganism would enter the system.

How the GEM is monitored

An entire range of methods is available to monitor the GEM in both terrestrial and aquatic microcosms. These include conventional antibiotic resistant markers with suitable selective media for culture [26] and various molecular probing methods, including PCR amplification to detect as few as 1 genome in 100 ml of water [6]. These methods are discussed in detail in other chapters.

References

1. Atlas RM (1984) Diversity of microbial communities. Adv Microb Ecol 7: 1–47.
2. Atlas RM, Bartha R (1993) Microbial Ecology: fundamentals and applications, 3rd ed., Benjamin/Cummings, Redwood City, CA.
3. Armstrong JL, Knudsen GR, Seidler RJ (1987) Microcosm method to assess survival of recombinant bacteria associated with plants and herbivorous insects. Curr Microbiol 15: 229–232.
4. Armstrong JL (1989) Assessing the persistence of recombinant bacteria in microcosms. In: Fredrickson JK, Seidler RJ (eds) Evaluation of terrestrial microcosms for detection, fate, and survival analysis of genetically engineered micro-organisms and their recombinant genetic material, pp. 3–89/043. US EPA ERL Corvallis EPA 600.
5. Awong J, Bitton G, Chaudhry GR (1990) Microcosm for assessing survival of genetically engineered micro-organisms in aquatic environments. Appl Environ Micro 56: 977–983.
6. Bej AK, Mahbubani MH, Dicesare JL, Atlas RM (1991) Polymerase chain reaction-gene probe detection of micro-organisms by using filter-concentrated samples. Appl Environ Microbial 57: 3529–3534.
7. Bolton HJ, Fredrickson JK, Bentjen SA, Workman DJ, Shu-mei WL, Thomas JM (1991) Field calibrations of soil-core microcosms: fate of a genetically altered rhizobacterium. Microbial Ecol 21: 163–173.
8. Bolton HJ, Fredrickson JK, Thomas JM, Li SW, Workman DJ, Bentjen SA, Smith JL (1991) Field calibration of soil-core microcosms: ecosystem structural and functional comparisons. Microbial Ecol 21: 175–189.
9. Brettar I, Hofle M (1992) Influence of ecosystematic factors on survival of *E. coli* after large-scale release into lake water mesocosms. Appl Environ Microbial 58: 2201–2210.
10. Colwell RR (1988) Engineering marine micro-organisms for biodegradation and waste control in the sea. J Shellfish Res 7: 552–553.
11. Cripe CR, Pritchard PH (1992) Site-specific aquatic microcosms as test systems for fate and effects of micro-organisms. In: Levin MA, Seidler RJ, Rogul M (eds) Microbial Ecology: Principles, Methods, and Applications, pp. 467–493. McGraw Hill, New York.
12. Cripe CR, Pritchard PH (1992) Application of microcosms for assessing the risk of microbial biotechnology products. In: Stern AM (ed) EPA 600r-92/066. US EPA Office of Res/Dev. Washington, DC.
13. Donegan K, Matyac C, Seidler RJ, Porteous A (1991) Evaluation of methods for sampling, recovery, and enumeration of bacteria applied to the phylloplane. Appl Environ Microbial 57: 51–56.
14. Federle TW, Livingston RJ, Wolfe LW, White DC (1986) A quantitative comparison of microbial community structure of estuarine sediments from microcosms and the field. Can J Microbial 32: 319–325.
15. Fredrickson JK, Seidler RJ (1989) Evaluation of terrestrial microcosms for detection, fate, and survival analysis of genetically engineered micro-organisms and their recombinant genetic material. In: Fredrickson JK, Seidler RJ (eds), DOE Report, 600/3-89/043. EPA.
16. Fredrickson JK, Bolton HJ, Bentjen SA, McFadden KM, Ward LS, Van Voris P (1990) Evaluation of intact soil-core microcosms for determining potential impacts on nutrient dynamics by genetically engineered micro-organisms. Environ Toxic Chem. 9: 551–558.
17. Gillett JW, Witt JM, Wyatt CJ (eds) (1978) Symposium of Terrestrial Microcosms and Environmental Chemistry. June 1977, Corvalis, Oregon, NSF/RA79–0026. National Science Foundation, Washington, DC.
18. Haas LM (1992) Improve epifluorescent microscopy for observing planktonic micro-organisms. Ann Inst Oceanogr 58: 261–266.
19. Halvorson HO, Pramer D, Rogul M (1985) Engineered organisms in the environment: scientific issues. American Society for Microbiology, Washington, D.C.
20. Hobbie JE, Daley RJ, Jasper S (1977) Use of Nucleopore filters for counting bacteria by

fluorescence microscopy. Appl Environ Micro 33: 1225–1228.

21. Hood MA (1991) Comparison of four methods for measuring chitinase activity and the application of the 4-MUF assay in aquatic environments. J Micro Methods 13: 151–160.
22. Jones RD, Hood MA (1980) The effects of organophosphorus pesticides on ammonium oxidation in estuarine sediments. Can J Microbial 26: 1296–1299.
23. Kandel A, Nybroe O, Rasmussen OF (1992) Survival of 2,4-dichlorophenoxyacetic acid degrading *Alcaligenes eutrophus* AEO106 (pR0101) in lake water microcosms. Microbial Ecol 24: 291–303.
24. Kogure K, Simidu U, Taga N (1979) A tentative direct microscopic method for counting living marine bacteria. Can J Microbial 25: 415–420.
25. Kroer N, Coffin RB (1992) Microbial trophic interactions in aquatic microcosms designed for testing genetically engineered micro-organisms: a field comparison. Microbial Ecol 23: 143–157.
26. Levin MA, Seidler RJ, Borquin AW, Fowle JR III, Barkay T (1987) EPA developing methods to assess environmental release. BioTechnol 5: 38–45.
27. Martin K, Parsons LL, Murray RE (1988) Dynamics of soil denitrifier populations: relationships between enzyme activity, most probable number counts and actual N gas loss. Appl Environ Microbial 54: 2711–2716.
28. Mellon MJ (1993) The Gene Exchange. Union of Concerned Scientists. 1616 P St NJW Suite 310, Washington, DC 20036.
29. Nublein K, Maris D, Timmis K, Dwyer DF (1992) Expression and transfer of engineered catabolic pathways harbored by *Pseudomonas* spp introduced into activated sludge microcosms. Appl Environ Microbial 58: 3380–3386.
30. Nybroe O, Christoffersen K, Riemann B (1992) Survival of *Bacillus licheniformis* in seawater model ecosystem. Appl Environ Microbial 58: 252–259.
31. Odum EP (1984) The mesocosm. BioScien 34: 558–562.
32. O'Neill E, Hood MA, Cripe CR, Pritchard PH (1988). Field calibrations of aquatic microcosms. Abstracts of the annual meeting of the American Society for Microbiology, Washington, DC., pp. 297.
33. Parson TR, Maita Y, Lalli CM (1984) A manual of chemical and biological methods for seawater analysis. Pergamon, New York.
34. Paul JH, David AW (1989) Production of extracellular nucleic acids by genetically altered bacteria in aquatic environment microcosm. Appl Environ Microbial 55: 1865–1869.
35. Peck RL, Hern JL (1988) Rapid introduction of acetylene into large assay chambers for acetylene reduction experiments. Soil Sci Soc Am J 52: 1624–1625.
36. Portier RJ (1985) Comparison of environmental effect and biotransformation of toxicants on laboratory microcosm and field microbial communities. In: Boyle TP (ed) Validation and Predictability of Laboratory Methods for Assessing the Fate and Effects of Contaminants in Aquatic Ecosystems, pp. 14–30. ASTM STP 865, Am Soc Testing Materials, Philadelphia, PA.
37. Pritchard PH, Borquin AW (1984) The use of microcosms for evaluation of interactions between pollutants and micro-organism. In: Marshall KC (ed) Adv in Microbial Ecology. Vol.7, pp. 133–215. Plenum, New York.
38. Sands DC, Rovira AD (1970) Isolation of fluorescent pseudomonads with a selective medium. Appl Microbial 20: 513–514.
39. Scanferlato VS, Lacy GH, Cairns JJ (1990) Persistence of genetically engineered *Erwinia cartovora* in perturbed and unperturbed aquatic microcosm and effects on recovery of indigenous bacteria. Microbial Ecol 20: 11–20.
40. Seidler RJ (1992) Evaluation of methods for detecting ecological effects from genetically engineered micro-organisms and microbial pest control agents in terrestrial systems. Biotech Adv 10: 149–178.
41. Sobecky PA, Schell MA, Moran MA, Hodson RE, (1992) Adaption of model genetically engineered micro-organisms to lake water: growth rate enhancements and plasmid loss. Appl Environ Microbial 58: 3630–3637.

42. Standard Methods for the Examination of Water and Wastewater (1980). American Public Health Association. Washington, DC.

43. Taub FB (1992) Synthetic microcosms as test systems for survival and effects of genetically engineered micro-organisms. In: Levin MA, Seidler RJ, Rogul R (eds) Microbial Ecology: Principles, Methods and Applications, pp. 643–661. McGraw Hill, New York.

44. Van Voris P (1988) Standard guide for conducting a terrestrial soil-core microcosm test. Standard No. E-1197-87. In: Annual Book of ASTM Standards, Vol. 1104. American Society for Testing and Materials, Philadelphia, PA.

45. Wagner-Dobler I, Pipke R, Timmis KN, Dwyer DF (1992) Evaluation of aquatic sediment microcosms and their use in assessing possible effects of introduced micro-organisms on ecosystem parameters. Appl Environ Microbial 58: 1249–1258.

46. Wimpenny JWT (1988) Models and Microcosms. In: Handbook of Laboratory Model Systems for Microbial Ecosystems, Vol 1. pp 1–17. CRC Press, Inc, Boca Raton, FL.

47. Wright RT, Coffin RB (1984) Measuring microzooplankton grazing on planktonic marine bacteria by its impact on bacterial production. Microbial Ecol 10: 137–149.

Molecular Microbial Ecology Manual **6.2.3**: 1-8, 1995

Designing release experiments with GEM's in foods:

Risk assessment of the use of genetically modified Lactococcus lactis *strains in fermented milk products; a case study*

NICOLETTE KLIJN, ANTON H. WEERKAMP and
WILLEM M. DE VOS
Netherlands Institute for Dairy Research, P.O. Box 20 6710 BA Ede, The Netherlands

Introduction

The enormous progress in the understanding of the genetics of micro-organisms has resulted in the ability to genetically engineer micro-organisms for the production of economically important products, including food and feed. It is important to distinguish at this point the use of genetically engineered microorganisms (GEM) as production organisms of food additives and as food ingredients such as in starter cultures. Until now the use of GEM in food production has been restricted to the contained production of food additives and enzymes (e.g., chymosin, high fructose corn syrup) and to the application in food of micro-organisms with homologous transformations [3].

In order to assess the risk associated with the release of GEM in the environment a number of studies have been initiated in recent years which were focused on the uncontained use of microorganisms in pharmaceutical, environmental and agricultural applications [10,17]. Because the application of GEM in food involves the intentional ingestion of usually high numbers of live microorganisms by the consumer [3], assessment of the associated risks is especially complicated.

A rational starting point for design of release experiments would be an analysis of the flow of the carrier food product through the environment, including the consumer, and the identification of the critical point factors for the release (Fig. 1). Such flow diagrams may also form a basis for the design of decision schemes aiding clearance procedures for novel foods by competent authorities [9]. To aid in the identification of critical points, hypothetical quantitative estimates of certain events may be included, based on theoretical and reference studies (see Fig. 1). The aim of risk assessment field studies is to provide objective quantification of the critical points.

Most flow schemes for the release of GEM through food products will contain the following critical points:
– growth and survival of the GEM during the production process,

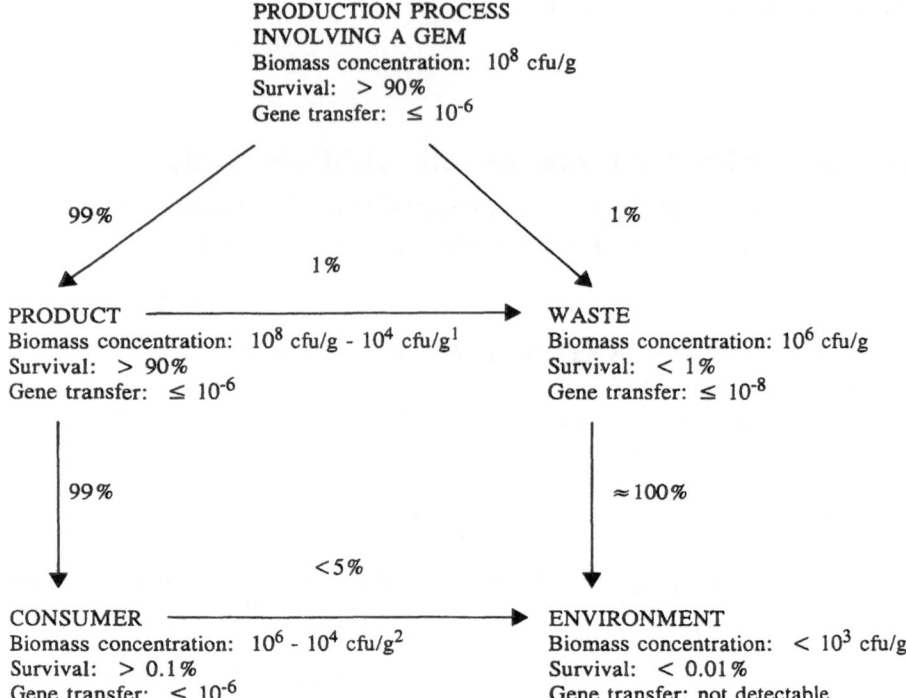

PRODUCTION PROCESS
INVOLVING A GEM
Biomass concentration: 10^8 cfu/g
Survival: > 90%
Gene transfer: ≤ 10^{-6}

99%

1%

1%

PRODUCT
Biomass concentration: 10^8 cfu/g - 10^4 cfu/g[1]
Survival: > 90%
Gene transfer: ≤ 10^{-6}

WASTE
Biomass concentration: 10^6 cfu/g
Survival: < 1%
Gene transfer: ≤ 10^{-8}

99%

≈100%

<5%

CONSUMER
Biomass concentration: 10^6 - 10^4 cfu/g[2]
Survival: > 0.1%
Gene transfer: ≤ 10^{-6}

ENVIRONMENT
Biomass concentration: < 10^3 cfu/g
Survival: < 0.01%
Gene transfer: not detectable

Figure 1. Flow scheme for a food product. The quantifications of the product flow, biomass concentration and the survival rate's are based on the results of the case study with *Lactococcus lactis*. For the gene transfer frequencies we used the values reported in the literature for self-transmissible plasmids.

[1] Depending on the kind of product.
[2] Depending on the kind of product consumed.

— growth and survival of the GEM in the waste flow and persistence in the environment,
— growth and survival of the GEM during passage of the gastro-intestinal tract of the consumer and potential colonization of the host,
— transfer of the (genetically modified) genes during these stages to other microorganisms.

Organisms intended for use in food products must have formal clearance for safe use in foods (GRAS, generally regarded as safe; STIDAFF, safe tradition in deliberate application in food fermentation). In addition, it is essential that their phylogenetic position is well established, also because knowledge of the properties of strains belonging to the same species or genus will be helpful in estimating the chances for survival and transfer of genes of the GEM. Since conventional microorganisms similar to the GEM are often already released in the environment, much information relevant to the risk assessment can be obtained in retrospective from studying their dissemination, survival, persistence and natural communities in the environment. Feeding trials with product made with a suitable wild type organism may provide relevant information about the potential of survival and behaviour of GEM in the human gastro-intestinal tract. Sensitive and reliable detection methods have recently become available to carry out such studies [12]. Nevertheless, it seems characteristic for industrial starter bacteria that little is known as yet about their behaviour outside the industrial environment.

In the authors laboratory a case study on the use of GEM in food production was performed with *Lactococcus* spp. as model organisms. These bacteria are widely applied as starter culture for the production of fermented foods, in particular in dairy industry. Although the genetics of lactococci has been extensively studied, little is known about their survival and genetic stability in the environment, and about the natural communities of lactococci. The study attempted to quantify the main events occurring during the release of a GEM in food products and to identify the knowledge required to allow tracing of genetically modified genes in the environment. Retrospective as well as progressive model studies were applied.

Experimental approach

The experimental approach taken includes the following aspects:
- The development of sensitive techniques for identification and characterization of *Lactococcus* spp., based on Polymerase Chain Reaction (PCR) [20] and DNA-hybridization [15].
- The retrospective study of the presence and survival of *Lactococcus* spp. in the environment.
- Feeding trials with human volunteers to determine the survival of lactococci in the gastrointestinal tract.
- The study of the capacity of *Lactococcus* spp. to transfer genes in vitro and in vivo model systems.

Identification and characterization of Lactococcus *spp.*

In recent years the use of ribosomal RNA sequences for identification and detection has been generally accepted [2]. DNA probes based on highly variable rRNA regions have been successfully applied for identification and detection of micro-organisms. By comparing published 16S rRNA sequences of *Lactococcus* spp. we identified the regions containing the highest variability, and designed specific DNA probes for those regions [11]. In order to increase the sensitivity of the procedure, PCR-amplification of the variable regions with primers based on the conserved flanking sequences were used. PCR-amplification in combination with techniques allowing DNA isolation from small amounts of bacteria (one colony on an agar-plate) allowed the development of a rapid method for the identification of *Lactococcus* spp. [11]. Based on this method, a screening strategy was developed for isolates from environmental samples. This included a first screening of the isolates with a genus-specific probe based on the first part of the 16S rRNA, followed by an identification on (sub)species level using specific probes based on the V1-region of the 16S rRNA [11]. Identification of organisms based on molecular methods can however show major discrepancies with identification based on conventional phenotypical methods. This is clearly illustrated in this study: although both subspecies of *Lactococcus lactis* were found outside the dairy environment based on genotypic identification, analysis of the phenotypical properties showed that isolates assigned to *L. lactis* subsp. *cremoris*, the most important species in starter cultures, had a quite different phenotype compared to industrially used organisms [7]. This indicates that in general, identification at only the species level is not specific enough and that additional characterization is required, based on phenotypical properties or by DNA fingerprinting methods.

Retrospective study

A retrospective study was carried out to locate the natural niches for lactococci. The choice of sampling sites was based on the hypothesis that starter organisms originate from spontaneous milk fermentations and thus would be present in the surroundings of cattle. The samples were plated out on semi-specific agar plates and colonies were randomly picked for DNA extraction after which they were identified with PCR-amplification and hybridization with species-specific DNA probes [11].

In this way the presence of lactococci could be demonstrated on the hide and in saliva of cattle and in soil and on grass of meadows. In order to evaluate the relatedness of industrially used strains and environmental isolates we additionally probed for the presence of genes coding for industrial relevant proteins like protease, citrate permease and nisin, and for the ability to hydrolyse arginine. The differences found between the industrial strains and the environmental isolates suggest that most starter organisms do not survive outside the dairy environment although natural niches are available [12].

Survival of Lactococcus lactis *in the gastro-intestinal tract*

There are no reports in the literature on the isolation of *L. lactis* from human faeces, although the isolation of this organism and of *L. garvieae* from blood and other clinical samples has been described [19]. Isolation of specific microorganisms (i.e., lower than 10^6 per gram faeces) from faeces is very complicated especially when suitable selective media are not available for the target organism. To allow low-level detection molecular methods can also be applied. A method to isolate microbial DNA from human faeces was therefore developed, which enables screening for the presence of lactococcal DNA using PCR-amplification with specific primers.

The isolation procedure is based on the following principles:
- isolation of the fraction of human faeces containing the microorganisms,
- lysis of the microbial cells after the formation of protoplasts using lysozyme,
- phenol/chloroform extraction of the DNA,
- purification of the DNA with Quiagen columns.

In order to validate the molecular detection method we developed a specific plate assay and a probe detection method for a suitable indicator organism. *L. lactis* subsp. *lactis* TC165.1, which has been marked with conventional genetic methods [18] was chosen for this purpose. This strain is a derivative of MG1614 that has acquired the conjugal transposon Tn*5276* coding for nisin production and sucrose fermentation. Strain MG1614 is a natural rifampicin and streptomycin resistant mutant of strain NCDO 712. The indicator strain was used to prepare a dairy product containing high levels of the indicator organism (10^8c.f.u./ml). A similar product, not containing *L. lactis* was prepared as a placebo. The products were used in nutritional studies.

The results of these experiments surprisingly showed that *L. lactis* was able to survive passage of the gastro-intestinal tract. Comparison with faecal counts of simultaneously ingested spores of *Bacillus stearothermophilus* allows the quantification of the transit time and the percentage of survival [13]. The results showed that approximately 0.1% of ingested indicator organisms survived transit through the intestine, but all viable organisms were recovered within three days after consumption. The clearance time of spores was eight days. The experiment clearly illustrates the potential of the methodology to obtain quantitative data suitable for assessing the risk of GEM in food products. Also, the results indicate the need for appropriate methods which allow the quantification of gene transfer in vivo or in vitro models of the gastro-intestinal tract [14].

Gene transfer

Recently a number of research groups have studied gene transfer between bacteria including *L. lactis* under environmental conditions [5]. It is well known that *L. lactis* strains transfer genes via plasmids and phage very easily [6]. Since phages usually show a narrow host range, transduction will not be the main mechanism of gene transfer in the environment. The environment in which the organisms are present is very important for the level of gene transfer. In liquid environments transfer frequencies are relatively low, but in environments where close cell to cell contact is possible, such as on the surface of particles or in the gastrointestinal tract, the frequency may be relatively high. In addition, the transfer frequency is directly related to the number of donor and receptors present. Besides these environmental factors also the degree of relatedness of the strains involved and the characteristics of the plasmid strongly affect the transfer frequency. For *Lactococcus* spp. it has been demonstrated that self-transmissible plasmids can spread easily in a bacterial population even in the absence of selective pressure, but non-self transmissible plasmids disappear [8]. The same has been shown to occur in gnotobiotic mice containing a human intestinal microflora and fed with an *E.coli* strain containing a self-transmissible plasmid. Although the donor strain disappeared rather quickly, transconjugants (*E.coli* strains indigenously present in the microflora) persisted at a constant level, without selective pressure [4]. The relatively high gene transfer in the GI-tract of animals in relation to the high concentrations of donors and receptors indicates that gene transfer experiments in in vitro models of the GI-tract [14] or animal models are highly relevant.

Conclusive remarks

The research approach described here is not only relevant for the introduction of GEM's but also for large scale introductions of new wild type organisms. The knowledge gained from this kind of research allows an estimation of the level of the risk involved in the use of a GEM in a food product. It does not provide any data on the possible *hazard*, a potential harmful event involving the consumer and/or the environment, involved with such an introduction. The latter obviously depends on the nature of the genetic modification in relation to the environments and the bacterial populations in which the GEM would be released. Only genetic information providing a selective advantage in a particular niche will lead to effects on the ecological level by changing the composition of the population or disturbing the chemical balance. With results gained from this kind of studies, situations can be identified in which the risk of dispersal of genetically modified material is to considered relatively high or low.

For example when the release of genetically modified *L. lactis* is considered, the highest chance on the occurrence of gene dispersal from a genetically modified lactococcal strain can be expected when the gene is present on a self-transmissible plasmid and the product is consumed in large quantities and contains a high number of viable cells (10^8/g). In such a case the probability of a gene being transferred to the intestinal flora must be considered high, since the transfer frequencies are relatively high in the gastrointestinal tract (see Fig. 1). In the waste flow of such a product the probability for transfer is lower, mainly because of the low numbers of endogenous microorganisms found in whey, waste water (10^6/g) and soil (10^3/g) and the unfavourable environmental conditions.

On the contrary, if a lactococcal strain which contains the modified gene on a non-self transmissible plasmid or in the chromosome is used for manufacturing a product like cheese (10^4/g), that does not contain high numbers of viable cells and is not consumed in large quantities, then the risk of the dispersal of the gene is considerably lower.

Based on the implementation of this kind of results in decision schemes, information is gained which is highly relevant in aiding clearance procedures for novel foods by competent authorities.

References

1. Aguirre M, Collins MD (1993) Lactic acid bacteria in human clinical infections. J Appl Bacteriol 75: 95–107.
2. Barry T, Powell R, Gannon F (1990) A general method to generate DNA probes for micro-organisms. BioTechnology 8: 233–236.
3. Bol J, de Vos WM (1991) In: Biotechnological Innovations in Food Processing, pp. 45–77. Butterworth Heinemann Ltd., Oxford.
4. Duval-Yflah Y (1994) Homologous recombinant DNA transfer to *E. coli* of human faecal origin in digestive tract of mice. Proc 4th Symp on the Bacterial Genetics and Ecology in Wageningen, 21–24 november 1993.
5. Final report Biotechnology Action Programme: Stability, survival and horizontal gene transfer of genetically engineered lactic streptococci 1991.
6. Fitzgerald GF, Gasson MJ (1988) In vivo gene transfer systems and transposons. Biochimie 70: 489–502.
7. Godon, J-J, Delrome C, Ehrlich SD, Renault P (1992) Divergence of genomic sequences between *Lactococcus lactis* subsp. *lactis* and *Lactococcus lactis* subsp. *cremoris*. Appl Environ Microbiol 58: 4045–4047.
8. Gruzza M, Duval-Yflah Y, Ducluzeau R (1992) Colonization of the digestive tract of germ-free mice by genetically engineered strains of *Lactococcus lactis*: study of recombinant DNA stability. Microbial Releases 1(3): 165–173.
9. International Food Biotechnology Council (1990) Regul Toxicol Pharmacol 12, S1–S196.
10. Jones RA, Broder MW, Stotzky G (1991) Effects of genetically engineered micro-organisms on nitrogen transformations and nitrogen-transforming microbial populations in soil. Appl Environ Micribiol 57: 3212–3219.
11. Klijn N, Weerkamp AH, de Vos WM (1991) Identification of mesophilic lactic acid bacteria using PCR-amplified variable regions of 16S rRNA and specific DNA probes. Appl Environ Microbiol 57: 3390–3393.
12. Klijn N, Weerkamp AH, de Vos WM (1992) Development of sensitive techniques based on DNA probes and PCR-amplification, for the study of the survival and genetic stability of *Lactococcus* spp. in the environment and the gastrointestinal tract. In: Gauthier MJ (ed) Gene transfer and environment, pp. 191–195. Springer Verlag, Berlin, Heidelberg.
13. Marteau P, Pochart P, Bisetti N, Godorel I, Bourlioux P, Rambaud JC (1990) Isolement des bifidobacteries dans les selles après ingestion prolongée de lait au bifidus. Méd Mol Infect 20: 75–78.
14. Marteau P, Minekus M, Havenaar R, Huis in 't Veld JHJ (1993) Survival of lactic acid bacteria in a computer controlled in vitro model of the gastrointestinal tract. FEMS Microbiol Rev 12: P179–P188 (Special Issue).

15. Matthews JA, Kricka JL (1988) Analytical strategies for the use of DNA probes. Anal Biochem 169: 1–25.
16. Mazodier P, Davies J (1991) Gene transfer between distantly related bacteria. Ann Rev Genet 25: 147–171.
17. Pike R, Wagner-Döbler I, Timmis KN, Dwyer DF (1992) Survival and function of a genetically engineered pseudonomad in aquatic sediment microcosms. Appl Environ Microbiol 58: 1259–1265.
18. Rauch PJG, de Vos WM (1992) Characterisation of the novel nisin-sucrose conjugative transposon Tn 276 and its insertion in *Lactococcus lactis*. J Bact 174: 1280–1287.
19. Rodrigues UM, Aguirre M, Facklam RR, Collins MD (1991) Specific and intraspecific molecular typing of lactococci based on polymorphism of DNA encoding rRNA. J Appl Biol 71: 509–516.
20. Saiki RK, Gelfand DH, Stoffel S, Schauf SJ, Highueni RG, Horn GT, Hulks KB, Erlich HA (1988) Primer-directed enzymatic amplification of DNA with a thermostable DNA polymerase. Science 239: 487–491.